THE PRINCIPLES & PRACTICE OF GEOPHYSICAL PROSPECTING

AUSTRALIA

Areas investigated by the Imperial Geophysical Experimental Survey

(G1)	Gravimetric Survey		No. 1
(E6)	Electrical	,,	No. 6
(S2)	Seismic	,,	No. 2
(M3)	Magnetic	,,	No. 3

Compiled by D. F. McGregor

THE PRINCIPLES & PRACTICE OF GEOPHYSICAL PROSPECTING

BEING THE

REPORT

OF THE

IMPERIAL GEOPHYSICAL EXPERIMENTAL SURVEY

Edited by

A. B. BROUGHTON EDGE, B.Sc., A.R.S.M., F.G.S., M.Inst.M.M.

Director, Imperial Geophysical Experimental Survey

and

T. H. LABY, M.A., Sc.D., F.Inst.P., F.R.S.

Professor of Natural Philosophy, University of Melbourne

CAMBRIDGE

AT THE UNIVERSITY PRESS

1931

CAMBRIDGE UNIVERSITY PRESS
Cambridge, New York, Melbourne, Madrid, Cape Town,
Singapore, São Paulo, Delhi, Tokyo, Mexico City

Cambridge University Press
The Edinburgh Building, Cambridge CB2 8RU, UK

Published in the United States of America by Cambridge University Press, New York

www.cambridge.org
Information on this title: www.cambridge.org/9781107612020

First published 1931
First paperback edition 2011

A catalogue record for this publication is available from the British Library

ISBN 978-1-107-61202-0 Paperback

PREFACE

The work which is recorded in these pages was carried out under the joint auspices of the British Empire Marketing Board and the Commonwealth Government, each contributing £16,000 towards its cost. The object was not to prospect for minerals but to test the applicability of various geophysical methods under field conditions in Australia which, it was considered, might be regarded as fairly characteristic of considerable portions of the British Empire.

The Executive Committee, which undertook responsibility in Australia for the work, included Mr H. W. Gepp (with whom the whole idea of the Survey originated in 1927), representing the Development and Migration Commission, Mr W. E. Wainwright, representing the Australasian Institute of Mining and Metallurgy, Mr E. C. Andrews (N.S.W.) and Dr L. Keith Ward (S.A.), representing the several States, and Professor T. H. Laby of the University of Melbourne, with Mr G. A. Cook as Secretary. To these gentlemen the Survey is under deep obligation. Early in 1929 the Geophysical Survey Research Committee was formed in London through the Department of Scientific and Industrial Research and direct contact was effected by a visit from one of its members, Professor A. O. Rankine: also an officer of the Department was attached to the Survey during the second year. The Australian Committee desires to express its appreciation of the ready advice and assistance thus provided. The lines followed by the Survey have developed from those broadly suggested by the Geophysical Sub-Committee of the Committee of Civil Research, whose report was published for the Empire Marketing Board by H.M. Stationery Office in 1927.

Under the title of the Imperial Geophysical Experimental Survey, field parties have worked in selected, widely separated areas in the six Australian States. Everywhere they have been accorded assistance by the State Departments of Mines and to these the Survey is indebted for the carrying out of confirmatory work by drilling and shaft sinking, without which the value of the geophysical observations and predictions could not have been assessed accurately. The Universities, particularly those of Sydney and Melbourne, have helped in many ways. In the Physics Department of the University of Melbourne, laboratory and workshop facilities have been generously provided throughout the investigations.

To the Director of the Survey, Mr A. Broughton Edge, and his enthusiastic staff, the Executive Committee offers its warm thanks and congratulations. Their keenness has never flagged, even in circumstances often somewhat trying. The death of the brilliant Deputy-Director, Dr E. S. Bieler, of McGill University, Canada, in July 1929, was a serious blow to the investigation. In particular, the preparation of the electromagnetic sections of this Report has been severely handicapped, since this branch of the investigation was one to which Dr Bieler had brought special knowledge and for the exposition of which he was exceptionally well qualified. In this emergency Professor Laby, at the request of the Australian Committee, came to the aid of the Survey, placing his personal services as consulting physicist unreservedly at its disposal.

It is hoped and believed that the results now made public will be considered to provide justification for the enterprise. Electrical, gravimetric, magnetic, and to a limited extent, seismic, methods have been examined in country carrying sulphide and other ores, brown coal, graphite, saline waters, etc. Apart from the general report upon the field work, full descriptions have been given of the instruments used by the Survey and of the field procedure that has been followed. Accounts of the principles upon which the several geophysical methods are based have also been included and these may be welcomed by students of practical geophysics.

The several sections of the Report have been submitted for comment to leading authorities in Britain nominated by the Department of Scientific and Industrial Research. While the Editors take full responsibility for any errors which may be found in the work as now presented, both they and the Australian Committee gratefully acknowledge the value of the constructive criticism given by:

S. BUTTERWORTH, Esq., M.Sc., Admiralty Research Laboratory, Teddington (Electrical section);

A. F. HALLIMOND, Esq., M.A., Geological Survey of Great Britain (Magnetic section);

H. L. P. JOLLY, Esq., M.A., Ordnance Survey Department (Magnetic section);

E. LANCASTER JONES, Esq., M.A., Science Museum, South Kensington (Electrical section);

Dr W. McLINTOCK, F.R.S.E., F.G.S., Curator, Museum of Practical Geology (Gravimetric section);

Dr H. MOORE, F.Inst.P., Asst.-Director of Research, British Scientific Instrument Research Association (All sections);

Professor A. O. RANKINE, O.B.E., D.Sc., Department of Physics, Imperial College of Science and Technology, South Kensington (Seismic section); and

Capt. H. SHAW, M.Sc., Science Museum, South Kensington (Gravimetric section).

In some cases the referees' comments have been added as footnotes to the text.

The help of officers of the Empire Marketing Board in completing arrangements for publication is also cordially acknowledged.

That there is ample opportunity within the Empire for the co-operation of physicists and geologists in the search for ore-bodies and the determination of earth structure is fully recognized to-day. A more general realization of the limitations of geophysical methods may also be desirable in those parts where their exploitation by untrained people is not unknown. This report should lead to a more accurate estimation of the present possibilities of applied geophysics.

A. C. D. RIVETT
Chairman of the Australian Geophysical Executive Committee

Council for Scientific and Industrial Research
May 1931 MELBOURNE

CONTENTS

PART I

PART II

CHAPTER VII. ELECTRICAL METHODS

CHAPTER VIII. GRAVIMETRIC METHODS

CHAPTER IX. MAGNETIC METHODS

CHAPTER X. SEISMIC METHODS

PLATE

FOLDING PLANS OR FIGURES (FOLDING)

IMPERIAL GEOPHYSICAL EXPERIMENTAL SURVEY

Director A. B. BROUGHTON EDGE, B.Sc., A.R.S.M., M.Inst.M.M.
Deputy-Director E. S. BIELER[a], M.Sc., Ph.D. (Cantab.).
Field Secretary W. H. SCOTT[b], later M. A. ELLIOTT[c].

ELECTRICAL SECTIONS

No. 1

Leader J. C. FERGUSON, B.Sc.
Field Assistants L. A. RICHARDSON, Assoc.I.S.
R. F. THYER[d].

No. 2

Leader S. H. SHAW, B.Sc., A.R.S.M., Assoc.Inst.M.M.
Field Assistants I. W. MORLEY[e], B.M.E., B.Met.E., Assoc.Inst.M.M.
J. R. HARPER[f], B.E.E.

GRAVIMETRIC SECTION

Leader N. B. LEWIS, B.Sc., D.Phil. (Oxon.).
Field Assistants A. J. MOORE, B.C.E.
E. L. BLAZEY, B.E.E.

SEISMIC SECTION

Leader R. L. ASTON[g], B.Sc., B.E., M.Sc. (Cantab.), A.M.I.E. (Aust.).
Field Assistants[h] N. H. HILL[i], B.Sc.

Consultant Seismic Section E. H. BOOTH, M.C., B.Sc., F.Inst.P.

ATTACHED TO SURVEY

J. McG. BRUCKSHAW[k], M.Sc. For the Department of Scientific and Industrial Research, London.
J. M. RAYNER[l], B.Sc. For the Department of Mines, New South Wales.

TEMPORARY ASSISTANTS (one month or less)

E. W. GREGORY, B.Sc., J. L. PAWSEY, B.Sc., N. H. FISHER, D. M. SUTHERLAND.

[a] Died in Australia 25. 7. 29. [b] Resigned 16. 10. 29. [c] Appointed 30. 9. 29.
[d] Appointed 17. 6. 29. [e] Served 14. 1. 29 until 31. 3. 29 and 1. 6. 29 until 15. 2. 30.
[f] Appointed 30. 10. 29. [g] Appointed 16. 1. 29. [h] Assistants provided by other sections as required.
[i] Served 27. 1. 29 until 8. 3. 29. [k] Attached 1. 1. 29. [l] Attached 21. 1. 29.

LIST OF CONTRIBUTORS

INTRODUCTION	A. B. BROUGHTON EDGE
ELECTRICAL METHODS	
Section 1. Principles and Applications . . .	A. B. BROUGHTON EDGE
2. Surface Potential Methods	
(*a*) Spontaneous Polarization Method .	A. B. BROUGHTON EDGE
(*b*) Resistivity Methods . . .	J. McG. BRUCKSHAW
(*c*) Equipotential Line Method . .	A. B. BROUGHTON EDGE
(*d*) A.C. Potential Ratio Method . .	A. B. BROUGHTON EDGE
3. Electromagnetic Method . . .	J. M. RAYNER
4. Surveys in Australia	J. C. FERGUSON, S. H. SHAW, and A. B. BROUGHTON EDGE
GRAVIMETRIC METHOD	N. B. LEWIS
MAGNETIC METHOD	N. B. LEWIS & E. L. BLAZEY
SEISMIC METHOD	E. H. BOOTH & R. L. ASTON

The Chairman

THE GEOPHYSICAL EXECUTIVE COMMITTEE

MELBOURNE, AUSTRALIA

Sir,

This final Report of the Imperial Geophysical Experimental Survey, which I now have the honour to submit to the Geophysical Executive Committee, includes a full account of the activities and findings of the Survey, from the date of its inception in London in February 1928, until its close in February 1930; after having completed eighteen months of field work in Australia.

The Report has been prepared in two parts in the belief that by adopting this arrangement the work will prove of value to a wider range of readers than might otherwise be the case. In Part I the field investigations made by the Survey, together with elementary expositions of the different geophysical methods employed, are given in such a manner that they may be followed without serious difficulty, whether the reader be physicist, geologist, or mining engineer. Part II, on the other hand, is devoted to the more detailed explanations of field procedure, design of apparatus, interpretation of results and such theoretical considerations as are necessary for a more complete understanding of the principles upon which the methods are based. Since much of the information regarding electrical instruments has never before been made public it is confidently hoped that this section of the Report will prove of material assistance to those who may contemplate the application of electrical methods of prospecting.

The large proportion of space which has been allocated to the electrical and electromagnetic methods may occasion some comment, but it is thought that this course is fully justified in view of the paucity of information of any practical value which is available regarding these methods, and the fact that the procedure and theory of gravimetric and magnetic investigations are already well established. The position is clearly expressed by the Sub-Committee of the Committee of Civil Research in its Report on Geophysical Surveying[a]. On page 15 of this Report the following statement is made:

> In particular, the electrical method has throughout been treated, by the companies employing it, as a jealously-guarded secret trade process. In the result, little information is available to the general scientific world regarding the methods employed...the apparatus required, the field operations, or the interpretation of results. We believe that...a full disclosure of the scientific facts would tend, more than anything else, to stimulate the natural development of this method of geophysical surveying, by placing it on a scientific footing, similar to that of the gravimetric method.

In submitting this Report it is necessary for me to refer to the serious loss sustained by the Survey in the death of Dr E. S. Bieler, which took place with such tragic suddenness in July, 1929, exactly a year after his arrival in Australia as Deputy-Director of the

[a] *Geophysical Surveying. Report of Sub-Committee of the Committee of Civil Research.* H.M. Stationery Office, London, November, 1927.

Survey. From the inception of the Survey it was realized that the services of a senior physicist of considerable experience were required and the Survey could not have been more fortunate than it was, when Dr Bieler was induced to obtain temporary leave from his post as Associate Professor of Physics at McGill University, in order that he might fill this position. Quite apart from his able guidance in physical matters, his services as Deputy-Director of the Survey were of the greatest value—particularly so during the period of my own absence from Australia, when the technical administration of the Survey was left entirely in his hands.

During the twelve months which preceded his death, Dr Bieler had devoted a large amount of his time and attention to the testing and further development of electromagnetic methods of prospecting; with which he had considerable experience prior to his leaving Canada. Much of this work, and the theoretical aspects of it in particular, had not been formulated in such a manner as to permit publication but, fortunately, it has been possible for Mr J. M. Rayner, who had taken part in the field tests, to prepare an account of the work which Dr Bieler had initiated, and to interpret, to some extent, the rather disconnected series of notes and calculations which were found amongst his papers.

The effects of Dr Bieler's death were serious; particularly so since, in his capacity of senior physicist, he had accepted responsibility for a large proportion of the theoretical aspects of the Survey's investigations. His services during the preparation of the final report had been regarded as indispensable, as there were certain features with which he alone was fully competent to deal. To fill his place during the few months that remained was quite impracticable but, very fortunately, the Geophysical Executive Committee was able to prevail upon Professor Laby of Melbourne University to act as consultant physicist to the Survey until its conclusion. As a member of the Executive Committee, Professor Laby had followed the work of the Survey with great interest and he had already proved an unfailing source of help on many occasions, and since Dr Bieler's death, and during the preparation of this report in particular, his services have been invaluable. Not only has he devoted a large part of his time to advising the several writers on the physical aspects of their subjects, but he has taken an active part in editing the work and in arranging for its publication.

I must also draw attention to the generous assistance and exceptional facilities that have been afforded to the Survey in the many parts of Australia that have been visited. This co-operation and help has been given by so great a number of government departments, mining companies, and other bodies, that to make individual acknowledgments is hardly possible. It is necessary, however, to make special mention of the services rendered by the several State Mines Departments and Geological Surveys, and also by the individual State Geologists and the Geological Adviser to the Federal Government from whom the officers of the Survey have sought advice on so many occasions. Geophysical investigations have been carried out in each of the six States, and in every instance the closest co-operation has been experienced. Guarantees were given by the Departments of Mines that any important results obtained would be followed up and tested by drilling operations and in many cases the field parties were provided with camp equipment, transport, etc.

The Survey also owes a great debt of gratitude to the Chancellor and Council of Melbourne University for the special facilities that have been put at the disposal of the Survey. Not only have the laboratories and workshop of the Natural Philosophy Department been made fully available for purposes of testing instruments, and effecting repairs, but the preparation of this Report was largely carried out within the University buildings.

In conclusion I should mention that, owing to the immediate dispersal of the personnel of the Survey at the close of the work in Australia, it has been impossible for several of the contributors to this report to receive and correct proofs of their work. The completion of the report and its preparation for publication have been carried out at Australia House (London) where accommodation and every facility have been provided. In this connection I particularly wish to acknowledge the generous help given to me by Mr F. L. McDougall, C.M.G., Mr A. W. Stuart Smith and Dr A. S. Fitzpatrick, and in proof reading by Captain H. Shaw, Mr S. H. Shaw and Mr J. McG. Bruckshaw. An expression of gratitude is also due to the Cambridge University Press for the ready assistance and advice it has given during the final stages of this work.

I am, Sir,

Yours faithfully,

(Signed) A. BROUGHTON EDGE
DIRECTOR
Imperial Geophysical Experimental Survey

Australia House
LONDON
May 1931

PART I

CHAPTER I

INTRODUCTION

Objects of the Investigations. The duties of the Imperial Geophysical Experimental Survey (I.G.E.S.), as laid down by the Sub-Committee of the Committee of Civil Research[a], and as further determined from time to time by the Geophysical Executive Committee, may be enumerated as follows:

(i) To conduct thorough trials of the principal geophysical methods, and to determine their practical value and limitations under a variety of geological conditions.

(ii) Priority to be given to test areas in which two or more different geophysical methods can be used in combination, since a knowledge of their relative merits is required, and also of the practical advantages to be gained by their conjoint application.

(iii) The work of the Survey to be restricted to areas in which the geophysical results obtained may be checked in a satisfactory manner. In cases where this cannot be done from existing mine or drilling records, arrangements to be made for confirmatory underground exploration, should important geophysical evidence be recorded.

(iv) These trial surveys to be carried out on a normal field scale in specially selected parts of Australia, the choice of areas to be decided, primarily on their suitability for the purposes stated in (i)–(iii) above, but also with due regard to their economic possibilities.

(v) These tests to be carried out by the gravimetric, electrical, and magnetic methods[b] and the scientific information and experience so acquired to be published in full at the close of the Survey.

(vi) Instruction in the use of instruments, and in the application of the methods in the field, to be given to such Australian students and others who may be nominated for this purpose.

With regard to personnel, the recommendations of the Sub-Committee were that the survey party should consist of the following principal members:

(i) The leader, who should have a special knowledge of the electrical method, and who should undertake to report in full detail on the method, apparatus, and every part of the procedure and, at the close of the survey, to prepare the whole for publication.

(ii) One selected Australian physicist to undertake the gravimetric and magnetic surveys. The Department of Scientific and Industrial Research to be asked to provide for his training in England, in consultation with the Science Museum.

(iii) One Australian physicist, who has specialized in electricity, to assist the leader in the electrical surveys.

(iv) Two surveyors and one laboratory assistant.

[a] *Report on Geophysical Surveying*, Published by H.M. Stationery Office, Nov. 1927.

[b] The investigation of seismic methods was initiated at a later date.

It was also recommended that the Survey should be provided with the following equipment at an estimated cost of £4000:

Electrical apparatus.	One magnetometer.
Two torsion balances and accessory equipment.	Survey instruments, etc.
Two motor lorries.	Camp equipment.

Preliminary Organization. The preparatory work in connection with the Survey was carried out at Australia House, London, in February 1928, in close consultation with the Empire Marketing Board, the Department of Scientific and Industrial Research, and the Geological Survey of Great Britain. Instructions were also received from the Geophysical Executive Committee that had recently been formed in Melbourne for the purpose of controlling the activities of the Survey as soon as it commenced operations in Australia. Throughout this preliminary work the advice and guidance of Mr F. L. McDougall, C.M.G., and his staff at Australia House, was indispensable, and to their knowledge of Australian conditions the avoidance of many mistakes is to be attributed.

In the selection of personnel, in accordance with the recommendations of the Sub-Committee, it was apparent that the previous experience of the party would be limited to the electrical methods almost entirely, and that the requisite knowledge of the other branches must perforce be gained after the establishment of the Survey in the field. The appointment of the physicist in charge of gravimetric and magnetic work had already been made, however, in the person of Dr N. B. Lewis of Melbourne and Oxford Universities and, for some months, he had been undergoing his preliminary training in England.

As a result of arrangements which had been made by the Department of Scientific and Industrial Research, Dr Lewis was able to study the Oertling Torsion Balance and the general field procedure of gravimetric surveying under the guidance of the Geological Survey which, at that time, had a gravimetric field party operating in Scotland. Grateful acknowledgment is due to Drs McLintock and Phemister, the two officers of the Geological Survey who were conducting this work, and who also gave valuable advice to the I.G.E.S. on many subsequent occasions. Dr Lewis was also able to spend some time in Cumberland with Messrs Shaw and Lancaster-Jones of the Science Museum, London, who were carrying out a gravimetric survey there over iron-ore deposits. On this occasion he obtained experience with the Gradiometer, a small type of balance that had recently been developed by Messrs Shaw and Lancaster-Jones. As the performance of this instrument was considered to be satisfactory, arrangements were made to secure one for the use of the Imperial Geophysical Experimental Survey. On the advice of the officers of the Geological Survey a large Oertling Torsion Balance and accessory equipment, similar to that employed by the Geological Survey, was also ordered for use in Australia.

In the case of certain of the electrical methods the Survey was in a rather stronger position, since the advantage of several years practical experience in Spain, Portugal, and

South Africa was available, and arrangements could be made to acquire the necessary equipment which, contrary to the cases of the gravimetric and magnetic methods, cannot be purchased in the open market from instrument manufacturers.

Although the officers of the Survey had some previous experience of electrical prospecting operations it should be explained that the position with regard to this branch of applied geophysics was, and still remains, entirely different from that of the gravimetric and magnetic methods, the theory and practice of which have been fully described in published works. The electrical methods have been described in the Report of the Sub-Committee of the Committee of Civil Research[a] as "having been treated by the companies employing them as jealously-guarded secret trade processes," and it must be agreed that there are grounds for this indictment. Although numerous papers on electrical prospecting methods have appeared during the past few years it is most unusual to find information in any of them which can be of the slightest assistance in the design of instruments, or in the use of the methods in the field; in fact, the sincerity of some of these publications is open to considerable doubt. As a result of this difficulty in obtaining information about instruments and procedure many of the electrical investigations carried out by the Survey were necessarily of an experimental character.

In order that the electrical investigations might be commenced in Australia without delay, the appointments of Mr S. H. Shaw, A.R.S.M., B.Sc., and Mr J. C. Ferguson, B.Sc., were approved, both of whom had two years' previous experience in South Africa. At a later stage Dr E. S. Bieler of McGill University, Canada, was appointed Deputy-Director of the Survey, and his previous experience and keen interest in electromagnetic methods of prospecting gave additional support to this section of the work.

In view of the comparative simplicity which attends the use of the magnetic methods and since full details of the procedure are readily available from published works, it was not considered that the services of a specialist were required. This decision was amply justified by subsequent experience in Australia, where it was found that the two vertical variometers of the Schmidt pattern[b], which had been purchased by the Survey, could be used efficiently by any of the technically trained staff.

During these early preparations in London the testing of seismic methods by the Survey was thought to be impracticable, since neither experienced operators nor the necessary equipment could be secured. This branch of geophysical work was entirely in the hands of certain geophysical companies, and not until nearly a year later was it considered possible for the Survey to carry out such investigations. They were then included in the programme of the Survey upon the recommendation of the Australian branch of the Institute of Physics, which drew attention to the fact that, following on their war experiences of seismic methods, the late Professor J. A. Pollock, F.R.S., and Major E. H. Booth of Sydney University had, for some years, been carrying out experiments by an electrical recording system, which is very similar to that now being

[a] *Geophysical Surveying*, H.M. Stationery Office, London, November, 1927.

[b] Made by the Askania Werke of Berlin.

employed in geophysical investigations. Major Booth was available and consented to act as consultant to the Survey in this branch, and on his recommendations the necessary microphones and electromagnetic recording apparatus were obtained. At the same time a standard seismograph of the mechanically recording type was purchased. This instrument[a] had recently been designed by Professor Schweydar of Potsdam and, although a single recording unit only was used by the Survey, it proved of considerable value for purposes of comparison. The field party, which was placed under the charge of Mr R. L. Aston, M.Sc., B.E., conducted field tests over two of the most suitable areas that could be found in New South Wales, and a large amount of valuable information was obtained. A more extensive programme of seismic work was not possible in view of the late date at which the investigations were commenced. Owing to the paucity of published information of a practical character regarding the application of seismic methods to geological problems, the Survey's experience of instruments, methods, and field procedure had to be gained entirely *ab initio*.

In connection with the organization and general policy of the Survey it should be mentioned that several offers of co-operation were received from continental geophysical companies, both directly and through their London representatives, and that in one or two instances the terms that were suggested were most generous. In every case, however, a difficulty existed with regard to the requirement for full publication at the close of the work. Although these companies were agreeable to the publication of results obtained, and even to disclose the details of their methods to authorized members of the Survey, they would not go so far as to concede the right to publish descriptions of the instruments employed, or the methods used in the interpretation of results. This difficulty could not be overcome, since the conditions under which the Survey had been authorized made it essential that the published report should include full accounts of the field procedure, and detailed descriptions of the apparatus employed, without reservations of any kind.

Selection of Areas in Australia. In accordance with the recommendations of the Sub-Committee, the Director of the Survey arrived in Australia several weeks in advance of the other members, in order to make a preliminary examination (without instruments) of areas likely to prove suitable for the purposes of the Survey. After consultation with the Geophysical Executive Committee, and with the State Geologists and the geological adviser to the Federal Government, the several States were visited in turn; every facility being afforded him by the Mines Departments and Geological Surveys. In this manner a large number of promising areas were examined in company with the Government Geologists concerned and, prior to the assembly of the field parties, it was possible, by careful selection, to submit a provisional programme of work for the approval of the Executive Committee.

In spite of the vast area which Australia presented for this purpose of selection, the

[a] Made by the Askania Werke of Berlin.

matter proved to be one of some difficulty; more particularly so in view of those special purposes of the I.G.E.S. which are enumerated on p. 1. Of these several requirements, the most difficult to comply with was that which relates to the confirmation of results; although in a few cases it was possible to select areas in which the mineral deposits, or geological structures, had previously been proved by drilling. Such instances are rare, however, since ore-deposits do not usually remain intact for long after their discovery, and those in course of exploitation are so disturbed, as a rule, that they are quite unsuitable for the testing of geophysical methods. The difficulties attendant on the application of electrical, magnetic, or gravimetric methods in the immediate neighbourhood of mine workings, with their complex systems of electrical cables, iron rails, pipe lines, and underground excavations, will be readily appreciated.

In the majority of the areas which were examined by the electrical methods it was impossible to find standing ore-bodies over which to conduct the special tests in question, and in many instances it was necessary to arrange for systematic surveys of the adjoining country, in the hope that new bodies of mineral might be located. The discovery of mineral was not the primary objective of the I.G.E.S., but in some districts the only way to determine the value of the methods was to see if this could be done. In such cases the most favourable ground was selected in consultation with the Government Geologists, and arrangements were made with the Mines Departments concerned that underground exploration, by drilling or otherwise, would be undertaken in the event of there being any important geophysical results. In many instances there was no alternative to this procedure, since the majority of Australia's base metal mining fields are inactive; the known mines having been worked out and abandoned long ago. In the case of new fields, in which vigorous exploration is in progress, this course might not have been necessary, since newly discovered and only slightly disturbed ore-bodies would be available. It happens, however, that the one or two fields in Australia which answer this description are, on other grounds, not very suitable for geophysical work.

The programme carried out by the Survey is shown below in Table I; the sequence in which the various examinations were made being indicated by numerals, prefaced by the initial letter of the method employed. Thus by E. 10 is meant that the investigation made at Gulgong, in New South Wales, was the tenth carried out by the Electrical Section, and reference to the map of Australia (frontispiece) will show, not only the position of the Gulgong district (E. 10), but also the other geophysical methods that were used there.

The preparation of this programme of work necessitated preliminary visits to a large number of districts, in all parts of Australia, about half of which were rejected, for one reason or another, as unsuitable for the purposes of the Survey. From the experience gained during this process of selection it was apparent that the work of the Survey would mostly take place in the Eastern States, since New South Wales and Victoria presented special problems for gravimetric, seismic, and magnetic investigation, and in Tasmania and Queensland there were favourable opportunities for the application of the electrical methods. The Survey was able to devote considerable attention to the west coast of Tasmania, but, unfortunately, for reasons outside the control of the I.G.E.S., arrangements could not be made for work in Queensland until shortly before the termination of its activities.

TABLE I. *I.G.E.S. Programme of Field Work*

State	Method used			
	Electrical	Gravimetric	Seismic	Magnetic
New South Wales	E. 1 Anembo *sulphide ores*	—	—	—
	E. 2 Captain's Flat *sulphide ores*	—	S. 2 Tallong *surface velocities and underground structure*	—
	E. 3 Leadville *sulphide ores*	—		—
	E. 10 Gulgong *gold deep-leads*	G. 3 Gulgong *gold deep-leads*	S. 1 Gulgong *gold deep-leads*	M. 5 Gulgong *gold sub-basaltic deep-leads*
Victoria	E. 4 Cooper's Creek *copper sulphides*	—	—	M. 2 Cooper's Creek *basic dyke*
	E. 7a Mallee *water horizons*	G. 1 Gelliondale *brown coal*	—	M. 1 Gelliondale *brown coal*
	E. 9 Laverton *brown coal*	G. 2 Lakes Entrance *oil structures*	—	—
Tasmania	E. 5 Zeehan *silver-lead sulphides*	—	—	—
	E. 6 Zeehan *copper-nickel sulphides*	—	—	—
	E. 6a Renison Bell *stanniferous pyrrhotite*	—	—	M. 3 Renison Bell *stanniferous pyrrhotite*
South Australia	E. 8a Moonta and Wallaroo *copper sulphides*	—	—	M. 4 Kadina *magnetic body*
	E. 11 Port Lincoln *graphite*	—	—	—
Western Australia	E. 7 Northampton *lead sulphide*	—	—	—
Queensland	E. 8 Chillagoe *copper-lead sulphides*	—	—	M. 6 Chillagoe *magnetite bodies*

Organization of Field Work. The original conception of the Survey was that of a single mobile unit provided with gravimetric, electrical, and magnetic instruments, together with the personnel required for their use. The party was to be supplied with camp equipment and motor transport, so that it could move freely from one district to another, applying suitable combinations of geophysical methods to the particular problems that had to be dealt with. In theory such an arrangement has much to commend it, but within a week or so of arrival in Australia it was agreed that this was entirely impracticable, for the following reasons:

Firstly, the gravimetric and the electrical were the two principal methods to be employed, and since their practical applications are so different, it was not surprising that no areas could be found in which they might be used in combination. The necessity for separating the two methods was apparent from the start, since to keep both personnels and equipments together in a district suitable only to one of them was out of the question. It was also realized that at any time, should the occasion arise, the two parties could be brought together again for as long as might prove necessary.

Secondly, it was apparent that the field of operations was the Australian continent as a whole, and that great distances had to be covered by rail in moving between the different States. These moves would take place at frequent intervals and, for reasons of efficiency and economy, it would be necessary that living accommodation and transport should be obtained locally. In most instances the areas selected for work were situated on established mining fields, and within a mile or two of house accommodation of some kind. Actually, in having nothing but instruments to transport from State to State, the Survey parties were far more mobile than would have been the case had they been encumbered with camp equipment and road transport of their own.

At a somewhat later stage the formation of a seismic party, and the sub-division of the electrical section into two independent units, resulted in a still greater decentralization of the work, and the Director and Deputy Director were then occupied in a continuous round of visits, spending a week or two at a time with each field party.

The administration of the field parties was carried out from the offices of the Council for Scientific and Industrial Research in Melbourne, where the Field Secretary to the Survey was occupied with supply and transport arrangements, cost and store keeping, and a variety of other general office work.

General Remarks. From Table I it will be observed that the number of areas examined electrically is far in excess of those dealt with by the gravimetric method. This is largely due to the greater speed of working which is possible with the former system, but also to the fact that the mining fields of Australia offered a greater number of opportunities for electrical investigations; so much so in fact that it was decided, early in 1929, to subdivide the electrical section into two independent field parties, so that a more extensive programme might be undertaken. Gravimetric surveying, on the other hand, is relatively a slow procedure and only occasionally do opportunities arise for employing it on metalliferous mining fields. Unquestionably it is more suited to investigations of geological structure, such as may arise in coal and oil fields, than to the direct location of ore-bodies, although opportunities for work of the latter kind sometimes do occur. Nevertheless, the comparatively small programme of work completed by the gravimetric section is to be attributed, very largely, to the serious instrumental difficulties that were experienced, rather than to the slowness of the method or absence of suitable problems. These instrumental troubles became so serious in January, 1929, that the party had to be withdrawn from the field altogether and was unable to resume until the following May.

It has already been explained that the seismic investigations were initiated at a late period in the life of the Survey, and that operations in the field were not commenced until January, 1929. The work of the section, although conducted on a field scale and with a view to structural investigations, was largely of an experimental character.

Throughout the period of the Survey the magnetic work was considered as subsidiary to the gravimetric and electrical investigations, the instruments being used by these sections whenever it was considered that additional information might be gained by doing so. In two instances the magnetic results proved to be of such importance that the method assumed priority over the other systems. In these cases observers were appointed for the purpose of making detailed magnetic surveys.

Although the investigations made by the I.G.E.S. were mostly carried out on long-established mining fields in which prospecting operations by the more usual methods had often been very thorough, a satisfactory proportion of positive results was obtained. Out of eighteen districts which were visited by the Survey, fifteen were examined for the purpose of testing various geophysical methods over mineral deposits, or in order to locate particular structures with which mineral might be expected to occur. Two were selected for electrical investigations in connection with underground water problems, and one for the determination of velocities by the seismic method. In eleven of these districts satisfactory results were obtained by one or more of the methods employed, and in several instances important discoveries of mineral were made. In the remaining cases the surveys were either unsuccessful or gave indications that were not up to expectations. It should be mentioned, however, that in several of the districts which were visited the chances of success were known to be rather remote. For example, saline underground water is prevalent in many of the Australian mining fields, and is so adverse a factor in the cases of the electrical methods that special investigations of the problem were carried out by the Survey in several of the affected areas. The possibility should also be borne in mind that, in some cases, the failure to locate mineral might well have been due to the fact that no mineral was present in the ground that was examined.

CHAPTER II

ELECTRICAL METHODS

SECTION I. PRINCIPLES AND APPLICATIONS

Introduction. There are numerous geo-electrical methods of prospecting, differing one from another in various respects. In some cases the differences are of a fundamental character, but in many they are notable merely as minor variations in the manner of application and in the design of apparatus. Five representative systems have been investigated in Australia by the I.G.E.S., the general principles of which are explained in the following sections of this chapter, and are followed by an account of practical results obtained in the field. In Part II, the theoretical details are discussed at greater length and full descriptions are given of the instruments employed.

Whether applied to the discovery of mineral deposits, or to the elucidation of geological structures, the successful operation of these methods is primarily dependent upon the existence of well-marked differences in electrical conductivity, either between ore-bodies and the rocks enclosing them, or between contiguous geological formations when structural problems are to be investigated. Their practical applications are largely concerned with the search for metalliferous deposits, since those ore-minerals which possess a metallic lustre, such as the sulphides of copper, iron, and lead, are with few exceptions good conductors of electricity which, in relation to the country rocks enclosing them, may afford conductivity ratios ranging from several thousands to several millions to one. More recently, however, it has become evident that electrical methods have a much wider range of practical application, since they are now being employed with considerable effect in a variety of geological investigations[a]. For example, there is ample evidence to show that, under favourable conditions, it is possible to make satisfactory determinations of the depth to the water table, or to bed-rock in country covered by recent and unconsolidated formations.

Resistance Values of Rocks and Minerals. From what has been said above it will be understood that relative electrical conductivities are the primary considerations in determining the practicability of these geo-electrical investigations. In practice, it is usual, and more convenient, to express the conductive properties of rocks and minerals in the form of the reciprocals of their specific conductivities. These values are known as the specific resistances, or resistivities, of the materials in question and are given in ohm cm.[b]. Resistivity tables appear in most of those geophysical publications that deal with the electrical methods, but they are often more misleading than useful, since they are mostly compiled from determinations made in the laboratory on specially prepared rock specimens which, owing to loss of moisture and to other factors, are in no way representative of the naturally occurring formations from which they have been taken.

[a] The value of electrical methods in this connection was indicated by C. Schlumberger as far back as 1920, see *Étude sur la Prospection Électrique du Sous-Sol*, Gauthiers Villars et Cie, Paris, 1920.

[b] An ohm cm. is the resistivity of a one centimetre cube, the resistance of which to a current passing between two opposite faces is one ohm.

That this is the case has been established by several workers[a] who have shown quite conclusively that the electrical resistivity of any normal rock, as it occurs undisturbed in nature, is determined almost entirely by the moisture it holds, both on account of its porosity and as a result of the numerous fissures and joint planes which traverse it. At depths of a few hundred feet from the surface, with which most electrical prospecting operations are concerned, this moisture content is the predominant factor, and country rocks which may give resistivity values of the order of 10^8 to 10^{11} ohm cm. when examined in a dry state in the laboratory will rarely show more than a few hundred thousand ohm cm. when tested *in situ*.

It is apparent, therefore, that in rock formations the passage of current is largely electrolytic in character, whereas in the case of ore-minerals of good conductivity the conditions are similar to those that obtain in metallic conductors. Nevertheless, even in the latter cases, electrolytic conductivity may play a very important part.

The values given in Table II will serve as a general indication of the order of resistivities of rocks and minerals met in practice. In the case of rocks these are mean values for large volumes of the undisturbed formations as they occur in place within a few hundred feet of the surface[b], but in areas where the underground waters are free from any abnormal content of soluble salts.

In determining the suitability of an area for electrical prospecting operations, one of the first points to be investigated is that of the resistivity of the country rocks in which the mineral deposits are likely to be found. For reasons given above, these tests must be carried out upon the ground itself and not upon specimens, however carefully the latter may have been collected. The resistivity methods described in Section 2 of this chapter are recommended for this purpose, since they are simple in application and give a more accurate record of the mean value of the ground resistivity, and of the change in this value with increasing depth from the surface, than is possible by any other method. In the case of the mineral itself it is often impracticable to make *in situ* tests. Resistivity determinations should then be made upon ore-samples considered to be characteristic of the district to be surveyed.

Particular importance is to be attached to the resistivities of the overburden and surface soils, as several of the electrical methods may be at a serious disadvantage if these are low (i.e. less than 500 ohm cm.). In some cases, where the overburden is thick and of low resistivity, the screening effect will be so great that the prospects of success will not be sufficiently good to warrant an electrical survey. In other instances, where the ground resistivities are found to be on the low side, it may be possible to take special steps to minimise the adverse effects; for example, by using a direct current method, or frequencies below 100 cycles per second, in place of the higher frequencies of 500–60,000 cycles that are more commonly employed, and which result in a serious degree of absorption by highly conducting strata.

[a] *Geophysical Prospecting 1929*, pp. 62 and 221–4, A.I.M.E., New York, 1929.

[b] Determinations made on surface exposures and involving relatively small volumes of rock may yield values several times greater than those given in Table II.

TABLE II. *Resistivities of Rocks and Minerals*

Material	Resistivity ohm cm.
A. Crystalline rocks:	
Igneous rocks, gneisses, and crystalline schists, of normal physical character	2×10^4—10^6
B. Consolidated sedimentary rocks*:	
Slates, shales, mudstones, sandstones, limestones, etc.	10^3—5×10^4
C. Unconsolidated and recent formations:	
Marls, clays, sands, alluvial deposits, etc. (excluding surface formations in arid districts)	50—10^4
D. Ore minerals†:	
Zincblende, haematite, limonite, stibnite, wulframite, sulphur, etc. ...	10^3—10^6
Molybdenite	1—50
Chalcopyrite, bornite, chalcocite, pyrite, pyrrhotite, galena, magnetite, specular haematite, etc.	10^{-4}—1
E. Underground waters:	
Normal water (potable)	10^3—10^5
Saline waters‡:	
1 % sodium chloride	75·0
5 „	15·0
10 „	8·25
20 „	5·10

* Some compact sandstones, quartzites and recrystallized limestones have resistivities of the order of 10^5 ohm cm., and fall into group *A* with the crystalline rocks.

† The values given are for picked mineral specimens. In many types of deposit the ore will contain a large proportion of poorly conductive gangue material, thus increasing the mean resistivity of the lode filling.

‡ The electrical resistivity of naturally occurring waters is dependent more on the number of free ions present than upon the actual nature of the dissolved salt.

Resistivity Conditions in Cleaved and Jointed Rocks. It has also been pointed out, and established experimentally by Koenigsberger[a], that rock formations are in many cases far from being isotropic in their resistivity characteristics, and that differences as high as 20 per cent. may be observed in the values for some cleaved or foliated rocks in which determinations are made in directions parallel and normal to these structural features. In most cases this is due to the retention of moisture along the cleavage planes, although it may sometimes be accounted for by the presence of conducting laminae of carbonaceous or graphitic material. Under certain conditions, use may be made of this circumstance for determining the strike direction of a concealed rock formation.

THE ELECTRICAL METHODS

Although numerous, electrical methods of prospecting may be grouped quite simply into the following three classes, according to the nature of the field determinations.

1. *Measurement of Natural Earth Potentials* due to currents set up spontaneously in the ground by natural electro-chemical activity, such as the oxidation processes that

[a] *Geophysical Prospecting 1929*, p. 224, A.I.M.E., New York, 1929.

take place where certain types of sulphide ore lie near to the earth's surface. In some cases the circulation of current is so well defined that the positions of concealed mineral deposits may be located accurately. The field procedure is quite simple and consists of potential measurements at the surface, using a portable potentiometer connected to a pair of non-polarizing electrodes. This is known as the *Spontaneous Polarization Method*, and is described in detail in this chapter.

2. *Determination of Surface Potentials* set up by direct or alternating current passed conductively through the ground between a pair of earthed electrodes. The current field may be represented by a series of current lines passing between the two electrodes and radiating from each of them in all directions beneath the ground surface. Where a body of relatively high electrical conductivity is present these current lines will be drawn together, producing an area of greater current density.

Normal to the current lines an infinite number of equipotential surfaces may be visualised (see Fig. 4). Reference to Fig. 6 will show that just as the current lines are attracted towards a conducting body, so will the equipotential surfaces be repelled from it. Providing that the conductive body does not lie at too great a depth its position may be located by means of potential observations made at the surface.

The surface potential methods used by the I.G.E.S. were (*a*) *The Resistivity Methods* in which actual measurements of the mean earth resistivity are made between particular equipotential surfaces; (*b*) *The Equipotential Line Method* which consists simply in mapping the traces or outcrops of the equipotential surfaces; and (*c*) *The A.C. Potential Ratio Method* in which measurements of the potential gradient are made over the ground surface, phase effects being properly recognised.

3. *Determination of Magnetic Fields* set up by alternating currents, either passed conductively through the earth or applied inductively to the ground by means of horizontal or vertical transmitting loops. In the neighbourhood of conductive bodies the magnetic field immediately above the earth's surface consists of the applied primary field upon which an inductive secondary field is superposed. As these two fields differ in phase from each other the resultant field is elliptically polarized. There are several practical methods by which the secondary, or out-of-phase, field due to a concealed conductor may be determined and used for prospecting purposes.

Two such methods have been used by the I.G.E.S. and are described in later sections of this chapter; namely, the *Bieler-Watson Method*, in which alternating current at 500 cycles per second is applied inductively by means of a large horizontal loop laid upon the ground, and a *High Frequency Method* (60,000 cycles per sec.) in which a comparatively small transmitting loop is used in a vertical position. In both of these methods the resultant magnetic fields are investigated by specially designed search coils.

Resistivity Ratios and Saturation Effect. It has been explained above that the location of a mineral body by electrical prospecting methods is primarily dependent upon the resistivity of the ore, relative to that of the rock formations enclosing it. It will also be seen from Table I that resistivity ratios of the order of several thousands may be met with in the cases of certain types of deposit. In considering the effect of these ratios a distinction must be made between some of the surface potential methods, such as those in which potential observations alone are made, and the electromagnetic methods in which the field is applied inductively. In the former it can be shown that a state of saturation is reached where resistivity ratios ranging from 25 to 50 to one exist between mineral and country rock[a]. This means that no practical advantage is gained where larger ratios exist, since the current distribution will not be affected to any appreciable extent. Difficulties may also arise in view of the fact that ratios approaching the saturation limit may exist between two contiguous geological formations, and produce potential anomalies at the surface just as strong as those due to mineral. This low saturation level is an adverse feature which, fortunately, does not apply in the cases of the electromagnetic methods, since the reaction of a buried conductor to an inductive field is largely dependent on the absolute value of its resistivity.

Depths to which Electrical Methods are Effective. This is perhaps the most frequent enquiry in connection with geo-electrical methods and, just as with the other geophysical systems, it is impossible to make a satisfactory answer without entering into a detailed consideration of each type of problem that is encountered in practice. Just as with the gravimetric and magnetic methods, described later in this volume, so the electrical methods are governed by an inverse cube law, from which it will be readily understood that the detectability of a conductive body diminishes with increasing depth at a most disconcerting rate.

The case of highly conductive mineral deposits has been considered by Max Mason[b], who shows that an ore-body lying at a depth of 500 ft. from the surface must be of unusually large dimensions in order to be capable of detection. This conclusion is certainly supported by practical experience, since few, if any, properly authenticated cases have been recorded in which metalliferous deposits have been *discovered*[c] at depths exceeding a few hundred feet below the surface. It is the opinion of the I.G.E.S. that there are very few mineral fields in which the expense of an electrical survey would be justified unless substantial ore-bodies are to be expected within 250 ft. of the surface.

The more important factors which influence the effective depth of these electrical operations may be enumerated as follows:

(1) *The mean resistivity values of the lode filling* (*i.e. ore and gangue together*) *and of the country rock enclosing the lode.* Although a large proportion of highly conductive ore-minerals may be present in a lode it sometimes happens that the conductivity of the lode as a whole is low, owing to isolation of the conductive constituents by quartz, calcite, or other gangue minerals of high resistivity. This is a common feature in dis-

[a] Max Mason, *Geophysical Prospecting 1929*, p. 17, A.I.M.E., New York, 1929. Warren Weaver, *ibid.*, p. 72, A.I.M.E., New York, 1929.

[b] *Geophysical Prospecting 1929*, p. 29, A.I.M.E., New York, 1929.

[c] By *discovered* is meant the location of a mineral body previously unknown. This excludes the very different case of confirmatory results obtained over known deposits.

seminated ores and, in such instances, the resistivity ratio between lode and country rock may not be large enough to yield a distinctive electrical indication.

(2) *Dimensions and depth of the conductive body below the surface.* These two factors are mutually dependent in accordance with an inverse cube law. For example, a spherical body detectable at the surface when lying with its centre at a depth of 100 ft., would have to have its diameter doubled in order to be equally detectable at a depth of 200 ft.

This is a general statement which, for alternating current methods, is inadequate, since it neglects the effects of increased absorption at greater depths. In the example given, the frequency would have to be changed in order that the ratio of the wave length to the dimensions of the body might remain the same at the two depths mentioned.

(3) *Shape and attitude of the ore-body.* The shape and the lie of the conductive mass are important factors which will largely determine the particular electrical method to be employed, and the depth to which it will be effective. Thus, a flat lying lens or sheet of mineral might be detected quite readily by an inductive or resistivity method but yield no distinctive response to an equipotential line survey which, on the other hand, is well suited to steeply dipping lodes. Similarly, a steeply pitching shoot or pipe of mineral of small horizontal dimensions will usually be difficult to detect unless lying at a very shallow depth. Generally speaking, the vertical extent of a body, although a significant factor[a], does not play quite so important a part as do its horizontal measurements.

(4) *The resistivity of the overburden and surface soils.* The serious effect of a highly conductive overburden has already been discussed above.

(5) *Homogeneity of the enclosing and overlying formations.* Great difficulty is experienced in recognising indications due to an ore-body in areas where other zones of relatively high conductivity are present. For example, a satisfactory interpretation of results may be quite impossible in cases where mineral deposits occur in rocks which also contain bands of graphitic schist, which is usually an excellent conductor of electricity. Similarly, a patch of highly conductive clay lying close to the surface, may completely mask the effects of an ore-body lying at a greater depth.

The above considerations refer particularly to the application of electrical methods to the search for metalliferous deposits which, in the majority of cases, occur in the form of lenticular or sheet-like bodies or lodes, usually inclined from the horizontal at steep angles. The use of electrical methods for other purposes, such as the determination of geological structures and depths to particular interfaces[b], is now receiving wide attention, and under certain conditions it is possible to carry out effective work at depths appreciably greater than those which are attainable when prospecting for mineral. This is due to the fact that these structural investigations are large-scale problems depending on differences in resistivity between extensive geological formations, hundreds

[a] Particularly so in the case of flat-lying bodies when resistivity methods are employed.

[b] The term *interface* is used in this work to indicate the common boundary or contact of two geological formations.

of feet or so in thickness, whereas metalliferous deposits seldom have dimensions[a] of this order. In this connection special mention may be made of the resistivity methods described later in this chapter, which have proved of great value in connection with depth determinations and in underground water investigations.

A further question that is often raised in connection with electrical prospecting operations concerns the possibility of determining the vertical extent of a conductive ore-body. In the case of flat-lying sheets or beds this is sometimes possible[b], but in the case of steeply dipping lodes there are no generally known methods of obtaining such information.

SECTION 2. SURFACE POTENTIAL METHODS

(A) SPONTANEOUS POLARIZATION METHOD[c]

This method is dependent on the fact that electro-chemical differences of potential are generated naturally in certain types of deposit in which oxidation processes are in active operation. When this electrical activity is sufficiently pronounced a well-defined circulation of current is set up within the deposit and throughout the surrounding country and, by mapping its distribution at the surface, the position of the concealed ore-body may often be determined with a high degree of accuracy. The method differs fundamentally, therefore, from all of the other geo-electrical systems of prospecting, in that the procedure is confined entirely to an investigation of naturally occurring electrical phenomena; no artificial source of current whatever being applied to the ground.

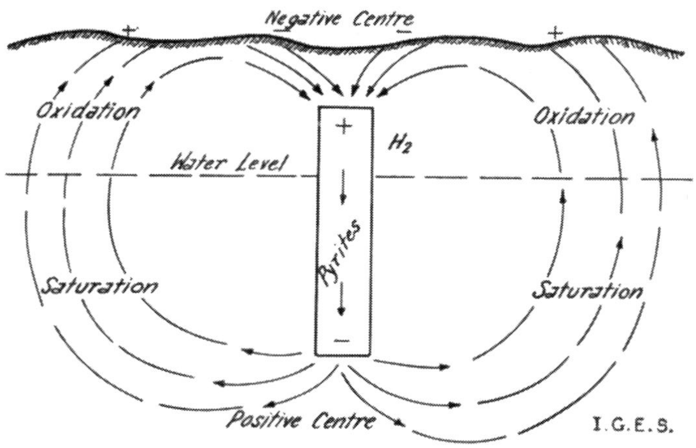

Fig. 1. Diagrammatic representation of spontaneous polarization

Since a vigorous chemical activity is essential to the generation of these earth currents it follows that the method will only prove effective when applied to the discovery of such mineral deposits as will readily undergo oxidation. In practice it has been found that the best results are obtained with sulphide ore-bodies of a pyritic character—provided, of course, that their upper limits extend within the zone of oxidation (Fig. 1). Spontaneous polarization effects are also associated, in some instances, with anthracitic coals and graphite bearing formations but, in general, it may be said

[a] Large iron-ore deposits sometimes present favourable conditions for exploration at considerable depths. See Eve and Keys, *Applied Geophysics*, p. 96, University Press, Cambridge, 1929.

[b] In such cases it is the thickness of the bed which is determined. Under the best of conditions it would hardly be possible to obtain reliable results if the thickness of the bed were less than 25 per cent. of that of the overburden.

[c] Schlumberger, *Étude sur la Prospection Électrique du Sous-Sol*, Gauthiers Villars et Cie, Paris, 1920.

that the practical applications of the method are restricted, almost entirely, to the discovery of pyritic ores. Ore-bodies containing pyrrhotite often show strong electro-chemical activity owing to the instability of that mineral. On the other hand there are many mining fields in which pyritic lodes fail to respond to this method.

The principles of the method are simple and may be followed in Fig. 1 which indicates, in a diagrammatic manner, the electrical conditions which arise in the case of a body of pyrites, the upper extremity of which is being subjected to an oxidizing environment. The zone of oxidation, through which water in an aerated condition tends to percolate downwards from the surface, will vary in depth in different districts, from a few feet up to several hundreds of feet, according to the topographical and other local conditions. In general, however, it is limited in depth by a well defined ground water table which, although subject to slight seasonal variations in level, is a very stable horizon. It marks the transition from the zone of percolation (oxidizing conditions) above, to the zone of saturation beneath, in which there is little, or no tendency for oxidation of the mineral. Under these circumstances the oxidation of the sulphides takes place at, and above, the level of the water table, and the resulting electro-chemical conditions are such that the upper portion of the pyritic ore-body assumes positive polarity. The current passes downwards through the deposit to its lower extremity, which becomes the negative pole of the system, and the circuit is then completed through the electrolyte which, in this case, consists of the rocks and soil enclosing the ore-body. It is apparent then that the ground surface, at which the current and potential observations are made, must be regarded as a plane of observation within the medium of the electrolyte. It will also be understood that the actual potential values observed at the surface will be dependent on the effective resistances of the various rocks and surface formations which occur in the area.

From Fig. 1 it will be seen that the ground immediately above the ore-body is an area of negative potential with respect to all distant points at the surface. The point of maximum negative potential in such an area is known as the *negative centre* and it is the location of these negative centres which is the main objective when the method is used for prospecting purposes.

Effect of Local Geology. In most areas where this method can be employed an important part is played by local topographical and geological conditions, a full appreciation of which is essential if serious mistakes in the interpretation of results are to be avoided. A typical case is illustrated in Fig. 2 in which two parallel sulphide bodies, lying a few hundred feet apart, are shown. The uppermost of these inclined sheet-like bodies, which are here shown in cross section, outcrops on the hillside; the oxidation products derived from the sulphides extending to the surface in the form of the usual porous gossan capping. The second body which lies at a lower level is concealed beneath a shallow cover of alluvium which, in this case, consists largely of an impervious clay. Now, the degree to which these two bodies will show the effects of spontaneous polarization will be determined by the presence, or absence, of factors which will excite oxidation at the base of the gossan capping. In this case (Fig. 2) the two dominant factors are firstly, the means of access of aerated water from the surface, and secondly, the seasonal variations in the level of the water table. Since the seasonal change in the water table is greater, as a rule, beneath a hill than elsewhere, it follows that the oxidation processes in the upper body (*A*) are maintained in a far more active state than are those in the lower body (*B*). In the case of the outcropping deposit this is assisted still further by downward percolation of aerated surface waters, which can take place quite freely through the overlying body of gossan which, owing to its high porosity, serves as a direct channel between the ground surface and the unstable sulphides below. In the case of the lower ore-body, the superficial covering of alluvial clays precludes the direct access of water from above, and the fluctuations in the water table at this point will be so slight that oxidation processes will be quiescent. Under circumstances such as these it may happen that an important ore-body will give no response to the method under discussion, whilst a second body, lying close by, may reveal itself in a most striking manner. It follows, therefore, that the method is ineffective for purposes of an exhaustive survey for mineral. On the other hand, as an auxiliary to other geo-electrical methods it will occasionally prove of great value.

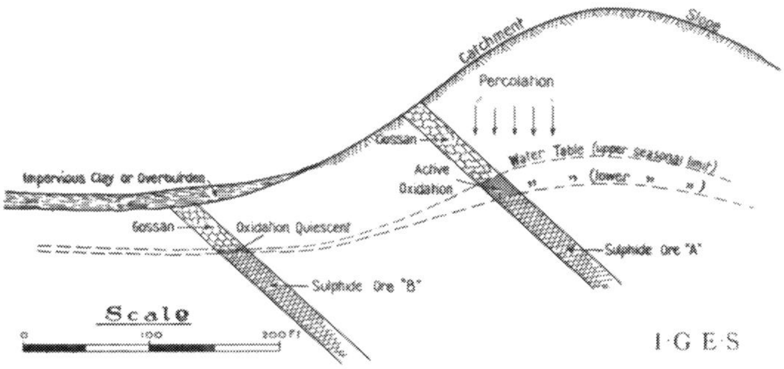

Fig. 2. Influence of topography on spontaneous polarization

Method of Application. The apparatus employed, and the field procedure which is followed in applying the method of spontaneous polarization is comparatively simple, but, nevertheless, close attention to detail is required if satisfactory results are to be obtained. The usual practice is to prepare a plan of the area (scale 1 : 1000) under examination, showing the current distribution at the surface by means of a series of equipotential lines. The negative centres are thus revealed and, in most cases, the electrically active bodies to which they are due will be found immediately beneath them. Thus, in simple cases, a general interpretation of results does not present any serious difficulties. It sometimes happens, however, that electrical anomalies are encountered which are due to causes other than the presence of mineral[a], and in these cases previous experience of the method under a variety of geological conditions will be the only safeguard against misinterpretation.

Equipment. The complete electrical equipment required for this purpose is shown in Fig. 3. It consists of a simple form of portable potentiometer, mounted on a tripod and

[a] For example: vigorous electro-chemical activity occurs in most peat-bogs and gives rise to well defined electrical anomalies.

connected by long insulated wire leads to a pair of specially constructed non-polarizing electrodes, with which the necessary contacts are made with the ground. As full descriptions of this apparatus are given on p. 236 it is unnecessary to do more than mention here that the potentiometer must be robust, and should not require special care in levelling; also that it should be capable of determining differences of potential up to 0·1 volt, with a 1 per cent. accuracy. The non-polarizing electrodes require special attention. They consist, as a rule, of porous pots filled with a saturated solution of copper sulphate into which copper rods or tubes are inserted; the quality of the contacts they make with the ground being a matter of importance. In wet country the operator may often prepare the ground contact surface with the heel of his boot, but when the surface is hard and dry it is necessary to scrape a small hole a few inches in diameter and an inch or so in depth. A small quantity of water is poured into the hole before the base of the porous-pot is placed in position.

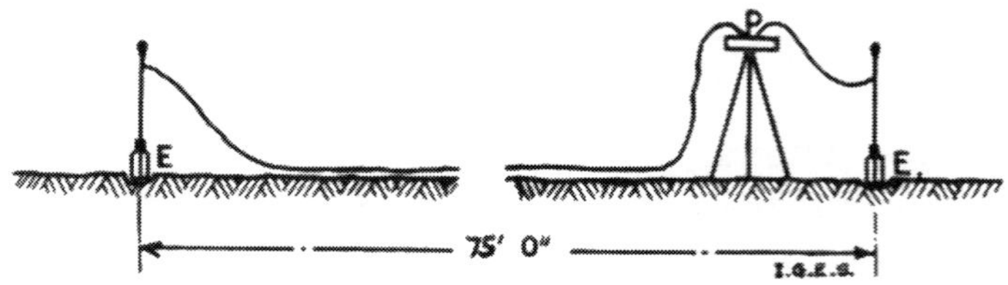

Fig. 3. Field equipment for spontaneous polarization method

Field Procedure. When commencing work in an unknown area it is usual to make a series of straight line reconnaissance traverses in a direction at right angles to the supposed strike of the mineralized zones, and so disposed as to intersect the most promising surface indications, if such exist. In the absence of surface evidence, a series of intersecting or radial traverses should be laid out in such a manner that any hidden ore-body of important dimensions may be crossed, or closely approached, by at least one of them. These lines are usually laid out by compass and tape, numbered wooden pegs being fixed along them at 100-ft. intervals. The apparatus is then placed in position as shown in Fig. 3, with the potentiometer and operator at the forward station. Having determined and booked the potential difference between the first two pegs on the traverse, and noted the polarity of the forward station, the operator then picks up the instrument and electrode and he and his assistant, who holds the rear electrode, move forward along the traverse and occupy the next pair of stations. The assistant then places the rear electrode in position at the point just vacated by the operator and the procedure of measuring and booking the potential value is repeated. Similar observations are made at every pair of stations along the line of traverse, the results then being summed and plotted with distances as abscissae and the potential values in millivolts as ordinates. From the reconnaissance curves thus obtained it may be seen at a glance if any notable gradients have been encountered. The potential values recorded over normal ground, free from mineral, are small and irregular as a rule, but even when large they are usually of so erratic a character that they are not readily confused with the steady gradients that characterise the more important indications.

When one or more of these reconnaissance traverses has proved successful in locating an important anomaly a more detailed survey is made over the area in question. If the reconnaissance work shows a well defined gradient of the order of half a millivolt or more per foot, which persists over a distance of several hundred feet, then time may be saved by making a direct determination of the equipotential lines[a]. To do this, the potentiometer (set at zero) and one electrode are placed at a point on the reconnaissance traverse at which the gradient is steep and regular. The assistant then takes the second electrode, connected to the potentiometer by a 100-ft. length of insulated cable, and applies it to the ground at a succession of trial points situated approximately on the circumference of a circle, the radius of which is determined by the 100-ft. length of connecting wire. The operator, by watching the direction and amount of deflection in the galvanometer for each of these trial contacts, will then be able to direct the assistant to an equipotential point at which no deflection of the needle is observed. A stake is driven into the ground at this point, which is then occupied by the operator with his instrument and the rear electrode. The assistant moves forward as before and a third equipotential point is located, and by continued repetition of this procedure there should be no difficulty in tracing out a line which eventually will close on the starting-point without any serious error. This, of course, will depend largely upon the behaviour of the non-polarizing electrodes, but also upon the care with which the work is conducted. As soon as the position of this equipotential line has been surveyed and plotted the disposition of the electrical field and the approximate position of the negative centre will be known. The precise position of the latter may then be determined by means of further equipotential lines which, with the assistance of the potentiometer, can be spaced at convenient potential intervals from each other.

As a rule, however, it will be more convenient and satisfactory to carry out the detailed investigations by means of a further series of straight line traverses. In rugged country, and when good contacts with the ground are difficult to obtain, this is by far the better practice and it should always be adopted in cases where the potential gradients are small. As far as possible the traverse lines should be so disposed that they will intersect the promising ground at right angles to its strike direction and care should be taken to connect and close the system by cross or diagonal traverses. The accuracy of the work can then be checked and, after a suitable distribution of closing errors, the potential values throughout the traverse system should be so adjusted that they all have reference to some point in the area to which an arbitrary zero potential is given. When the results of such a survey are plotted, the insertion of the equipotential lines by interpolation is a simple matter.

(B) RESISTIVITY METHODS

An inspection of the table of rock and mineral resistivities, which appears on p. 11, will show what a wide range of values is covered by the ordinary rocks and minerals which constitute the earth's crust. Between the consolidated rocks, such as quartzites and granites, and the more recent formations, such as alluvial clays and sands, there

[a] This is a very slow procedure excepting in open country with a moist ground surface.

may be conductivity ratios of the order of several thousands, whilst ratios of millions can exist between some of the ore-minerals and the rocks which enclose them.

Some of the earliest experiments in electrical prospecting were based on these considerations, and they consisted of simple resistance measurements carried out between pairs of electrodes placed in contact with the ground at fixed intervals apart. It was thought that the presence of a body of relatively high conductivity in the intervening ground could be detected in this way. In practice, however, it was found that the resistance values so obtained were determined, almost entirely, by the nature of the electrode contacts with the ground. This method was abandoned long ago but, more recently, the use of resistance measurements has received renewed attention, notably by C. and M. Schlumberger of Paris[a], Gish and Rooney of the Carnegie Institution at Washington, and Professor Koenigsberger of Freiberg University[b]. In their early experiments Gish and Rooney were investigating ground resistivities in connection with natural earth current phenomena but, from their results, they were able to show that the four-electrode system they had developed was also applicable to the determination of depths. Since then numerous trials by this method have been carried out in America and elsewhere, and it has been shown that under favourable circumstances the depth to any extensive geological discontinuity may be found, provided that the conductivity ratio between the contiguous formations is sufficiently marked.

In the Gish and Rooney method an average value of the resistivity is actually obtained, and it is by observing the variations of this value with increasing depth that the positions of underground discontinuities may be determined. The practical value of a method of this kind is unquestionable, and particularly so since reliable estimates of depth are so difficult to obtain by the majority of electrical prospecting methods. Its greatest uses will be in connection with known structural problems, such as depths to bed-rock in areas covered by great thicknesses of alluvium or other recent formations. As far as ore-deposits are concerned, it may prove of value in determining the depth of a body already located by some other method but, excepting for flat lying ore-bodies of large dimensions, it is not very suitable for ordinary prospecting purposes.

The resistivity investigations carried out by the I.G.E.S. were conducted in accordance with the principles described by Gish and Rooney and special acknowledgment is due to the Carnegie Institution for the detailed information, regarding the design of instruments, which it provided at the request of the Survey.

The Principles of the Four-Electrode Methods

Current Distribution. In this method assumptions are made that the surface of the earth is a plane, that it extends to infinity in every direction below its surface, and that the resistivity of the country rock is everywhere the same. In other words the ground is assumed to be a semi-infinite homogeneous conductor. The observations made are directed to detecting departures from this assumption; that is, to detecting bodies of

[a] *Technical Publication*, No. 315, p. 4, A.I.M.E., New York, 1930.
[b] *Geophysical Prospecting 1929*, p. 221, A.I.M.E., New York, 1929.

greater, or less, conductivity than that of the surrounding medium. If two electrodes are connected to points on the surface of such a conductor and to an external electric source, an electric current will pass through the medium from one electrode to the other, and will be distributed throughout the whole region surrounding the electrodes as shown in Fig. 4. There are two main features of the system, the lines of flow and the equipotential surfaces. The former are shown by the broken curves and give the direction of the current at all points in the conductor, while the equipotential surfaces, the continuous curves, join points of similar potential and are everywhere at right angles to the direction of the current. In the immediate neighbourhood of the electrodes the equipotential surfaces are hemispheres with the electrodes at the centre (see Fig. 12, p. 29). At greater distances from the electrode the hemispheres become distorted, and finally at a point midway between the two current electrodes the equipotential surface is a plane perpendicular to the line joining the electrodes.

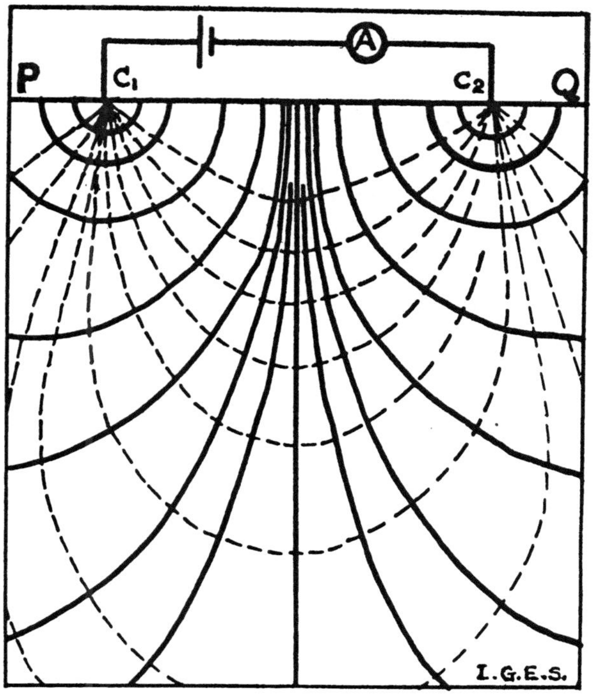

Fig. 4. Distribution of current and potential in a homogeneous medium
PQ surface of medium. C_1C_2 current electrodes.
Continuous curves—equipotential surfaces.
Broken curves—lines of current flow.

Resistivity Measurements. Wenner[a] has given a simple relation which enables the resistivity of the ground to be measured. To make this determination two potential electrodes are placed between, and on line with the current electrodes so that the three intervals between the electrodes are equal, being a cm. By measuring the potential, V volts, between the two inner electrodes and the total current, I amperes, flowing between the current electrodes, the resistance, R ohms, between the two equipotential surfaces which are in contact with the inner electrodes, may be obtained. Wenner's relation (see p. 250) connecting the above quantities is

$$\rho = 2\pi aV/I = 2\pi aR$$

where ρ ohm cm. is the resistivity of the ground. Every part of the conductive medium, however remote, contributes to the total resistance R, although the regions close to the electrodes play a far more important part than those lying at distances which are large compared with the interval a between the electrodes. For all practical purposes the resistivity determined may be taken as referring to a block of ground whose dimensions

[a] *Bureau of Standards, Bull.* vol. 12, p. 469 (1916).

are comparable with the electrode separation, and it may be regarded as the average value to a depth *a*.

Depth Determinations. As an illustration, the case of a layer of overburden of uniform thickness, resting on a semi-infinite medium of greater resistivity, may be taken. For electrode separations which are small compared with the thickness of the overburden, the contribution of the lower medium is negligible and the resistivity measured will be that of the overburden. Conversely, for large separations the measured value will be the resistivity of the lower medium, for the effect of the comparatively thin surface layer will be small. These two cases are illustrated in Fig. 5, in which the line joining $C_1 C_2$ is the ground surface and AB is the interface between the two adjoining media. The shaded areas represent those portions of the ground which contribute the major part of the resistance; the outside areas being responsible for a small fraction only of the total resistance. It will be seen from Fig. 5 (*a*) that when the electrode separation is

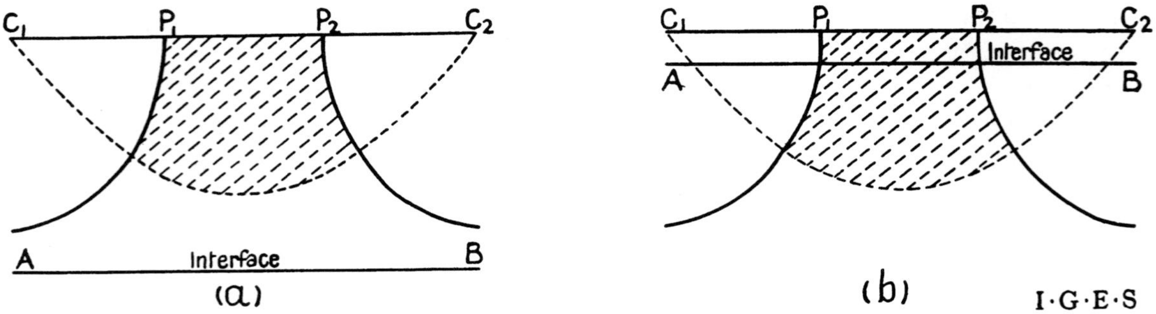

Fig. 5. Effect of separation on measured resistivity
$C_1 C_2$ ground surface. AB interface between media.
(*a*) Electrode separation small compared with depth of overburden.
(*b*) Electrode separation large compared with depth of overburden.

small compared with the thickness of the overburden the lower medium is too remote from the electrodes to have any appreciable effect, while for the relatively large separation shown in Fig. 5 (*b*) the resistance is mainly due to the lower layer. It is apparent, therefore, that in passing from one stage to the other, intermediate resistivity values will be obtained, which will gradually change as the separation is increased from that of the top layer to that of the medium lying beneath. It should be realised, however, that the above picture is a very crude outline of the actual conditions that exist; a more detailed analysis of the problem being given on p. 255.

Except in certain simple cases in which the resistivity changes abruptly across sharply defined boundaries, it is impossible to calculate the variation of effective resistivity with electrode separation. Fortunately, as a result of their investigations, Gish and Rooney[a] were able to give the simple empirical rule that *the depth at which a discontinuity occurs is equal to the electrode-separation corresponding to the turning point on the resistivity—electrode-separation curve.* Hence, under favourable circumstances, a

[a] *Geophysical Prospecting 1929*, p. 53, A.I.M.E., New York, 1929.

superficial examination of the plotted results is sufficient for an estimate of the depth of an interface, but this rough rule must be used with some caution in view of the heterogeneous nature of many areas and the consequent irregularities which may appear on the resistivity curves. In many cases, however, it is possible to determine whether a turning point on the curve is due to a local geological anomaly or to a feature in the main structure of the area.

It is important to notice that Wenner's formula, as given on p. 21, is true for a semi-infinite homogeneous conductor. In practice these conditions are never exactly realized, since the ground is not homogeneous and its surface is rarely flat. So far, no method of making topographical corrections has been devised, but in many areas where structural conditions are favourable for the use of the method, the correction would be small. For small scale problems in rugged country it is unlikely that the method would give satisfactory results, on account of the complications arising from local topographical detail, but when the electrode separations are large, compared with the deviations of the surface from the mean plane, little error would be introduced. It has already been pointed out that the departure of a region from homogeneity may enable some estimate to be made of details of the subsurface structure, but at present the problems actually capable of solution by the resistivity methods are limited to the determination of depths, and dips, of buried formations, or to the location of other extensive discontinuities involving changes in resistivity.

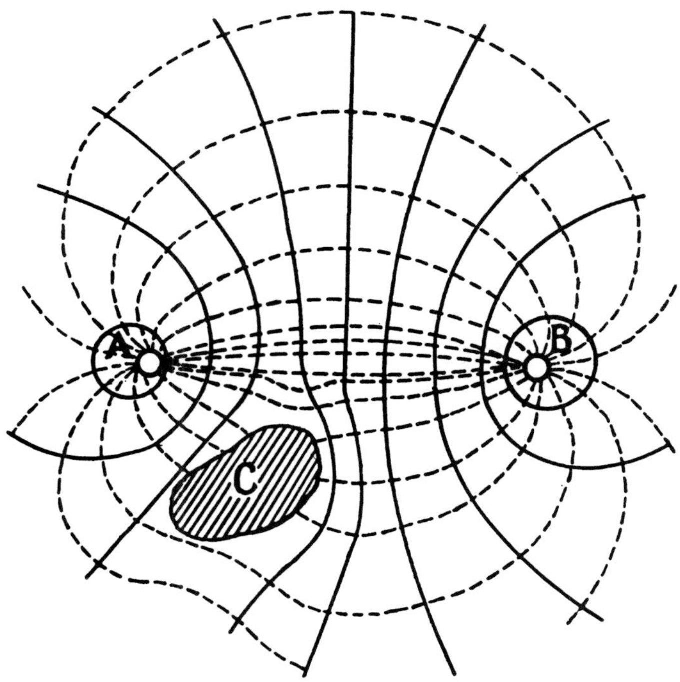

Fig. 6. Distortion of equipotential surfaces by a conductive body *C*. *AB* current electrodes

Before any interpretation can be placed on the values found for the resistivity it is necessary to ensure, as far as possible, that they are truly representative of the resistivity to a depth equal to the electrode separation. This is a matter of importance since the resistivity of the formation may vary considerably throughout its volume, on account of varying water content or local lithological changes. A more serious source of error will be produced by a disturbing body of different resistivity, lying near the surface and in the neighbourhood of the line of observation. Such a case is shown in Fig. 6 in which serious distortion of the equipotential surfaces is caused by the body *C*. As this body *C* has less resistivity than the country rock the surfaces diverge from it. On the other hand, for a body of greater resistivity the surfaces would tend to crowd into the disturbing region. If measurements were made in the neighbourhood of such a body the modified value of the resistivity would not be representative of the mean resistivity due to the main vertical features of the structure, and an indication, quite indistinguishable from that due to a horizontal discontinuity in resistivity might be recorded on the resistivity curve. To guard against errors due to such causes, a second line of measurements is necessary, preferably either at right angles to the first, or parallel to it at some distance away. The indications due to the vertical structural details will be present on both curves while those due to local surface anomalies will occur on one only.

Fig. 7 is a practical illustration in which resistances were measured for regularly increasing electrode separations; each calculated resistivity being plotted against its corresponding electrode separation[a]. The determinations were carried out with the electrode separation increasing by steps of 20 ft. and the general tendency of both curves indicates an increase of resistivity with depth. The peculiar flattening of the curve *A* at a separation of about 50 ft. is not repeated on the second line *B*, which was laid out at right angles to line *A*, and therefore it may be regarded as the effect of a local surface variation.

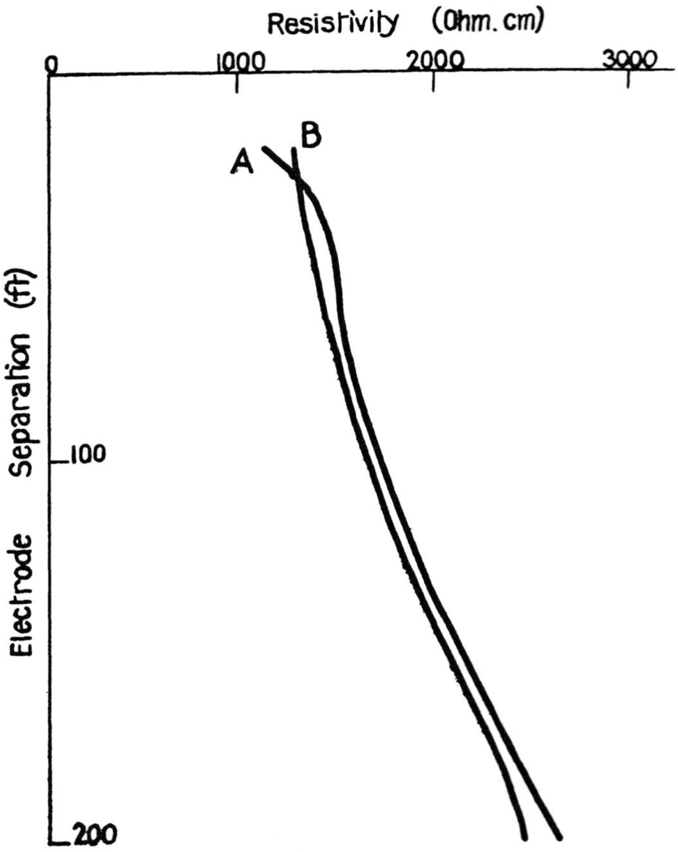

Fig. 7. Field results at Murrayville, Victoria

To ensure good contact in areas of high resistivity the electrodes are driven well into the ground and in exceptional cases groups of electrodes are necessary to reduce contact resistances. Under these conditions corrections for the finite size of the electrode may be necessary.

GROUND RESISTIVITY METHOD IN PRACTICE[b]

It has been shown that the determination of ground resistivity requires the measurement of a length, a current, and a potential. The four electrodes are placed at equal intervals in a straight line; the two outer terminals are connected in series with an

[a] Since electrode separations are equivalent to depths from the surface it is convenient to plot them as ordinates.

[b] *Terrest. Magn.* vol. 30, p. 161 (1925).

ammeter and a source of current, while the two inner electrodes are joined to a potentiometer. Simultaneous observations of the current and potential yield the necessary data for the calculation of the effective resistance for the electrode separation used. Certain difficulties arise in practice which seriously affect the measurement unless special precautions are taken. *Natural earth currents* are always present and these will set up natural potentials between the inner electrodes, in many cases of the same order of magnitude as the potentials due to the artificial currents passed through the ground. Further, *polarization* electromotive forces are set up between the two inner electrodes in the same way as in a voltaic cell, for each of the electrodes in contact with the wet soil behaves as a metal immersed in a dilute electrolyte, the concentrations in general being different for the two electrodes. In some cases this may give rise to polarization potentials between the inner electrodes as high as 200 millivolts.

Two methods are used to overcome these difficulties. In the first of these the polarization is reduced to a minimum by means of non-polarizing electrodes, while the effect of the natural earth potential is eliminated by reversing the current and measuring the potential drop between the inner electrodes for both directions. The second method is that employed by Gish and Rooney, which is shown in a simplified form in Fig. 8. It involves the use of a double commutator which reverses the direction of the current through the ground, and at the same time transposes the leads to the potentiometer. The two current electrodes C_1 and C_2 are connected through the commutator C to the battery B and the ammeter A, while the potential electrodes P_1, P_2, are joined to the potentiometer P through the other section of the commutator. The resultant of the natural and polarization potentials produces a steady potential between the inner electrodes and the effect of the commutator is to apply this potential to the galvanometer first in one direction and then in the other. If the number of reversals per second is sufficiently high, no potential due to these causes can be detected, while that due to the artificial current will be rectified and can be measured.

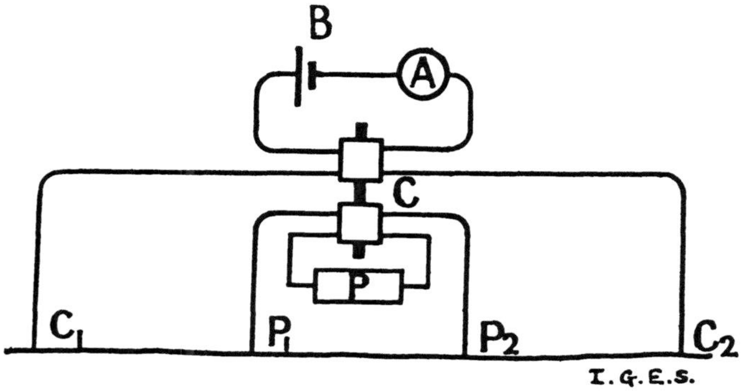

Fig. 8. Simplified circuit of Gish and Rooney apparatus. A, ammeter. B, battery. C, commutator. P, potentiometer. C_1C_2, current electrodes. P_1P_2, potential electrodes

These methods are only capable of dealing with a certain range of resistivities, but fortunately the values encountered in the field usually lie within this range. From the equation given on p. 21 it will be seen that the measured resistance decreases rapidly as the separation increases. If the ground has a low resistivity the resistance becomes small and, with the currents available in the field, the potential may assume a value too small to be measured accurately. On the other hand, if the ground has a high resistivity the current through it will be small and although a large potential may exist between the inner electrodes, the high contact resistances reduce the sensitivity of the potentiometer circuit to such an extent that the potential cannot be determined with sufficient accuracy.

The instrument known as the "Megger" Earth-tester[a], made by Evershed and Vignoles, embodies in a very compact form the fundamental features of the commutator used in the Gish and Rooney apparatus. It contains a generator, a commutator, and two coils rigidly connected together, moving in a magnetic field. The two coils are so arranged that the position they take up depends on the ratio of the currents in them. The current entering the ground is passed through one coil and the current in the other is proportional to the potential between the inner electrodes. Hence the position of equilibrium depends on the resistance between the two potential electrodes and this value may be read directly. It has the advantages that the readings are independent of the current used, and that the instantaneous values of the current and potential influence the coil system simultaneously, thus overcoming one of the weaknesses of the Gish and Rooney system. The instrument is calibrated for a fixed resistance in the potential circuit, and in the field this is changed by the value of the contact resistances at the inner electrodes. If these are small a correction factor can be applied, but if large the correction will be greater than the resistance being measured and the results will not be reliable. Further, the smallest range of the instrument at present available is 0–3 ohms, and assuming that 0·5 ohm is the smallest value that may be read with accuracy, this places a lower limit on the range of resistivities which may be measured by the instrument. For example, at a separation of 300 ft. an effective resistance of 0·5 ohm would give a ground resistivity of 30,000 ohm cm. Resistivities of less than this value would give an effective resistance too small to be measured accurately at the above electrode separation, while for larger separations higher resistivities would be required for the same reason. An additional instrument is required, designed on the same lines as the "Megger" Earth-tester, but capable of dealing with a range of resistivities intermediate between those of the Earth-tester and Ducter[b]. With an instrument of this kind as an auxiliary to the present Earth-tester, it should be possible to cope with the full range of resistances met in the field; which vary from 0·005 ohm up to several hundreds of ohms.

Field Procedure. A resistivity survey, which consists in determining the variations of resistivity throughout an area, may be carried out in two ways. In the first, the electrode separation is kept constant and observations are made along a predetermined system of traverses, in order to determine the variation in resistivity over the area for a fixed depth[c]. The second and more usual procedure is to determine the change of resistivity with depth at one or more selected points within the area. In this case a survey line is laid

[a] Evershed and Vignoles, Catalogue No. 163b (1930). *Bureau of Mines*, Techn. paper 440 (1928).

[b] Also manufactured by Evershed and Vignoles. It is understood that this firm is now designing an instrument to comply with the special requirements of the geophysicist.

[c] The values obtained may then be plotted on a plan and, by interpolation, lines of equal resistivity may be drawn to show the variation in resistivity throughout the area. A series of such plans for different depths may prove of great value in revealing underground structures.

out through the point of observation and is pegged at suitable intervals. The four electrodes are placed in position as shown in Fig. 9, two on each side of the datum point O, so that $c_1O = c_2O = 3p_1O = 3p_2O$ where O is the observation point, $c_1 c_2$ the current electrodes and $p_1 p_2$ the potential electrodes. It is found in practice that if the electrodes are driven into the ground to a depth less than 5 per cent. of the electrode interval the error due to the current source not being confined to a point is less than the experimental error[a]. The measuring apparatus is set up at the central point O and cables connect it to the electrodes. The resistance is determined and checked by using three or four different current values through the ground. The electrodes are then moved to the new positions, C_1, C_2, P_1, P_2, and the process repeated. In changing the electrode positions both sets are moved either towards or away from the central point O, the outer electrodes being moved three times the distance by which the potential electrodes are changed, that is

$$c_1C_1 = 3p_1P_1.$$

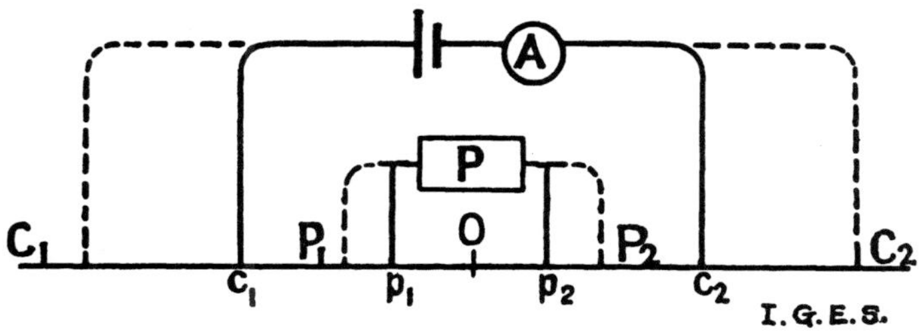

Fig. 9. Method of changing electrode positions
C_1C_2, c_1c_2, current electrodes. P_1P_2, p_1p_2, potential electrodes.
$3p_1O = 3p_2O = c_1O = c_2O$; $3p_1P_1 = c_1C_1$.

Apparatus. Some idea of the nature and layout of the apparatus may be obtained from Figs. 10 and 11. In Fig. 10 the commutator[b], ammeter, and the batteries used to supply the current can be seen, while in Fig. 11 the potentiometer circuit is shown. The apparatus appears rather cumbersome for field work but no serious inconvenience was experienced while using it. This may possibly be due to the fact that the second of the two methods described above was employed by the Survey, for in this case the apparatus can remain in position for a complete series of observations. If work had been carried out by the first method, which would have necessitated the complete dismantling and reassembling of the apparatus after every three or four observations, it is likely that the work would have been very slow. The complete apparatus weighs 200 lb. but, if redesigned, this could be reduced very materially.

Speed of Working. The choice of electrode intervals depends entirely upon the nature of the problem to be solved. When the depths to be determined are in the neighbourhood of 100 ft., intervals increasing by 20 ft. may be used while, for a depth of several hundreds of feet, increments of 50 ft. are more suitable. Under favourable conditions, employing two observers and one assistant, it is possible to perform some 25 measurements in a day of 8 hours, including the erecting and dismantling of the apparatus. This corresponds to measurements of resistivity to a depth of 500 ft. in a day, at intervals of 20 ft. With the measurements taken at the larger intervals the number of measurements per

[a] *Terrest. Magn.* vol. 30, p. 163 (1925).

[b] It will be seen that the commutator belonging to a "Megger" Earth-tester was adapted for use with this apparatus.

I.G.E.S.

I.G.E.S.

Figs. 10 and 11. Commutator and potentiometer arrangements for resistivity apparatus (Gish and Rooney principle) day would be less on account of the longer time taken in changing over from one electrode system to the other. In the above figures allowance has been made for the time taken for calculation and graphing of the results.

"Megger" Earth-tester. The procedure in using the Earth-tester is precisely the same as described above excepting that, in this case, the values of the contact resistance at the potential electrodes must be specially determined. Since the instrument is primarily designed for such determinations no further apparatus is required. The Megger has a number of advantages from the point of view of field efficiency. Only one operator is required, a greater speed of working is possible (about 60 measurements can be made in a day) and, further, the instrument is readily portable, weighing only 21 lb. It has already been pointed out that in areas of high resistivity with a dry surface soil the instrument is at a grave disadvantage on account of the high values of the contact resistances.

THE ELECTRODE PROBE METHOD[a]

It has already been stated that in the immediate neighbourhood of a current electrode the equipotential surfaces in a homogeneous medium assume the form of a series of concentric hemispheres with the electrode as a centre. This property is used for the determination of change of resistivity with depth by the electrode probe method. Two electrodes are connected to a source of direct current, one of them being placed at the point on the ground at which the resistivity values are required, while the other is at a large distance from it. Observations are then carried out in the vicinity of the former. Two potential electrodes are placed in line with the current electrode and the effective resistance between two equipotential surfaces is measured as before. The two potential electrodes are then moved by successive equal steps away from the current electrode, so that they will be situated on hemispherical surfaces which penetrate deeper and deeper into the ground. In this way information can be obtained about the resistivity to gradually increasing depths (Fig. 12). The relation between resistivity and resistance is given in this case by

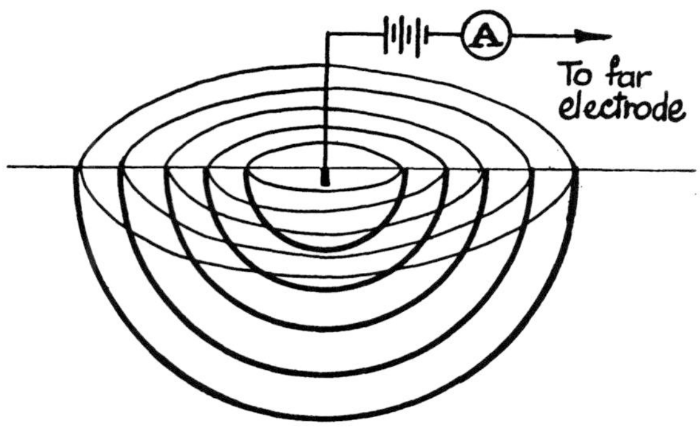

Fig. 12. Equipotential surfaces in neighbourhood of one electrode

$$\rho = 2\pi R \left(\frac{ab}{a-b}\right),$$

where ρ and R have the same significance as on p. 21, and a and b cm. are the distances of the two potential electrodes from the current electrode. In the case of a uniform overburden above a medium of different resistivity, when the mean of a and b is equal

[a] Eve and Keys, *Applied Geophysics*, p. 107, University Press, Cambridge, 1929.

to the thickness of the overburden, the two hemispheres will enclose a portion of the lower medium and a change in effective resistivity will be produced. Here again the interpretation of observations will be based on abrupt changes of resistivity, the greatest rate of change of slope in the resistivity curve giving the depth to the discontinuity. It must be understood that the above is a simplified and incomplete picture of the phenomenon, since the existence of a layer of different conductivity necessarily results in a distortion of the equipotential surfaces as the interface is approached, so that a change in resistivity is obtained before the depth to the lower medium is reached. Nevertheless, it is claimed that the simple rule given above yields accurate values for depths when applied to practical problems[a]. Fig. 13 shows the distortion near the interface of two semi-infinite media of differing resistivity, and although this is not the precise problem dealt with above it gives some idea of the effect. Further cases of distorted current distributions due to the presence of a second medium can be seen in a paper by W. Weaver[b].

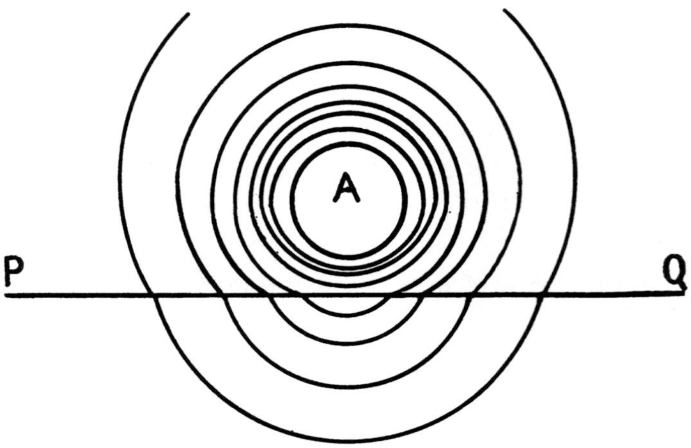

Fig. 13. Distortion of equipotential surfaces due to a second medium
A, source of current in medium of higher resistivity.
PQ, interface between media.

Field Procedure. The apparatus devised by Gish and Rooney for the four-electrode method can also be used for the electrode probe. The field work is carried out along similar lines and the same precautions against error have to be taken. The only difference lies in the arrangement of the current electrodes. The near current electrode is placed over the point at which the structure is required, while the far electrode is situated at a distance equal to about ten times the greatest depth to which it is intended to probe. The far current electrode will influence the measurements to some extent, especially for the greater distances of the potential electrodes from the near current electrode. If the potential electrode traverse is carried out at right angles to the line joining the current electrodes the effect will be reduced to an approximate minimum[c]. The separation

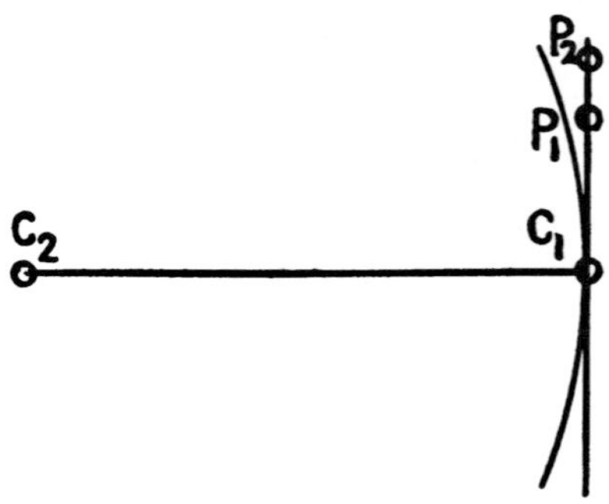

Fig. 14

[a] *Geophysical Prospecting 1929*, p. 199, A.I.M.E., New York, 1929.
[b] *Ibid.*, p. 68, A.I.M.E., New York, 1929.
[c] If the potential electrodes are placed along an arc of a circle with the distant electrode C_2 as centre, and passing through the near electrode C_1, that is, on an equipotential line due to the far electrode, no error will be introduced by the far electrode (Fig. 14). In practice such a curved traverse is impracticable, but in using the tangent to the above circle the effect will be negligible.

of the potential electrodes depends on the nature of the problem and may vary from 10 to 50 ft. After a measurement has been completed the potential electrodes are moved along the traverse to the next pair of stations, the near electrode being placed in position at the point just vacated by the leading electrode. In this way all the ground around the current electrode is included in the measurements.

Electrode probe work was carried out in two areas, at Laverton in Victoria and at Gulgong in New South Wales (p. 42), which had previously been tested by the Gish and Rooney method. At Gulgong, agreement was obtained by the two methods, but at Laverton it proved impossible to correlate the two sets of results. An examination of the data gave no evidence as to the cause of the disagreement and, although little work was done with the method, it is believed that there are certain factors which limit its use.

RESISTIVITY INVESTIGATIONS IN AUSTRALIA

Determinations of water levels, and depths and thicknesses of geological formations, were carried out in the Victorian Mallee, at Laverton (near Melbourne), and also on the Gulgong gold fields in New South Wales, using resistivity methods. As the investigations were of a purely experimental character, with the object of testing the methods, all work was carried out in the neighbourhood of pre-existing bore-holes, so that the results could be checked.

TESTS IN THE VICTORIAN MALLEE (E. 7a)

The district in which the work was carried out lies some 350 miles north-west of Melbourne on the Ouyen-Pinnaroo branch railway line. It was selected as an experimental field in which the application of resistivity methods to underground water problems might be investigated. It was hoped that the depth to the water table, and the relative salinities of water horizons lying at greater depths, might be determined. The work was undertaken by the I.G.E.S. at the suggestion of Mr H. W. Gepp[a], who emphasised that any effective method of obtaining such information in the more arid parts of Australia would be of the greatest economic value.

The Mallee is an arid area, occupying about one-fifth of the State of Victoria, with an average rainfall of 14 in. Wheat is cultivated on a large scale and the water supply is obtained from numerous bore-holes. As a result of the boring operations the geology of the district is known in detail. The surface soils, some 30 ft. thick, are entirely wind distributed and overlie about 200 ft. of lacustrine clays and drifts. Beneath these there are marine formations, interbedded with estuarine blue clays, and followed by some 700 ft. of terrestrial fluviatile deposits containing lignite and pyrites, which rest on palaeozoic rocks. The water, which is sub-artesian, is encountered in the marine formations beneath the blue clays, at depths varying from 100 to 600 ft. below the surface. To the north and east of the district all the upper waters are highly saline, but become

[a] Chairman, Development and Migration Commission.

fresher to the west and south. Even in the areas where the first water is saline other water layers may be encountered by penetrating to a sufficient depth; the salt content decreasing as a rule with increasing depth.

Three tests in all were undertaken at the points shown in Fig. 15, one at the Manpy bore, some 8 miles from Underbool, the second at the bore in the township of Murrayville, and the third at the Koonda bore which is about 20 miles north-east of Murrayville.

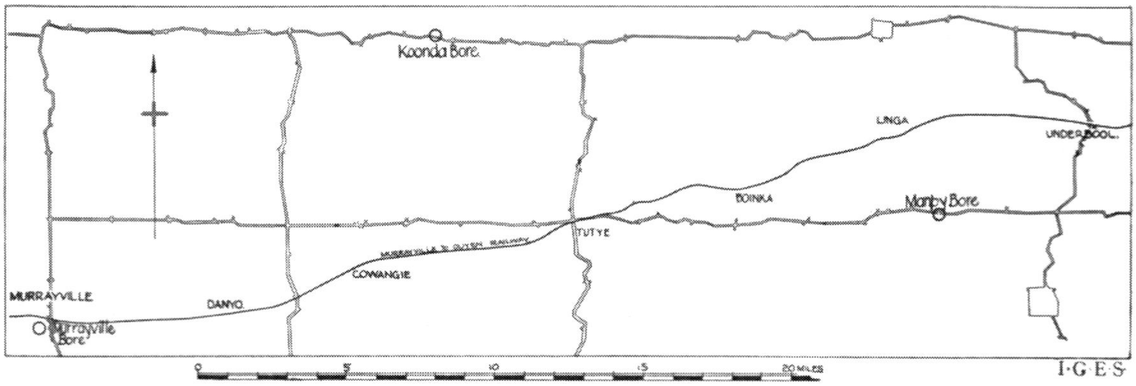

Fig. 15. Map of Mallee showing resistivity test points

Manpy. The line along which the measurements were made at Manpy is shown in Fig. 16. Three sets of observations were made with the centre stations at S_6, S_2 and N_2. Most of the work carried out here was experimental and many of the measurements proved useless on account of various difficulties which were experienced, chiefly induction effects between the current and potential leads which at first were not fully appreciated. Measurements at each point were carried out with electrode separations ranging from 20 ft. up to 600 ft., and in most cases the determinations were repeated three or four times. Up to separations of 300 ft. good agreement was obtained between the various repetitions and here induction effects were small, but beyond this range the results recorded for one separation varied by as much as 20 per cent. This, no doubt, was due to the low resistances measured (0·003 ohm), which is probably beyond the range of the apparatus employed. The three curves obtained are shown in Fig. 17 and it will be seen that they all show the same characteristics. In all cases the resistivity rises until a separation of 50 ft. is reached, and from this point there is a continuous drop to 200 ft., beyond which the resistivity remains constant, except for a slight anomaly at 280 ft. The curves beyond the 300 ft. separations are not shown since their behaviour is too uncertain to permit any satisfactory interpretation.

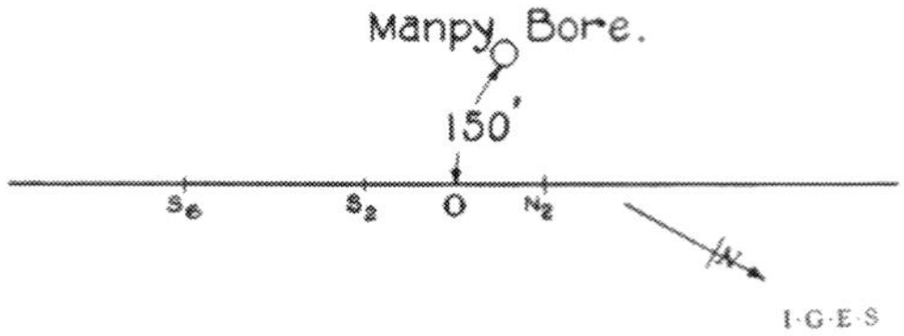

Fig. 16. Resistivity observations at Manpy, Victoria

The bore log, which is shown on the graph, exhibits sand and sandy clays only down to 200 ft., and in all probability there is no abrupt change in the resistivity of the various layers down to this depth, since there is no marked lithological variation. In the vicinity of 200 ft. an abrupt change is to be expected, due to the water level recorded

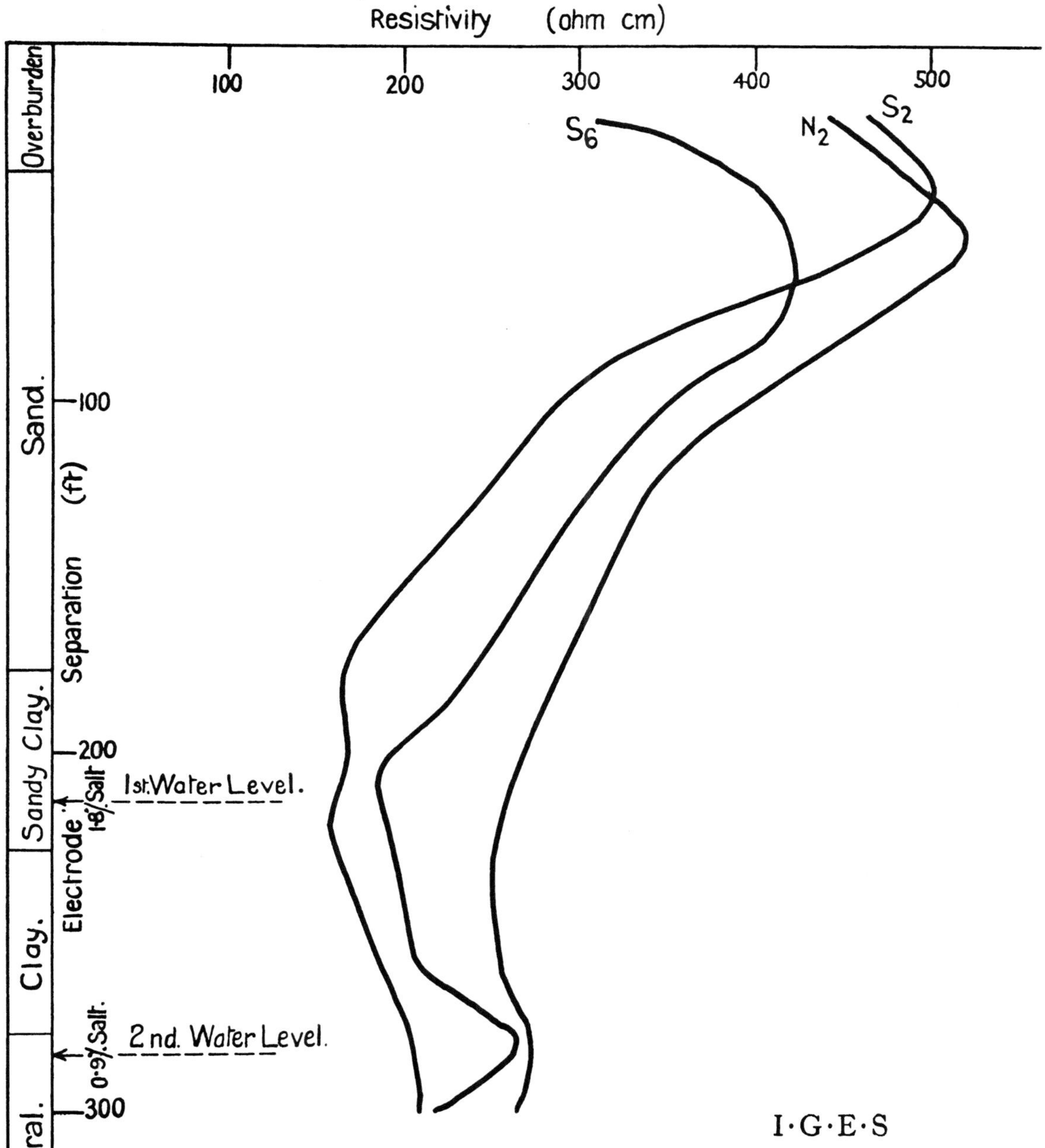

Fig. 17. Resistivity curve and bore section at Manpy

at 214 ft., samples from which contained 1·8 per cent. of salt. The rise in the curves up to the 50 ft. separation may be attributed to a diminution in surface humidity towards the base of the overburden, while there is little doubt that the gradual fall to the 200 ft. interval is largely due to the increasing influence of the low resistivity water layer. Applying the Gish and Rooney rule to this portion of each of the three curves N_2, S_2 and S_6, the greatest rates of change of slope occur at 230 ft., 180 ft. and 205 ft. respectively. These give a mean value of 200 ft. for the depth of the first water, which agrees very closely with the 214 ft. recorded in the bore log. Just below the first water level there is a bed of impervious clay, presumably with a higher resistivity, since there is a general rise up to 280 ft. The drops that follow may be associated with the second saline water level which was recorded in the bore at a depth of 285 ft., on entering the coral formation. An examination of the curves shows that the greatest rates of change of slope take place at 285 ft., 295 ft. and 280 ft., giving a mean of 287 ft. which is also in close agreement with the bore-hole evidence.

Koonda. One set of observations was carried out along the line shown in Fig. 18. In this case the effects of induction on the shape of the curve were investigated and, by taking measurements at each electrode setting with three different commutator speeds, a correction to the readings was applied. In Fig. 19, curve *A* shows the corrected values, and curve *B* the values for a commutator speed of 36 revs. per second. It will be seen that for short electrode intervals the induction effects are small, but that

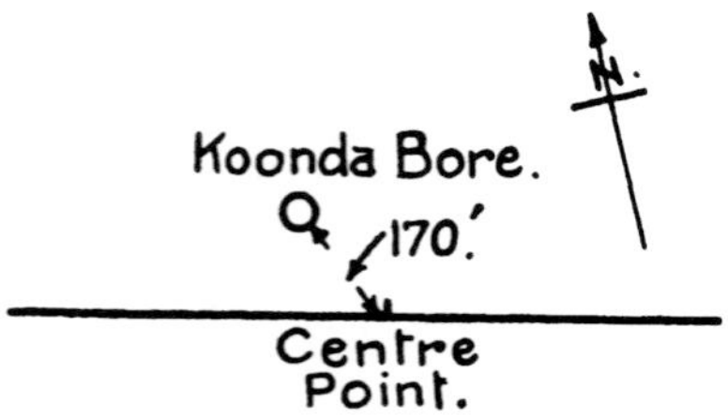

Fig. 18. Layout at Koonda

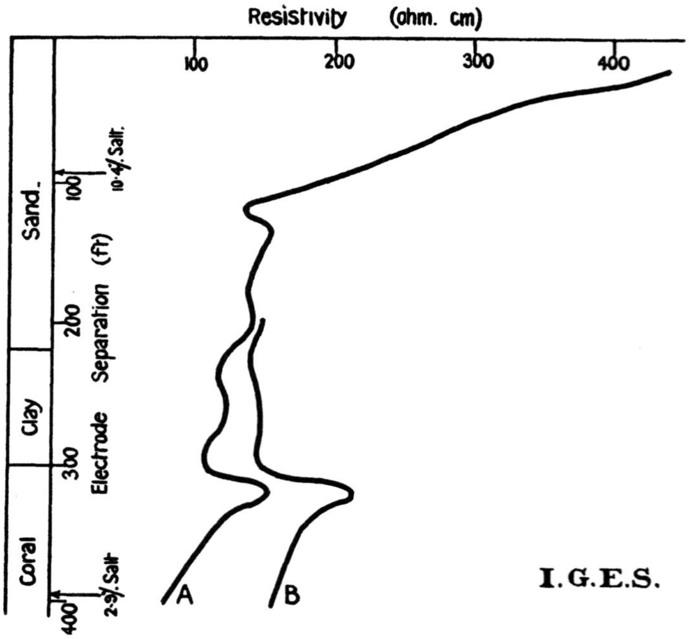

Fig. 19. Resistivity curve and bore section at Koonda

they increase with increasing separation until at 400 ft. one value is 100 per cent. greater than the other. In the case shown the two curves have the same characteristics, and in practice, since the absolute value of the resistivity is not required, the structure could be obtained from the uncorrected curve. For larger separations, however, the induction effects become more important and probably would mask the real shape of the curve, but the resistances measured would then be so low as to be outside the range of the apparatus in its present form. The measurements at Koonda were made with increments of 20 ft. up to a separation of 400 ft. and in general the resistivity decreased with increasing depth; the slope of the curve being greater for the small separations. The bore log in

Fig. 19 shows that up to a depth of 300 ft. sand and clay are present, and that below this level there are layers of soft and hard coral. Down to a depth of 400 ft. two water horizons were met, one at 94 ft. containing 10·4 per cent. of salt, and the second at 394 ft. with 2·9 per cent. of salt. It seems permissible to associate the marked change of slope at 120 ft. with the first water level[a], although the agreement in position is not as good as might be expected. It is possible that the peak at 320 ft. may be due to some local surface anomaly, but as no check measurement was made this is not certain. On the other hand a band of hard coral some 40 ft. thick occurs at 300 ft. and may perhaps be responsible for this feature. The continuous fall from 320 ft. up to 400 ft. may well be due to the influence of the second water horizon at 394 ft.

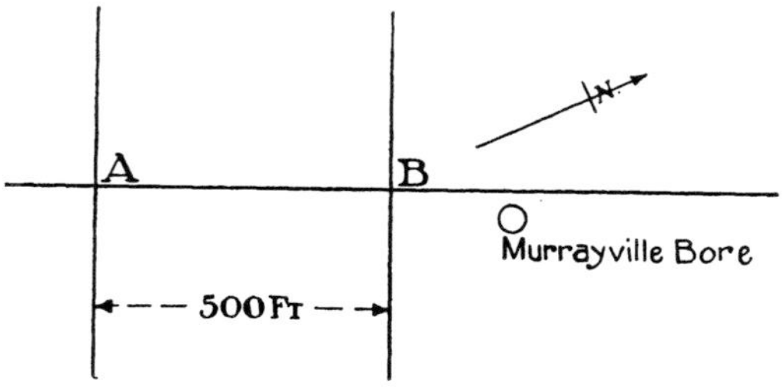

Fig. 20. Layout at Murrayville

Murrayville. The lines shown in Fig. 20 were laid out and measurements made along them with the centres of the electrode systems at the points of intersection. The values obtained in the field for the two pairs of resistivity curves about *A* and *B* are shown in

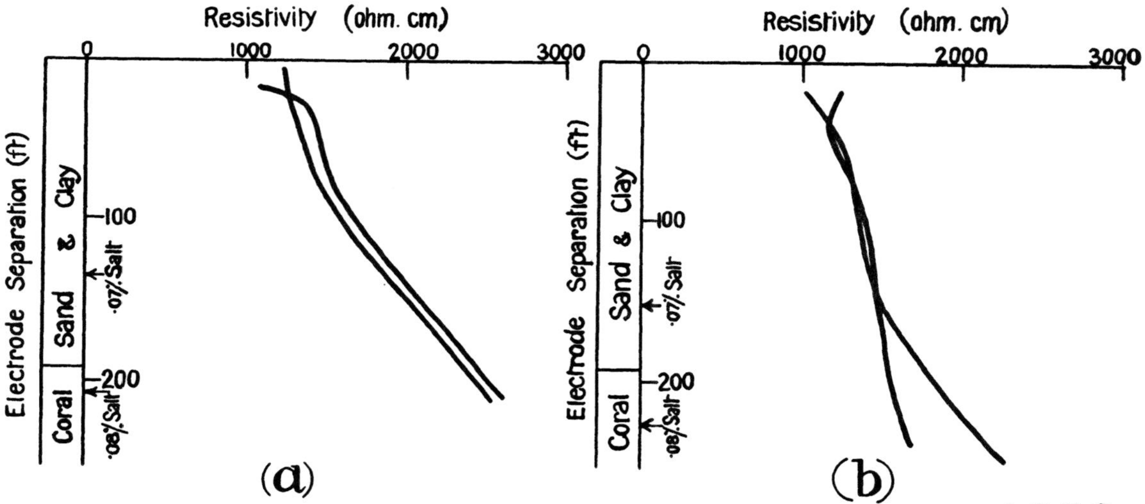

Fig. 21. Resistivity curves and bore section at Murrayville
(*a*) Curves about *A*. (*b*) Curves about *B*.

Figs. 21 (*a*) and 21 (*b*) respectively. The resistances here were sufficiently large for induction effects to be neglected, but in all cases the commutator was rotated at a constant speed of 18 revolutions per second.

The results at Murrayville differed from those at Manpy and Koonda both in absolute values of the resistivity and in the shape of the curves. In the latter areas, in

[a] There were no means of checking these levels at the time the resistivity test was made. It is possible that they have fallen somewhat since the bore-hole was put down.

which the bores contain water of high saline content, values ranging from 500 ohm cm. to 100 ohm cm. were recorded and in all cases they decreased with increasing separation, while at the fresh water bore at Murrayville the values varied from 1000 to 3000 ohm cm. and the resistivity increased with electrode separation. The agreement between the two sets of determinations at Murrayville about *A* and *B* affords a good check on the general character of the curves, but those at *A* are considerably steeper.

The log of the bore shows a series of thin layers of sand and clay down to 200 ft. followed by 50 ft. of coral. A weak flow of water (0·07 per cent. of salt) was recorded at 135 ft. It will be seen that the resistivity curves do not show any marked anomalies which may be associated with an abrupt change in resistivity. This may be due to the numerous alternations of beds of clay and sand which merely yield a mean value of their separate resistivities; or possibly the whole region is saturated with water, which would also tend to mask the effects of individual strata.

Conclusions. Although these tests were not extensive, and were confined to a single type of geological structure and water occurrence, the observations obtained suggest that the method employed has important possibilities in saline areas, as a means of determining the depths to ground water levels. From Fig. 17 it will have been noted that an abrupt change in the resistivity occurred at the depth of the water table, and there is some reason for believing that, with further field experience, the absolute values of the measurements in any area would provide a useful indication of the salt content of the water; low resistivities corresponding to high salt contents.

In areas such as the Mallee, where water of various salinities may occur at three or four different horizons, it should be possible, by this method, to make a distinction between them and, possibly, to locate the horizon at which the salinity is at a minimum. It is considered that the provision of such information, together with the depth to the first water level, would prove of immense value in many of the arid parts of Australia and South Africa. In the case of fresh water areas the evidence obtained is far from conclusive, since only one fresh water bore was investigated and there it appeared that the resistivity change at the water table was insufficient to produce any indication on the curve. If this is found to be a characteristic feature, then the method might be applied to a search for fresh water areas in arid districts, by the indirect procedure of eliminating those areas in which a drop in resistivity was recorded. It should be pointed out, however, that in these saline areas and with the apparatus usually employed in these resistivity determinations, some difficulty arises on account of the low resistance measurements that have to be made. Nevertheless, it should be possible, by further experiment and slight modifications of the method and apparatus, to overcome this trouble and it is recommended that the investigations commenced by the I.G.E.S. should be continued in other areas where underground water conditions afford suitable opportunities.

TEST AT LAVERTON (E. 9)

For the purpose of carrying out further tests by the Gish and Rooney and other resistivity methods advantage was taken of an area lying 12 miles to the west of Melbourne, in which a series of bores had been put down by the Geological Survey of Victoria to prove the thickness and character of the brown coal deposits that occur there. The general structure of the district is illustrated by the line of bore-holes shown in Fig. 22[a], and the general uniformity of the geological sequence may be concluded from the fact that these bores lie two and two and a half miles apart respectively.

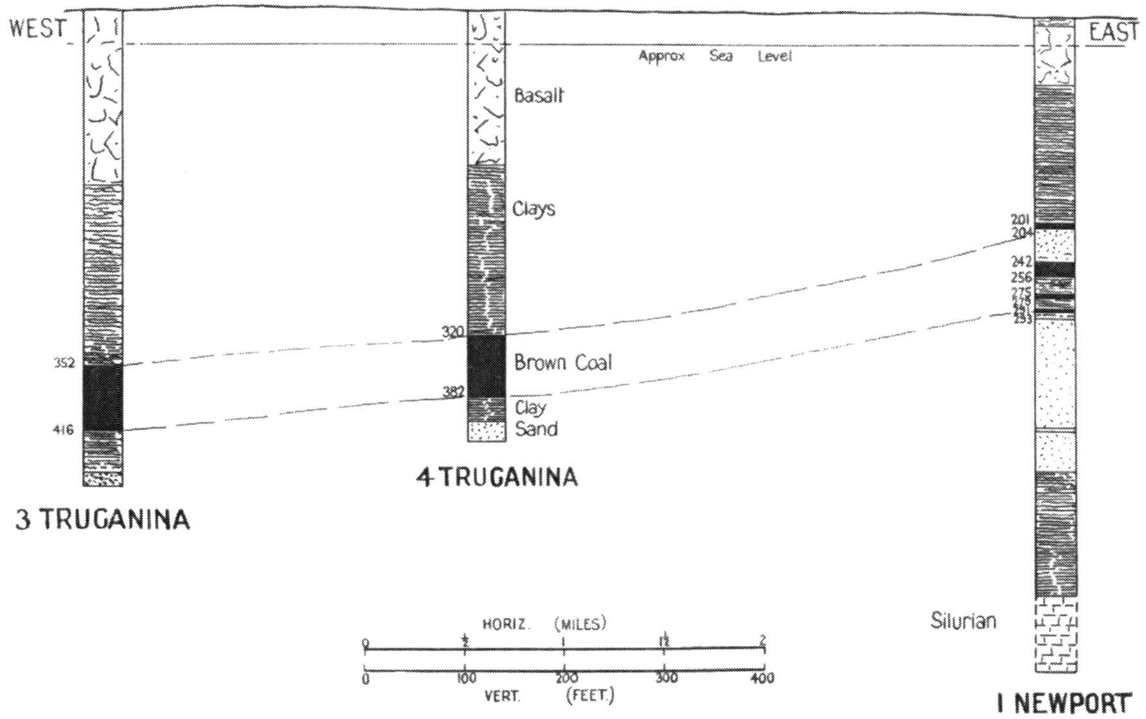

Fig. 22. Line of bores at Laverton

The tests in question were carried out over the No. 3 Truganina Bore which lies one and a half miles to the N.N.E. of Laverton Railway Station. The surrounding country, which is a flat open plain, entirely free from timber, provided excellent conditions for field work. From Fig. 22 it will be seen that the surface soil, some two feet thick, rests directly upon a thick sheet of basalt which extends down to a depth of 175 ft. from the surface. This rests in turn upon a series of clays and mudstones, within which a 64 ft. bed of brown coal occurs at a depth of 352 ft. This series of unconsolidated formations is saturated with brackish water, containing about 0·3 per cent. of salt (about 250 ohm cm.), which was first struck in the bore at a depth of 32 ft., and now stands

[a] From data kindly supplied by the Director, Victorian Geological Survey.

28 ft. From the evidence provided by the No. 1 Newport Bore it may be assumed that the silurian bed-rock would also have been encountered at No. 3 Truganina had the bore been continued a few hundred feet deeper.

In selecting this area for tests it was hoped that a large resistivity ratio would exist between the basalt and the saline formations underlying it, and that the thickness of the basalt sheet might, therefore, be determined.

The observations were made at the No. 3 Truganina Bore on the three parallel lines *A*, *B*, and *C*, shown in Fig. 23, the electrical contacts[a] being exceptionally good owing to the dampness of the surface soil. The tests were made by the Gish and Rooney method and the results obtained are shown in Fig. 24 by the three curves *A*, *B*, and *C*, which for convenience are plotted against the geological section obtained in the bore. It will be seen that the curves have no marked characteristic form, and that at small separations the resistances vary from 2000 to 7000 ohm cm. For larger separations, in which the surface layers do not play such an important part, the curves show more consistent values and, excluding the zone of surface irregularity, it may be concluded that the resistivity of the ground at a given depth is fairly constant throughout the area.

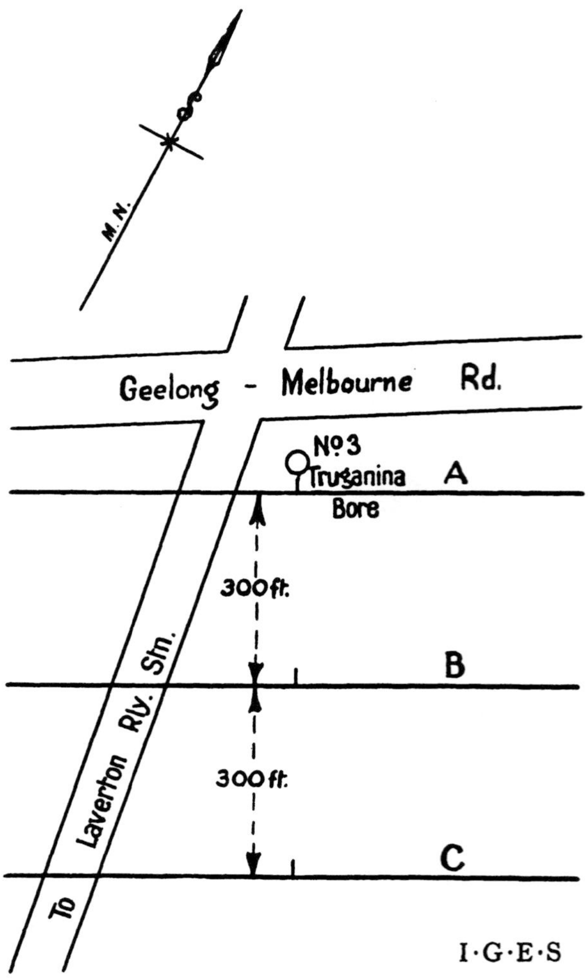

Fig. 23. Layout at Laverton
Measurements carried out along lines *A*, *B*, and *C*.

In spite of the lack of similarity between the curves, a careful examination shows that points of greatest curvature occur on all three of them at approximately the same separation and that these points may be correlated with the various interfaces. On the graph these points are shown by circles, and corresponding points on each curve are denoted by the same letter. The values recorded are given below Fig. 24 in the form of a table. Although the agreement between the last two columns is well within the accuracy to be expected from the method it would be necessary to carry out further measurements in the same neighbourhood before certain of these correlations could be accepted.

[a] Measurements showed that the contact resistances were less than 200 ohms and in some cases as low as 20 ohms.

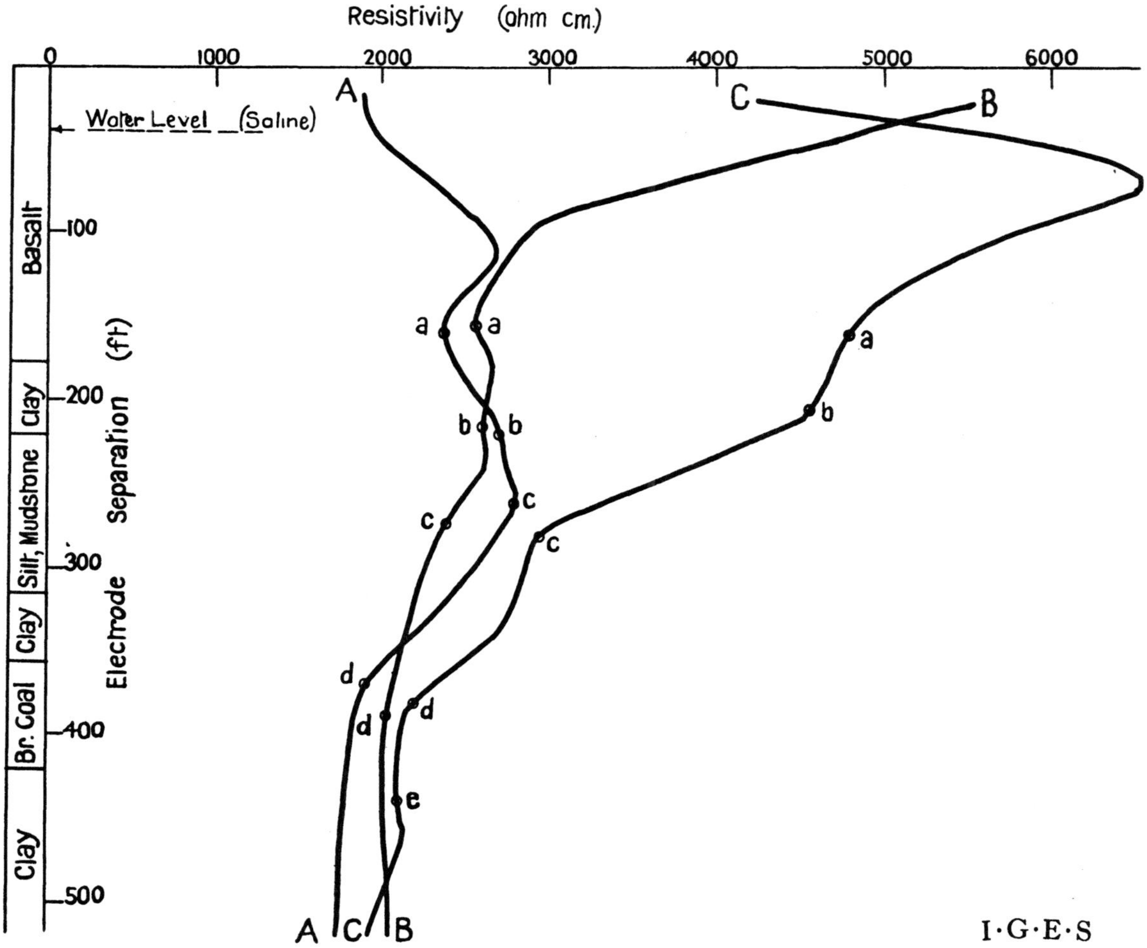

Fig. 24. Resistivity curves at Laverton

Point on curve	Interface associated with point	Depth in ft. from			Depth	
		A	*B*	*C*	Mean depth ft.	From bore ft.
a	Basalt—clay	160	155	160	160	175
b	Clay—mudstone	220	215	200	210	220
c	Mudstone—clay	260	270	280	270	315
d	Clay—brown coal	370	390	350	370	355
e	Brown coal—clay	—	—	440	440	420

The differences between the curves for small separations, and the low resistivity values obtained for the basalt, may be attributed to certain factors, the influence of which was not realised until after the work had been completed. The first of these concerns the physical character of the basalt. Although this has been recorded as 175 ft. in thickness it is apparent from the drill log that it is far from being homogeneous. It is mostly recorded as broken basalt interspersed by layers of basaltic clay from 1 to 14 ft. in thickness. Also from the fact that salt water was met with in the bore at 37 ft. there can be little doubt, when the electrical evidence is also considered, that the basalt sheet is largely water-logged. Owing to the saline character of the water, and the presence of these clay bands, low and irregular resistivity values would be expected. It is also conceivable that the discrepancies in value at short separations may be due, to some extent, to the proximity of the bore-hole. The salt water rose in this hole to within 28 ft. of the surface and it is possible that it has penetrated the surrounding rock in such a manner as to cause local anomalies.

TEST AT GULGONG, N.S.W. (E. 10)

The work was carried out some two miles to the north-east of Gulgong township and within the area that had already been surveyed magnetically. The geology of the district is described in detail in the magnetic section of this work (see p. 188), and it is unnecessary, therefore, to do more than mention the position and structural features of the ground that was subjected to these resistivity tests. The area chosen was on the Gulgong Gold and Diamond Lease, since some definite structural information was available there from a series of bore-holes, and from the New 44 Shaft. At this point a gold bearing deep-lead lies at a depth of 180 ft. from the surface, on a granite bedrock and is overlain by a sheet of basalt which extends within 20–40 ft. of the surface. The surface cover consists of dry sandy soil, which was found to afford very poor electrode contacts.

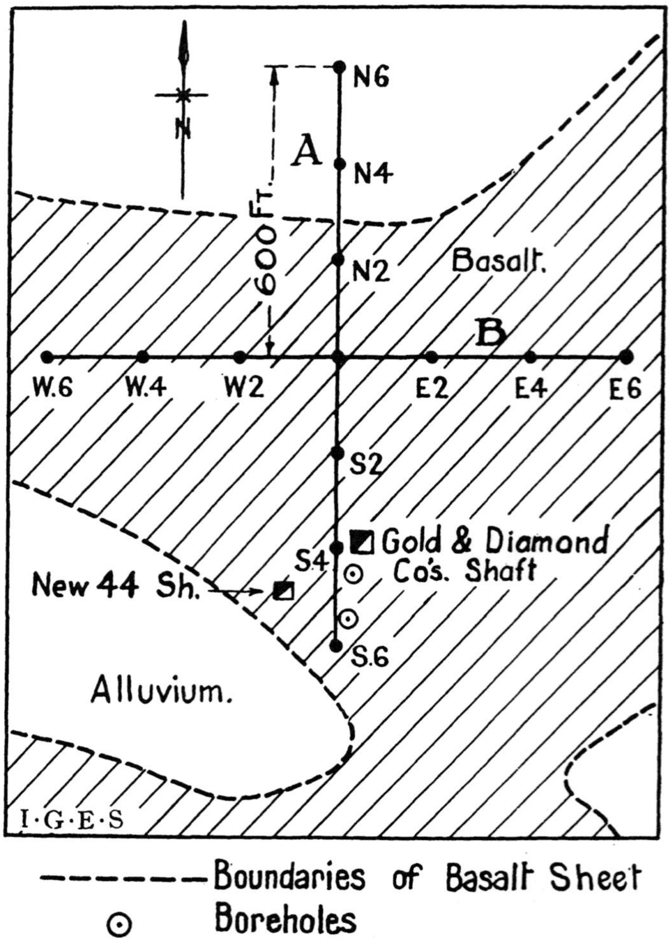

Fig. 25. Resistivity layout at Gulgong. Basalt covered area shown by shading

The object of the investigation was to determine the depth and thickness of the basalt and possibly the depth to bed-rock. The selection of the site of the operations was rendered much easier by a knowledge of the basalt boundaries that had previously been determined by magnetic methods (Fig. 133). The datum point was fixed about 400 ft. due north of the Gold and Diamond Company's shaft and two lines, 1200 ft. in length, were run out in the directions shown in Fig. 25. Unfortunately the datum point was fixed before the completion of the magnetic survey and it was found later that one of the lines extended beyond the limits of the basalt sheet.

The contact resistances at the electrodes were of the order of 1500 ohms and as a consequence small currents only could be used. Although the sensitivity of the potentiometer circuit was decreased it was found that the measurements could be repeated to within 3 per cent. The results obtained by the Gish and Rooney method are shown in Fig. 26; curve *A* showing the results along the north-south line and curve *B* along the east-west line. Both curves show a sharp drop at the commencement (the resistivity for a separation of 10 ft. was 19,000 ohm cm.) followed by a continual rise. At a separation of 25 ft. each curve shows a decided indication of a discontinuity which is attributed to the top of the basalt sheet. A second inflection occurs on both curves at 140 ft. which may be correlated with the lower surface of the basalt. The slopes of the curves at 140 ft. decrease showing that the medium which is entering into the measurements has a lower resistivity than the basalt. At this depth the sands and gravels underlying the basalt will be saturated with water and their resistivities may well be of the order of 4000 ohm cm. or less. Immediately after the decrease in slope at 140 ft., an increase occurs at 160 ft. which is probably due to the granite bed-rock. Beyond this point the similarity between the two curves ceases.

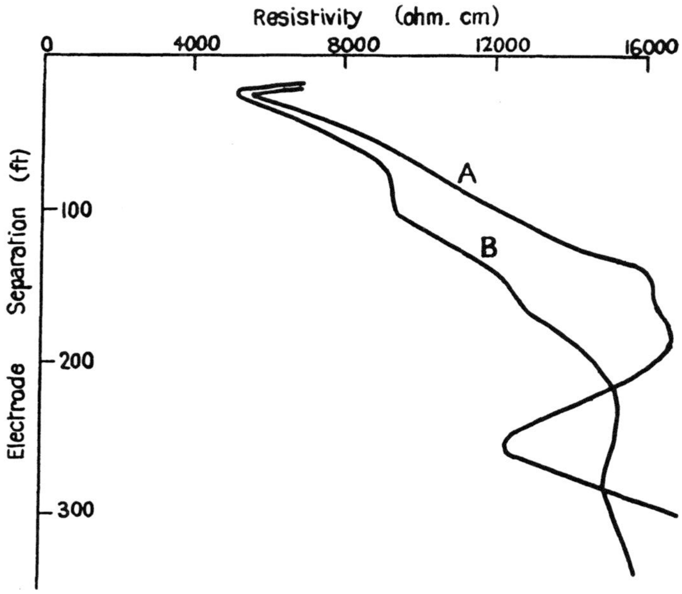

Fig. 26. Resistivity curves at Gulgong

It should be remembered that these depth determinations are average values for the area surrounding the observation point, and since the thickness of the basalt decreases towards the edges of the flow it follows that the depths determined at the datum point will be somewhat less than the actual depths. The bores, some of which are situated near the edges of the flow, show that the depth of the lower surface of the basalt varies from 120 ft. to 150 ft. while the bed-rock occurs at 180 ft.; which figures are in good agreement with the experimental determinations. The two curves provide an example of the effect of surface anomalies, for curve *B* shows an indication at 80 ft. which is absent on curve *A*. The peculiar behaviour of curve *A* for electrode separations of 200 ft. and greater may be due to the effect of the edge of the basalt. At 200 ft. separation the current electrodes are 300 ft. on each side of the observation point and it will be seen that on the north-south line, in this neighbourhood, the northern electrode is passing beyond the edge of the basalt. If this electrode moves into a region where the average resistivity is higher, the measured resistivity will drop (see p. 255), the higher resistance in this case being due to the increased effect of the granite bed-rock which approaches nearer to the surface towards the boundary of the basalt sheet. An attempt was made to carry out a survey over the same ground with the "Megger" Earth-tester, but the high contact resistances made it impossible to use this instrument.

Determinations by the electrode probe method were carried out along the same southern and western lines from the datum point shown in Fig. 25, the far current electrode being placed 2000 ft. due east and north respectively. The resistivity curves *A* and *B* are shown in Fig. 27[a]. On curve *A*, which refers to measurements with the potential electrodes due south, an indication occurs at 30 ft. which agrees well with the depth of the overburden. On *B*, with the electrodes west, the first indication occurs at 40 ft., and this agrees with the surface anomaly observed on the Gish and Rooney curve *B* at the 80 ft. separation (corresponding to potential electrode positions 40 ft. on each side of the datum). In both cases this local effect, which is near the surface and 40 ft. west of the datum point, tends to reduce the effective resistivity. The only other resemblance between the two curves is the definite change of slope on both at 140 ft.; which corresponds with the depth of the basalt sheet. No indication of bed-rock occurs but, as the interval between the bottom of the basalt and the surface of the granite is only 20 ft. while the potential electrode separation was 30 ft., it is hardly to be expected that the effect on the curves would be noticeable.

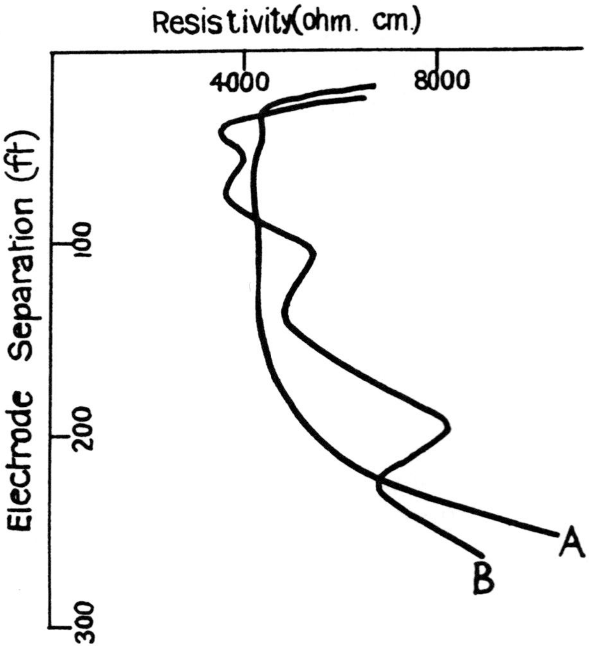

Fig. 27. Electrode probe results at Gulgong

The practical results of these tests in the Gulgong field may be tabulated as follows:

Geological feature	Gish and Rooney	Electrode probe	Bore-hole evidence
	ft.	ft.	ft.
Thickness of overburden	25	30	24–40
Depth to bottom of basalt	140	140	120–160
Depth to bed-rock	160	—	180

[a] In the electrode probe method the electrode-separation value is the mean distance of the two potential electrodes from the near current electrode.

General Conclusions

The above series of resistivity tests affords a satisfactory confirmation of the conclusions of Gish and Rooney and others that this electrical method is capable of determining depths to discontinuities to within 15 per cent., and at the same time demonstrates the practical value of the empirical rule enunciated by Gish and Rooney. Although the resistivities measured cover a range from 100 ohm cm. in the Victorian Mallee to 16,000 ohm cm. at Gulgong, these values are small compared with those obtained by other observers in America and elsewhere, who have recorded resistivities of several million ohm cm. The low resistivities encountered in Australia by the I.G.E.S. present special problems on account of the low resistances measured. Such conditions require an increased sensitivity of the measuring instruments; but in many cases sensitivity is seriously limited by the high electrode contact resistances encountered in the field.

In conclusion it should be said that although the interpretation of resistivity curves may be quite convincing, in the light of a known geological structure established by boring, it is another matter to apply the method to the investigation of an unknown structure. Fortunately, however, it is frequently the case in practice that some boreholes, or shafts, are present in the area to be examined. These will afford favourable opportunities for obtaining characteristic curves and for correlating them with the known geological sequence at these points. With the aid of such curves it will often be possible to extend the investigations into unexplored ground and to make satisfactory interpretations of the results obtained.

(C) EQUIPOTENTIAL LINE METHOD

Introduction. Although one of the earliest of electrical prospecting methods, the tracing of equipotential lines is still practised and in many types of problem it can be recommended for general reconnaissance purposes. In simple cases, such as sulphide lodes extending to within 50 ft. or so of the surface, it may prove fully effective, giving the positions of the conductive bodies with all the accuracy required for practical purposes. For deeper deposits, occurring under less favourable conditions, the equipotential method will do little more than provide ill-defined indications, and electrical methods of greater precision must then be applied in order to obtain more detailed information regarding the positions of the conductors.

Surveys by the equipotential line method are carried out by passing either direct or alternating current conductively through the ground between two widely separated electrodes. A detailed discussion of the conditions that arise when direct current is used in this way has already been given on pp. 20–24 in connection with the resistivity methods and, from Fig. 4, it will have been seen that the potential distribution around each electrode, and throughout the intervening country, can be represented by a series of equipotential surfaces, the traces or outcrops of which, at the ground level, are known as equipotential lines. These lines can be traced out on the ground with a simple portable circuit, consisting of two electrodes for making contact with the ground, connected by 100 ft. or so of insulated wire to a galvanometer or, when alternating current is used, to a pair of telephones[a]. In homogeneous ground the configuration of these lines, over each half of the field, will be precisely as shown in Fig. 4, in which the line *PQ* is an axis of revolution. Characteristic distortions in the equipotential lines will occur, however, in the neighbourhood of any local body of widely different resistivity; the behaviour near a body of lower resistivity having already been shown in Fig. 6. In this case the distortion of the lines is of a very mild character, owing to the round shape of the body *C* and to the fact that the distance between the electrodes *A* and *B* is not great enough in relation to the size of *C*.

Layout of Equipotential Line Survey[b]. In practice the electrodes are spaced, up to a mile or more apart, so that over the section of ground under examination the equipotential lines will be approximately straight and parallel to one another. A field layout of this kind is shown diagrammatically in Fig. 28, from which it will be seen that the actual equipotential line survey is restricted to the rectangle *abcd*, and that work is not undertaken within $\frac{1}{4}ab$ of the electrodes, where the lines are too curved to be really effective in revealing the position of conducting bodies, and where there is an insufficient penetration of the equipotential surfaces. Consideration of this figure will also show the importance of having some previous knowledge of the main geological features of the district and of the general character of the mineral deposits to be located. In this case the conducting bodies C_1 and C_2 have an east-west line of strike and, in horizontal section, are elongated in form. As a rule, the general trend of the mineralized zones will be known beforehand; in which case the electrodes *A* and *B* should be placed so that the general bearing of the equipotential lines will be at right angles to the strike direction. In Fig. 28 the distortions caused by the conducting bodies C_1 and C_2 in the neighbouring equipotential surfaces are typical of those sometimes experienced in actual practice, but it will be easily understood that they would not occur if the electrodes *A* and *B* had been laid out along a north-south bearing; since the equipotential lines would then be running parallel to long axes of C_1 and C_2. It

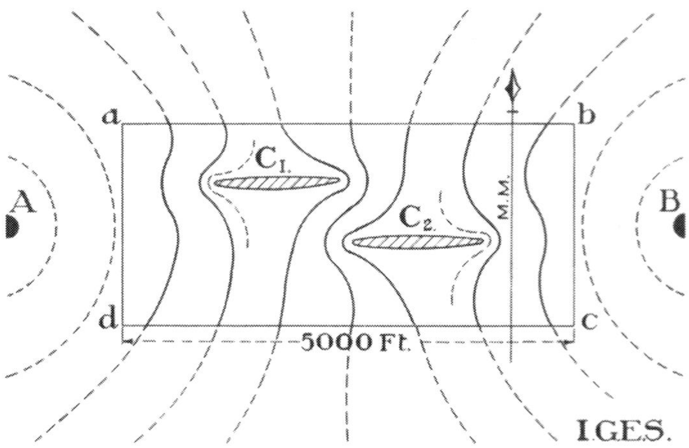

Fig. 28. Equipotential line survey with point electrodes

[a] When direct current is used the apparatus is the same as that described on p. 17 for the spontaneous polarization method. The procedure with alternating current and the instruments employed are described in detail on p. 261.

[b] Described in greater detail on p. 260.

will also be understood that, owing to their elongated form, the best results are to be obtained with lodes or lenticular ore-bodies, rather than with rounded masses or steeply pitching shoots or pipes which are equidimensional in horizontal section.

Direct or Alternating Current. The above description of the equipotential line method is strictly true only in the case of direct current. When alternating current is used[a] the electrical conditions are more complex, owing to the inductive effects of the alternating field upon all conductive materials (e.g. the ground, mineral bodies, etc.) lying within the area under investigation. Secondary fields are generated within these conductors and differ in phase with respect to the primary field upon which they are superposed. It follows, therefore, that in the strict sense, equipotential lines cannot exist in a compound field of this type[b], since the equipotential points can only have an instantaneous existence[c]. These out-of-phase conditions are not so pronounced, however, as to preclude the tracing of lines over the ground surface and, in actual practice, this procedure is found to be quite effective in revealing the general current distribution. In spite of theoretical objections they are generally known as equipotential lines and the fact that alternating current is usually preferred to direct current indicates that no serious disadvantages arise. Although alternating current was used exclusively by the I.G.E.S. it is possible, from previous experience with direct current, to enumerate the following conclusions regarding their relative merits.

Advantages of Direct Current.

1. The equipotential points can be located with greater precision than is possible in the case of alternating current.

2. The galvanometer gives a clear indication of the direction in which to move the electrode to find the equipotential point.

3. The theoretical basis of the method is satisfactory and comparatively simple.

4. Measurements of potential difference in absolute units are easily made, and values can be assigned to the equipotential lines.

Disadvantages of Direct Current.

1. Periodical reversals of the applied current are necessary in order to distinguish the applied potentials from those due to natural earth currents, electrode polarization effects, etc. This operation seriously delays the work.

2. Non-polarizing electrodes must be used, and to make good earth contacts with them, when searching for equipotential points, is a slow and tedious procedure, except in open country where the ground is very moist.

3. Serious interference is sometimes caused by leakage from a local direct current supply.

4. The primary electrodes by which the current is passed into the ground are rapidly eaten away by electrolytic action. Serious trouble is experienced from this cause if these electrodes are made from long lengths of wire, as is often the case.

[a] A frequency of 500 cycles per second is commonly employed.

[b] Such a field has an elliptically polarized character, see p. 59.

[c] Ambronn, *Elements of Geophysics*, pp. 136–9. McGraw Hill, New York, 1928.

Advantages of Alternating Current.

1. The portability and convenience of the movable circuit, consisting of pointed steel electrodes, headphones, and an amplifier. This is decidedly more mobile than the galvanometer and non-polarizing electrodes required when working with direct current. (Compare Figs. 3 and 194.)

2. Owing to the use of a valve amplifier, the power of the generator may be greatly reduced. A 350 watt alternator is as effective in the field as a kilowatt dynamo.

3. The same alternating current generator can be used for a variety of surface potential and electromagnetic methods.

4. No interference from local electrical supplies will be experienced since the frequency is so different from those used commercially.

Disadvantages of Alternating Current.

1. Potential values cannot be given to the equipotential lines and, therefore, potential profiles cannot be prepared[a].

2. Lack of precision owing to phase displacements, which become so pronounced in highly conductive areas that it is sometimes impossible to locate any definite minima[b]. In some cases such conditions will lead directly to the discovery of a conductive body and, therefore, may perhaps be considered as an advantage in the use of alternating current rather than as a defect.

3. In windy weather the strain of listening with headphones is considerable.

4. Interference with local telephone systems is sometimes caused, especially those having earth returns.

Although it is possible to locate equipotential points using direct current the work would be too slow and tedious to be practicable under average field conditions. As a rule a direct current equipotential line plan would be prepared by interpolation from a large number of potential determinations made over the area along a closed series of straight line traverses. With alternating current it is usually possible to trace out the equipotential lines directly on the ground without serious difficulty. As the direct location of the lines is an essential feature of any equipotential line method it is apparent that the use of alternating current has very substantial advantages, and in all further references to the method it will be assumed that alternating current is employed.

Electrodes. There are two forms of electrode used for passing current into the ground. They are generally known respectively as the *point* and the *linear electrode systems*. With the former the equipotential lines, over the working area, take the form of arcs of circles of large radii, as shown in Fig. 28, whilst the production of a uniform series of parallel straight lines is the object of the linear type of electrode (Fig. 29).

[a] This can now be done by the A.C. potential ratio method described on pp. 50–53.

[b] Gall has shown that this difficulty may be overcome by including a small search coil in series with the detecting circuit. Since the E.M.F. induced in this coil is in quadrature with the current, whereas that in the ground is approximately in phase with the current, it is possible, by suitably orientating the search coil, to balance the out-of-phase component and so locate the *in-phase* equipotential points with precision. (See *Nature*, 21st March 1931.)

A point electrode in its simplest form would consist of a metal spike driven into the ground but, in practice, the resistance introduced by a single contact of this kind is so great in relation to that of the ground portion of the circuit that a very inefficient system results. For example, if a pair of single point contacts is employed it will usually be found that from 75 per cent. to 90 per cent. of the total fall in potential through the circuit takes place at, and in the immediate neighbourhood of, the electrodes. In order to reduce this contact resistance the point becomes in practice a multipoint arrangement and, therefore, the term *multipoint electrode* is more descriptive and is to be preferred for general usage. Effective electrodes of this type can be made in many ways and may have any shape, providing that good contacts are made with the ground at a sufficient number of suitably spaced points. It has been the practice of the I.G.E.S. to use long lengths of stranded galvanized iron wire, pegged firmly to the ground by metal spikes[a] at intervals of 20 ft. or so. Various arrangements used by the Survey are illustrated in Fig. 30, in which the dimensions given must be regarded as a rough guide only. The actual extent and number of contact points required will vary in different districts according to the moisture in the surface formations and other local conditions. It is usual to test the resistance of each electrode[b] so that the requisite number of contacts may be determined.

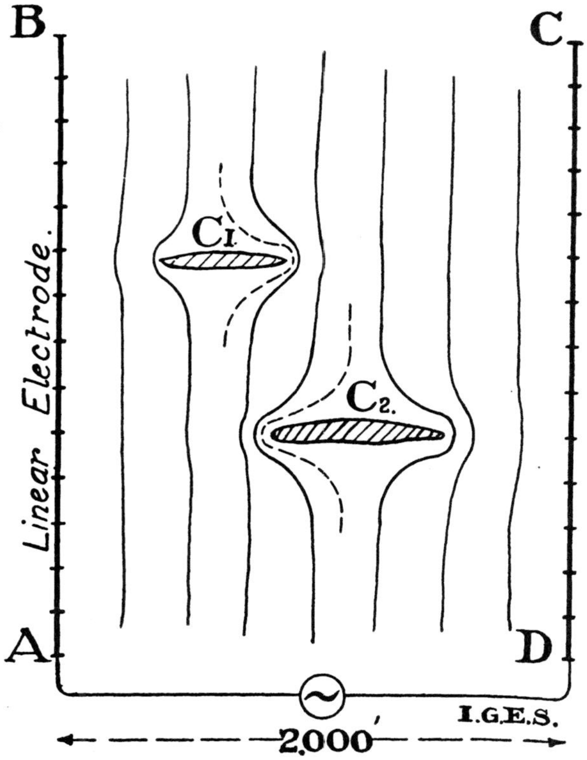

Fig. 29. Equipotential line survey with linear electrodes

Types *a* and *b* in Fig. 30 are suitable where open areas of moist ground are available. Type *c* has been used in forest country in which narrow glades were the only places where damp ground was found, and type *d* represents the case of a bare wire pegged to the bed of a creek in dry country where no better facilities exist[c].

Although the area covered by each of these multipoint electrodes may be of the order of several thousand square yards, the dimensions are actually so small in relation to the distance of a mile or so which separates them that, for all practical purposes, they behave and may be considered as point contacts. Any irregularities in the shape of these electrodes may also be disregarded, since they will have no appreciable effect on the

[a] These spikes are described on p. 264.
[b] The Evershed and Vignoles "Megger" Earth-tester is convenient for this purpose.
[c] See p. 264 for special methods of improving electrode contacts.

trend of the equipotential lines in the central part of the area where the prospecting work is carried out.

The linear electrode system[a] is a special case of the multipoint electrode in which each electrode consists of a single length of bare copper or iron wire laid out in a straight line at right angles to the probable strike direction of the conductive bodies to be located. The distinctive feature of this system, however, is that the length of each of these linear electrodes is of the same order as that of the interval between them (see Fig. 29). These long lengths of bare wire are pegged to the ground in the same manner as the multipoint electrodes already described, the spacing of the metal pegs depending on local conditions (e.g. humidity of the surface soil etc.). With this arrangement the equipotential surfaces become long semi-cylinders, coaxial with the linear electrodes, of which the traces at the ground surface appear as a series of parallel straight lines—provided that the ground is homogeneous and that the electrode wire is maintained at the same potential throughout its length.

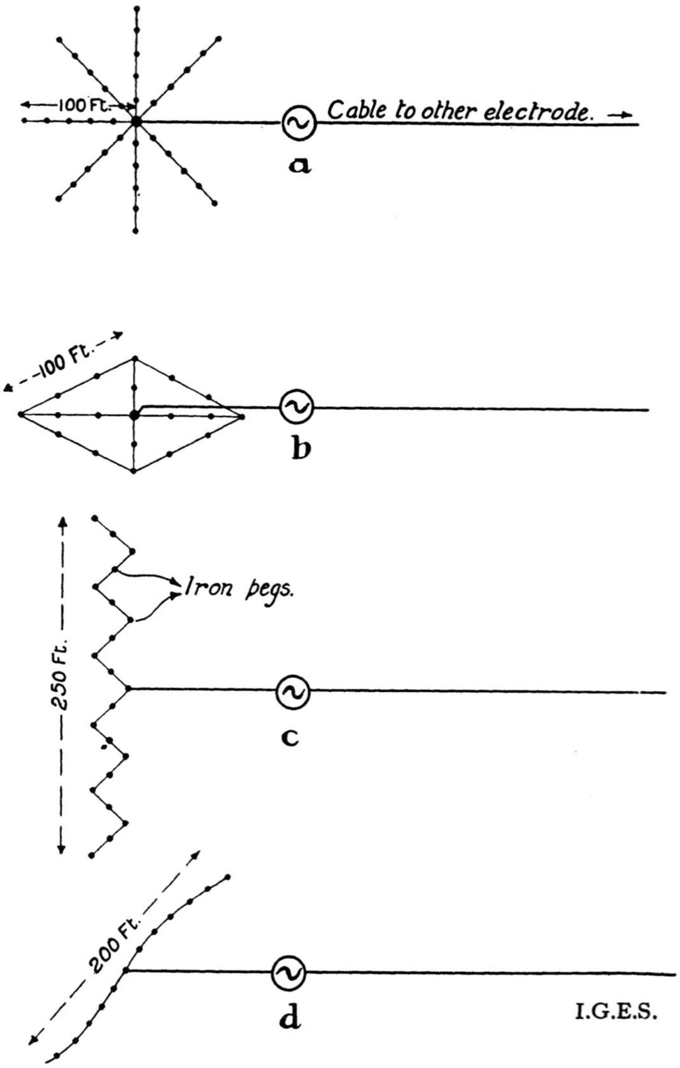

Fig. 30. Types of multipoint electrodes

From the appearance of Fig. 29 it might be concluded that the linear electrode system has marked advantages over the ordinary multipoint arrangement, since the effective working area over which the equipotential lines maintain their direction and parallelism is far greater. This may be the case in areas where the linear arrangement will operate satisfactorily; for example, in flat swampy country where the moisture in the ground is uniform over the whole length of the electrode. In other types of country, where marked variations in surface humidity may exist between the hill tops and valley bottoms, and over different types of surface

[a] Sundberg, Lundberg, and Eklund, "Electrical Prospecting in Sweden," *Sveriges Geologiska Undersökning*, Årsbok 17 (1923), No. 8, Stockholm, 1925.

formations, it will usually be found that the potential distribution along the line of the electrode is very irregular[a]. It will also be subject to sudden changes if the quality of the contacts at any part of the line is altered; for example, as a result of a rain storm following a long spell of dry weather. Under such circumstances the conformation of the neighbouring equipotential lines may be very unsatisfactory and will not always remain constant from day to day.

In typically dry countries, such as Australia, it is only occasionally that local circumstances will permit the use of linear electrodes and, even so, it is more convenient as a rule to select a piece of marshy ground, or a creek bed, in which to place a multipoint electrode of the type shown in Fig. 30. It is possible, however, to regulate the potential drop along the length of a linear electrode by increasing the number of contacts over the dry areas, and reducing them over wet patches. Further control can also be effected by connecting the insulated cable from the alternator to a number of points along the line of the electrode, instead of to one end only, as shown in Fig. 29. As a rule, however, it is impracticable to devote time to adjustments of this kind, which are by no means satisfactory.

A further trouble experienced with linear electrodes arises as a result of their great length (1000 yards in some cases) and the frequent necessity of laying them across roads and railways, and other places where the wire may be severed by wheeled traffic, or where portions may be stolen or interfered with by ill-disposed persons. Damage of this kind to a localised multipoint electrode has no appreciable effect on the configuration of the electrical field, but in the case of linear electrodes the distribution of potential throughout the area may be changed to a serious extent, while the tracing of the equipotential lines is in progress; possibly without the fact becoming known until some time later, when the equipotential points have been surveyed and plotted on paper.

Considering all circumstances, it has been found that under average field conditions the localised multipoint electrode is more serviceable than the linear system, the use of which is restricted in practice to special types of country. It is clear, however, that the former also has its weak points; largely on account of the steepness of the potential gradient in the neighbourhood of the earth contacts and owing to the fact that the electrodes must be placed so far from the area to be surveyed. For example, in the cases of small mining claims, it may sometimes be necessary to ask permission to place the electrodes on adjoining properties.

Interpretation of Results. In most cases it is relatively easy to make a general qualitative interpretation of the results obtained by an equipotential line survey, by marking off the axes of the conductive zones upon the plan, and determining by inspection which of them are the best defined and are the most in accordance with geological possibilities. It is impossible, however, without the use of auxiliary electrical methods, to subject

[a] Even where the ground contact along the line of the electrode is uniform, but of very low resistance, the greater proportion of the current will enter the ground in the neighbourhood of the connection between the electrode and the lead from the generator. Hence, each electrode will tend to behave as a point source rather than as a uniform line source of current.

the results to any graphical or mathematical treatment which will enable the depth of the conductive body, or its strike length and width, to be determined. It is apparent, however, that the nearer the conductive body lies to the ground surface the more pronounced and acute will be the flexures in the equipotential lines, and with a little experience there is no great difficulty in recognising an indication of outstanding importance. On the other hand, there is great difficulty in assessing the value of mild indications, and especially those that extend over a wide area. They may be due to a deep-seated conductive body of economic value or, equally well, to some inhomogeneity in the geological formations lying nearer to the surface. It will be understood, therefore, that the interpretation of an equipotential line survey is largely empirical in character, based on previous experience of the method and on a satisfactory knowledge of the geological features of the area under examination. Owing to the speed with which the ground can be covered, this method will sometimes prove of great value for general reconnaissance purposes, but the conductive zones that are located by it should then be re-examined by other electrical methods which are capable of making more exact determinations.

(D) A.C. POTENTIAL RATIO METHOD[a]

Principle of Method. This method has been developed and employed by the I.G.E.S. for making more accurate investigations of surface potentials and phase relations than are possible by the equipotential line or other generally known electrical methods. The system, which is shown in Fig. 31, is simple in application and in several instances it has proved of great value, both for direct prospecting purposes and for making detailed examinations of indications obtained by other methods. It consists in principle of a bridge, in which *A*, *B*, and *C* are contacts made with the ground, and *AO* and *CO* are ratio arms in each of which a condenser and a resistance are connected in series. Alternating current, at about 500 cycles per second, is passing through the ground between distant earthed electrodes, and the purpose of the device is to compare the potential drops V_1 and V_2 between the equidistant pairs of contacts *AB* and *BC*, and also to determine the difference in phase angle between them. To accomplish this, the capacity and resistance in each arm are adjusted until an exact balance of potential and phase is indicated by silence in the headphones. When this condition has been obtained the potential drops V_1 and V_2 are directly proportional to the total impedances in the arms *AO* and *CO* respectively (see p. 268) or, in terms of

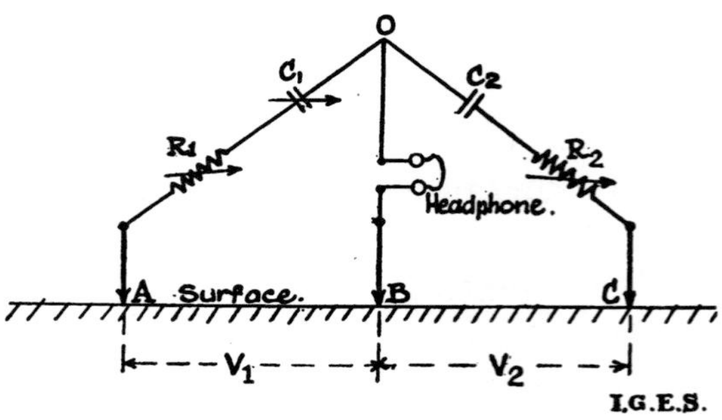

Fig. 31. Principle of A.C. potential ratio method

[a] British Patent Application No. 19120/30 by A. Broughton Edge. The three contact ratio arm bridge method (A.C.), from which this system has been developed, was devised by A. Broughton Edge in 1925, and was first used in N. Rhodesia during that year. For special applications see p. 352.

the resistances R_1 and R_2 and capacitative reactances X_1 and X_2, which are read off directly from the instruments:

$$V_2 = \frac{R_2 \sin \tan^{-1} (R_1/X_1)}{R_1 \sin \tan^{-1} (R_2/X_2)} V_1$$

and

$$\theta_2 - \theta_1 = \tan^{-1} R_2/X_2 - \tan^{-1} R_1/X_1,$$

where $\theta_2 - \theta_1$ is the difference in phase angle between V_2 and V_1; being of positive value when V_2 is leading V_1, and vice versa[a].

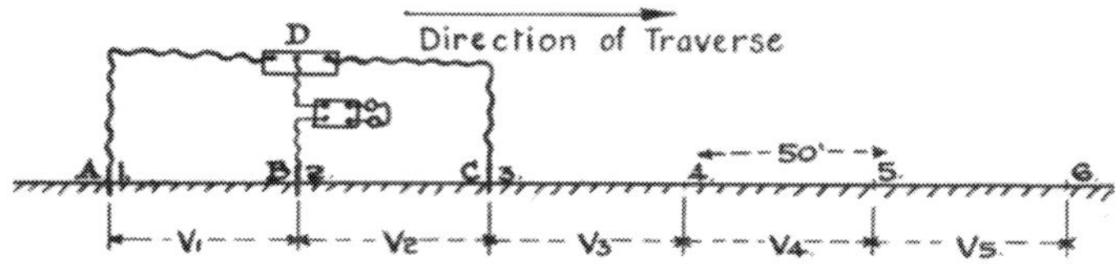

I.G.E.S.

Fig. 32. Field procedure in A.C. potential ratio method

Field Operations. As a rule, the observations are made along straight line traverses, in some cases using the same linear or point electrode system that has previously been employed for an equipotential line reconnaissance survey. In order to illustrate the field procedure, and the form of the equipment employed, the first few stations of such a traverse are shown in Fig. 32. The intervals between stations are usually of the order of 25 to 100 ft. or more, depending upon the nature and scale of the problem under investigation, and these distances are marked out by pegs along the line which has been selected.

The apparatus consists of three pointed steel rods, A, B, and C, driven into the ground and connected by insulated leads to the ratiometer D, which is carried by the operator, together with the headphones and amplifier as shown in Fig. 33. The latter is connected into the lead from the central contact B—thus completing the bridge system already shown in Fig. 31. Observations are made along the length of the traverse; the operator and his assistants moving forwards, one station at a time, so that the potential ratios and differences in phase angle are obtained for every successive pair of stations.

I.G.E.S.

Fig. 33. Ratiometer in use, with amplifier on the operator's back

A brief consideration of the system illustrated in Fig. 31 will show that, in order to obtain a balance, it is essential that the earth at contact B should be intermediate in potential with respect to the contacts at A and C. It may happen, therefore, that, even on straight line traverses, points will be met with beyond which it is impossible to proceed in the ordinary way. For example, if two traverses

[a] In Australia, these potential and phase values were obtained from the sliding scale described on p. 274, but a direct reading ratiometer has now been designed (see p. 272 note a).

P and *Q* are laid out across an area in which the potential distribution is as shown by the equipotential lines in Fig. 34, it is apparent that the current direction is consistent along the traverse *P*, but that a reversal takes place at the point *y* on traverse *Q*. It will happen, therefore, that the survey party will be brought to a halt on traverse *Q* when the apparatus is applied to the stations *x*, *y*, *z*. It will always be possible[a], however, to find an auxiliary point *y*′, such that balances are obtainable both at *xyy*′ and at *zyy*′, and by including this additional step the traverse may be continued in the ordinary manner. The exact location of these reversal points is a matter of importance, since they mark the electrical axis of the indication. The two forms of potential variation curve, represented by traverses *P* and *Q*, are shown in Fig. 35; from which it will be seen that in the case of the latter the potential values change sign at the point *y*.

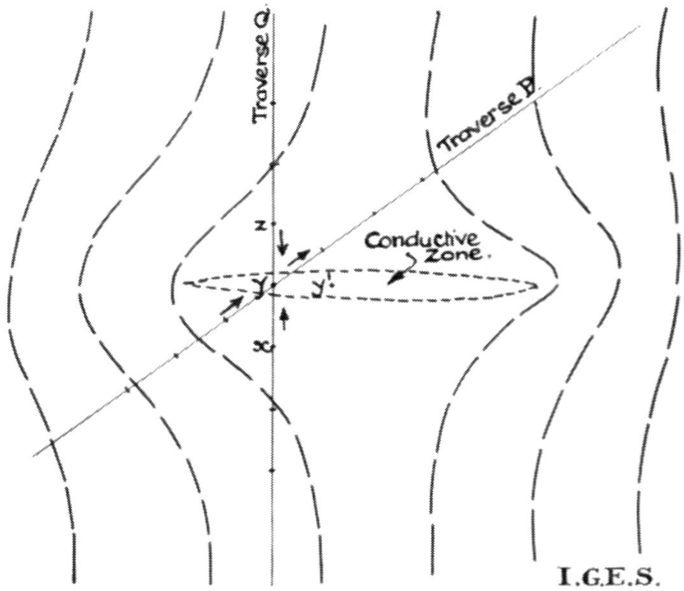

Fig. 34. Potential ratio traverses *P* and *Q* in relation to potential distribution indicated by equipotential lines

Plotting Results. From these observations potential and phase variation curves can be plotted which, in many cases, will reveal important features that may not even be suspected from the usual type of equipotential line survey. The method of preparing these curves is simple. In the case of the traverse shown in Fig. 32 the potential difference V_1 would be taken as an arbitrary unit and V_2, V_3, ..., etc. would then be referred to it by successive multiplication of the individual ratios V_2/V_1, V_3/V_2, ..., etc.[b] that have been obtained right along the line of traverse. The multiplied ratios so obtained

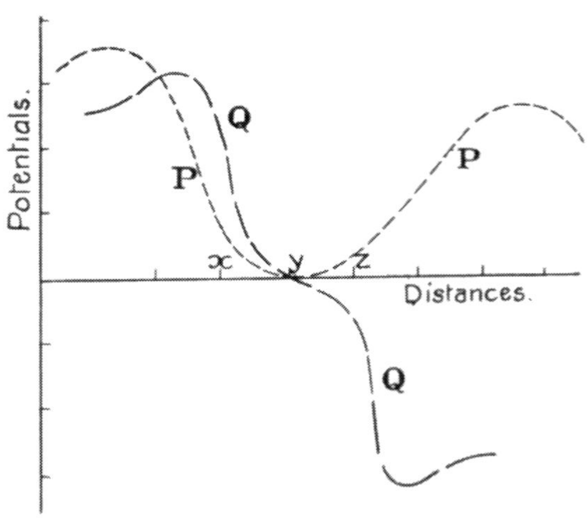

Fig. 35. Potential variation curves along traverses *P* and *Q* (Fig. 34)

[a] This is not always an easy matter, and in some cases two auxiliary stations must be occupied before the *blind spot* can be passed.

[b] This multiplication is of complex numbers, i.e. V_2/V_1 and V_3/V_2 are multiplied to obtain the magnitude of the product and their phases are added to obtain the phase of the product.

are then plotted as shown in Fig. 36, above the mid-points of the station intervals to which they refer. The changes in phase angle are dealt with in a similar manner by simple addition or subtraction according to whether they lead or lag from one pair of stations to the next. As in the case of the potentials, the phase angle of V_1 may be regarded as the datum phase to which the subsequent determinations are referred. It will be observed, of course, that these curves show nothing more than variations in potential and phase *per unit distance* on the earth's surface. They must be regarded, therefore, as anomaly curves and it should be understood that comparisons of potential values at points along the traverse cannot be made excepting by vectorial addition of the values that are shown.

When this method is being used in conjunction with an electrical survey by some other system, merely as a check test, or in order to obtain more detailed information of the phase anomalies over some particular area, it will usually be unnecessary to do more than carry out a few selected straight line traverses and to prepare potential and phase variation curves in the manner indicated above. If, on the other hand, a complete survey of an area is to be made by the A.C. potential ratio method a different procedure is necessary. Owing to the close of the I.G.E.S. operations in the field it was not possible to conduct any extensive surveys of this kind, but the general procedure was as follows. The working area (e.g. the rectangle *abcd* in Fig. 28) is marked out by two sets of intersecting diagonal traverse lines, with stations pegged along them at 100 ft. intervals[a]. Using the ratiometer in the usual way, observations are made along each line so that potential and phase values may be assigned to every station within the area. These values are determined by *vectorial addition* and must all have reference to a common potential; preferably that of a point lying towards the centre of the survey area. After adjustment of closing errors[b], equipotential and iso-phase lines are drawn by interpolation.

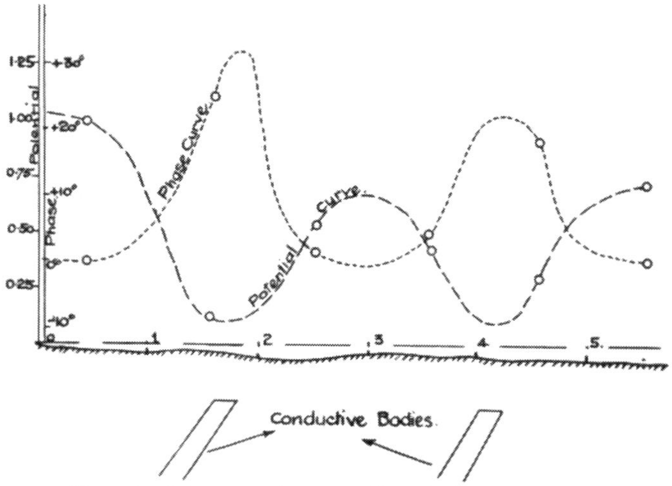

Fig. 36. Diagram of potential and phase variation curves over conductive bodies

Interpretation of Results and Applications of the Method. Field work with this method has not been sufficiently extensive, as yet, to permit any definite account of the potential and phase changes that are met with in the neighbourhood of conductive bodies of various shapes. The fact has been definitely established, however, that variations in

[a] A separation of 300 ft. between parallel traverse lines is suitable for ordinary purposes and gives a sufficient number of stations that are common to both sets of traverses.

[b] In certain cases these were found to be large, possibly on account of some factor that has not yet been properly appreciated.

ground conductivity along lines traversed with the ratiometer are indicated very faithfully by anomaly curves plotted in the manner indicated above. In passing over the more conductive areas the potential drop is recorded in a far more detailed manner than is possible with the equipotential line method, and the axes of conductive bodies may be located with precision. In the case of a steeply dipping lode of good conductivity the curve will show a well-defined peak over each wall, where the current density is relatively high, and an abrupt fall is observed over the body itself, where the high conductivity results in a relatively small potential gradient.

In most cases the curve showing the changes in phase angle per unit distance will exhibit a close relation to the potential curve, as shown in Fig. 36, but important departures from this relationship have been recorded; notably in the cases of conductive bodies dipping at flat angles. It is believed that the phase anomalies have greater individuality and diagnostic possibilities than have the potential values, which are at a serious disadvantage in that they are recorded in purely arbitrary units that do not allow any satisfactory comparison of results obtained in different districts. The full interpretation of the phase changes recorded by this method is a matter which calls for theoretical investigation, accompanied by experiments on a model scale.

Although the system may be employed as an independent prospecting method, in the manner explained above, it is also of particular service as an auxiliary to the equipotential line method which, although of admitted value for reconnaissance purposes, is too crude for detailed investigations. For example, owing to out-of-phase conditions it often happens that the plotting of equipotential lines becomes increasingly difficult as a conductive body is approached, and in extreme cases it is actually impossible to continue them beyond certain points. Under such circumstances the equipotential method does no more than indicate the presence of a conductor in an indefinite manner, and even under the most favourable circumstances there is a certain vagueness about the indications; particularly so when they are of great linear extent. In cases such as these, valuable information may be obtained by making traverses across the ground with the ratiometer[a].

Various other useful applications of the method have been tried out in the field. For example, in an equipotential line survey it is usual to commence the lines at equidistant points along a surveyed base line, and to complete the work without any knowledge of the actual potential drops between successive lines. With the ratiometer it is possible to carry out a calibration traverse along the base line and so assign relative values to each of the equipotential lines. The method has also been applied to the investigation of underground structure by the electrode probe principle described on page 29.

In some districts the method will be at a disadvantage owing to its extreme sensitivity to variations in surface conductivity, but in areas where the overburden is reasonably uniform in character there should be no serious trouble on this account.

[a] A special arrangement of the electrode system may be necessary for this purpose. A pair of linear electrodes laid parallel to the strike of a conductive zone, and on opposite sides of it, will sometimes yield very distinctive indications, not only of the strike axis of the conductor but also of its dip direction.

Application of Ratiometer to Comparison of Magnetic Fields. Although properly belonging to the electromagnetic section of this chapter, a useful application of the ratiometer may more conveniently be mentioned at this point. This relates to the simultaneous comparison of alternating magnetic fields at different observation points in the neighbourhood of conductive deposits. The use of the instrument for this purpose is shown in Fig. 37. In this particular case a comparison between the vertical components of the magnetic fields at the points A and B is illustrated. The two coils C_1 and C_2 each have an area of about 4 sq. ft., and are wound with several hundred turns of insulated wire. They are supported in a horizontal position immediately above the points A and B and are connected together by a pair of insulated wires[a], so that the electromotive forces induced in them will *assist each other*[b]. The ratiometer is then inserted into the circuit in the manner shown, and is operated in precisely the same manner as in the surface potential application of the instrument. By this means the field intensities may be compared and the differences in phase angle determined. Measurements may

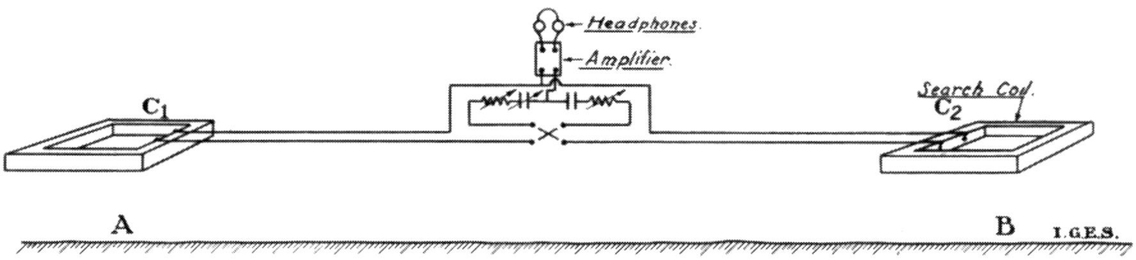

Fig. 37. Comparison of magnetic fields with ratio-arm bridge

be made for other positions of the coils than those shown in Fig. 37 and, in fact, the apparatus may be employed in various ways for investigating the character of the elliptical field; providing that the differences in phase angle do not lie outside the range of the bridge[c].

However carefully the coils C_1 and C_2 are made to match each other, it may be found that a correction factor must be applied in order to allow for slight differences in construction. This factor is determined in the usual way by taking a double set of observations; the first with the coil C_1 at station A (Fig. 37), and the second with coil C_2 at A. Thus, if E is the potential set up in coil C_1 when it is placed in a unit magnetic field, then the potential set up in coil C_2 by the same field will be KE, where K is the required correction factor. Now, if R_1 is the measured ratio of the e.m.f. set up in C_1 with reference to the e.m.f. set up in C_2, when the coil C_1 is at A and C_2 at B, then

$$R_1 = \frac{EH_1}{KEH_2} = \frac{H_1}{KH_2},$$

[a] These wires should be twisted together in order to reduce inductive effects.

[b] If found to be opposed, rotation of one coil through 180° on a horizontal axis will produce the required relation. The same result may be obtained by including a reversing switch in the leads from one of the coils.

[c] It has been suggested by D. C. Gall that this difficulty may be overcome by using a mutual inductance bridge, which for other reasons as well would be more suitable for this kind of measurement. An instrument on these lines is now being made.

where H_1 and H_2 are the field strengths at A and B respectively. Similarly, if R_2 is the ratio of the e.m.f. in C_1 to that in C_2 when C_1 is at B and C_2 at A, then

$$R_2 = \frac{H_2}{KH_1},$$

and
$$K = \sqrt{\frac{1}{R_1 R_2}}.$$

As a rule the comparison of field intensities by this method will be carried out between successive pairs of equidistant stations, marked out along prepared traverse lines. If any marked anomalies occur in the magnetic field at the surface, due to conductive bodies that happen to be crossed by these traverses, the fact will be observed on plotting intensity and phase variation curves[a] in the manner already described for the A.C. potential ratio method[b].

The method described above resembles in some respects the two coil balancing system, due to Sundberg and Lundberg[c], in which two coils are joined *in opposition*, with an amplifier and telephone in the circuit. In this method, when comparing the strengths of the vertical field at two points, the coils are first held in a horizontal position as shown in Fig. 37. One of them is then tilted out of the horizontal until a minimum of sound is observed in the headphones. The relative strengths of the vertical fields at the two points are then given approximately by the secant of the angle of tilt. This method is open to two objections. Firstly, that a horizontal component will usually be introduced into the determination at the point where the coil is tilted out of the horizontal plane; and secondly, that it is sometimes very difficult, or even impossible, to obtain a balance, owing to a difference in phase in the currents induced in the two coils. As these out-of-phase conditions are usually at maximum in the neighbourhood of highly conductive bodies, the use of the ratio-arm bridge is to be preferred. Not only does it enable a satisfactory balance to be obtained, but the determination of the difference in phase angle is an essential part of the prospecting operations.

SECTION 3. ELECTROMAGNETIC METHODS

GENERAL PRINCIPLES

Introduction. In order to prospect an area by the electromagnetic methods an alternating current is maintained in the ground, the distribution of that current being dependent on the electrical conductivities of the various rock and mineral formations. The methods to be described in this section determine this distribution by investigating the magnetic field which is set up at the surface. Conclusions can then be drawn

[a] Lines of equal vertical or horizontal magnetic intensity and iso-phase lines may also be drawn on a plan of the area.

[b] In this application of the ratiometer it is preferable that the magnetic field to be examined should be produced inductively by means of loops as explained on the opposite page.

[c] Eve and Keys, *Applied Geophysics*, p. 122, University Press, Cambridge, 1929.

concerning the geological structure of the area; particularly with regard to the presence of highly conductive zones such as ore-bodies or salt water horizons.

Inductive and Conductive Methods. The alternating current may be caused to flow in the ground either by connecting the poles of a suitable generator to a pair of earthed electrodes, or by induction from insulated loops suitably disposed on the ground surface, and excited by an alternator or valve oscillator. It is with the latter methods that this section is concerned. The electromagnetic field above the area to be examined, which results from the currents flowing in the ground as well as from those in the exciting loop or electrode system, is investigated by means of portable search coils.

The inductive methods of applying current to the ground are of particular advantage in rocky or arid districts where the electrode system often presents difficulties owing to high contact resistances. Another advantage of the inductive over the conductive methods is that the fields set up by the former can reach a body lying beneath an insulating layer[a], whereas the currents applied in the latter may be unable to do this. On the other hand, a highly conductive surface layer acts as a screen to regions below it and gives rise to extremely unfavourable conditions for both methods. The ability of electrical methods to detect mineral deposits depends upon the ratio that the electrical conductivity of the mineral bears to that of the surrounding formations. For conductive methods it is found that a saturation effect exists whereby a ratio of say a million is of no more practical advantage than one of a thousand. Inductive methods, however, make full use of large conductivity ratios and, in general, the greater the ratio the more certain are the results obtained[b].

Induction of Currents in the Ground from a Loop. To form some picture of the induction phenomena taking place, consider a loop laid on the ground and carrying an alternating current. Actually, either horizontal or vertical loops are used, depending on the nature of the problem, but the response of a conductive body is fundamentally the same in whatever manner it has been excited. Within the horizontal loop (Fig. 38) the magnetic field due to it will be practically vertical at all points on the ground surface and, according to well-known physical laws, currents will be set up in any conductor in the vicinity.

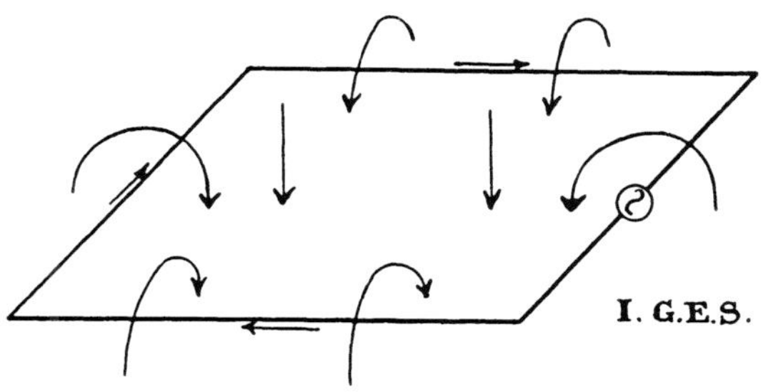

Fig. 38. The magnetic field set up by current flowing in a horizontal loop

The strength of these will be determined by a number of factors such as frequency and strength of the energizing current, impedance of the conductor considered, and its mutual inductance relative to the energizing current. Since the ground also has a definite conductivity, depending chiefly on the nature and quantity of its moisture content, a system of eddy currents will be induced in it, which will give rise to a magnetic

[a] In practice it is unusual to find layers of such extremely high resistivity.

[b] Mason, *Geophysical Prospecting 1929*, p. 17, A.I.M.E., New York, 1929.

field. If in the neighbourhood of the loop there are also regions of high conductivity such as ore-bodies, saline waters, graphitic schists, etc., then further induced currents of greater magnitude will flow around their peripheries and give rise to fields at the surface having both horizontal and vertical components. Other small currents and fields, such as those due to the capacity effects of the cable, will also arise in the neighbourhood for various reasons which it is unnecessary to discuss here. Thus we have to investigate the combined effect of a primary and a secondary field. The secondary field is the total effect of the magnetic fields due to the currents which are induced throughout the region by the primary field. What must be particularly noted about these secondary fields is that at the surface they have a horizontal component and, more important still, they do not rise to their maximum values at the same instant as the primary or energizing field; in other words, they have a phase lag. The higher the frequency used the greater would be the current induced into the conductive body if the surrounding material were a perfect insulator. Actually, a practical limit exists after which any such advantage is offset by the losses which take place owing to the absorption of the energy of the electromagnetic fields in the overburden. Thus, if too high a frequency is used, large bodies at depth might well be overlooked owing to their weak response, whereas small conductors lying near the surface but of no economic importance would give rise to very large fields.

The Horizontal Component. If the energizing loop is laid out in flat country the primary field is everywhere vertical, so that the approach to a buried conductor is marked by the appearance of a horizontal component in the field, which will attain its maximum values near the boundaries of the body where the induced currents are flowing. The intensity and direction of the horizontal component of the field may be measured by holding a search coil in a vertical plane, and turning it about a vertical axis till the reading on a sensitive voltmeter is a maximum. When in this position the plane of the coil is perpendicular to the direction of the field[a]. In Fig. 39 the change in the value of the horizontal component of the secondary field, as found when a traverse is made across a body (*d*), is shown by the curve (*c*). The horizontal field due to the current in the loop is small since measurements are not made close to the cable. That due to currents induced in the overburden may, however, be large. It is a maximum near the sides of the loop and is a minimum at the centre of the loop. The way in which the intensity of this field varies with distance from the cable is shown in Fig. 39(*b*), and as the search coil measures the combined field it is necessary to subtract the effect of the former in order to obtain the effect of the body alone. If, however, the prospecting is being carried out in hilly country, horizontal components due to topography would be introduced and would not necessarily indicate the presence of conductive bodies.

[a] This statement is not exact, since it neglects the fact that the field is elliptically polarized. It would only be true, therefore, if the major axis of the ellipse happened to be vertical. As there may be considerable departure from the vertical, it follows that the horizontal component of the elliptical field is still an ellipse (see next paragraph), and the plane of the coil, when recording a maximum potential, will be perpendicular to its major axis.

As mentioned above, the secondary fields will lag in phase behind the primary field, and this condition provides the most reliable evidence as to the presence of conductors. It follows, therefore, that the search for conductors becomes a search for out-of-phase conditions in the magnetic field.

Elliptically Polarized Fields. It must be recognzied that the out-of-phase condition gives to the field a characteristic structure. Consider the conditions, which we will simplify slightly, at some point on the ground surface. There is the primary field which is roughly vertical[a], and the secondary field which in the neighbourhood of conductors may be approximately horizontal, and between these a phase difference exists. The intensities of these fields are alternating quantities, but we may represent their maximum values or amplitudes by vectors, which, if no phase difference existed, could be compounded by the familiar parallelogram of forces[b]. As it is, the vectors can be combined by another geometrical construction and it is found that the resultant has no fixed direction, but revolves about the point considered once during each cycle of the alternating field with its extremity tracing out an ellipse[c]. This is termed the representative ellipse[d] since it completely represents the conditions existing at any point of the field.

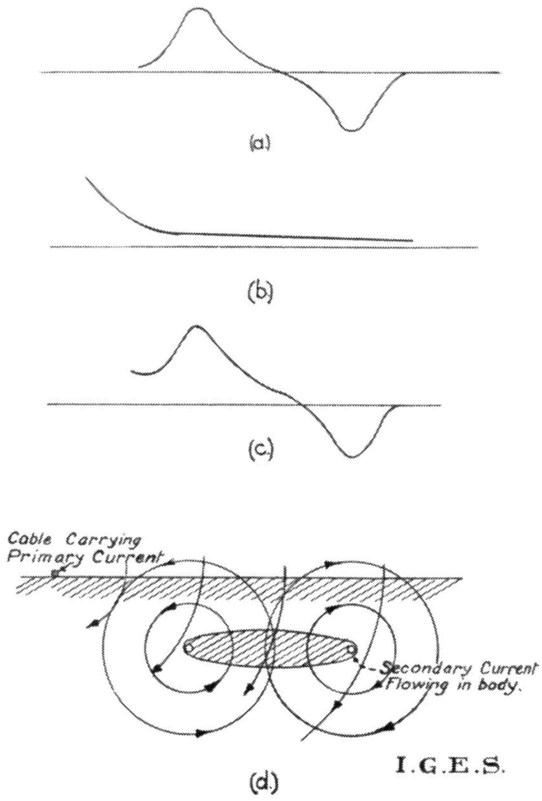

Fig. 39. A conductor gives rise to a horizontal component in the electromagnetic field above it. (*a*) Horizontal field due to body. (*b*) Horizontal field due to current in cable and to general ground conductivity. (*c*) Resultant horizontal field from (*a*) and (*b*). (*d*) Section of the conducting body showing primary and secondary fields

The magnetic field which must be investigated is said to be elliptically polarized, and the more satisfactory of the known electromagnetic methods of prospecting consist essentially of measuring the size and shape of this ellipse at a number of points at the

[a] Still supposing, for purposes of illustration, that a horizontal loop is used.

[b] In this case the field is plane polarized.

[c] The conditions existing at any one point in the field are more complicated than this, although for many practical purposes the above treatment is adequate. An alternating magnetic vector in free space is always accompanied by an alternating electric vector, which differs in phase and does not coincide in direction with the magnetic vector. The combination of these two is the electromagnetic field, and it can be shown that a transmission of energy takes place in a direction perpendicular to the plane containing these two vectors. This flow of energy may be regarded as a propagation through space of the electromagnetic field (see Jean's *Magnetism and Electricity*, or Poynting's *Collected Papers*). In the case of an elliptically polarized magnetic field an elliptically polarized electric field, lying in the same plane, will also be present, and the direction of propagation will be perpendicular to the plane containing these two ellipses.

[d] Ambronn, *Elements of Geophysics*, p. 181, McGraw Hill, New York, 1928.

ground surface. To give a complete specification of the ellipse at a point it is necessary to state:

(1 and 2) The direction of one axis (defined by two angles).

(3) The magnitude of the axis.

(4) The direction of the second axis which, being at right angles to the first, only requires the statement of one angle.

(5) The magnitude of the second axis

(6) The phase of one of the axes of the ellipse with reference to that at some datum point.

Other methods of definition are used but they are equivalent to the above. Fortunately, in the application of the electromagnetic method, a complete specification of the ellipse is rarely required. If measurements are taken in the vicinity of a highly conducting body then a large out-of-phase secondary field will occur. This gives rise to a large minor axis in the representative ellipse; whereas, if no secondary field is present, the minor axis vanishes and only the primary vertical field is left. Thus one way of determining the presence of conductors is by measuring the ratio of the minor to the major axis of the ellipse, a large ratio indicating the presence of a conductor. Actually, many fields may be present at any point, but whatever their directions, intensities, and phases, the resultant can be represented by a single rotating vector as described above.

Investigation of Elliptical Fields. Suppose a magnetic field is explored with a search coil of many turns of fine wire connected in series with a sensitive voltmeter which will give a reading proportional to the intensity of the field threading the coil. In the absence of conductive bodies there is no secondary field, and the magnetic field under investigation is plane polarized. Under such circumstances, it will be found that the voltmeter will continue to show a zero reading whilst the coil is rotated into an infinite number of planes round an axis corresponding to the direction of the field. On the other hand, the presence of a conductor will cause the magnetic field to be elliptically polarized, which can be recognized by there being only one plane in which the coil can be placed, so that it is not threaded by any lines of force; a condition indicated by zero reading on the voltmeter. This is the plane of the ellipse. Its location is most quickly carried out by setting the coil in a vertical plane, and rotating it about a vertical axis until a position of minimum intensity is reached. The coil is then rotated through 90° about the same axis and is then tilted about a horizontal axis in its own plane until the reading on the voltmeter is again at a minimum. A small final adjustment of the coil is made, until the reading on the voltmeter is zero. This is done by tilting it slightly sideways, to one side or the other, to allow for the fact that the minor axis of the ellipse is not, in general, perpendicular to the vertical plane containing the major axis[a]. The coil then lies in the plane of the ellipse as illustrated in Fig. 40(*a*)[b], and the dip

[a] In other words the minor axis of the ellipse may not be horizontal.

[b] In Fig. 40 the plane of the ellipse is vertical. For other cases see p. 278.

and strike[a] of this plane may then be determined. If the coil is now turned through 90° into a plane perpendicular to the ellipse and is rotated about an axis also perpendicular to the ellipse, then in a complete revolution the voltmeter will show two maximum readings and two minimum readings[b]. The maximum reading (Fig. 40(*c*)) gives the value of the field represented by the major axis, and the minimum reading (Fig. 40(*b*)) that represented by the minor axis. Thus the ratio of the axes of the ellipse is determined.

This method is given to illustrate the nature of the problem encountered in examining an elliptical field. There are more convenient ways of obtaining the same information. For example, the coil may be placed in turn in three mutually perpendicular planes at any field point, and the amplitudes and phases of the fields threading it determined, thus giving six independent quantities which are sufficient for the construction of the ellipse. A practical field method, which will be examined in detail in a later section, employs two coils fixed at right angles to each other and placed perpendicular to the two axes of the ellipse. By this means the vertical and horizontal fields are directly compared with each other.

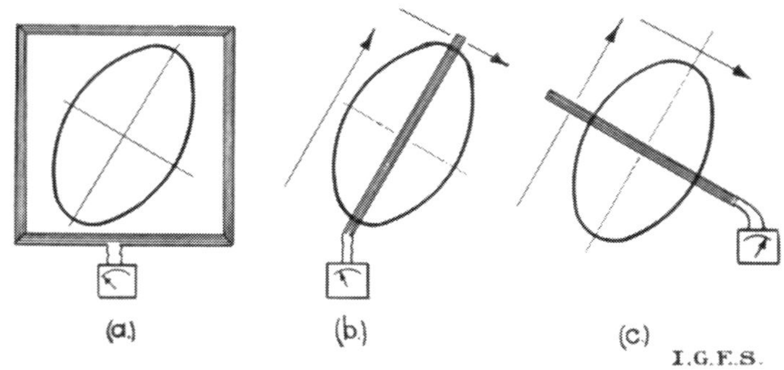

Fig. 40. The investigation of a field ellipse by a search coil

When the secondary field is small the minor axis is small and the ellipse is attenuated. In the absence of conductors it degenerates into the straight line representing the primary field. It may happen that at parts of the field the minor axis becomes equal to the major axis, so that at these points the electromagnetic field is circularly polarized; under these conditions the search coil shown in Figs. 40(*b*) and 40(*c*) can be turned through a complete revolution without changing the reading on the voltmeter.

If the coil, with sensitive voltmeter connected, is set in a plane containing the primary field, which will not be far from the vertical, then the quantity measured may be termed the horizontal component of the out-of-phase field. Under most field conditions, at the lower frequencies, it has been found that the ratio of the axes of the ellipse is approximately the same as the ratio of the out-of-phase horizontal component to the in-phase vertical component. This result is of considerable practical importance

[a] *The dip* is the angle between a horizontal plane and the plane of the ellipse, i.e. the plane in which both the electric and magnetic fields vibrate. For computational purposes this definition leaves an ambiguity as to whether the angle is acute or obtuse. This is removed if the dip is defined as the angle between the direction of propagation of the electromagnetic field and the vertical.

The strike is the direction of the line of intersection of the plane of the ellipse and the horizontal plane. It is measured with respect to the magnetic north or some other specified direction. The ambiguity in this definition is eliminated by defining the strike as the angle between the magnetic east and the projection in the horizontal plane of the direction of propagation.

[b] This operation is similar to the examination of an elliptically polarized beam of light by means of a polariscope. As the eyepiece containing the analysing nicol prism is rotated through 360°, the light is never extinguished but waxes and wanes with an angle of 90° between maximum and minimum intensities.

and will be referred to again when the question of detailed interpretation is considered[a].

As already explained, at any point in an elliptically polarized field there is only one plane in which the search coil can be placed to give zero intensity on the voltmeter. If the coil is rotated about an axis lying in any other plane, then in general a position of minimum intensity will be found, but its determination may not be at all critical. When the phase difference between the primary and secondary fields is small, then it is often sufficiently accurate to suppose that they may be combined by the parallelogram of forces, and that the resultant field is linear. Then, if the search coil is turned into a plane parallel to this resultant, a sufficiently sharp minimum may be obtained and in this position the coil indicates the direction of the field.

Geological applications. The electromagnetic methods may be used for general reconnaissance work, but it has been the practice of the I.G.E.S. to reserve them for the more intensive examinations of areas suspected of mineralization, either as a result of indications obtained by other geophysical methods, or from promising geological evidence. The magnitude of the anomalies and the certainty of interpretation depend upon the electrical conductivity of the bodies sought for being much greater than that of the rocks enclosing them. Thus the sulphides and arsenides of the metals, together with some oxides, provide the best indications. Sphalerite (zinc sulphide) which is practically a non-conductor cannot be detected unless associated with some other conducting mineral. Saline waters, wet clays, and graphitic schists are also fairly good conductors, and in some cases will seriously complicate electrical prospecting operations.

The most unfavourable conditions for electromagnetic work occur in districts in which the solid rock surface is covered by a superficial layer of relatively high electrical conductivity, which in itself gives rise to large out-of-phase fields. Difficulties due to this cause arise in swampy country and in ground carrying any other form of wet overburden, such as damp clay or peat. Surface formations containing saline waters have a similar effect and if the conductivity is extremely high the application of the methods may be impracticable. A notable advantage of the electromagnetic methods is that they may be applied where the ground is frozen and snow-covered, and in some arid districts, where the contacts with the ground that are required for surface potential methods cannot be made. A non-conductive surface layer is not a drawback, since it is readily penetrated by the electromagnetic fields.

If any information is available regarding the strike and dip of the mineralized ground, the loop may be placed in the most favourable position and it is also possible to decide whether the greatest advantage will result from setting it in a horizontal or a vertical plane. Sheet-like bodies are much easier to locate than massive ones. Thus a buried sphere of metal would give a very small effect at the surface compared with that obtained from a disc of the same volume and material and lying flatly at the same depth. The maximum depth at which a body can be detected depends on a number of

[a] See p. 286.

conditions, but in most cases it will be of the order of a few hundred feet. Many workers agree with the conservative statement that, under normal conditions, a suitably conductive ore-body should be readily detectable if lying at a depth equal to about half of its strike extent. Under favourable conditions it is possible to obtain by indirect calculation useful information regarding the body, such as its dip, strike, extent, and depth below the surface.

CLASSIFICATION OF INDUCTIVE ELECTROMAGNETIC METHODS[a]

Horizontal Loop Methods. In this group of methods the exciting loop covers an area of about a quarter of a square mile and usually carries a current at a frequency of 500 cycles per second, although values as high as 10,000 have been used. The principles underlying the better known methods of investigating the electromagnetic field set up may be classified as follows:

1. The comparison of the vertical (or horizontal) fields at neighbouring points on the ground by means of a pair of horizontal (or vertical) coils connected in opposition through an amplifier and telephones[b], or connected to a ratio arm bridge in the manner described on p. 55.
2. The determination of equipotential lines in the magnetic field using a similar pair of coils.
3. The measurement of the intensities and phases of the vertical and horizontal components of the field along the strike of the representative ellipse[c].
4. The comparison of the axes of the representative ellipse by two coils fixed permanently at right angles to one another[d].

Vertical Loop Methods. The exciting loops are much smaller and frequencies ranging from 500 to 80,000 cycles per second may be used.

1. The investigation, by the double coil used in 4 above, of the fields associated with a large vertical loop carrying a 500 cycle per second current.
2. The determination of the strike and dip of the resultant field when a small exciting loop is used carrying currents at frequencies of the order of 30,000 to 60,000 cycles per second[e].

HORIZONTAL LOOP METHODS

Introduction. The excitation of a conducting body by means of current flowing in a horizontal loop on the ground surface has already been discussed in illustrating general induction phenomena. In the work carried out by the I.G.E.S. the resultant field was

[a] Eve and Keys, *Applied Geophysics*, p. 117, University Press, Cambridge, 1929.

[b] Lundberg, *Geophysical Prospecting 1929*, p. 96, A.I.M.E., New York, 1929. Broughton Edge, *Nature*, vol. 127, p. 37, Macmillan, London, 1931.

[c] Lundberg and Nordstrom, *Geophysical Prospecting 1929*, p. 134, A.I.M.E., New York, 1929.

[d] Bieler-Watson method, p. 64.

[e] Jakosky, *Geophysical Prospecting 1929*, p. 138, A.I.M.E., New York, 1929.

usually investigated by a double-coil system, which, however, could be applied in different ways according to the magnitude of the fields to be measured. Since, in general, the major axis of the ellipse is almost normal to the plane of the loop, the primary field and the horizontal component of the secondary field are about 90° out of phase, and are at right angles to each other.

Apparatus[a]. The energizing loops used by the I.G.E.S. have usually been laid out in the form of squares with sides from 1000 to 1700 ft. in length, and consisting of a single turn of insulated wire. A portable generator maintains a current of about 3 amps. in the loop at a frequency of 500 cycles per second.

I.G.E.S.

Fig. 41. The Bieler-Watson double detecting coil in use

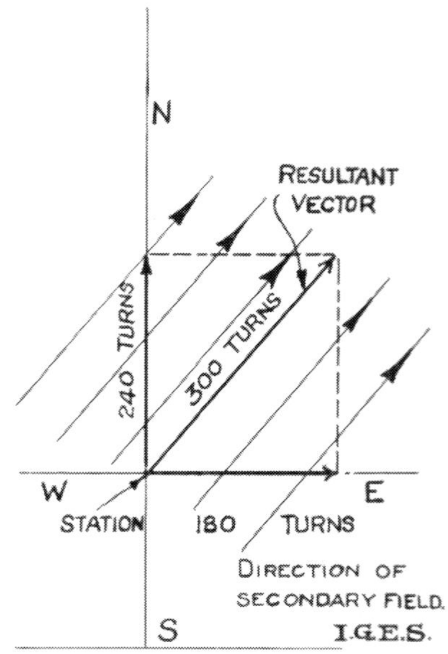

Fig. 42. Method of plotting and compounding field results

The apparatus for measuring the electromagnetic field is shown in Fig. 41, and consists essentially of two coils permanently fixed at right angles to each other, and joined in opposition through an amplifier[b] and telephones. The larger or vertical coil (so called because in operation it is usually in a vertical plane) has many turns of fine wire and is connected in parallel with a condenser of such a value as to tune the coil to the frequency of the alternator. On the other hand the number of turns in the smaller or horizontal coil may be varied over a wide range by the use of multipoint switches. This combination of coils, together with a control box containing the tuning condenser, switches, and a reversing key, is usually mounted on a staff, but it may also be supported on a tripod in such a way that it may be turned into any desired position. The amplifier is designed to amplify selectively the frequency used and

[a] This is described in greater detail on p. 282.
[b] This is on the back of the assistant and, therefore, cannot be seen.

thereby suppress harmonics which are extemely detrimental to the progress and accuracy of the survey.

Operation. To take a measurement at any point of the field the larger coil is held in an approximately vertical plane lying in a north and south direction. The apparatus is then rocked slowly backwards and forwards while the number of turns on the horizontal coil is varied by the switches until the greatest change in intensity is heard in the telephones, thus eliminating harmonics from the measurement[a]. Before a balance is obtained it may be necessary to push over the reversing switch. The apparatus is then turned through 90° about the vertical staff and the operation repeated. The number of turns in the horizontal coil necessary to obtain these balances is then entered into the field book against the name of the station.

Only one such reading would be required if the coils were set perpendicular to the axes of the ellipse, and this might seem to be the obvious procedure. The practical reason for taking two readings along perpendicular lines is that too much time would be occupied in searching by rotation for the plane of the ellipse. The same result is obtained, however, by compounding the two observations in the manner shown in Fig. 42.

In this way the results recorded in the field book are combined to give a number of turns of the horizontal coil. If the major axis is assumed vertical, this number is proportional to the ratio of the minor to the major axis of the representative ellipse. That this is the case can be seen by considering the physical conditions when a reading is being taken. Suppose for simplicity that the vertical coil is perpendicular to the minor axis of the ellipse and the horizontal coil is perpendicular to its major axis (Fig. 43). Then the electromotive force induced in the horizontal coil by the field threading it is altered by varying the number of turns until it balances that due to the vertical coil, when this latter electromotive force has been brought into phase by means of the condenser and reversing key. Then the number of turns of the horizontal coil becomes a measure of the field threading the vertical coil, using the field threading one turn of the horizontal coil as unit.

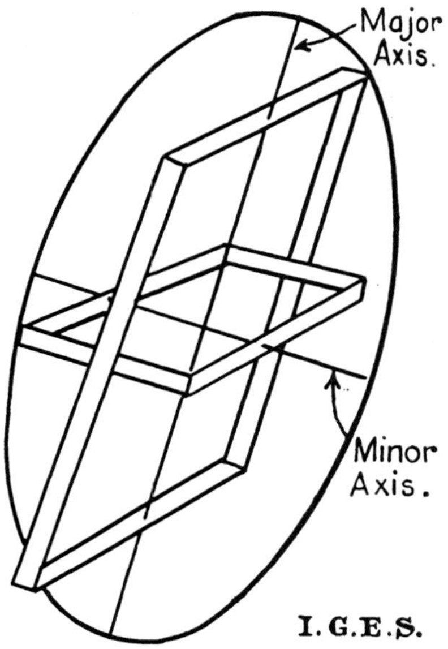

Fig. 43. Double coil in position to compare the axes of the representative ellipse

It sometimes happens that the out-of-phase field is so large that the maximum number of turns in the horizontal coils is insufficient to obtain a balance. In this case a variation of the method is used in which the double coil is mounted on a tripod in the

[a] The late Dr Bieler maintained that with practice it was possible to distinguish the harmonics from the fundamental and that he could always silence the latter and obtain a satisfactory balance. Great difficulty was experienced by other officers of the Survey in training their ears to this degree of proficiency.

manner shown in Fig. 44; provision being made for reading the strike and dip of the vertical coil in any position. A description of the procedure to be followed in making observations by this method is given on p. 285.

Field Procedure. For a close reconnaissance examination parallel lines are surveyed across the loop at intervals of 100 ft. and readings are taken at stations every 50 ft. along them. Over promising indications observations at intermediate stations are also taken. The operator carries the coil and takes the readings which are recorded in the field book by an assistant who also carries the amplifier (Fig. 41). Readings are not taken within about 150 ft. of the cable since the primary field then begins to depart markedly from the vertical. The whole question of interpretation is greatly simplified if it can be assumed that the loop lies in a plane and that the readings of the field are taken approximately in that plane. In hilly country this condition may be difficult to satisfy, but it can often be realized by laying one side of the loop along a ridge, and the opposite side along a neighbouring parallel valley. Placing the loop across a ridge or valley should be avoided.

In any case the area under examination should be contoured, say at intervals of 100 ft. for reconnaissance work or perhaps 25 ft. for intensive work, since irregularities may give rise to considerable local effects. Preliminary to the survey, resistivity tests of mineral, country rock, and surface formations, should be made.

I.G.E.S.

Fig. 44. Double coil mounted for measuring intense fields

Plotting and Interpretation of Results[a]. On a plan of the area under examination the two readings taken at any station are set off on a suitable scale in the directions in which they were measured in the field, and then combined by the parallelogram of forces. This resultant line is a vector quantity giving at each point on the map the ratio of the axes and the direction of the plane of the ellipse. Approximately it gives the direction and intensity of the horizontal out-of-phase component if the primary vertical field is assumed constant over the surveyed area. In the absence of any conductors of importance, these vectors are all directed towards the centre of the loop and decrease in magnitude fairly uniformly as the centre is approached, where they practically vanish. If, however, there are zones within the loop of higher conductivity than the normal ground, then the vector pattern is distorted so that vectors in the vicinity of the conductive zone all point towards it, and reach

[a] Bieler, *Canad. Min. Bul.* No. 193, pp. 638–644, Canad. Inst. M.M., Montreal, 1928.

their maximum values at points above its boundaries (Fig. 45). Each vector, of course, is partly due to general ground conductivity and other causes, but as a rule these are minor effects. Should they attain large values a correction for them may often be applied.

If points of equal vector magnitude are joined up by lines, then a plan of magnetic contours is obtained which will often illustrate the results of an electromagnetic survey in a very clear manner. As is shown in Fig. 46 the shaded conductive zones are indicated by lines of high value. The direction of dip of the bodies may also be deduced, since the lines will lie close together towards the foot-wall sides of the bodies.

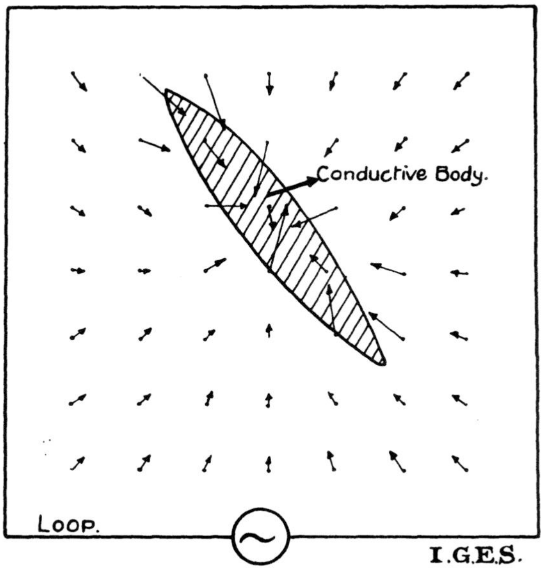

Fig. 45. The vectors point towards the conductor and reach their maximum values above its boundaries

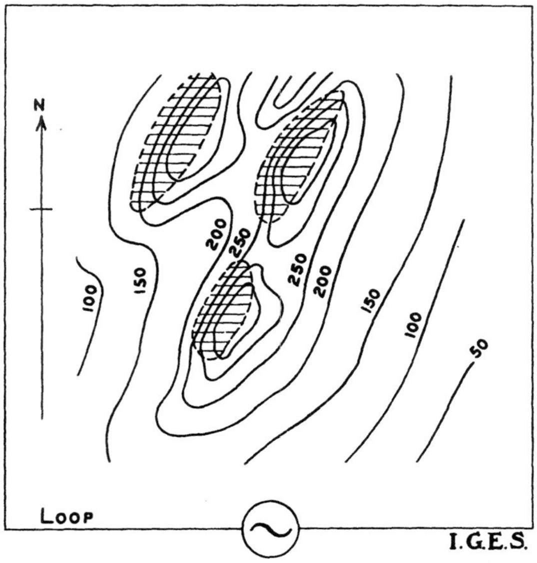

Fig. 46. Magnetic contours surrounding conductive bodies dipping towards the south-east

Profiles of this electromagnetic contour map may be prepared in a manner similar to that followed in preparing topographical sections and they will also assist in the interpretation of the survey results.

If the field conditions are at all favourable, then it is possible by indirect calculation to determine the dip, strike, and lateral extent of the body as well as its depth below the surface of the ground. The procedure adopted is to make reasonable assumptions as to the position of the conductors, and then to compare calculated results with results obtained in the field.

VERTICAL LOOP METHODS

Introduction. In order that a small generator may be used, and so that the electromagnetic fields set up by the loop may have a sufficient intensity over a suitable range, it is necessary to make the process of inducing current in the conductor as efficient as possible. If the body has the form of a thin sheet the maximum current will be induced in it by the energizing loop when the two are lying in parallel planes. Thus a vertical loop is preferable when prospecting for steeply dipping ore-bodies. By the same arrangement the adverse effects of a relatively high general ground conductivity may be minimized. In regions where these conditions are encountered, large out-of-phase fields are set up by the currents induced into the surface layers and may obscure the fields due to conducting ore-bodies. Since these layers of overburden are practically horizontal, then the use of a vertical loop will reduce the induction into them to a minimum. Both of these advantages have been demonstrated by field tests carried out by the I.G.E.S.

A general idea of the fields set up when the loop is vertical, and the body is dipping steeply, is given in Fig. 47. In this case the primary field due to the loop is horizontal[a], and in the neighbourhood of the conducting body a secondary field is set up, which in general will differ both in direction and phase from the primary.

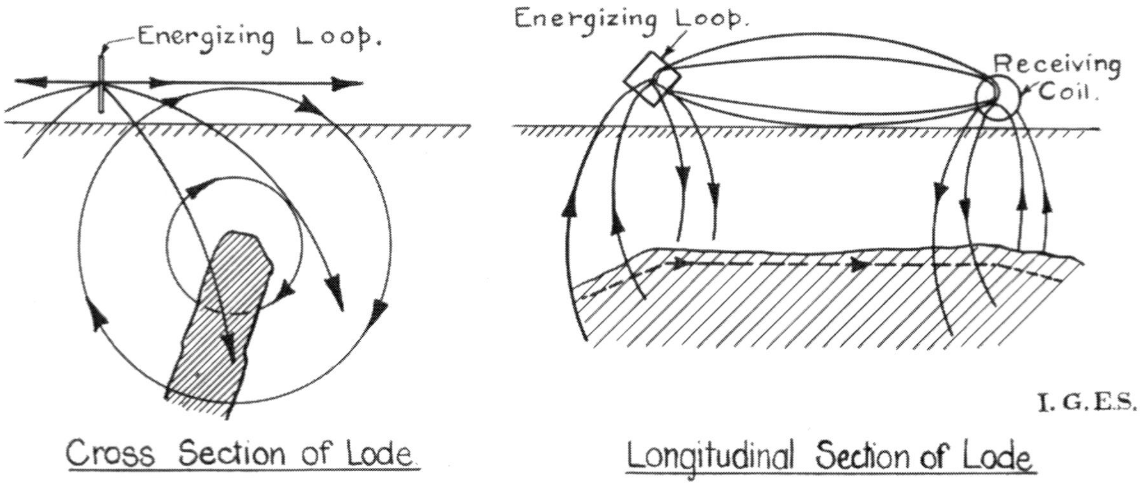

Fig. 47. Fields associated with steeply dipping bodies when a vertical loop is used

Work at Low Frequency. The early experiments conducted by the I.G.E.S. were carried out in an area of high ground conductivity in which the known ore-bodies have the general disposition indicated in Fig. 47. A large vertical energizing loop, carrying a 500 cycle current, was used. It consisted of ten turns of cable, approximately in the form of a triangle, 133 ft. along the base and 45 ft. in height, supported in the manner shown in Fig. 48. The elliptically polarized magnetic field set up in the vicinity of the conductors was investigated by the double search coil and, apart from a few necessary modifications, the procedure adopted was similar to that employed when working with horizontal loops. When the results were compared with those obtained over the same ground with a horizontal loop the advantage of the vertical loop in problems of this kind was very apparent[b].

At 500 cycles per second it was found that measurements could not be made beyond a distance of 500 ft. from the loop in spite of the high degree of amplification employed. As it was impracticable to increase the range by enlarging the loop, and

[a] This will be approximately true when the coil is set up at a greater distance from the lode than is shown in Fig. 47.

[b] See pp. 80–82.

since variations in the frequency of the generator were a source of difficulty, it was decided that this experimental equipment should be abandoned in favour of a triode oscillator giving a high frequency of 60,000 cycles per second, and connected to a small and portable transmitting loop.

Principles of High Frequency Work. In accordance with well-known physical principles the electromotive force induced in a conducting body by a neighbouring loop carrying a current increases with the frequency if other factors remain constant, and the resulting current flowing round the edges of the body will increase accordingly. Advantage of this fact has been taken in order to set up fields of sufficient intensity in the neighbourhood

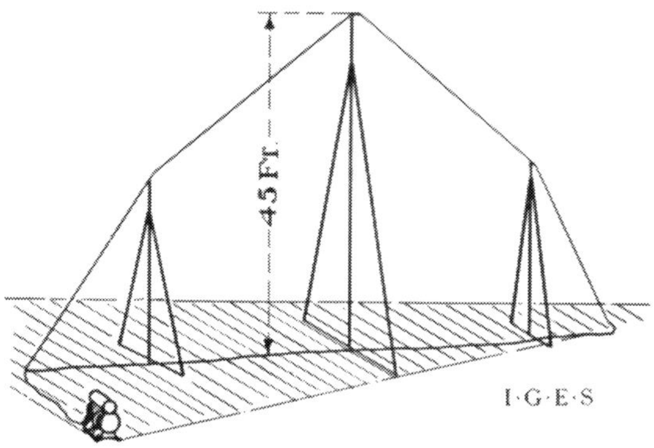

Fig. 48. Large vertical loop used for low frequency work

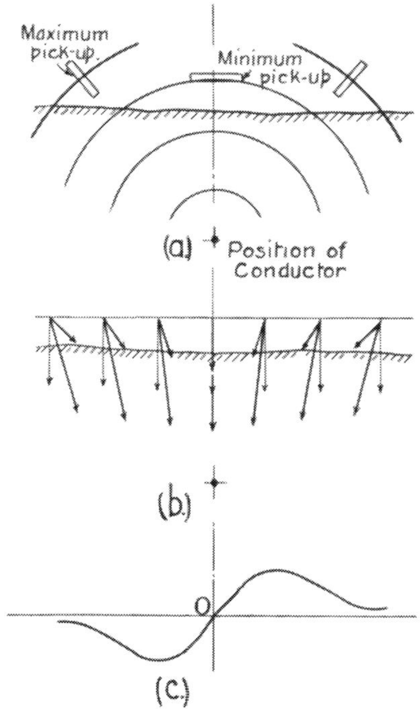

Fig. 49. Direction of fields associated with a linear conductor

of the body with a minimum expenditure of energy, and comparatively high frequencies of from 10,000 to 80,000 cycles per second have been used in geophysical prospecting. In practice, however, the frequency which may be employed is limited, since the energy losses due to the absorption of the electromagnetic fields in the overburden also increase with the frequency. The amount of these losses will depend upon the conductivity of the overburden, which is largely dependent on its moisture content. It follows that in some dry districts the higher frequencies may be used without suffering from this disadvantage. The skin effect should also be considered. It can be shown that the effective electrical resistance of any conductor increases with frequency and, owing to the large cross sectional areas of the conductive bodies met with in practice, this may become an important factor.

The field set up by a current flowing in a linear conductor spreads out from it concentrically as shown in section in Fig. 49 (*a*). At any point in the field a search coil will have an electromotive force induced into it which is dependent upon the frequency, the number of turns of wire in the coil, its area, and the angle it makes with the lines of force. In the diagram the coil is shown in positions in which its response is either a maximum or a minimum. In practice, however, conditions are more complicated since this secondary field due to the conductor is superimposed on the primary field due to the vertical loop.

A hypothetical traverse taken along the ground surface above and at right angles to such a conductor is shown in section in Fig. 49 (*b*). The normals to the primary and secondary fields are indicated by arrows whose lengths are proportional to their intensities. If the phase difference between the fields is small, then the combined effect may be obtained by compounding their representative arrows by the parallelogram of forces as shown. Thus these resultant arrows are normal to the plane of the coil when it is rotated into a position where it is not cut by any lines of force. The angle between the resultant arrow and the vertical at any point is termed the dip and its determination is the fundamental measurement in the search for conductors. A typical indication curve is given in Fig. 49 (*c*), in which the dips are plotted along a traverse running across a linear conductor, which lies vertically beneath the point *O*. Over barren ground the dips are practically zero, but the approach to a body is marked by increasing dips, which vanish directly above it, and then increase again but in the opposite sense, until far away from the body they finally become zero again.

If the plane of the energizing loop is directed towards the detecting coil, and the latter is kept in a vertical plane and rotated about a vertical axis into a position of minimum intensity, then the azimuth angle of the coil is termed the strike bearing. If only the primary field due to the loop is present, then for minimum intensity the detecting coil will be in a plane perpendicular to that of the loop. In the presence of a secondary field, however, the strike reading will differ from this value and thus will provide additional evidence of the proximity of a conductor.

The diagrams in Fig. 49 have been drawn on the assumption that the primary and secondary fields are in phase. This condition is not generally realized in practice and the resultant field is elliptically polarized. When a reading is being taken the search coil does not necessarily lie in the plane of the ellipse, so that the note does not completely vanish in the phones attached to the coil. As a rule, however, good minima may be obtained when the coil is rotated into what may be termed the average direction of the resultant field. If the minima are considered too indefinite to give sufficiently accurate results, then the fields may be brought into phase by adjusting the frequency of the energizing current[a].

Apparatus[b]. The energizing loop was designed to be collapsible and consisted of 10 turns of wire wound in the form of a square with sides 8 ft. in length. The oscillator, built

[a] Jakosky, *Geophysical Prospecting 1929*, p. 157, A.I.M.E., New York, 1929.

[b] Designed by W. G. Baker, B.Sc., B.E., according to the requirements specified by the I.G.E.S. It was built under his supervision in the workshops of Sydney University by kind permission of Professor Madsen.

into a case as shown in Fig. 50, was of the Hartley type and was directly connected to the loop. A single valve with an anode dissipation of 25 watts was used; the high tension being supplied from a small generator run by high capacity accumulators. The transmission is unmodulated and frequencies of 60,000 and over may be obtained.

A circular coil of 50 turns of wire mounted on a tripod as shown in Fig. 51 is used to investigate the electromagnetic field. This coil supplies the inductance in a portable

I.G.E.S.

Fig. 50. High frequency loop and oscillator

Reinartz circuit and the heterodyne note after being amplified can be heard in the earphones. The dip and strike of the plane of the coil in any position may be read from graduated circles.

Field procedure. This equipment has been used by the I.G.E.S. for intensive work over conductive zones which have been roughly indicated by one of the reconnaissance methods. As a preliminary to the investigation a base line is laid out along the strike of the conductive zone and traverse lines are set off at right angles to it at intervals of 100 ft. (Fig. 52). The transmitting (energizing) loop is then set up in a vertical plane and as nearly over the conductive zone as possible. Along each traverse readings are taken with the receiving coil at stations placed 50 ft. apart. Further observations are taken at intermediate points if more detailed information is required.

To take a reading of dip at any station the tripod is set up and its head levelled. The sighting axis, lying in the plane of the coil, is then directed towards the transmitting loop, which has already been rotated about a vertical axis until its plane passes through the receiving station. The transmitting and receiving sets are then put in operation,

I.G.E.S.

Fig. 51. High frequency detecting coil and receiving set

and the observer rotates the receiving coil about the sighting axis until it reaches a position where the induction into it is least, as indicated by minimum sound in the telephones (Fig. 51). The dip reading is then entered into the field book against the name of the station, together with another figure representing the breadth of the minimum. The field book and receiving set are carried by one assistant while another is left in charge of the transmitting unit.

A reading takes only a few minutes so that progress is fairly rapid. Although the transmitting unit may be taken down and set up in another position in a very short time, it is often possible to examine a conductive zone with one position of the transmitting loop. This could be done in the simple case illustrated in Fig. 52. In general, however, when an indication has been obtained at a certain point on a traverse, the aerial is then moved to this point and further traverses are run out, so that the conductor may be followed along its strike.

In thickly timbered country it may be impossible to see far enough to align the coils properly, so that the azimuth readings of the coil and loop for each station must be laid off on a map beforehand, and the figures entered into the field book for reference when setting up.

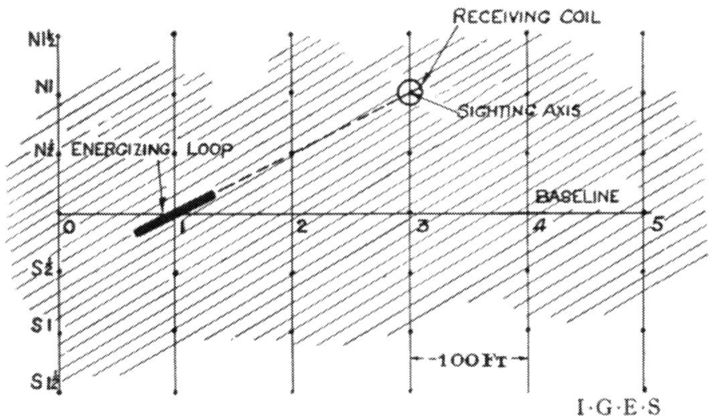

Fig. 52. Relation between the loop and the coil when a reading of dip is being taken

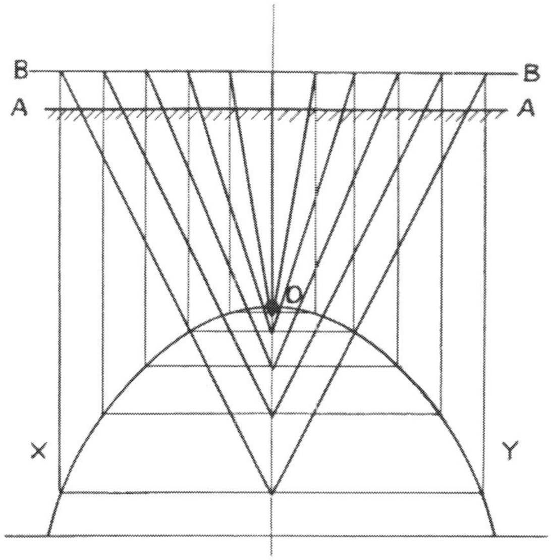

Fig. 53. Index curve corresponding to a long straight conductor of small radius

Plotting and Interpretation of Results. The curve of dips along a traverse as shown in Fig. 49 (*c*) is sufficient in itself to indicate points on the surface directly above a conductor. More specific information can be obtained by plotting the dips, and obtaining an *index curve*[a]. This procedure is shown in Fig. 53 in which *AA* is the level of the ground, *BB* is the level at which the coil is set up, and *XOY* is the index curve obtained by the graphical construction shown (see p. 296). The apex of the curve at *O* approximately marks the electrical axis of the conductor, so that an estimation of the depth of the body is possible. It has been found, however, that except in the simplest cases, the interpretation requires to be made in the light of considerable field experience. The shape of the curve also allows conclusions to be drawn regarding the dip and general configuration of the body. Thus, when field conditions are at all favourable, some picture of the conducting body may be formed by comparing the curves from a series of parallel traverses. In the interpretation of index curves, results that have been obtained for known bodies are of great assistance. Nevertheless, a general idea of the shape of the curve to be expected under certain typical field conditions may be computed from appropriate assumptions.

[a] Jakosky, *loc. cit.* p. 161.

CHAPTER III

ELECTRICAL SURVEYS IN AUSTRALIA

SECTION 1. ANEMBO, NEW SOUTH WALES (E. 1)

The Anembo area, which was the first to be investigated by the I.G.E.S., is situated in hilly country about 14 miles south of the Lake George Copper Mine at Captain's Flat. The surface indications there were thought to be very suggestive of sulphide mineralization and, as no serious exploratory work had been done, it was decided to carry out a reconnaissance survey by the equipotential line method, in order to see if any conductive bodies could be located. An area of 142 acres was included in this equipotential survey, two lines of gossans[a] being covered. The results were rather disappointing, however, since only one conductive zone was found. It will be seen from Fig. 54 that this lies on the eastern side of a large gossan outcrop which, on geological evidence, appears to dip towards the east.

Upon the completion of the equipotential line survey three of the gossan bearing areas were tested by the Bieler-Watson method[b], but no important anomalies were recorded. To some extent this may be accounted for, as it was found that the country rocks in this district have an unusually high conductivity. Large out-of-phase components were observed in the vicinity of the equipotential line indication shown in Fig. 54, but the high ground conductivity precluded any satisfactory interpretation.

The following general conclusions were drawn from these early investigations at Anembo:

(i) That the high ground conductivity of the shales, which form the normal country rock in the mineralized areas, was an adverse factor which may seriously have detracted from the effectiveness of the methods employed. This shows the importance of testing the ground resistivity prior to the final approval of an area for a detailed electrical survey. Apparatus for determining ground resistivities was not available, however, at the time the Anembo Survey was commenced.

(ii) Owing to this high ground conductivity the horizontal out-of-phase components recorded by the Bieler-Watson method were everywhere so large that it is doubtful if the effects of a deep-seated mineral body would have been sufficiently distinct for recognition. Although the equipotential line indication shown in Fig. 54 is not very striking in appearance it is better defined than that obtained over the same ground by the inductive method.

(iii) The negative results obtained by both methods over so great a proportion of the area surveyed are not attributed entirely to high ground conductivity. A close

[a] The word *gossan* is frequently used in this work. It is a well-established mining term for the outcrop material commonly associated with metalliferous deposits. In the case of sulphide ores it often consists of a porous red rock largely composed of ferric hydroxide and fragmental material, sometimes cemented by secondary silica.

[b] See p. 64.

geological examination was carried out concurrently with the electrical investigations, and there is reason to believe that the gossans may have been derived from mineral in which ferrous zincblende was the principal constituent. Since zincblende has a high specific resistance, there would be little hope of locating the mineral bodies, excepting in certain instances where the content of associated iron, copper, or lead sulphides, happened to be great enough to raise the electrical conductivity of the ore well above that of the surrounding country rocks.

(iv) The conductive zone, marked *electrical indication* in Fig. 54, is reasonably well defined for reconnaissance purposes and, from the manner in which it is related to the neighbouring gossan outcrop, which dips to the east, it may reasonably be concluded that it is due to the presence of mineral. For reasons already given it was impossible to make any satisfactory depth determination by the inductive method used (Bieler-Watson), but on geological evidence and from experience of the equipotential line method under similar conditions elsewhere, it was considered that the conductor lies between 75 and 100 ft. from the surface. A recommendation for a shaft to prove the ground at this point was made by the I.G.E.S., and it is to be regretted that circumstances have not yet permitted the work to be carried out.

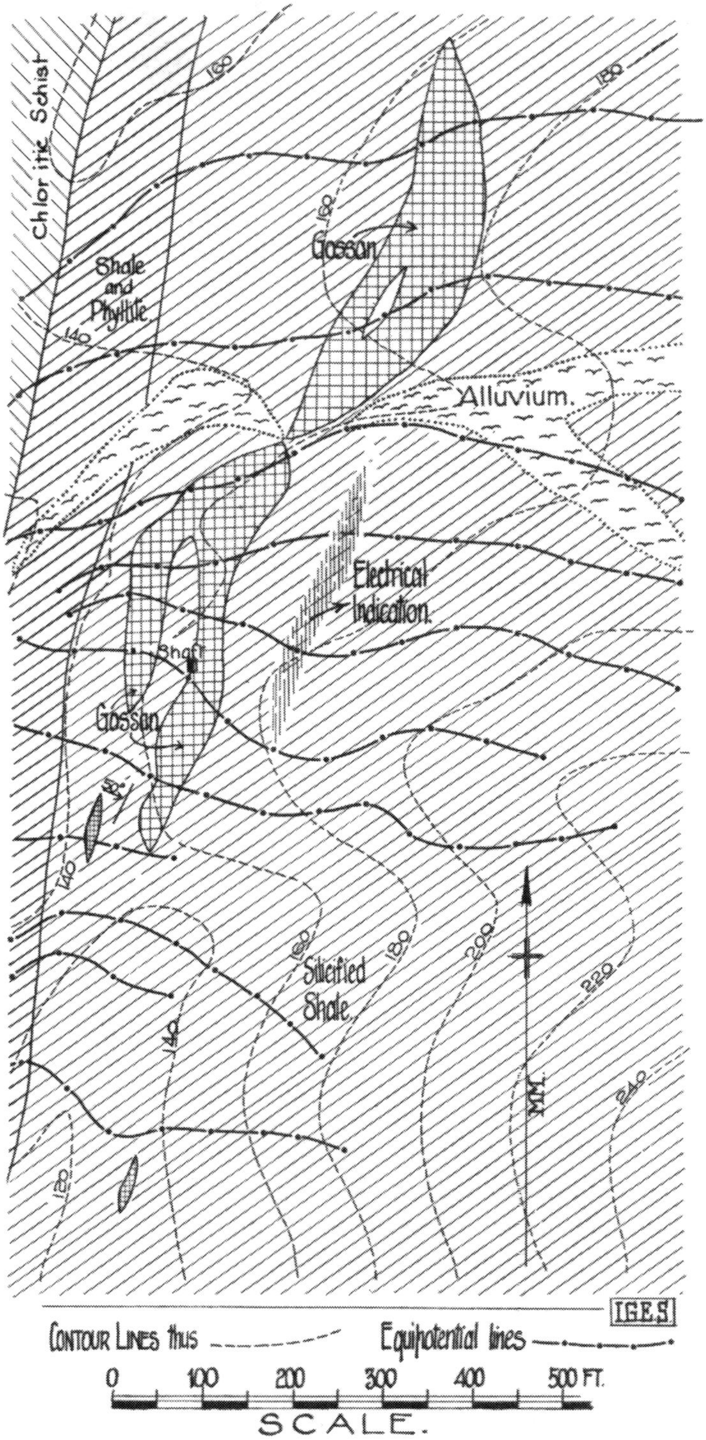

Fig. 54. Geological map and equipotential line survey, Anembo, N.S.W.

CAPTAIN'S FLAT, NEW SOUTH WALES (E. 2)

Upon completion of the work at Anembo a test was carried out over the known lode at the Lake George Mine, Captain's Flat, which lies about 14 miles to the north. The test was carried out by the Bieler-Watson method with a view to comparing the effects of the ground conductivity there with those experienced at Anembo, where it was thought that the unusually high values had been responsible, to some extent, for the negative results obtained over the greater part of the area. A test over a known ore-body occurring under similar conditions was very desirable.

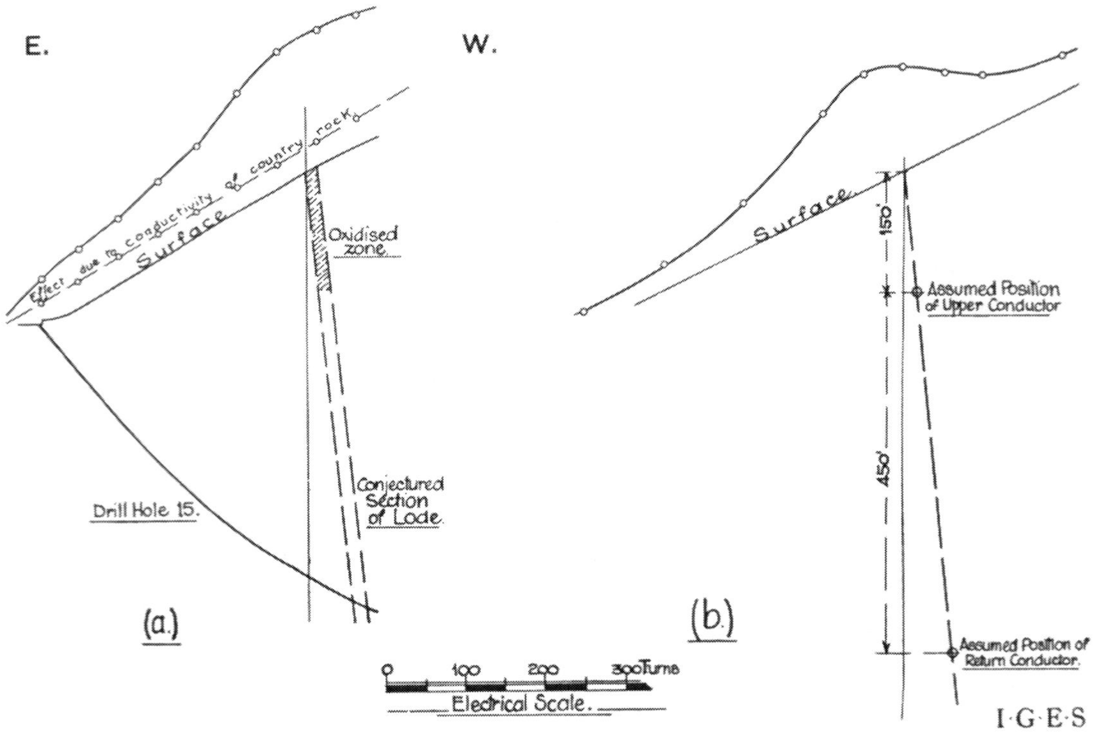

Fig. 55. Actual and calculated horizontal out-of-phase components over Lake George lode, Captain's Flat, N.S.W.

The Lake George lode[a] is a pyritic body in which the ore consists largely of the sulphides of iron, copper, lead, and zinc, and has a high electrical conductivity. It has an average width of 20 ft. and dips very steeply to the west. The portion investigated by the I.G.E.S. lay between the Elliot's and Keating's section of the mine, and was selected because it had been proved by drilling and by shallow workings, which were free from rails and pipe-lines. The depth of oxidation was here about 150 ft.

The lode strikes in a north and south direction, and a square loop was laid out so that two of its sides were roughly parallel with it. It was found that the horizontal out-

[a] Carne, *Mineral Resources Report*, *Geol. Survey*, *New South Wales*, No. 6, pp. 123–128 (1899). Kenny, *Bulletin*, *Geol. Survey*, *New South Wales*, No. 3, pp. 40–41 (1923).

of-phase components reached their maximum value over the line of the lode, although, in the absence of other information, the indication would hardly have been sufficiently strong for the existence of an ore-body to have been definitely inferred. Fig. 55 illustrates a profile (*a*) over a proved portion of the lode, plotted from the actual values of the east-west components of the vectors, and a second profile (*b*) calculated from the available geological data. It will be seen that the actual and calculated profiles agree fairly well, except that in the latter case there is a more marked fall in the readings on each side of the maximum. This is due to the influence upon the field results of the general conductivity of the country rock on each side of the lode, a factor which is not taken into account in the calculated profile. As at Anembo, the ground conductivity was found to be high; especially on the alluvial flat of the Molonglo River, which lies at the foot of the hill to the east of the lode. Although the results of this test did not throw very much light on the work carried out at Anembo, they were sufficient to show that inductive electromagnetic work, using vertical fields, is not very satisfactory over narrow, steeply-dipping lodes which do not extend within 150 ft. or so of the surface. It was concluded that horizontal fields would have two distinct advantages, namely:

(1) The current induced in the lode would be greater since the field would be approximately at right angles to it.

(2) The currents induced in the associated horizontal conductors such as swamps, alluvial beds, soils, and other types of highly conductive overburden, would be distinctly less.

As a result of this experience at Captain's Flat the construction of high frequency apparatus (60,000 cycles per sec.) with a vertical transmitting loop was undertaken by the Survey. The use of frequencies of this order is necessary when employing horizontal fields since the effective range at 500 cycles is insufficient for practical requirements.

LEADVILLE, NEW SOUTH WALES (E. 3)

The greater part of the work was carried out in the vicinity of the Extended and Grosvenor Workings of the Mount Stewart Mine (abandoned), which is situated upon a low ridge rising gently from an alluvial flat. From the old mine records it was known that some substantial sulphide bodies had been encountered and left untouched below the 170 ft. level. Although the main object of the Survey was to test electrical methods over these known bodies, a systematic survey was also conducted over neighbouring ground that had not been thoroughly prospected by ordinary methods. A general survey of the area was made by the equipotential method, in addition to which certain portions were investigated by the Bieler-Watson and spontaneous polarization methods.

Geology. The geology of the area has been described in a valuable paper by Willan[a], and a shorter account is given by Kenny[b]. The ground is occupied by a series of highly altered sediments, consisting mainly of slate, chert, and epidote rock, intruded by

[a] *Proc. Austr. Inst. Mining and Met.*, New Series, No. 57, 1925.
[b] *Bulletin, Geol. Survey, New South Wales*, No. 2, pp. 31, 32 (1923).

granophyre and quartz-porphyry (see map, Fig. 56); and the mineralization usually occurs at, or near, the contact of the latter with the sedimentary formations. The lodes trend in a general north-easterly direction, and have a westerly dip of from 60 to 75°. In the oxidized zone they consist of gossan carrying lead carbonate; whilst in the sulphide zone, pyrites, chalcopyrite, zincblende, and galena, are the characteristic ore minerals. The ore-bodies are roughly lenticular in shape and, as a rule, are from 100 to 200 ft. in length, and from 15 to 20 ft. in width.

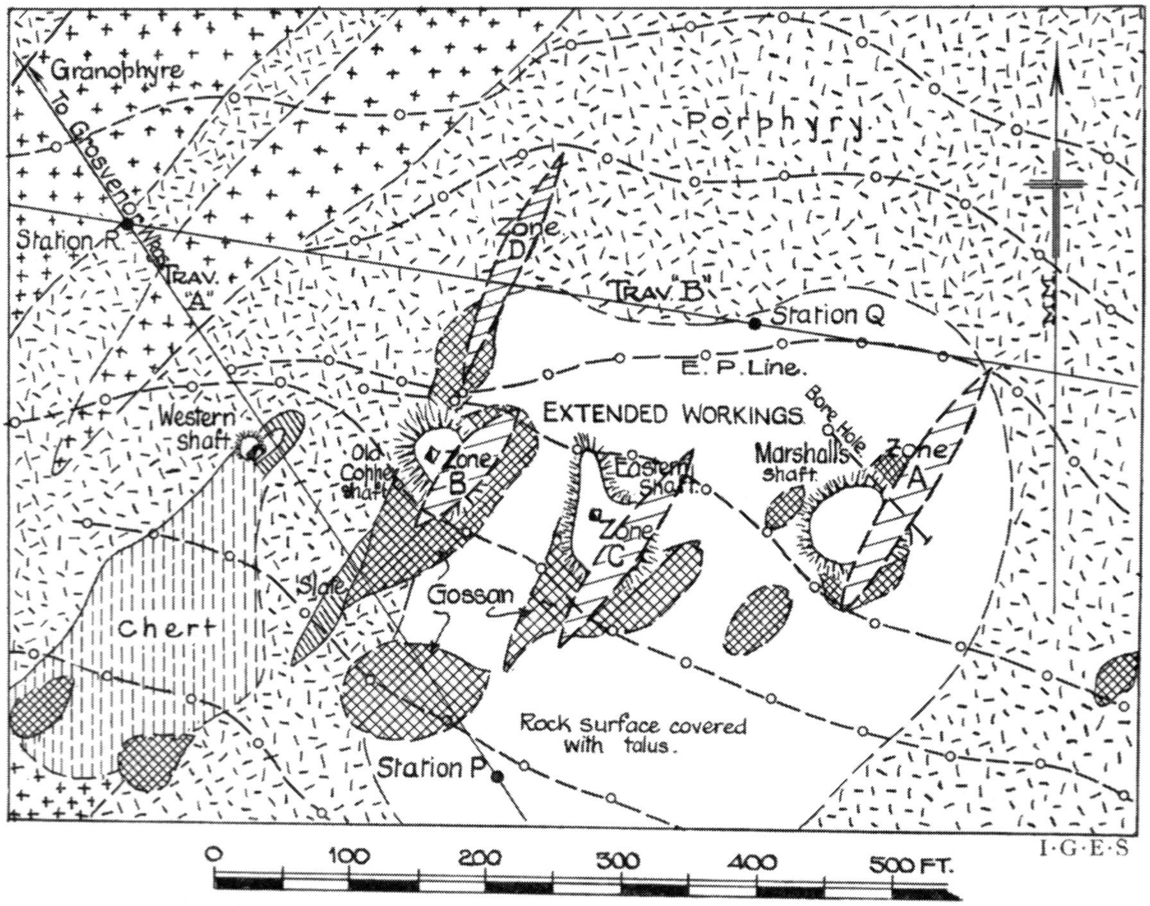

Fig. 56: Geological map and equipotential line-survey of the Extended Workings, Leadville, N.S.W.

In the area under consideration the lodes have been proved in the Extended and Grosvenor Workings. In the former group the sulphide zone commences at a depth of about 170 ft., and in the latter at from 80 to 100 ft. Some outlying gossans, known as the Latimer Group, upon which no mining has been done, were also included in the electrical survey.

Equipotential Survey. Four conductive zones were located in the neighbourhood of the Extended Workings (Fig. 56), but their general appearance is not suggestive of mineral bodies of important dimensions. It is reasonable, however, to correlate Zones A, B, and C with the sulphide bodies that have actually been proved in the three adjoining shafts. In view of the depth of oxidation it could hardly be expected that small bodies

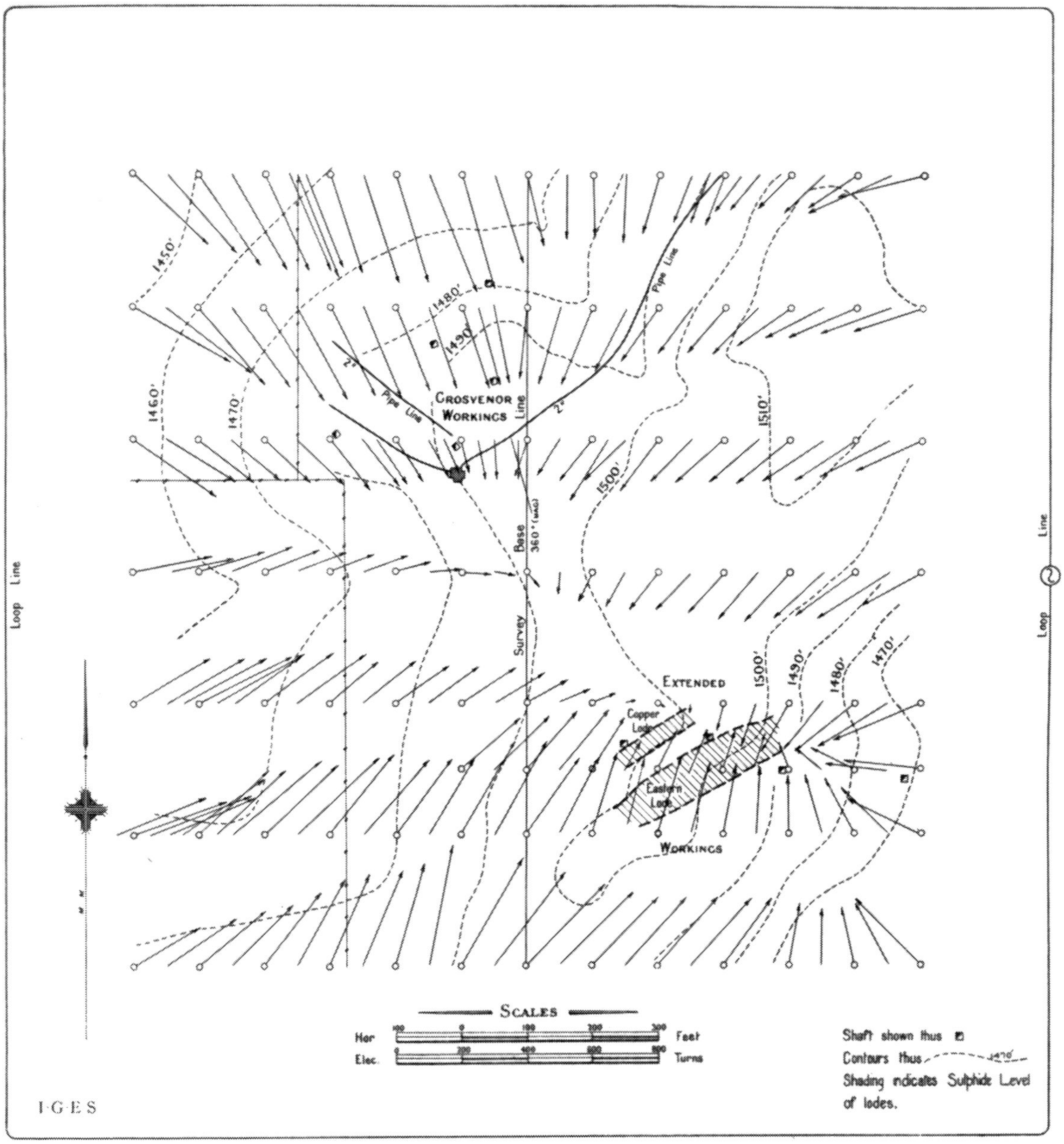

Fig. 57. Electromagnetic survey of the Extended Workings, with horizontal loop, Leadville, N.S.W.

of this kind would produce marked anomalies at the surface. Zone D lies in totally unproved ground, but may well be an extension of the line of mineralization to which Zone B is due. There is, however, a distinct displacement between the two lines of strike. Zone A is the best defined of the four indications.

Electromagnetic Survey. A survey of the Extended Workings by the Bieler-Watson method gave a well-defined indication, which can confidently be attributed to the Eastern Lode (Fig. 57). It is true that no other indications were obtained in this neighbourhood, but there can be little doubt that this is due to the fact that the Eastern Lode has a considerably greater width (sometimes exceeding 60 ft.) than any of the others which were covered by this survey.

A further test was carried out at the Extended Workings, using a vertical transmitting loop connected to the 500-cycle generator. This was erected in the position shown in Fig. 58, and consisted of 10 turns of 3-strand cable in the form of a triangle,

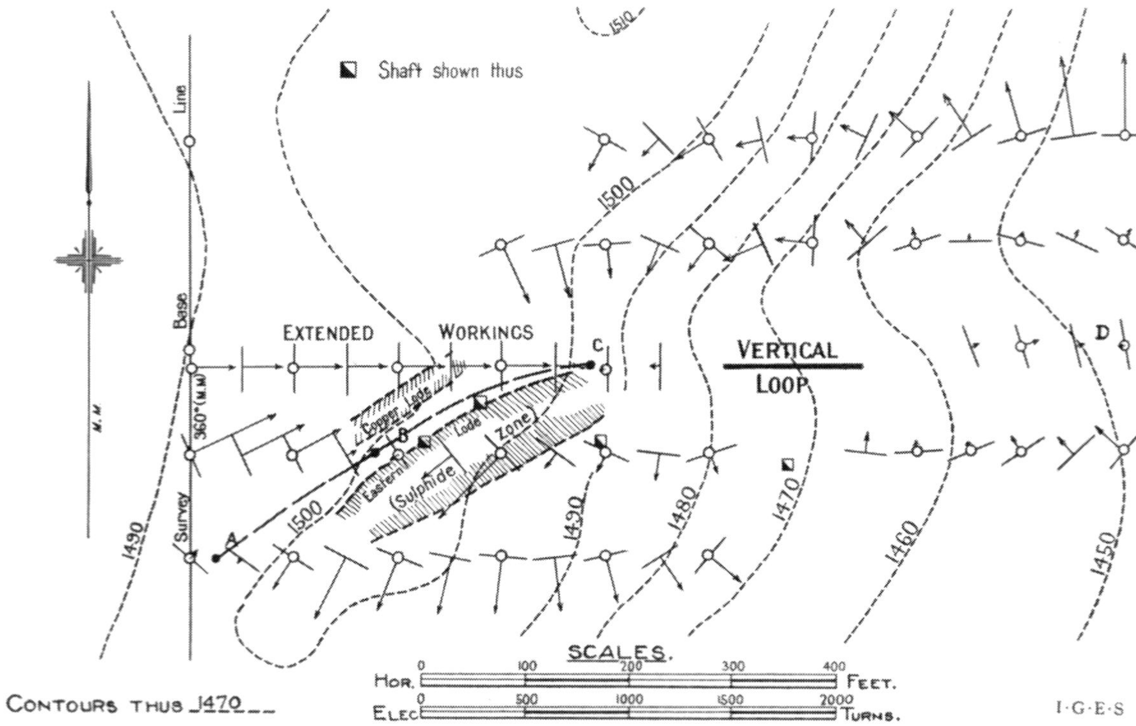

Fig. 58. Electromagnetic survey with vertical loop of the Extended Workings, Leadville, N.S.W.

measuring 45 ft. in height and 133 ft. along the base, with its plane in an east-west direction. The receiving apparatus consisted of the Bieler-Watson double coil, mounted on a tripod in such a way as to be capable of rotation about two axes; namely, a horizontal axis parallel to the common line of both coils, and the vertical axis of the tripod (see Fig. 44). A suitable compass-clinometer device was fitted to the apparatus for reading the dip of the large coil from the horizontal and the azimuth of the plane perpendicular to both coils.

Before the field work with the vertical loop was commenced the direction of the primary field due to it was calculated for each station, and in each case the receiving apparatus was set up so that the plane perpendicular to both coils was in the direction of the calculated primary field. The coils were then rotated about the horizontal axis and the number of turns in the variable coil was altered until silence was obtained in the telephones. The angle of dip of the large and the number of turns in the small coil were then read[a]. It was found that the range of the transmitting loop was limited to about 500 ft. on account of the weakness of the note at greater distances.

In Fig. 58 the result obtained at each station is shown by means of an arrow whose length is proportional to the out-of-phase component perpendicular to the primary field and whose direction is normal to that of the primary field. It will be seen that west of the transmitting loop there is a region where the magnitude of the out-of-phase component (as represented by the length of the arrows) diminishes rapidly to zero and then increases again, but with a change of direction. This change takes place at the points marked A, B, and C, on the plan, and it will be seen that a line joining these points coincides approximately with the western wall of the Eastern Lode. There can be little doubt, therefore, that this indication was due to the effects of the Eastern Lode, and that the displacement of the axis of the indication to the north-west, away from the centre of the Eastern Lode, may be accounted for by the presence of the Copper Lode, which is sufficiently close to influence the field. There is one other point, D, at which the out-of-phase component changes direction, but the variation in the magnitude of the component is far less striking than in the instance just described, and no corresponding changes of direction were recorded elsewhere in the vicinity. This, then, must be regarded as a minor effect due to some local increase in ground conductivity.

Spontaneous Polarization Method. This work was experimental in character and was carried out along traverse lines that were laid out across the known sulphide ore-bodies of the Extended and Grosvenor Workings. The two potential profiles in Fig. 59 wer measured along traverses A and B, shown in Fig. 56. It will be observed that no important variations in potential were recorded over the unmineralized portions of the ground, but that well-defined gradients occur in the immediate neighbourhood of the Grosvenor and Extended Workings. Over the former there is a series of weak negative centres which can tentatively be correlated with the several narrow lodes that occur there. Over the larger ore-bodies of the Extended Workings a more striking fall of potential will be observed, average gradients of the order of 0·26 millivolts per foot being recorded over distances of 300 to 400 ft. The trough in the potential curve on traverse B is peculiarly flat, and may be due to the combined effect of several bodies of mineral.

It is probable that the results obtained by the spontaneous polarization method in this area were accentuated by the increased vigour of oxidation brought about by the old mine workings, which penetrate the sulphide bodies at numerous points.

Conclusions. The main features at Leadville were high ground conductivity, a considerable depth of oxidation (80 to 170 ft.), the small dimensions of the lodes, and their steep dip. Under these conditions it was found that the equipotential and spontaneous polarization methods gave results which could be correlated with the known ore-bodies[b]. In

[a] For an explanation of the procedure followed in this test see p. 285.

[b] The series connected ratiometer, from which the instrument shown in Fig. 33 was designed, was first used by the I.G.E.S. at Leadville in October 1928.

the case of the Bieler-Watson method a positive result was obtained over the largest of the ore-bodies, but it was found necessary to employ a vertical transmitting loop in order to obtain a more satisfactory result. Electromagnetic methods are at a disadvantage in regions of high ground conductivity and, when the ore-bodies dip steeply and are of comparatively small horizontal extent, the use of a horizontal primary field is to be preferred. This was shown to be the case at Leadville, where the results obtained with the vertical loop were far more definite than those recorded with the horizontal layout.

Confirmatory Exploration. At the request of the I.G.E.S. a confirmatory drill-hole was put down on Zone A, by the New South Wales Department of Mines, in the position shown on Fig. 56. On completion of the work the Chief Inspector of Mines reported that a soft gossan was encountered at 50 ft. and continued to 154 ft. when 2 ft. of unoxidized lode material were passed through. At 156 ft. a band of iron pyrites was met with, and more lode material occurred between 158 ft. and 160 ft. From 160 ft. to 165 ft., where the hole was stopped, the drill appeared to be in footwall country.

The Chief Inspector of Mines states in his report (29/8989M) that "as the strata have a westerly dip of approximately 55°, the two bands of lode material met with in the drill-hole correspond with the surface position shown on the electrical survey plan supplied by the I.G.E.S. to the Department of Mines." He also states that "although the drill-hole has not developed any satisfactory lode, it is quite probable that a sulphide body of some description underlies the gossan material penetrated by the drill."

SECTION 2. COOPER'S CREEK, VICTORIA (E. 4)

The area examined centred about the old Thomson River copper mine[a] at Cooper's Creek. The lode formerly worked is a sulphide body consisting mainly of chalcopyrite, pyrites, and pyrrhotite, lying at the western contact of a dyke of hornblende-diorite. The ore carries platinum values up to 10 dwts. per ton.

The object of the geophysical survey was to determine whether other ore-bodies existed along the course of the dyke, beyond the limits of the old workings. The investigation was carried out by means of the spontaneous polarization, Bieler-Watson, and magnetic methods, but no indications were obtained of the existence of ore-bodies other than the known lode.

SECTION 3. SOUTH ZEEHAN, TASMANIA (E. 5)

The primary object of this survey was to determine whether lead lodes of the type common in the Zeehan field were amenable to detection by electrical prospecting methods. As the known lodes had all been worked out more or less completely it was necessary to find ground in which the geological evidence was considered favourable for further occurrences of mineral. For this purpose a section of the Silver King line

[a] Baragwanath, *Bulletin, Geol. Survey, Victoria*, No. 20 (1906).

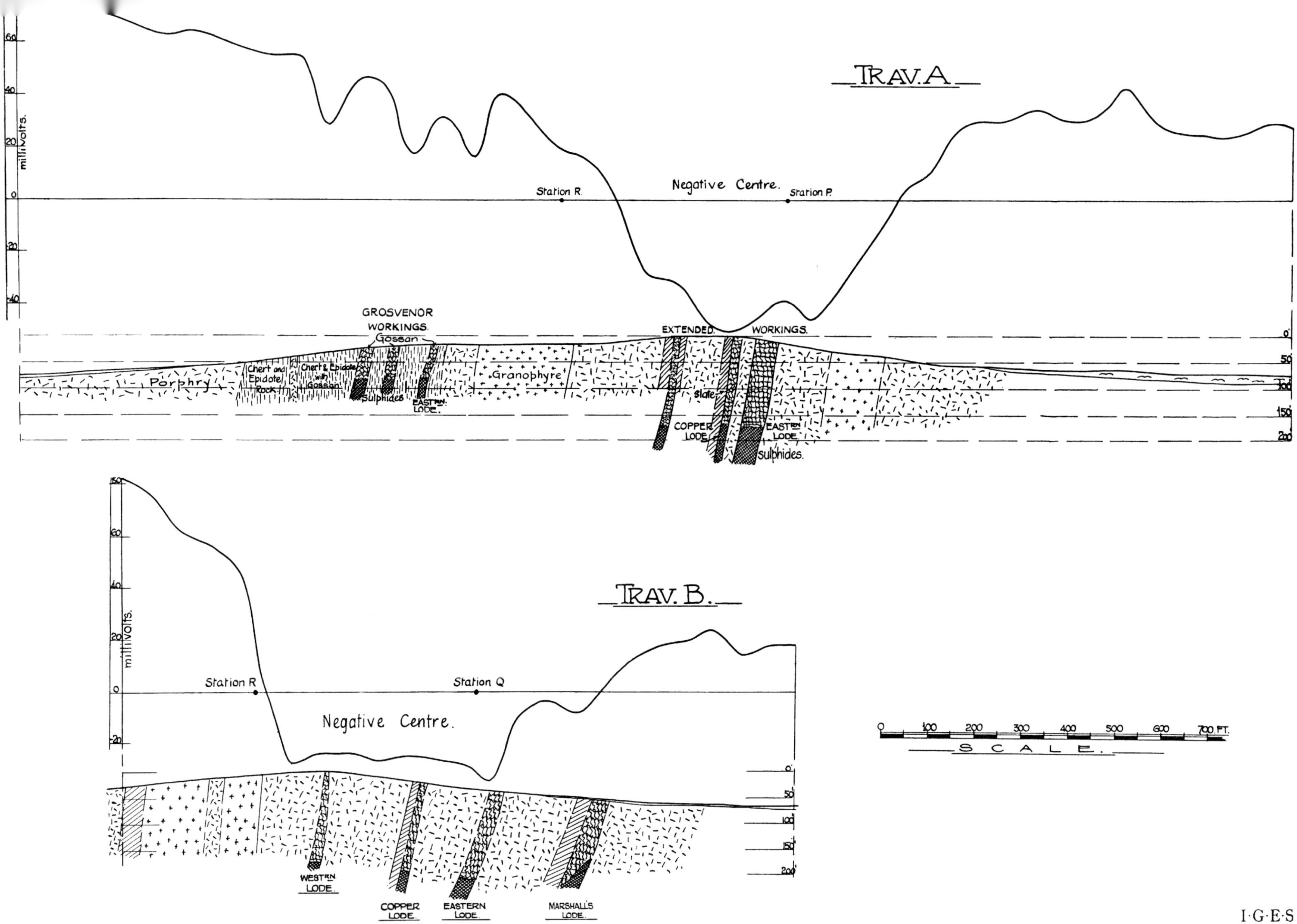

I·G·E·S

Fig. 59. Potential profiles recorded by spontaneous polarization method over known sulphide lodes at Leadville, N.S.W.

of lode[a] was selected, since this is one of the most persistent lodes in the district. It has been described as a well-defined fissure formation carrying abundant siderite, with argentiferous galena, and small quantities of pyrites, chalcopyrite, and zinc-blende.

The portion selected lay between the Silver King and South King mines, in which the lode has been worked to depths of 246 ft. and 258 ft. respectively. In the South King the lode averaged 3 to 4 ft. in width and carried two shoots of rich ore. In the ground between these two mines the line of mineralization may be traced by means of some old shallow workings but, apparently, it has been little prospected; mainly, no doubt, because it traverses low-lying swampy ground. Whether any substantial bodies of mineral occurred there was unknown, but on general evidence the chances of success appeared favourable.

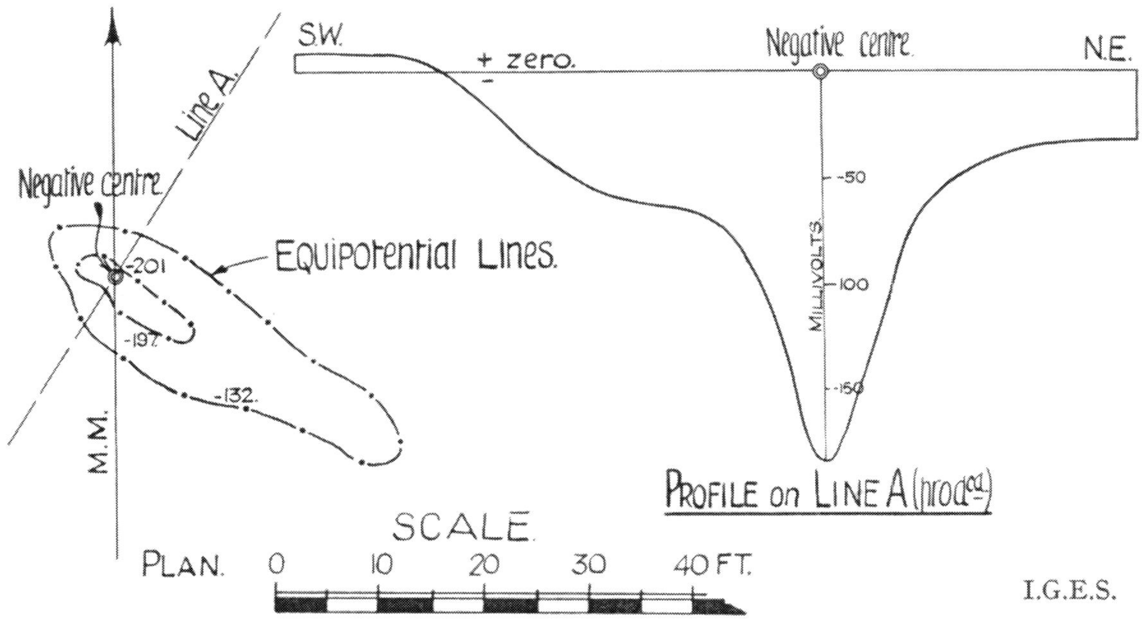

Fig. 60. Spontaneous polarization potential profile and equipotential lines, South Zeehan, Tasmania

A reconnaissance survey of this area by the equipotential method failed to reveal the presence of any anomalies, either along the line of the lode, or elsewhere. The workings of the South King mine were also included in this survey, but in view of the depth to which they had been mined, it was not surprising that the results obtained were negative. In addition, the northern part of the area was investigated by means of the spontaneous polarization method; traverses being made across the line of the lode at 200 ft. intervals. Again, no indication was obtained but an interesting negative centre was located elsewhere. This is illustrated by the plan and profiles given in Fig. 60.

Although this electrical disturbance was of a very local character, it is notable that the potential gradients were unusually steep (up to 12·5 millivolts per foot) and that the

[a] Waller, *Report on the Zeehan Silver-Lead Mining Field*, pp. 97–101, Mines Department, Tasmania, 1904.

total fall of potential to the negative centre, over a distance of 60 ft., is of the order of 150 millivolts. As so local an indication was suggestive of a cause lying close to the surface a trench was dug across the negative centre. A bed of strongly graphitic shale was exposed, and since the oxidation of material of this type is known to result in vigorous electrical activity the anomaly was satisfactorily accounted for.

The results of the work carried out at South Zeehan must be regarded as inconclusive. As is usually the case in abandoned mining fields there are no standing bodies of mineral over which satisfactory tests can be made, and an extension of the work at Zeehan, in the hope of locating bodies hitherto undiscovered, was not considered to be possible in view of the special duties of the I.G.E.S. and the programme of work that had to be completed.

ZEEHAN COPPER-NICKEL FIELD, TASMANIA (E. 6)

The Copper-Nickel field is a tract of comparatively low-lying country, largely swampy and covered by thick bush, situated about five miles north-east of Zeehan. Copper-nickel ores were mined here profitably before the outbreak of war in 1914 and, more recently, work has been resumed in the northern part of the field by the Copper Nickel Mining Company.

Geology[a]. The area is occupied by a series of slates and tuffs of cambro-ordovician age. These rocks are intruded by a number of ultrabasic dykes, of which gabbro, norite, peridotite, and pyroxenite, are the principal types. It is with these dykes that the copper-nickel ores are closely associated, and it is in the neighbourhood of the pyroxenite and peridotite that the important shoots of mineral occur. The ore is a mixture of pyrrhotite, pentlandite, and chalcopyrite, the chief accessory minerals being pyrites, marcasite, galena, and zinc-blende. As a rule the ore is but little oxidized, and in some parts of the field fairly fresh sulphides occur within a few feet of the surface. Below a depth of about 100 ft. the ores tend to become highly siliceous.

From Fig. 61 it will be seen that two principal dykes traverse the area in a north and south direction, one of pyroxenite and the other of gabbro. Ore-bodies occur along the footwall (or western) side of the former at several points; but no discoveries have been made along the latter.

Objects of Survey. Copper-nickel ores had been worked in this area at five principal points, spread over a distance of more than a mile, but there are long intervals in which no discoveries have been made. A considerable part of the ground is swampy, and obscured by alluvium and scrub, so that it seemed possible that undiscovered ore-bodies might exist along the line of the pyroxenite dyke. A reconnaissance survey was carried out by the equipotential line method, and this was followed by more detailed investigations by the A.C. potential ratio, high frequency, and magnetic methods;

[a] *Bulletin, Geol. Survey, Tasmania,* No. 36, pp. 23–38 (1925).

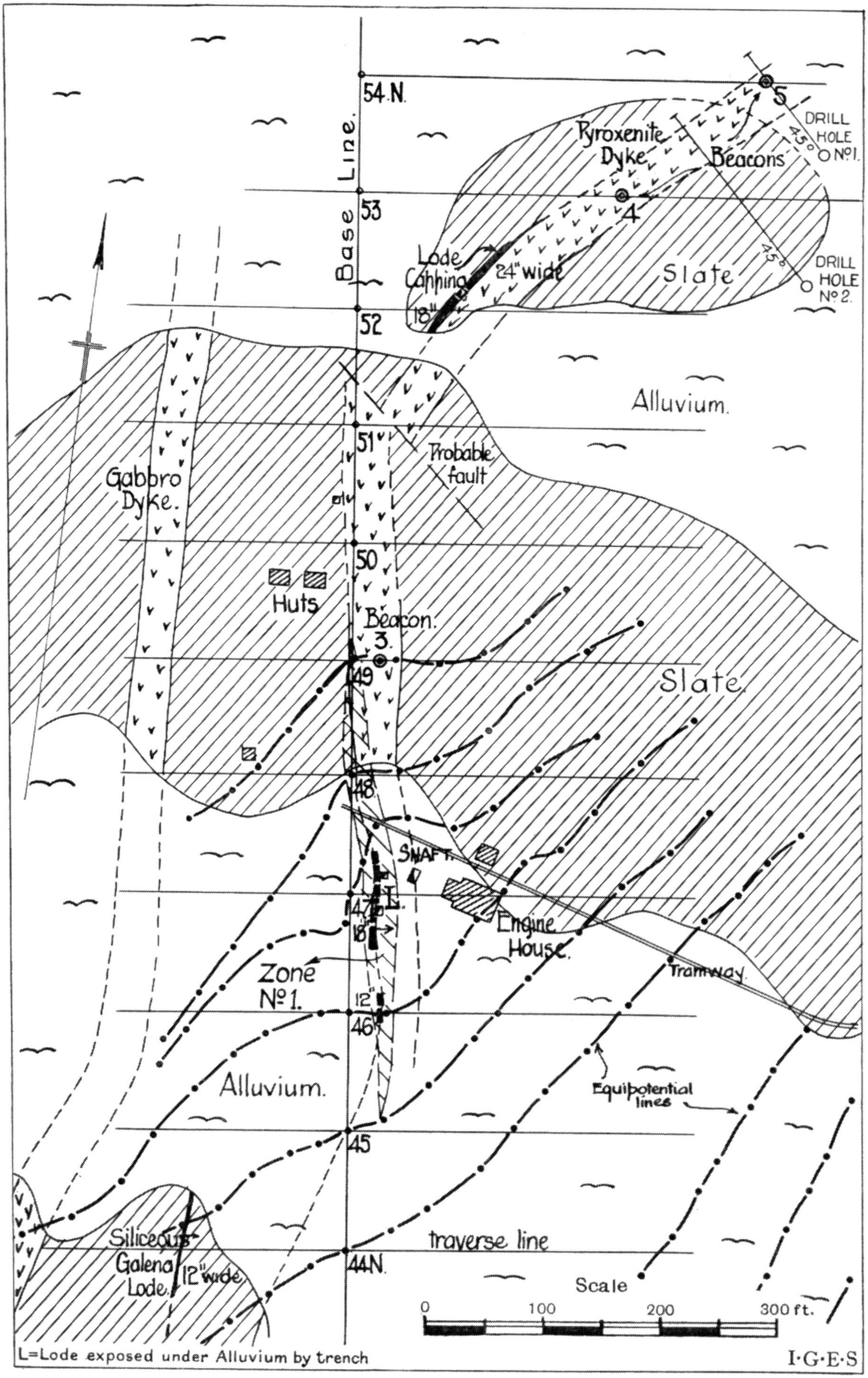

Fig. 61. Geological map and equipotential line survey of northern part of Copper-Nickel field, Tasmania

principally in the northern part of the field, where the reconnaissance results appeared to be most promising.

Equipotential Line Survey. Three indications were obtained, the principal one being over the workings of the Copper Nickel Mining Co., in the northern part of the field. This indication (Zone No. 1, Fig. 61), which has a length of about 300 ft. with a well-defined inner portion 200 ft. in length, corresponds exactly with the lode now being worked. The latter is a body of solid sulphide ore with an average width of about 2 ft. At the time the survey was made it had been stoped out for about 30 ft. above the 70 ft. level, over a length of 80 ft. Near the surface the lode is nearly vertical, but at a depth of 40 ft. it is dipping eastwards at 40°. The equipotential survey was not carried farther to the north, on account of the proximity of the northern electrode.

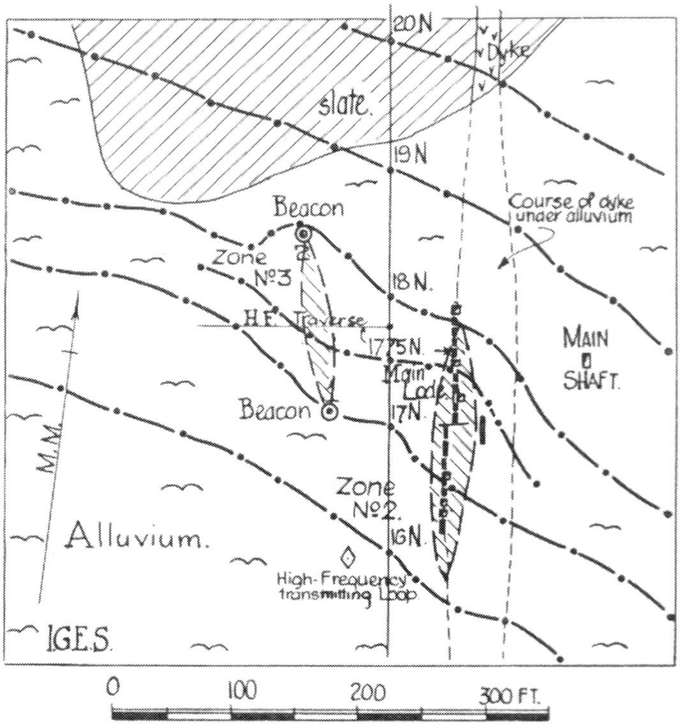

Fig. 62. Equipotential line survey, Dundas Cuni area, Tasmania

The two other conductive zones obtained (Nos. 2 and 3, Fig. 62) were of a minor character. Both were situated near the old Dundas Cuni Mining Company's Main Lode workings, 1000 yds. to the south of the area shown in Fig. 61. The axis of Zone No. 2, about 150 ft. in length, coincides with the Main Lode, in which the ore-shoot has been stoped out down to a depth of 75 ft. The other is of similar dimensions, but is situated about 100 ft. to the west of the Main Lode on unexplored ground occupied by slates of the Dundas series.

A.C. Potential Ratio Method. This work was carried out in the northern part of the area, along the east-west traverse lines shown in Fig. 61, and the potential profiles and phase angle curves obtained along certain of them are illustrated in Fig. 63. It will be noticed that on the southernmost traverse (45 N), which is beyond the limit of the equipotential indication, there are no important anomalies either in potential or phase angle. On the traverse through 47 N, however, a distinct trough is recorded in the potential variation curve, accompanied by a change in phase angle at a point corresponding closely with the axis of the equipotential indication. Similar features are shown by the traverse through 49 N, but in a more marked degree. On traverse 51 N a well-defined trough occurs in the potential variation curve, but the change in phase angle is small, and the same features occur in traverse 53 N. It will be seen that the indication in the latter case is displaced to the east, corresponding with the abrupt change in the strike of the lode.

High Frequency Method. Most of the observations were made in the northern part of the field along the lines laid out for the A.C. potential ratio method, but a traverse was also run across the western of the two equipotential line indications in the Dundas Cuni area, shown in Fig. 62.

The results of the work carried out in the north, which are illustrated in Fig. 64, agree quite well with those obtained by the equipotential line and potential ratio methods. The strongest indications were obtained on traverses 46 N, 53 N, and 54 N, the inclination of the receiving coil and the reversal of direction being very well marked in these instances. It should be noticed, however, that on traverse 51 N, where the potential ratio method gave a distinct indication, the high frequency results show very little of interest.

From Fig. 64 it will be seen that the index curves gave depths ranging from 50 to over 200 ft. These values serve as an indication of the depths at which the induced currents may be assumed to be concentrated, and they assist to some extent in locating the sites and inclinations of drill-holes.

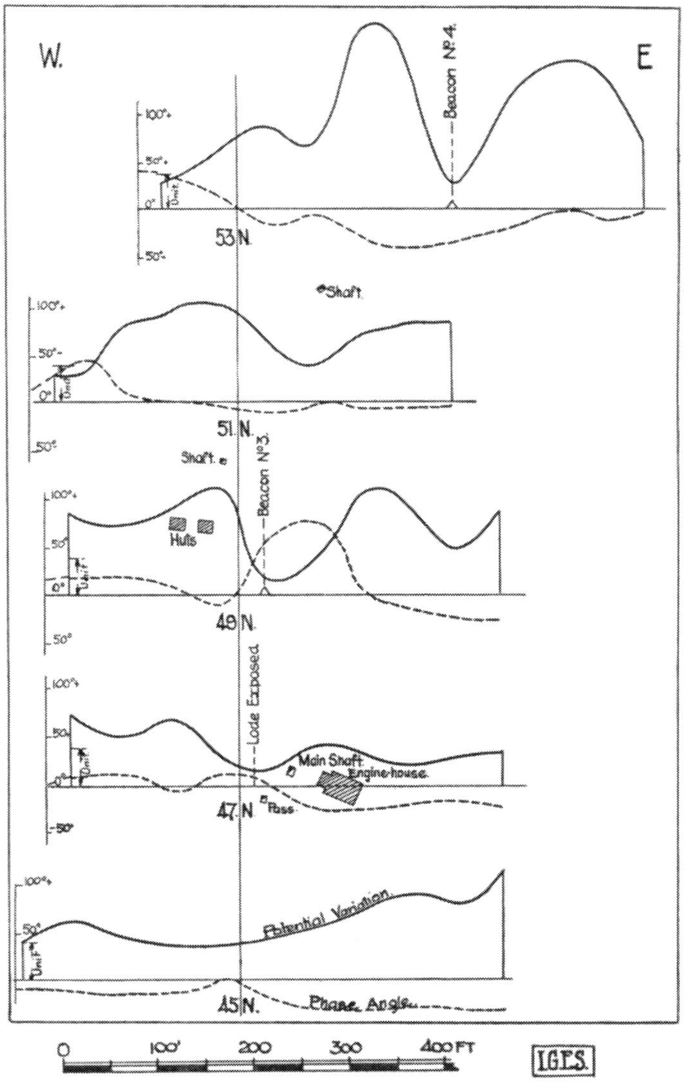

Fig. 63. Potential and phase angle curves over northern part of Copper-Nickel field, Tasmania

The results obtained along the traverse at $17\frac{3}{4}$ N in the Dundas Cuni area (see Fig. 62) are also shown in Fig. 64 and definitely confirm the equipotential line indication; the reversal in the direction of inclination of the receiving coil being well defined. In this case the depth as shown by the index curve is 75 ft.

Magnetic Survey. Several traverses were made in the northern part of the field (Fig. 61) using the Askania vertical field balance. The results are illustrated in Fig. 65 in the form of profiles superimposed upon a plan of the ground. Only one anomaly can be definitely attributed to geological causes; namely, the increase of about 200γ over the normal value at a point on traverse 46 N, at which sulphide ore outcrops at the surface. No effect was caused by the material of the basic dyke, but this may be explained by the fact that the rock is decomposed down to a considerable depth[a]. It will be observed that marked anomalies were recorded in the vicinity of the iron buildings on traverses 48 N and 50 N. It was for this reason that no traverse was made through 47 N, where the engine-house and tramway would have caused large disturbances.

Recommendations for Exploration. After comparing the results obtained by the several methods permanent beacons were erected to mark the more important of the electrical indications. The following recommendations for exploration work by trenches and drills were then submitted to the Director of the Mines Department in Hobart.

Beacons Nos. 1 *and* 2. These were placed to the west of the Dundas Cuni Main Shaft (Fig. 62), to mark the southern and northern ends of the conductive Zone No. 3 located by the equipotential line survey. It was recommended that a trench should first be dug midway between Beacons Nos. 1 and 2, in an east-west direction, in order to

[a] This dyke rock is completely decomposed at the 70 ft. level in the workings of the Copper Nickel Mining Co., and possibly the alteration continues to greater depths.

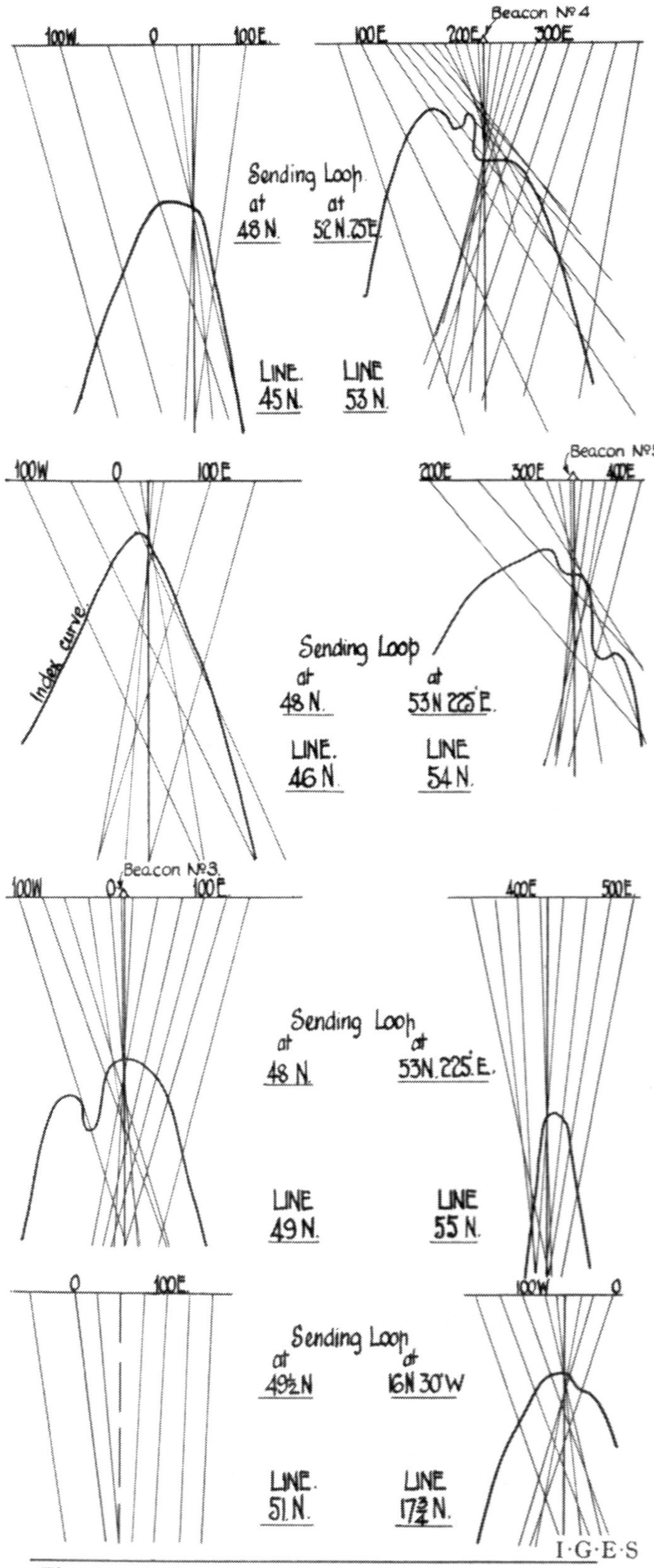

Fig. 64. Index curves for high frequency survey, Copper-Nickel field, Tasmania. (Scale 1″ = 200′)

intersect the axis of Zone No. 3. A second trench at Beacon No. 2 was proposed; also in an east-west direction. To follow this preliminary exploration an inclined diamond drill-hole was suggested.

Beacon No. 3. In the northern part of the field (Fig. 61), marking the strong indications obtained at this point, both by the A.C. potential ratio and the high frequency methods. Trenching at this point was recommended.

Beacons Nos. 4 *and* 5. In swamp country to the north-east of Beacon No. 3 (Fig. 61), marking the well-defined indications that were recorded in that part of the field. Trenching and drilling at both points were suggested.

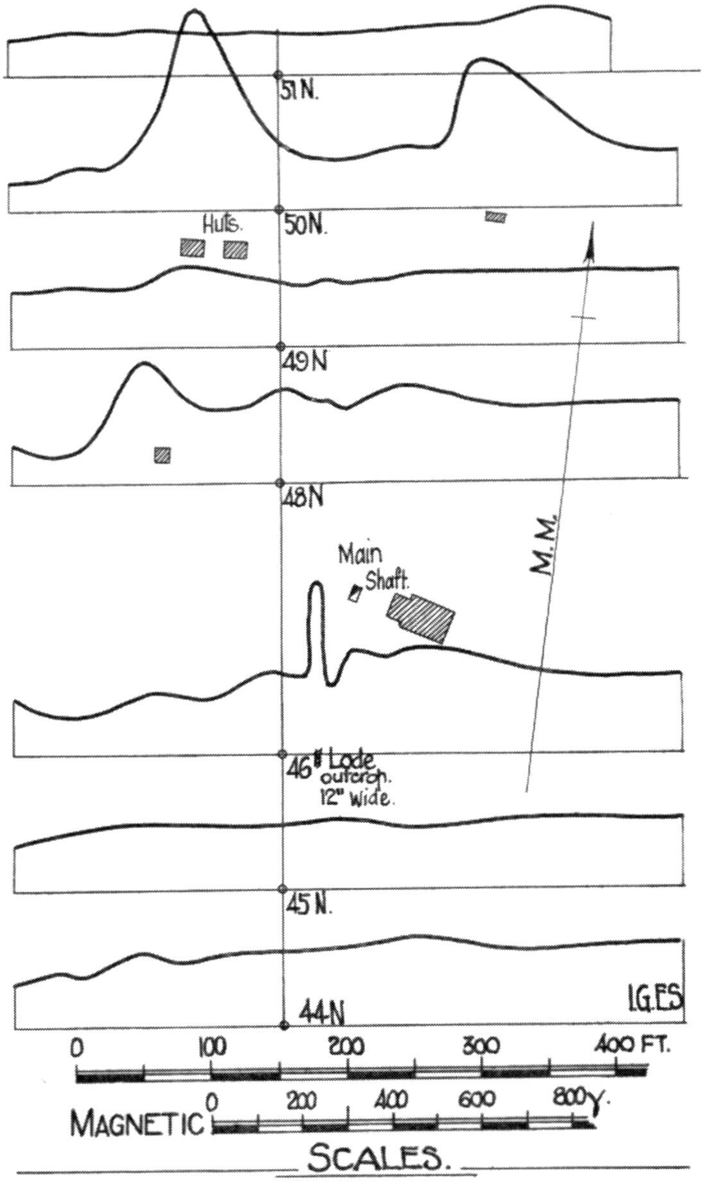

Fig. 65. Profiles of vertical magnetic anomaly, northern part of Copper-Nickel field, Tasmania

Results of Confirmatory Exploration. The recommendations made by the I.G.E.S. were executed under the direction of the Tasmanian Department of Mines in January, 1930, and the following results have been reported by the Director of Mines.

Beacons Nos. 1 *and* 2. A trench 10 ft. in length, and bisecting the line joining these two pegs, was sunk to a depth of 6 ft. in slate country, the slate being covered to a depth of 1 ft. by surface soil. At a depth of 6 ft. a definite mineralized formation 2 ft. wide was exposed, a vein of pyrites 1 in. wide being present directly below the indicator peg left by the I.G.E.S.

A trench of the same dimensions and direction was sunk at Beacon No. 2. Slate was encountered beneath 2 ft. of surface soil, and was found to be considerably stained with iron oxide, the staining increasing with depth. A well-defined wall was found below Beacon No. 2, dipping towards the east and forming the footwall of a definite oxidized formation 4 ft. wide. Owing to the heavy flood water the further exploration of this ground would require the use of drills or the equipment of a shaft. It was impossible, therefore, to continue the work down to the sulphide zone.

Beacon No. 3. A trench 18 ft. long and 5 ft. deep was cut through this point on an east-west bearing, and revealed a strong gossan body 4 to 5 ft. in width. A pit is being sunk to expose the sulphide ore.

Beacon No. 4. A trench 8 ft. in length and 6 ft. deep was cut through this point in a north-westerly direction. A definite lode formation 3 ft. wide was exposed at a depth of 6 ft. from the surface. A pit 8 ft. deep at this point exposed bunches of sulphide ore distributed through the gossan. The sulphide showed a marked increase in quantity as the depth from the surface increased and it appeared to be of good quality. A recent cable communication from the Director of Mines states that a diamond drill-hole (No. 2, Fig. 61) was subsequently put down by the Department of Mines, to prove this lode in depth at a point midway between Beacons Nos. 4 and 5 (Fig. 61). It showed 4 ft. of solid sulphide ore at a depth of 140 ft. from the surface; samples of which gave on assay, nickel 9·3 per cent. to 11 per cent. and copper 4 per cent. to 9·3 per cent.

Beacon No. 5. This point is in swamp country, the inflow of water making exploration work by trenching a matter of great difficulty. Nevertheless, sulphide ore (copper-nickel with pyrrhotite) of good quality was exposed in a lode formation which, at a depth of 5 ft. from the surface, appeared to be 1 or 2 ft. in width. This ore-body has now been intersected by a diamond drill-hole (No. 1, Fig. 61), placed in accordance with the recommendations of the I.G.E.S. The drill proved several feet of sulphide ore at this point, at a depth of 75 ft. below the surface. On assay, samples of the core gave nickel 3 per cent. to 6 per cent., and copper 2·5 per cent. to 3·5 per cent.

Conclusions. The results of the work carried out in this copper-nickel field were most successful in demonstrating the practical value of electrical methods as a means of locating sulphide ore-bodies lying near to the surface; especially in thickly-timbered and swamp-covered country where ordinary prospecting methods are at a disadvantage. The three electrical methods employed all gave satisfactory and concordant results. As regards their respective merits, it may be said that the equipotential line method was at an advantage for reconnaissance purposes in view of its simplicity and the speed with which the ground was covered. For the more exact determination of the positions of the lodes there was little to choose between the A.C. potential ratio and the high frequency methods. With the latter some evidence of depth was obtained by the index curve principle, but further field tests would be necessary before the values could be accepted with any degree of confidence. In spite of the pyrrhotite content of the ore the magnetic method was not found to be very helpful.

The exploration results given above have been summarized from a report by the State Mining Engineer and from cabled information subsequently received from the Director of Mines. In concluding his report the former draws attention to the difficult nature of the country and to the fact that there were no geological or other indications at the surface that would lead one to suspect the presence of the mineral bodies which were discovered at each of the points located by the I.G.E.S. The Director of Mines also emphasizes the importance of the discoveries that have been made, which, he considers, will prove of considerable economic value to those engaged in mining in this district.

RENISON BELL, TASMANIA (E. 6a)

The Renison Bell tinfield is situated in rugged country on the west coast of Tasmania and, on account of both climate and topography, considerable difficulties were experienced in carrying out geophysical field work. The average annual rainfall for the district is about 100 in., of which two-thirds may be expected to fall during the spring and winter months; thus, even under the most favourable conditions existing during the summer, field work was considerably hampered by rain. The country is covered with a dense growth of timber and scrub which in many places is practically impenetrable. Even where the heavy timber has been cleared, a strong secondary growth has usually sprung up which, frequently, is more difficult to pass through than the original vegetation. The rugged topography also impeded the work to some extent, the steep scrub-covered hillsides making the use and movement of field instruments an arduous task.

Geology. The principal sedimentary rocks of the district are slates of cambro-ordovician age, the name *Dundas Slates* being applied to the whole series, although coarser grained and more siliceous sediments are included among them. Rocks of the latter type actually predominate in that part of the ground examined by the I.G.E.S. at Renison Bell. During later palaeozoic times the sedimentary formations were intruded both by basic and acidic rocks, of which the principal representatives were gabbro and quartz-porphyry respectively. Although there were other periods of igneous activity this was the most important, since with it was associated the extensive mineralization of the district.

The chief geological features of the ground examined at Renison Bell are shown in Fig. 66. So far as can be ascertained the grits and shales overlie the cherts and silicified sediments, and these in turn overlie the arenaceous Dundas Slates. The strike of these beds is approximately north-west to south-east, and their general dip is to the north-east at angles less than 35°.

The relation of these beds to one another is not as simple a matter as the foregoing statement may suggest and in the vicinity of the ore-bodies it is particularly obscure. Intense local folding and faulting have occurred and, in the mineralized areas, the boundaries of the different formations are hidden by a ferruginous rubble and gossanous material which, no doubt, is due to the oxidation products derived from the sulphide bodies. A full description of the field, together with the various theories that have been advanced as to the origin of the ore-bodies, may be found by reference to the works cited below[a].

[a] Ward, *Bulletin, Geol. Survey, Tasmania*, No. 6 (1909). Conder, *Bulletin, Geol. Survey, Tasmania*, No. 26, pp. 1–75 (1918). Reid, *Bulletin, Geol. Survey, Tasmania*, No. 36, pp. 61–73 (1925).

The mineralization takes two main but related forms, characterized either by large flat-lying bodies of massive pyrrhotite, which usually carry small quantities of cassiterite, or by impregnations of quartz and pyrites (or pyrrhotite) in the slates. The mineral also occurs in steeply standing channels or "feeders" many of which are connected with the flat-lying sills of massive sulphide in the manner indicated by the cross-section in Fig. 66. Oxidation of these massive sulphide bodies has led to the liberation of the tin content which, at several points, has been profitably extracted from the gossans. So far, however, no systematic attempt has been made to mine the sulphides, since their tin values do not average more than 0·7 per cent. and their sulphur content is too low for profitable exploitation. Several bodies of massive pyrrhotite have been exposed at Renison Bell in the course of winning tin from the gossan outcrops; notably at the Old Mill Workings (see Fig. 66), where a large face of sulphide ore has been laid bare within a few feet of the surface.

The Purpose of the Geophysical Work. From the known geological evidence it was clear that the ground to be examined was highly mineralized, the mineralization being characterized by large bodies of sulphide of extremely complex form, which at certain points were known to lie within a few feet of the surface. Both pyrite and pyrrhotite possess high electrical conductivity, while the latter has also a high magnetic susceptibility; thus the field lent itself both to electrical and magnetic methods of prospecting. The underground waters are fresh and of comparatively high resistivity, so that no interference with electrical work was anticipated from this source. The main objects of the work were, therefore, to locate and, if possible, to determine the boundaries of hidden sulphide deposits, and also to make comparative tests of several methods by using them over the known bodies. It will be understood, of course, that the geophysical methods employed could only hope to locate the sulphides and would give no indication whether they carried tin. Nevertheless, by using both electrical and magnetic methods over the same ground there was the possibility of distinguishing between the magnetic bodies of pyrrhotite and the non-magnetic pyrites and, since the tin values are largely confined to the former, there was every hope of eliminating indications due to barren pyrites.

Equipotential Line Survey. An area of 114 acres was surveyed by this method and the results obtained are shown in Fig. 66. This figure includes only the more interesting portions of the ground and does not show the complete layout of electrodes, cable, and generator, although the distances to the two electrodes are indicated. The electrodes were placed 9000 ft. apart on a north-west bearing so that the normal current direction should be parallel to the general strike of the known deposits. Each electrode consisted of 200 yd. of galvanized iron wire; in one case buried in swampy ground, and in the other submerged in the bed of a creek. The insulated cable joining the electrodes was laid for a considerable distance along a disused tramway track; an arrangement which saved time spent on the layout but was not altogether desirable since it introduced a

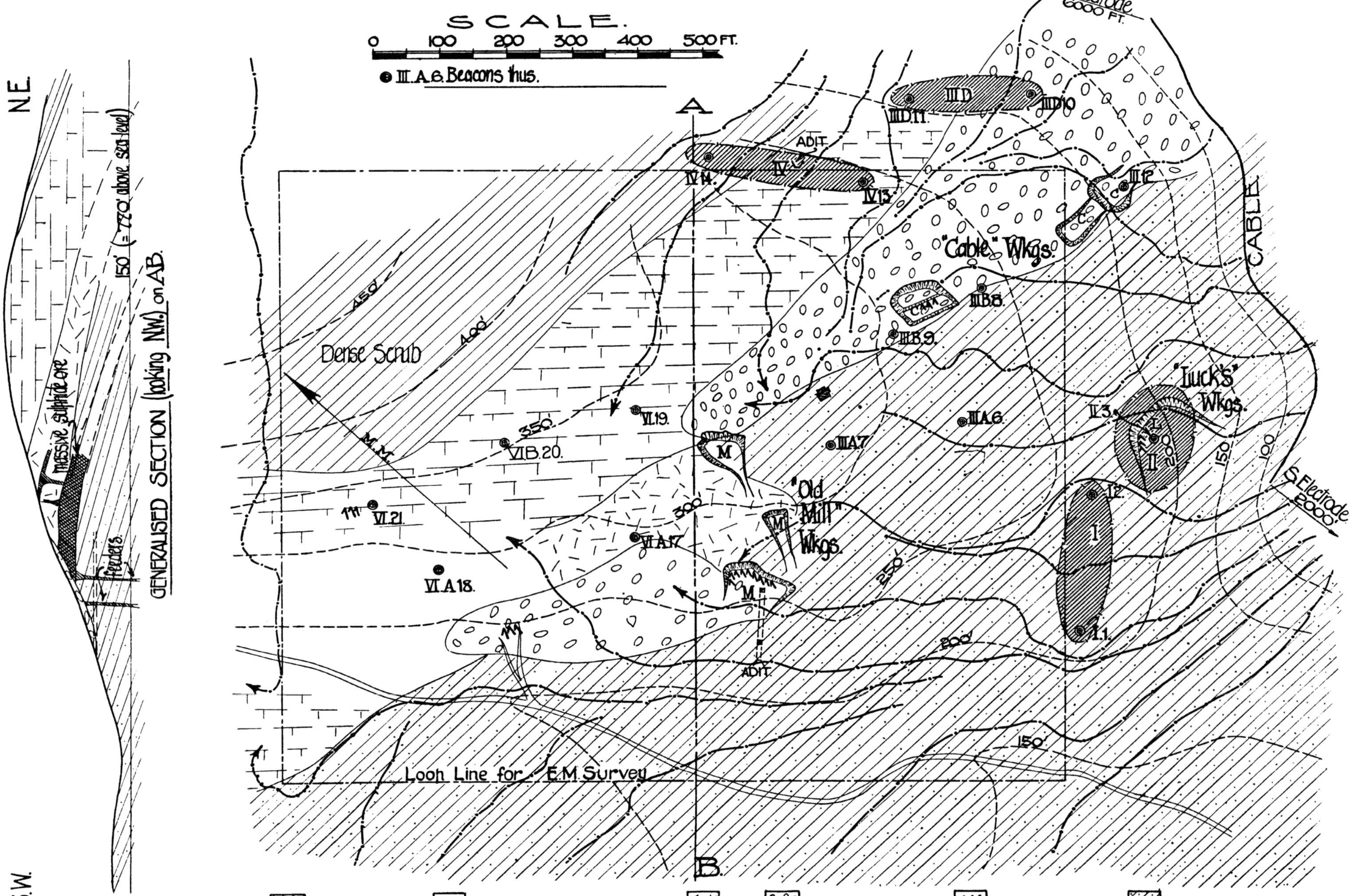

Grits and Shales; Chert and Silicified Sediments; Tuff; Ferruginous rubble; Massive sulphides; Dundas slate (arenaceous)

350' Contour lines; Equipotential lines (unable to continue beyond this point); I, II, III D, IV Electrical Indications

I.G.E.S.

Fig. 66. Geological map and equipotential line survey, Renison Bell, Tasmania

number of curves into the line of the cable. For the reasons discussed on p. 112, irregularities of this kind should be avoided as far as possible.

The general distribution and behaviour of the equipotential lines show considerable divergence from that which would normally occur on level ground free from conducting bodies, and it is apparent that the marked distortions in the current field must be due to disturbing features of a rather unusual character. To some extent they may be attributed to topographical effects and to the position of the cable, which, in the vicinity of Luck's and the Cable Workings, follows the contour of the hillside roughly in the form of an arc of a circle. The radial disposition of the lines is certainly due to this cause, since in this method there is always a natural tendency for the equipotentials to cross the cable line at right angles. These conditions, together with the local disturbances due to the presence of flat-lying sulphide bodies, are quite sufficient to account for the unusual features.

Over the larger part of the area there was no difficulty in following the equipotential lines, but on approaching the northern and north-western portions the minima as indicated by the headphones became increasingly indefinite, owing to the extreme out-of-phase conditions which were encountered there. Finally the work became impossible and the lines had to be abandoned at the points where they are shown terminated by arrows (Fig. 66). The purpose of these arrows is to draw attention to the fact that out-of-phase conditions of considerable magnitude occur at these points, and that the presence of secondary conductors in the ground is to be inferred. It follows, therefore, that considerable importance is to be attached to areas over which equipotential lines cannot be traced and that arrangements should always be made for their detailed examination by other electrical methods which can deal effectively with the phase conditions. In the case under discussion an electromagnetic method was applied, and a comparison of the results obtained (Fig. 68) will show that the inferences drawn from the equipotential line survey were well substantiated. Difficulties due to out-of-phase conditions are frequently experienced when applying the equipotential line method, and that they would arise at Renison Bell in an acute form was fully anticipated from the known mode of occurrence of the sulphide deposits. The survey was conducted, however, both as a demonstration of phase effects in equipotential line work and in order to show that, although prejudicial to the satisfactory operation of the method, they are sometimes of material diagnostic value when carrying out preliminary surveys.

In the eastern part of the area, where out-of-phase conditions did not prevent work, a more complete survey was made and four distinct indications were obtained. In Fig. 66 they are numbered I, II, III D, and IV respectively, the line-shaded areas being an indication of the limits within which the conducting bodies may reasonably be expected to occur. Indication No. I is the largest and best defined, the distortion of the equipotential lines being quite typical of a normal indication. Although there is no surface evidence of mineralization in the neighbourhood of Indication No. I, a strong magnetic anomaly (Fig. 72) was subsequently obtained at the south-west end of the indication, a fact that leaves little doubt that a sulphide body is present. Indication No. II, which occurs over Luck's open-cut workings and corresponds very closely with a body of sulphide which is visible at the foot of the northern face, is also confirmed by magnetic observations. In this case the equipotential lines have been somewhat distorted owing to the proximity of the cable, but nevertheless the position of the conductive zone may be inferred without any serious likelihood of error. Indication No. III D is not so well defined as I or II, and has also been influenced to some extent by the distorting effect of the cable. There is, however, a bold outcrop of gossan corresponding with the south-eastern end of the indication, and the presence of mineral is also inferred from magnetic disturbances at this point. In the case of Indication No. IV there is nothing on the surface to suggest mineralization, but it has been ascertained that a lode of some kind was cut in an old adit which was driven into the hillside at this point. As no other evidence is available, and since the adit is now inaccessible, it is impossible to correlate the position of this lode with the axis of the electrical indication, but there is reason to believe that they are in close agreement.

Electromagnetic Survey. A rectangular energizing loop 1200 by 1000 ft. was laid on the ground as shown in Fig. 67. The uneven nature of the topography prevented a true horizontal disposition of the loop, but as far as was possible it was laid out on the southern slope of the hill, care being taken that it should not extend over the crest of the ridge. In this manner a close approximation to a plane was obtained for the loop layout, and for practical purposes no great error was introduced by making the assumption that all observations were taken in that plane. A series of parallel traverses at intervals of 100 ft. was laid out within the rectangle and, after stations had been marked along them at 25 ft. intervals, the investigations of the electromagnetic field at each point were made with the double search coil[a]. The vector plan reproduced in Fig. 67 was drawn from the results of this survey, and shows to an arbitrary scale the direction and magnitude of the out-of-phase magnetic field at each point of observation, although for the sake of clearness some of the vectors have been omitted. The magnitude of the vector alone is a fairly good indication of the proximity and importance of conductors in the ground in its neighbourhood, but a convenient method of showing the distribution of the intensities of the out-of-phase field throughout the surveyed area is to prepare a plan in which curves are drawn through all points of the same value. In such a plan indications stand out clearly as areas surrounded by lines of high value. This *electromagnetic contour plan* for the Renison Bell area is shown in Fig. 68, and a series of profiles constructed from it in Fig. 69.

On all of these plans the unit used to represent the electromagnetic field is the flux of the primary field through a single turn of the horizontal coil of the apparatus, and the values are proportional to the ratio of the out-of-phase horizontal component to the primary vertical field. Although large readings indicate the presence of conductive material it does not necessarily follow that this will be mineral; and, in the case of

[a] Bieler-Watson Method.

barren ground of high conductivity, uniformly high readings may be obtained everywhere within the loop without having any economic significance. At Renison Bell, for example, where somewhat unusual conditions were encountered, it will be seen that there were very few stations at which readings of less than 400 units were registered. As this is the normal upper limit for which the apparatus is designed it was necessary

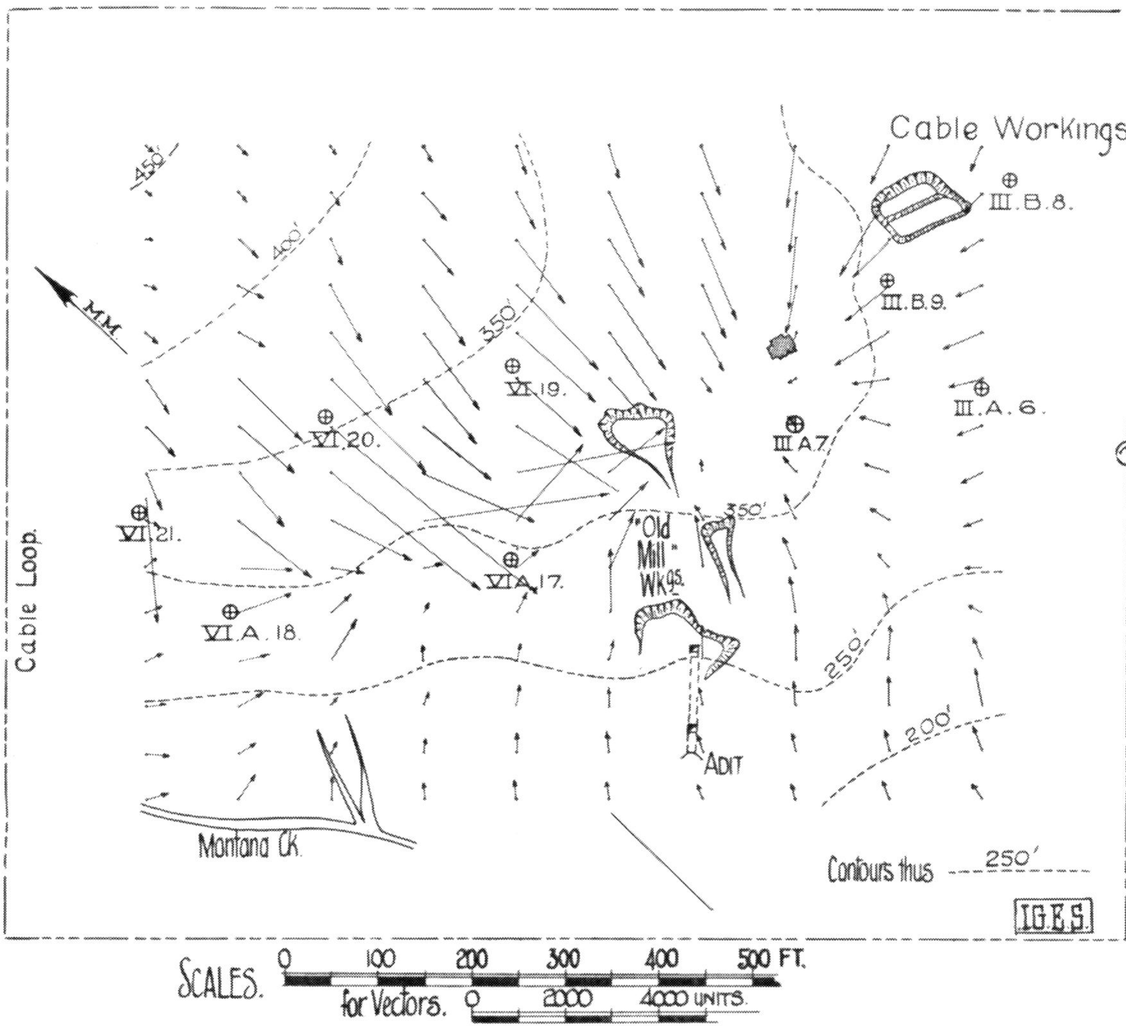

Fig. 67. Plan of electromagnetic survey, Renison Bell, Tasmania

to employ the special methods of making the observations that are described on p. 285. Fortunately, however, the readings in the vicinity of the sulphide bodies were so large that there was no difficulty in distinguishing important conductors, even though they were lying in ground of unusually high conductivity. The exceptionally large values recorded are best illustrated by comparison with indications obtained in other districts; such as Leadville in New South Wales (see p. 79), where the readings over the unmineralized ground were of the order of 200 units, and over the indications about 400.

At Renison Bell 400 units was the usual minimum, while over the main indications the readings rose to a maximum value of over 5000.

From the intensity of these indications it can be concluded that the conducting bodies must lie close to the surface, and from the distribution of the areas of greatest

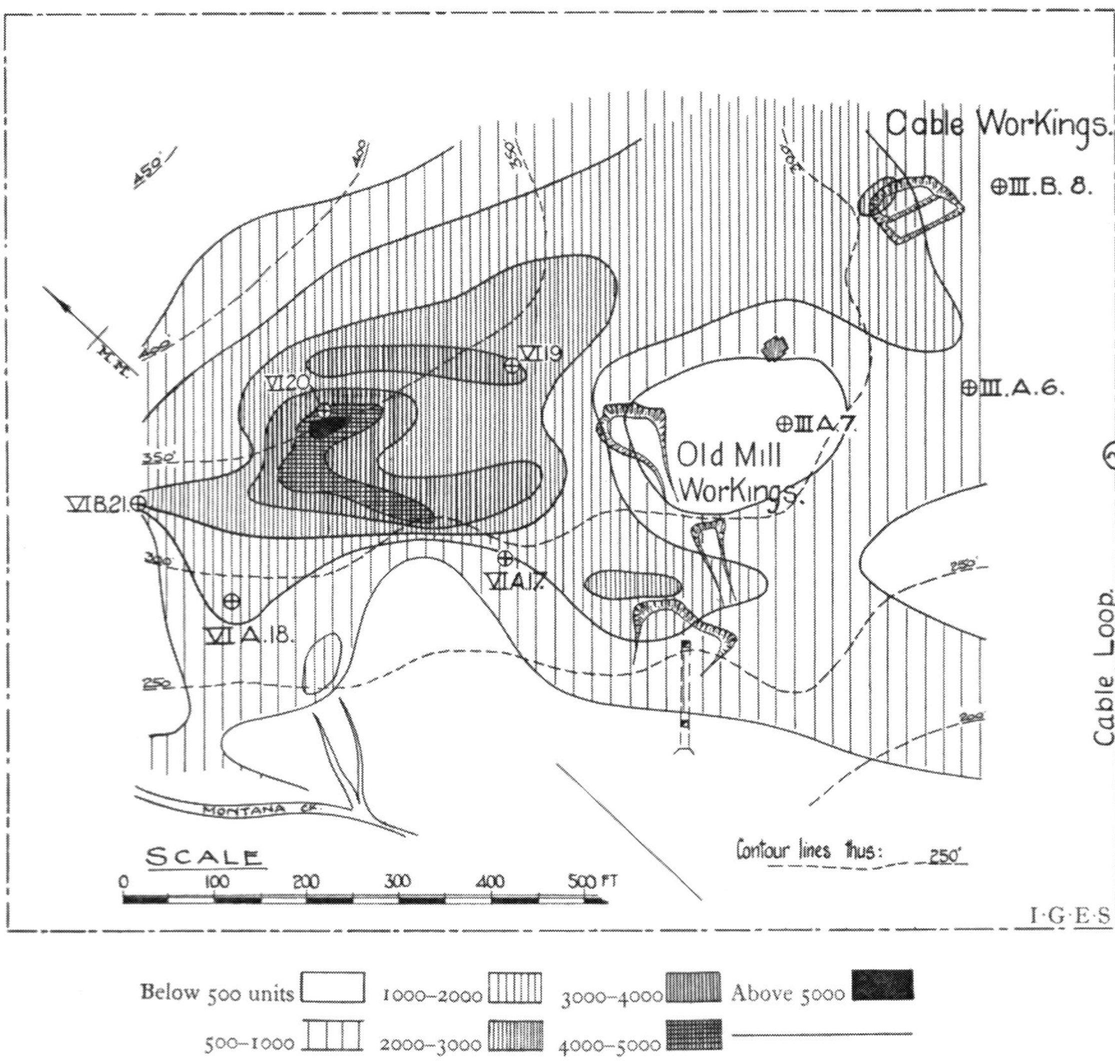

Fig. 68. Electromagnetic contour plan. Renison Bell, Tasmania

intensity (Fig. 68) it is apparent that the principal conductors lie in the unexplored ground to the north of the Old Mill Workings[a]. A southern extension of the indication occurs over the body of pyrrhotite exposed in the western open-cut of the Old Mill Workings, and another persists as far as the Cable open-cut. The known sulphide body in the Old Mill Workings serves a useful purpose in that it provides a standard by which the relative value of the other parts of the indication may be estimated. The pyrrhotite

[a] It will be observed that this is the ground over which equipotential lines could not be traced owing to extreme out-of-phase conditions.

here lies within 5 to 10 ft. of the surface, is 10 to 15 ft. thick, and extends for a length of nearly 100 ft. in the face. Over this considerable body the intensity was between 1000 and 2000 units, a value which is more than doubled in the unproved ground to the north where the indication reaches its maximum. It is reasonable, therefore, to assume that conductors of similar dimensions are widely distributed within the area enclosed by the 2000 contour line, and that this line represents, quite conservatively, the boun-

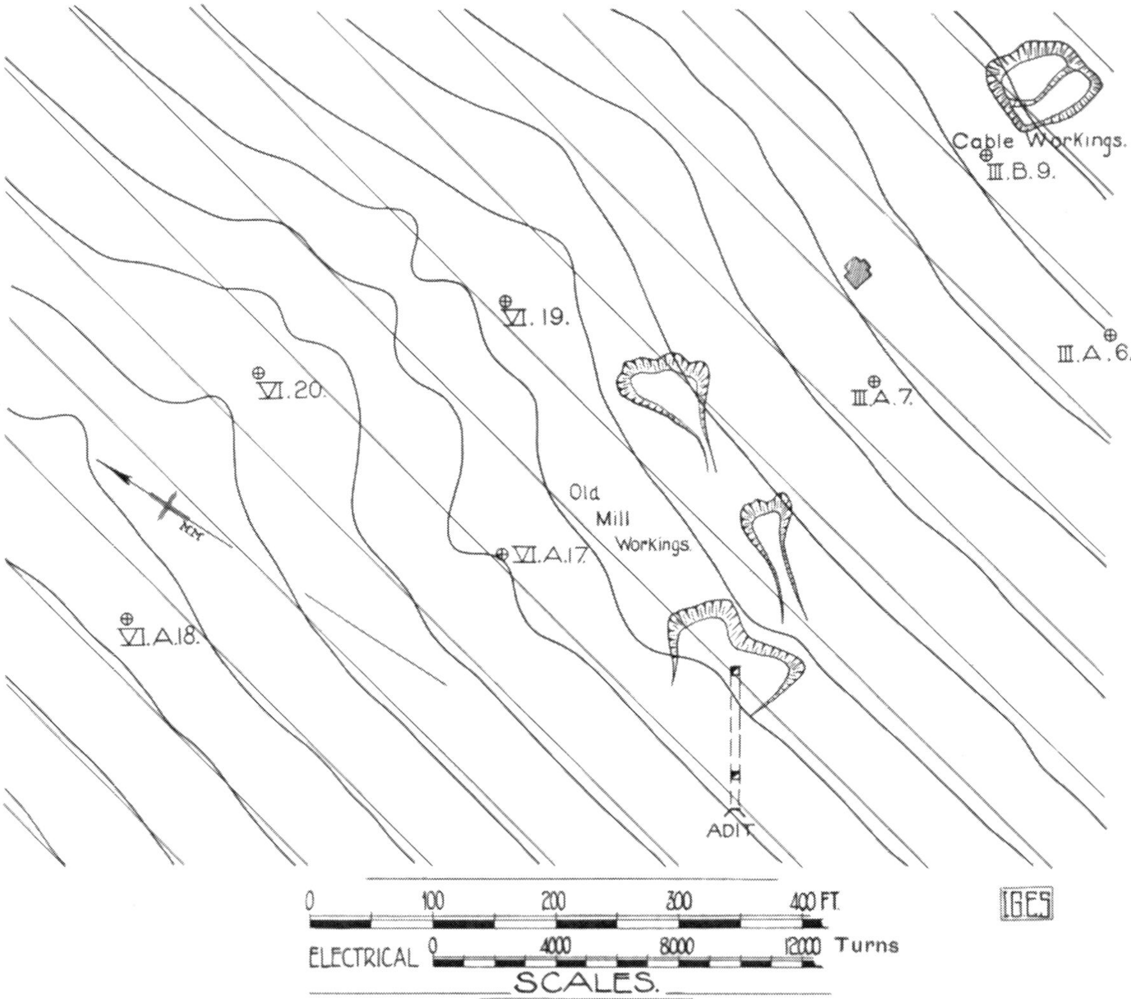

Fig. 69. Electromagnetic profiles. Renison Bell, Tasmania

dary of a highly conductive zone. Although there are no actual outcrops of mineral within the areas of maximum intensity it is thought that, at many points, the conductive bodies must lie very close to the surface, and that the indications are more likely to be due to massive sulphide than to mineralized slate.

Although a very large conducting area was revealed by this survey it would hardly have been wise to assume that it was due entirely to massive sulphides; therefore, before coming to any final decisions and making recommendations for exploration work, a careful comparison was made between these electromagnetic results and those obtained by the other geophysical methods that were employed over the same ground.

Spontaneous Polarization Method. An attempt was first made to locate the boundary of one of the known sulphide bodies by means of direct current, which was directly supplied to it by an electrode making contact with an exposed face of mineral in the western open-cut of the Old Mill Workings (see Fig. 66). The second electrode, consisting of some 200 yd. of bare wire, was buried in the ground at a point 2000 yd. distant. Equipotential lines were traced in the neighbourhood of the electrode which made contact with the mineral, but it was found that the potential gradients were due almost entirely to earth-currents resulting from the natural polarization of the sulphide bodies. These were large compared with those produced by the dynamo, whose available current was too low to be effective. The use of an artificial source of current was abandoned, therefore, and the work was continued by the normal spontaneous polarization method.

Equipotential lines were first plotted directly by means of a simple flying circuit containing a galvanometer and a pair of non-polarizing electrodes but, owing to the roughness of the country and the wandering character of the lines, this procedure was found to be too slow. It was abandoned, therefore, in favour of a system consisting of a series of intersecting straight line traverses along which potential measurements were made with a potentiometer.

The equipotential plan shown in Fig. 70 was constructed by interpolation from the potential values obtained on these traverses, and three marked negative centres will be observed. Two of them (Nos. 1 and 2) correspond closely with the known sulphide bodies in the Old Mill Workings, while the third (No. 3) lies in unexplored ground to the south-west of the Cable open-cut.

The potential distributions across these negative centres are shown in Fig. 71 by the two profiles along *AA* and *BB*, both of which are of interest in view of the magnitude of the gradients that they record. In the method of spontaneous polarization, gradients of 1 or 2 millivolts per foot are usually considered to be large, but in this case there are potential differences of nearly half a volt between the negative centres and points at a distance of 100 ft. from them. It should be remembered, however, that it is the activity of oxidation, rather than the size of an ore-body, which determines the strength of these natural earth currents, and that the conditions for vigorous electro-chemical activity are particularly favourable at Renison Bell owing to the wet climate and the character of the mineral which, in many cases, extends to within a few feet of the surface.

From the character of the indications in the neighbourhood of the Old Mill Workings it is believed that two bodies of mineral are present there, and that each of them is acting independently as a polarizing unit. It is always possible, however, for two or more negative centres to occur over the same ore-body.

Whichever of these structural possibilities may prove to be correct there can be little doubt that substantial sulphide bodies are present immediately beneath the three points which have been recorded as negative centres. The work carried out has

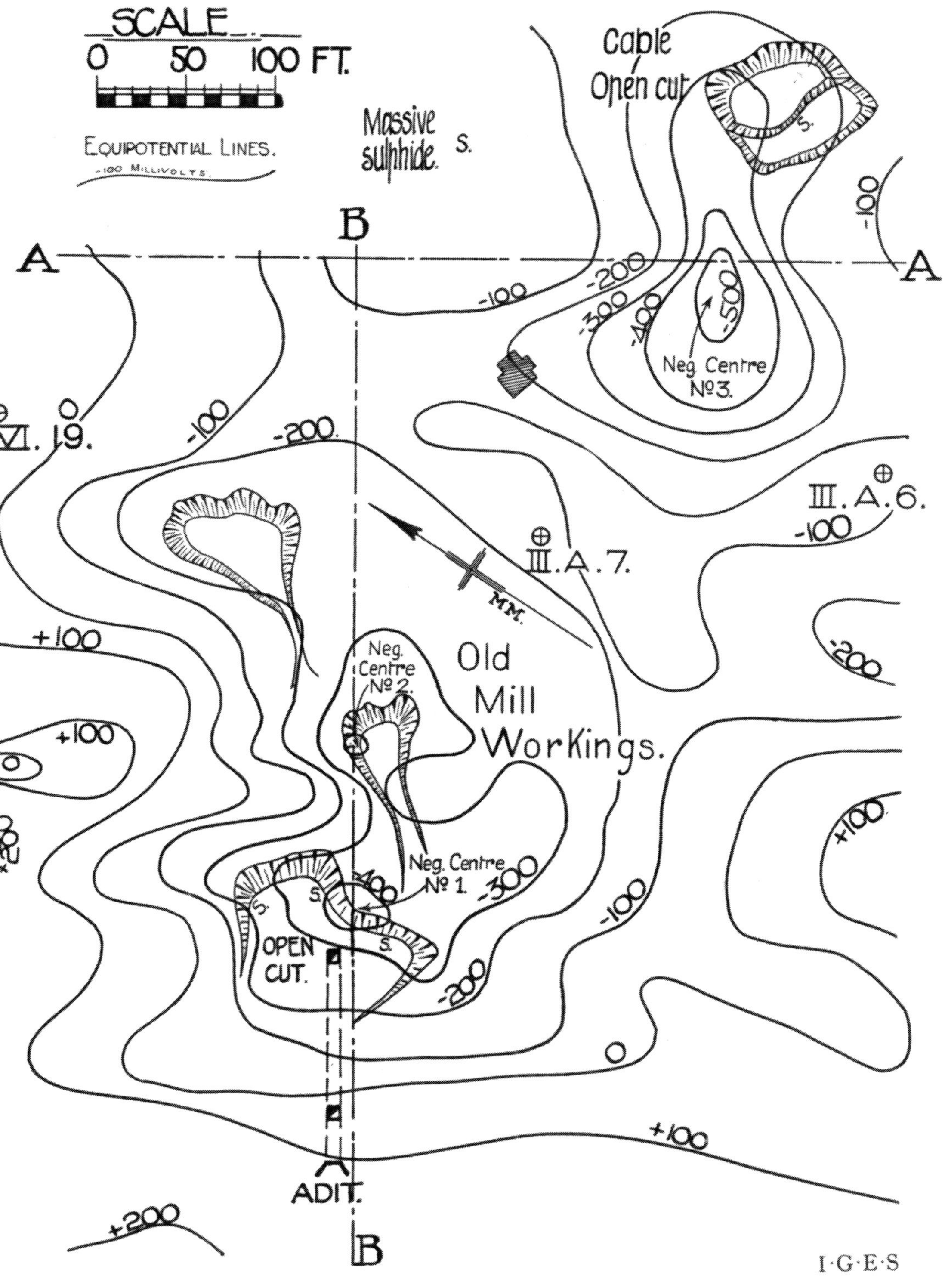

Fig. 70. Spontaneous polarization equipotential line plan, Renison Bell, Tasmania

also shown, very effectively, that the presence of these electro-chemically active bodies can be detected at considerable distances from them, and that to determine their position is quite a simple matter.

Magnetic Survey. Following the location of a number of conductors by the equipotential and electromagnetic methods a survey of the ground in which these occurred was made with the Thalen-Tiberg magnetometer. A large number of observations of vertical intensity was made over an area of some 39 acres, and a map showing lines of

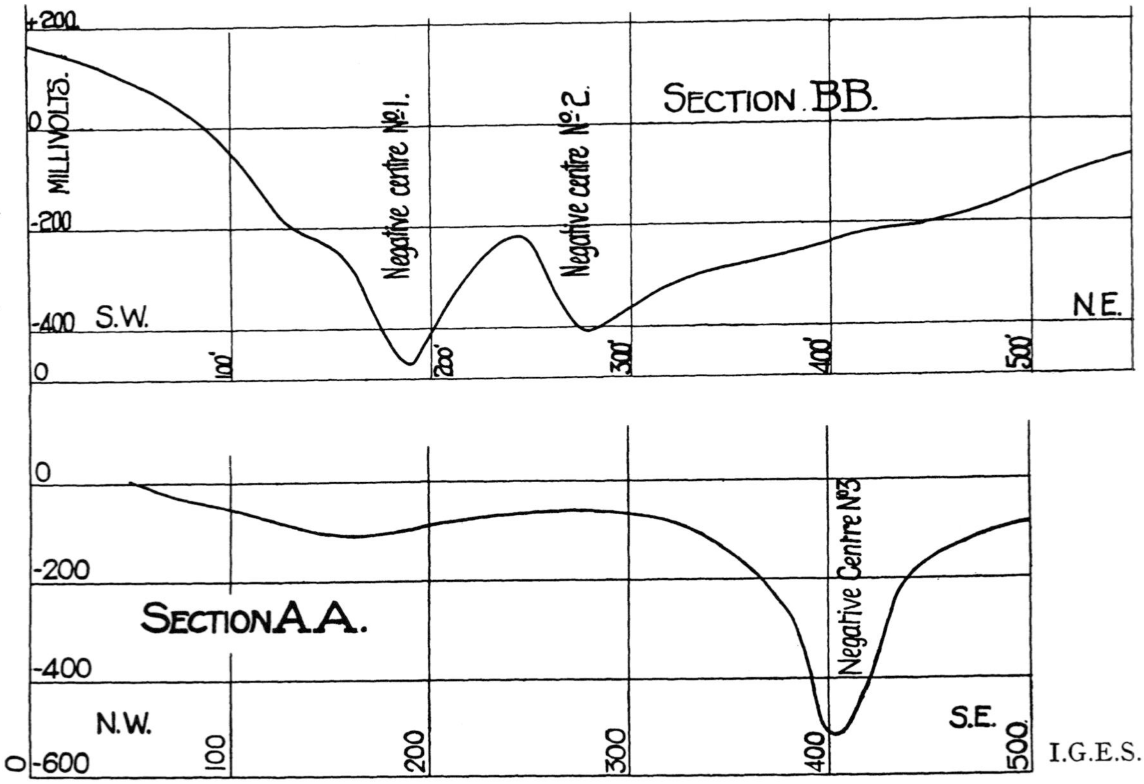

Fig. 71. Spontaneous polarization. Potential profiles along *AA* and *BB* in Fig. 70

equal anomaly in the vertical field was prepared. This map, together with profiles constructed from it, is reproduced in Fig. 72, and shows that the area is one in which there are well defined magnetic disturbances of a somewhat irregular character. The strengths of these magnetic anomalies are indicated by a system of line-shading; the earth's normal magnetic field being represented on the map by the datum or zero line, at every point on which there is no abnormality. As a general rule, the areas of positive anomaly represent the north seeking or north poles of magnetic-bodies, and the areas of negative anomaly their south seeking or south poles. From the inclination of the earth's field it follows that the north poles of magnetic-bodies will usually give more strongly marked anomalies than the south poles, since it is in the uppermost portions of the bodies that the north polarity is induced. This condition may of course be reversed according to the position of the body and its relation to the topography.

In all, eight magnetic indications were recorded and are shown on Figs. 72 and 73 by Beacons No. 1, 3, 6, 7, 8, 10, 12, 17, 18, 19 and 21. The two best defined areas of magnetic anomaly outlined on the indication plan (Fig. 73) are those numbered VI A and VI B respectively[a]. From Fig. 72 it will be seen that they form two parallel areas of positive magnetic intensity, extending for some 500 ft. in an east and west direction, and separated by a narrow strip of ground where the intensity falls to its normal value. To the south of these areas of positive intensity, and parallel to them, is a single area of well-marked negative intensity. There is no doubt that this group of anomalies constitutes a strong magnetic indication, the areas of positive intensity representing north poles of the magnetic body and the area of negative intensity its south pole. The field evidence confirms the magnetic work in a very satisfactory manner; the point of maximum positive intensity on profile *CD* being within a few feet of an occurrence of massive pyrrhotite which was discovered and exposed near Beacon No. 21 as a result of the magnetic survey, while the point of greatest negative intensity is located over an outcrop of the same class of mineral lower down the steep hillside. There is no showing of mineral corresponding to the intermediate peak of positive intensity, but its occurrence may be explained by assuming that the indication is caused by a flat sill of pyrrhotite (see profile *CD*, Fig. 72) which approaches the surface more closely at some parts than at others. Such an explanation, while not the only one possible, is the most simple, and for that reason is to be preferred. The profile along *AB* shows the same general features as that along *CD*, and the point of greatest negative intensity corresponds with a known body of pyrrhotite in the Old Mill Workings. The magnetic anomaly is very pronounced over the large area forming Indication VI B, the greatest positive reading giving a value of over 6000γ which is equivalent to about 10 per cent. of the normal vertical intensity. It is probable that several of the areas of maximum intensity (positive and negative) correspond closely with the more accessible portions of pyrrhotite bodies and, therefore, they have been delineated on the indication plan (Fig. 73).

Indications I and II are cases in which the negative anomaly is more strongly marked than the positive. In the former, the hill slopes upwards to the north-east so that the north pole of the body may well lie at a greater depth than the south pole. In Indication II an area of maximum negative intensity occurs some 30 ft. to the south-east of an exposure of pyrrhotite in the base of the face of Luck's Workings, and a closely related area of positive intensity occurs over the same exposure and extends northwards from it. The remaining indications, marked by Beacons No. 6, 7, 8, 10 and 12 all correspond to areas of positive intensity of 2000γ or more.

In an area of such intense and complex mineralization it was hardly to be expected that the results of this magnetic work would be of a simple character, or that exact interpretations would be possible. Such indeed was the case and although the location of the principal areas of magnetic disturbance proved of the greatest practical value

[a] The areas covered by the several indications are shown in Fig. 73 by Roman numerals.

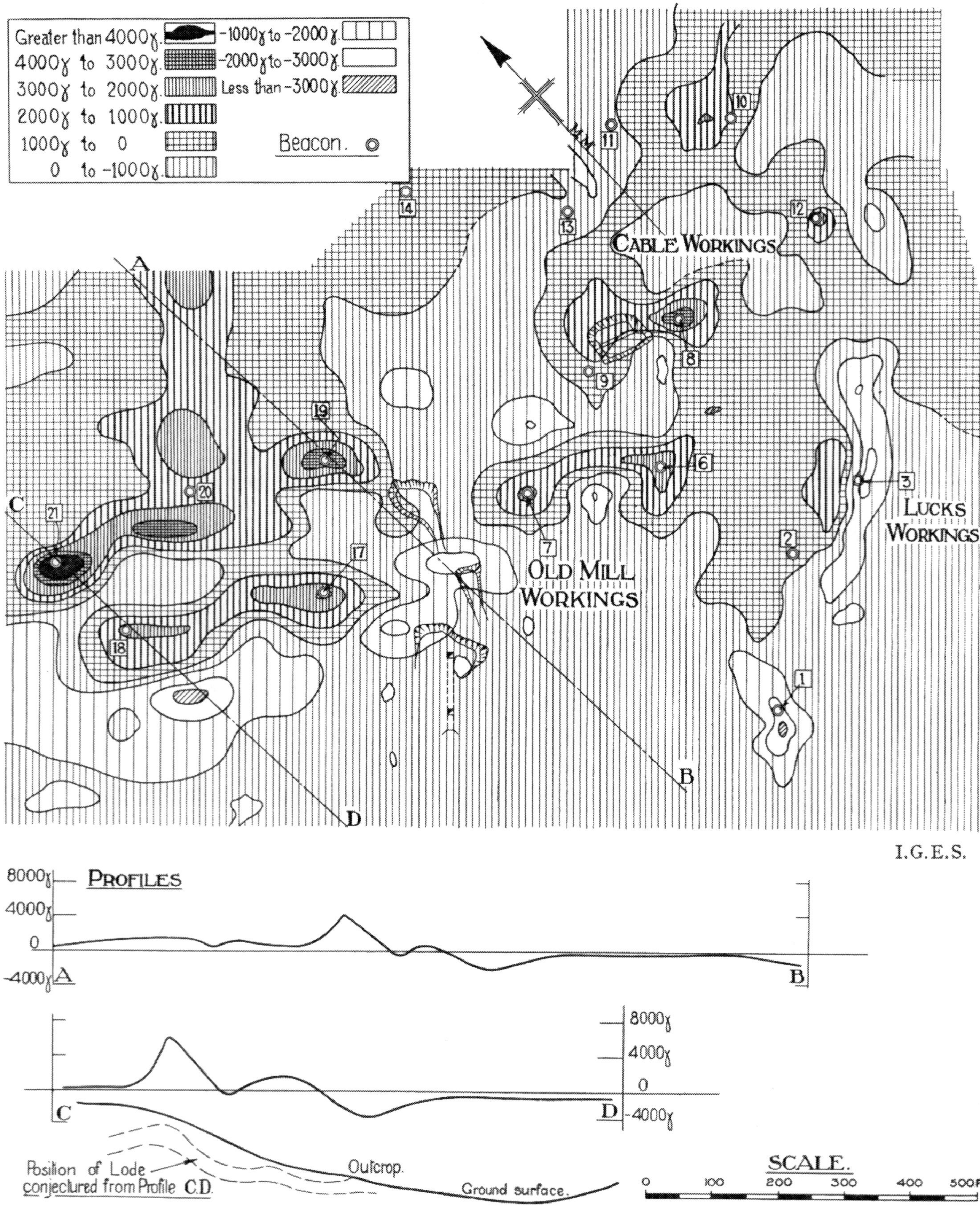

Fig. 72. Plan of magnetic survey, Renison Bell, Tasmania

when considered in conjunction with the results obtained by other methods, it was not possible to determine the distribution of the mineralization from the magnetic results alone. Vertical intensities only were measured since these give the most direct information regarding the position of hidden magnetic bodies, especially where the structural conditions are not simple. Although determinations of the horizontal and total

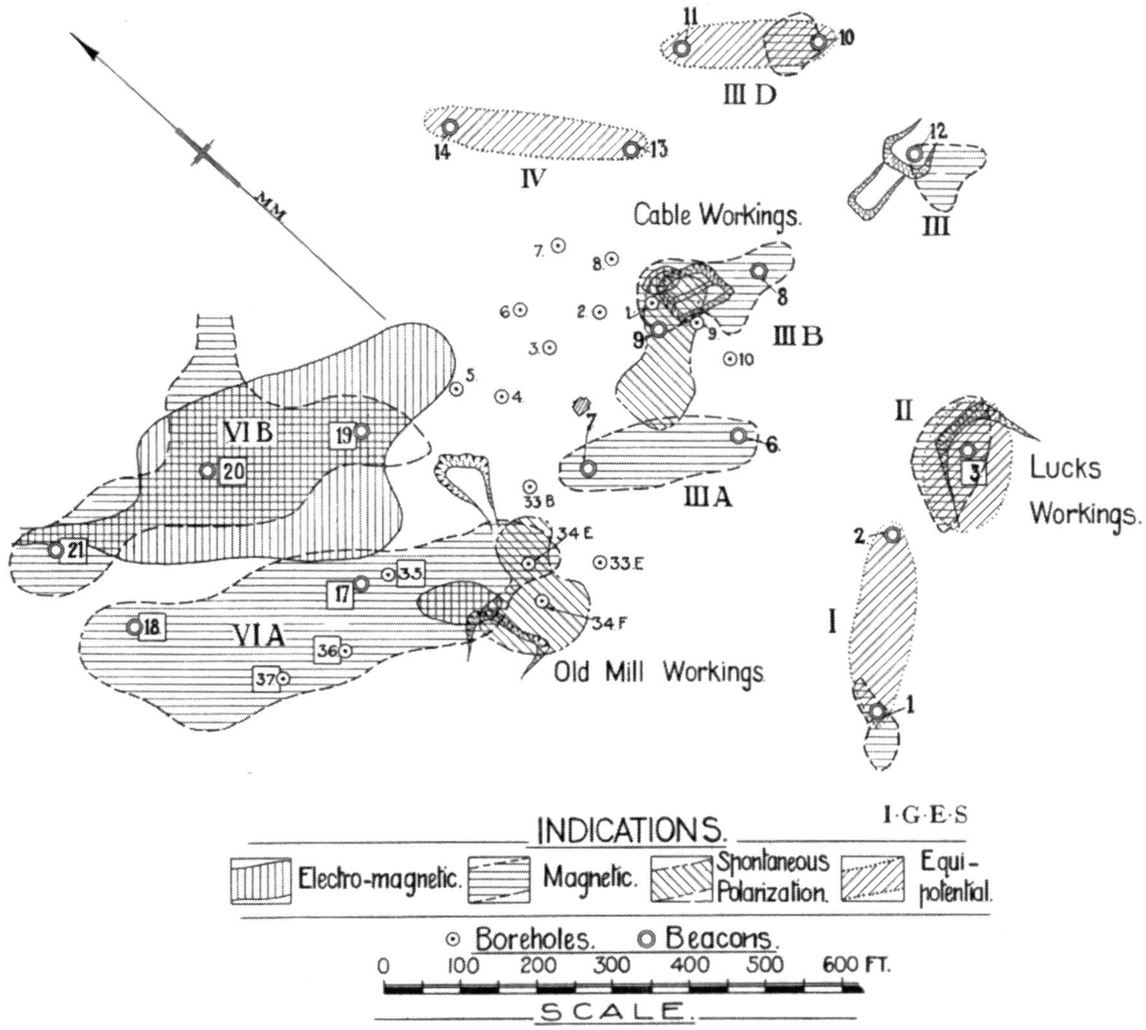

Fig. 73. Plan of geophysical indications, Renison Bell, Tasmania

intensities would have provided additional data it is doubtful if they would have carried the work of interpretation much farther. In any case it was the original intention that the magnetic results should be interpreted in conjunction with those obtained by the other methods. While magnetite is known to occur in some of the gossans of the district[a] and, possibly, may have contributed to some of the anomalies that were recorded, there can be little doubt that the majority, if not all, of them are due to bodies

[a] Conder, *Bulletin, Geol. Survey, Tasmania*, No. 26, p. 68 (1918).

TABLE III. *Comparison of results by different methods. Renison Bell, Tasmania*

Indication no. and reference beacons	Indication given by				Geology	Remarks
	E. P.	S. P.	E. M.	M.		
I 1 and 2	Strong	Not used	Not used	Strong negative anomaly	No surface indication of mineralization	The magnetic work confirms the south-western end of the equipotential indication
II 3	Fairly strong	Not used	Not used	Fairly strong	Pyrrhotite body visible in face of Luck's Workings	Good agreement between the two methods used. Indications correspond with known sulphide body
III 12	Indecisive	Not used	Not used	Fair	No sign of pyrrhotite. The anomaly occurs in old open-cut workings where oxidized ore existed	Indications very indefinite. The magnetic anomaly may be due to magnetite in the gossan
III A 6 and 7	Indecisive	No confirmation	No confirmation	Fair	No surface indication of mineralization	No satisfactory interpretation can be made
III B 8 and 9	Indecisive	Strong negative centre	Fairly strong	Strong	Pyrrhotite exposed in open-cut	Equipotential evidence indefinite but good agreement between the other methods
III D 10 and 11	Fair	Not used	Not used	Fair	Ironstone outcrop at south-east end of indication	Lodes are known to have been cut in this ground by an adit some 300 ft. below the surface
IV 13 and 14	Fair	Not used	Not used	Not used	No surface evidence of mineralization	An old adit is stated to have cut a lode here
VI A 17 and 18	Impossible to use owing to phase conditions	Not used	Fair	Strong	Pyrrhotite exposed in Old Mill Workings and at a point 100 ft. west-north-west of Bore-hole No. 37	These extensive indications, mostly in ground hitherto unexplored, are by far the strongest and most important obtained by the I.G.E.S. in this district
VI B 19, 20, and 21	Impossible to use owing to phase conditions	Not used	Very strong	Very strong	Little surface evidence. Pyrrhotite discovered near Beacon No. 21 as result of magnetic observations	

of pyrrhotite. Although detailed interpretations were not practicable the value of the magnetic method in locating mineral of this character, and in confirming the results obtained by other methods, is well shown by the surveys carried out at Renison Bell.

Summary of Results. Following the completion of the field work, a comparison was made of the indications obtained by the different geophysical methods employed, and after an examination of all the available evidence, a plan (Fig. 73) was prepared upon which the final practical conclusions were shown. A certain amount of boring had been carried out on the ground prior to the examination made by the I.G.E.S. but information gained from this work was not available and, in fact, no details of any kind were known until the work of interpretation had been completed. At a later date, however, the company concerned was asked to supply its boring records so that the geophysical conclusions might be checked. The ready acquiescence to this request, and the permission given to publish any necessary information have been of great value to the Survey. In appraising the value of the work, therefore, there are two things to be considered: firstly, the correlation of indications due to different methods and, secondly, the agreement between the geophysical and the bore-hole evidence.

Table III summarizes the evidence on which the value of each indication is based, and this, together with the final indication plan shown in Fig. 73, will be largely self-explanatory. Numbered beacons were placed on the ground to mark the indications and, as the positions of these are shown on all relevant plans (Figs. 66–73), they will serve as a guide to the reader who wishes to compare for himself the results obtained.

In Fig. 73 the individual indications obtained by each method have been shown by distinctive line-shading, and it follows that the best confirmed indications will be those in which the greatest superposition occurs. For the electromagnetic work a conservative estimate was made and only the ground within the 2000 unit line was included in the indication; also, in the case of the magnetic survey, only the principal areas of anomaly were included. The equipotential indications have been shown conventionally by shading the areas within which the conductors probably lie, and for the spontaneous polarization work a conservative view was again taken and only small areas around the negative centres have been delineated.

As a general rule it will be seen that agreement between the methods used was good, especially in the case of Indication VI B where the electromagnetic and magnetic results correspond closely. The pronounced out-of-phase conditions noted in this area during the course of the equipotential line survey also afford additional confirmation. The anomalies recorded over Indication VI B are so large, however, that each method independently provides a convincing indication. The least satisfactory indication is III A, because it is not confirmed either by the electromagnetic or the spontaneous polarization results. It is based, therefore, on magnetic evidence alone and, possibly, may be due to a cause other than the presence of mineral: a basic dyke rock for instance.

Bore-hole evidence. The locations of the bore-holes are shown in Fig. 73 and the information obtained from the Company's records is given in Table IV.

TABLE IV. *Records of vertical bore-holes. Renison Bell, Tasmania*

No.	ft.	
33 B	0– 90	Pyritic slate (5% pyrite)
	90–105	Unmineralized slate
33 E	0– 10	Massive pyrrhotite
	10– 55	Unmineralized slate
34 E	0– 20	Massive pyrrhotite
	20– 75	Pyritic slate
	75–100	Unmineralized slate
34 F	0– 20	Massive pyrrhotite
	20– 35	Unmineralized slate
35	0– 5	Slate
	5– 40	Massive pyrrhotite
	40– 50	Unmineralized slate
36	0– 5	Slate
	5– 45	Massive pyrrhotite
	45– 50	Unmineralized slate
37	0– 15	Ironstone gravel
	15– 20	Red rock
	20– 55	Massive pyrrhotite
	55– 60	Quartzite
1	0– 10	Ironstone
	10– 20	Massive pyrrhotite
	20– 25	Unmineralized slate
2	0– 15	Ferruginous clay (hole abandoned)
3	0– 25	Ferruginous clay (hole abandoned)
4	0– 30	Ferruginous clay
	30– 35	Unmineralized slate
5	0– 20	Ferruginous clay
	20– 35	Slate with quartz veins and stringers
	35– 40	Clay
	40– 50	Unmineralized slate
6	0– 40	Ferruginous clay
	40– 45	A little sulphide (not recorded whether massive or not)
7	0– 20	Ferruginous clay
	20– 30	Clay
	30– 40	Unmineralized slate
8	0– 20	Ferruginous clay
	20– 22	Unmineralized slate
9	0– 10	Massive pyrrhotite
	10– 15	Unmineralized slate
10	No record	

The series of percussion bore-holes (Nos. 33–37) proves a sheet of massive pyrrhotite extending north-westwards from Bore-hole No. 33 E as far as the outcrop of sulphide which lies to the south of Beacon No. 18. The mineral increases rapidly in thickness; from 10 ft. in Hole No. 33 E to 35 ft. in the ground lying immediately to the north-west of the Old Mill Workings. It will be seen that with the exception of No. 33 E all the holes fall within the limits of the geophysical indication VI A, and afford a very satisfactory confirmation of it. Both drilling and geophysical work are in agreement as regards the south-eastern end of the sheet, since the drilling shows it to be thinning and the magnetic indication becomes less intense in that direction. Hole No. 33 E, although showing 10 ft. of mineral, affords ample evidence of the thinning out of the pyrrhotite sheet to the east of the Old Mill Workings; while Hole No. 33 B provides a still more definite confirmation of the geophysical evidence since the mineral has cut out entirely at that point. The pyrite content of the slates recorded in this hole is, of course, too low to have given an electrical indication. It is possible, however, that to the south-east of Hole No. 33 B these pyritic slates give place to massive pyrrhotite, thus accounting for Indication III A; the geological characteristics of the local mineralization being quite in accordance with such a change. It will be observed that the north-western end of Indication III A lies within 40 ft. of this bore-hole.

In the ground lying to the east and north-east of Indication III A valuable evidence is afforded by the series of shallow holes Nos. 1–10. It will be seen that both No. 1 and No. 9 lie within Indication III B and encountered sulphide, while of the remainder only one (No. 6) showed any evidence of mineralization, and in that case the records do not indicate that any substantial sulphide body was met. It will be seen, therefore, that the geophysical results are closely in accord with the bore-hole evidence, and the fact that the latter information was entirely unknown to the officers of the Survey until some months after their report had been submitted to the Department of Mines, together with the positions of the indications as shown in Fig. 73, ensured that the interpretation of results was not influenced by it. It is true, however, that the work of interpretation was carried out more confidently in view of the satisfactory correlations that could be made between certain of the geophysical indications and the several surface exposures of mineral that occur in the area.

At the conclusion of the investigations in this district the Director of Mines for Tasmania was asked to undertake some additional boring operations that were required for further testing the geophysical results that had been obtained, and a list of thirteen shallow holes was submitted to him. Unfortunately, however, this full programme could not be carried out but, eventually, it was arranged that four holes should be put down in the neighbourhood of Indications VI A and B, at the positions marked by Beacons Nos. 18, 19, 20 and 21. Information has not yet been received from Tasmania that this series of holes has been completed, but the reports to date may be summarized as follows, together with such explanatory comments as are necessary.

Bore-hole No. 18. This hole is situated on a well-defined magnetic high. In correspondence with the I.G.E.S. the Director of Mines for Tasmania states that, on geological grounds, he has reason to believe that this hole will reveal another pyrrhotite bearing ore-body. So far, however, the results of this hole have not been received.

Bore-hole No. 19. This hole was also selected on magnetic evidence. It was carried to a depth of 47 ft. mostly through mineralized slate and quartzite. From 16 ft. to 35 ft. the rock carried from 5 per cent. to 8 per cent. of iron pyrites. From the bore-hole log there is no satisfactory indication of the cause of the magnetic anomaly.

Bore-hole No. 20. This is the location of the largest electromagnetic readings by the modified Bieler-Watson method. The bore log shows weakly mineralized ground similar to that met with in Bore-hole No. 19. No massive mineral was encountered.

Bore-hole No. 21. This hole is situated at the point of the maximum magnetic high on Indication VI B. After passing through 8 ft. of dolomitic rock it passed into dense sulphide ore (pyrrhotite) which continued to a depth of 25 ft. From 25 ft. to the bottom of the hole at 47 ft. the core showed grey slate with disseminated pyrites.

The absence of any substantial bodies of mineral in Bore-holes Nos. 19 and 20 is rather disappointing, particularly so in the case of the latter hole where the electromagnetic results were so impressive. There are good reasons for believing that a pyrrhotite sheet is present in this area, and it is distinctly possible that the two bore-hole sites lie just beyond its boundary.

Generally speaking the agreement between the bore-hole and the geophysical evidence has been very satisfactory in this area, which is largely to be attributed to the use of several methods over the same ground. The most successful of these was the magnetic, although the electromagnetic method also provided convincing evidence. In the case of the latter, however, there would have been considerable uncertainty in defining the limits of the indications had no other methods been used as a check. For example, it will be observed that at a point over hole No. 37, where the sulphide is only 20 ft. from the surface and is 35 ft. thick, readings of less than 500 units were obtained. Farther to the east, however, in the neighbourhood of Bore-holes Nos. 2–8, where no mineral has been found within 30 or 40 ft. of the surface, values of over 1000 units were observed. These must be attributed to the low resistivity of the thick clay overburden encountered in the latter bore-holes. Under these circumstances it is clear that the magnetic method, which is independent of ground conductivity conditions, is at a very great advantage.

The equipotential survey proved of material value as a reconnaissance method in spite of the fact that extreme out-of-phase conditions prevented its use over certain parts of the district where the sulphide bodies are believed to reach their maximum development. In itself this fact was of value in indicating important areas where detailed investigation by other methods was required. As an auxiliary, the spontaneous polarization method was also of great assistance in spite of its natural limitations for prospecting purposes. Although incapable of detecting deep seated mineral bodies it affords a very simple and expeditious means of locating concealed outcrops where oxidation processes are active.

SECTION 4. MOONTA, SOUTH AUSTRALIA (E. 8a)

The Wallaroo and Moonta mining district is situated on the eastern shores of Spencer's Gulf, South Australia, and from 1860, the date of discovery, up to 1923 it produced many millions of pounds worth of copper. Since the latter date the production has fallen to a very low figure, owing to the working out of the most payable lodes.

This mining field is one of many in Australia in which the character of the mineralization and its mode of occurrence are entirely favourable to electrical methods of prospecting—excepting for the serious factor presented by saline underground waters and highly conductive surface formations. Both at Wallaroo and at Moonta the salinity conditions are particularly bad, and, under normal circumstances, the application of electrical methods would have been ruled out completely. Nevertheless, it was decided that the Survey should carry out some experimental investigations in saline areas, in order to determine the extent to which the various methods are affected, and to see if useful work of any kind is possible under such adverse conditions. The field under discussion was selected for this purpose in view of there being important possibilities that undiscovered ore-bodies still exist within reasonable depths from the surface.

Topography and Geology. The area of one and three-quarter square miles examined by the I.G.E.S. lies a mile and a half to the east of the township of Moonta, and consists of flat, open country typical of the district. There are no streams of any kind, the only topographical features of note being caused by low sand hills of recent geological age. The rainfall averages only 15 in. per annum, and the absence of any definite drainage system appears to be largely responsible for the accumulation of soluble salts in the underground waters.

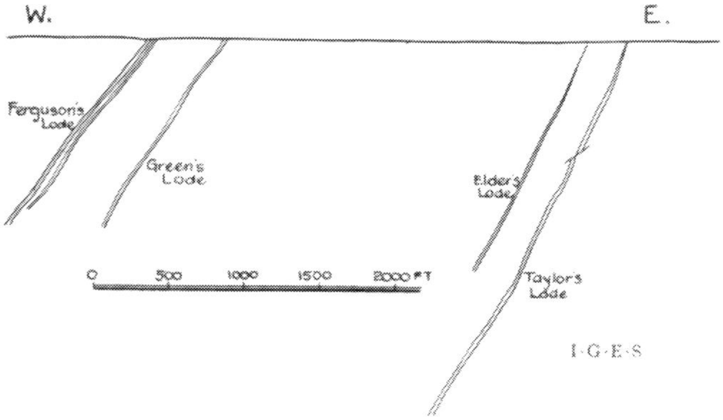

Fig. 74. Typical cross-section of lodes at Moonta, S.A.

The geological features of the district are comparatively simple, and have been described in the publications of the South Australian Mines Department[a]. The country rock consists of a felspar porphyry of pre-cambrian age in which the copper mineralization occurs in the form of well-defined lodes. These have a general strike somewhat north of north-east, and dip consistently to the west at angles from 45 to 65°. Several different lines of fracturing in the porphyry have given rise to a series of parallel lodes as illustrated in Fig. 74 by a typical cross-section. The lode channels persist for great distances across country, but the development of ore-bodies within them is somewhat intermittent and usually takes the form of steeply dipping lenticular shoots. The

[a] Jack, *Bulletin, Geol. Survey, South Australia*, No. 5 (1917).

principal copper minerals present are chalcopyrite and bornite, pyrites being uncommon. In the richest parts of the lodes these minerals were found massive over widths of a few inches to 10 ft.; elsewhere the lodes consist largely of pegmatitic material through which the sulphides are irregularly dispersed. The upper portions of the lodes have been oxidized, but the primary sulphide minerals are usually to be expected at a depth of 60 or 70 ft.

Although the felspar porphyry has been planed down to a level surface by successive periods of denudation the lodes do not outcrop, since a thin covering of recent formations extends over the whole district. This overburden rarely exceeds 10 or 15 ft. in thickness, and consists of a sheet of travertine limestone[a] underlain by clay—the former either outcropping or lying within a few inches of the surface. A typical section is shown in Fig. 76 from which it will be seen that the underlying clays give place to the solid porphyry at a depth of about 10 ft.

Resistivity Conditions. Before commencing work in this district the local resistivity conditions were investigated by means of laboratory tests on a series of rock and mineral specimens and on samples of underground water collected both from the Wallaroo and the Moonta sections of the field[b]. The results obtained are shown in Table V.

TABLE V. *Resistivity values for Moonta and Wallaroo copper field.*

Locality	Material	Description	Resistivity in ohm cm.
Moonta	Country rock	Unsheared felspar porphyry	4×10^5
,,	,, ,,	Felspar porphyry (schistose facies)	3×10^5
,,	Mineral	Rich bornite ore	0·3
,,	,,	Bornite-chalcopyrite ore	4·0
,,	Water	Mean of five samples from flooded mine workings	56·0
Wallaroo	Country rock	Mica schist	1×10^7
,,	,, ,,	Schist slightly mineralized	$1{\cdot}5 \times 10^6$
,,	Mineral	Rich chalcopyrite ore	0·12
,,	,,	Low grade veinstuff (sulphides)	13·0
,,	Water	Mean of three samples from flooded mine workings	36·0

Although resistivity determinations made in the laboratory upon specimens of rocks and minerals cannot be relied on to give a true indication of the actual ground resistivity conditions it would appear from the above table that the ratio between mineral and country rock is quite satisfactory. It was feared, however, that the low values recorded for the underground waters might prove to be the controlling factor in this district, although, on geological grounds, there were reasons for believing that these highly conductive saline waters were largely confined to the lode-channels, in which case they might actually assist in the electrical prospecting operations.

[a] This type of calcareous surface formation extends over vast areas in many of the arid and semi-arid parts of Australia. The terms travertine, calcrete, and hard-pan are commonly applied to it, but the first two names are not very appropriate. The material slowly accumulates *in situ*, apparently by deposition from moisture caused to ascend through the sub-soil as a result of the intense desiccation of the ground surface.

[b] The Survey is indebted to Professor Kerr Grant, of Adelaide University, who kindly arranged for the examination of many of these specimens in his department.

After consideration of the relative merits of the Moonta and Wallaroo sections of the field for the purpose of the proposed experimental tests, it was decided that the difficulties to be contended with would prove rather less formidable in the former area. The mean resistivity of the underground waters at Moonta is appreciably higher than at Wallaroo, and this was the principal factor which determined the site of the work.

Equipotential Work. Equipotential line reconnaissance surveys were carried out over an area of one and three-quarter square miles, but before the work had proceeded very far it became apparent that the electrical conditions were of an unusual character. Not only were there widespread out-of-phase conditions but at distances from a quarter to half a mile from the cable it was almost impossible to work owing to the

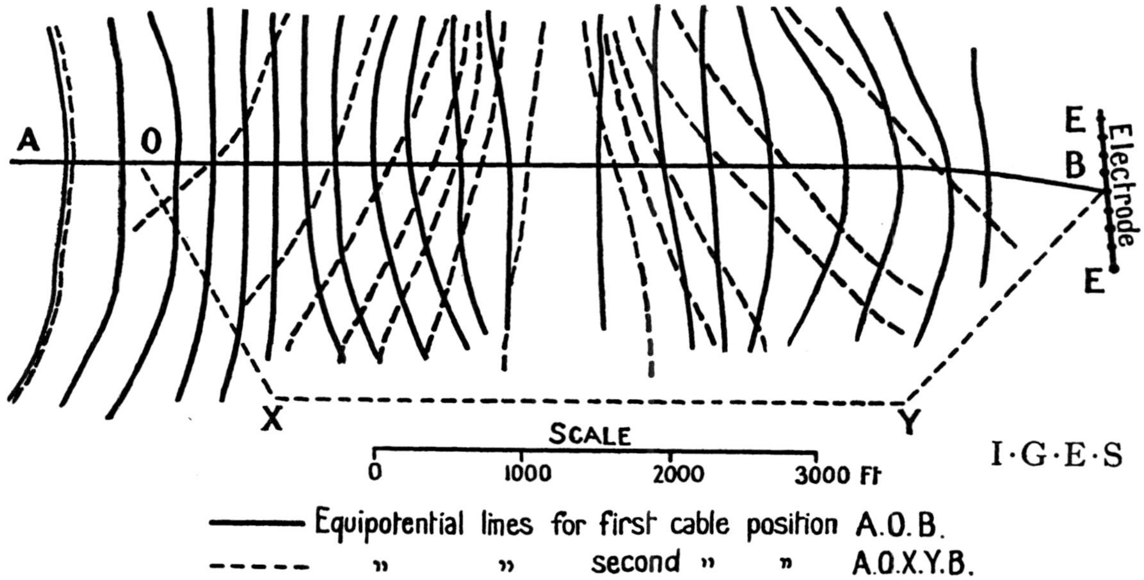

Fig. 75. Example of cable control, equipotential line survey, Moonta, S.A.

weakness of the note in the headphones. Under normal conditions there is no serious diminution in intensity of the note even at distances of a mile or more on each side of the cable. At first this was attributed to an abnormally high ground conductivity due to the salt content of the underground waters, and although several indications were recorded by the equipotential lines it was thought, from their peculiar nature, that they must be due to variations in this sub-surface salinity, rather than to the presence of mineral bodies lying at greater depths.

In one area the large anomaly indicated by the full lines in Fig. 75 was obtained, which suggested that a conductive body of considerable extent might be present. It will be observed, however, that its axis coincides exactly with the line of cable, *AOB*, which connects the alternator to the two electrodes. In order to eliminate any local effects due to the cable a re-examination of the ground was made with the cable diverted laterally to a distance of 1600 ft. along the line *AOXYB*. This was considered to be necessary since, even under normal conditions, the cable always exerts an appreciable control on the equipotential lines where they cross it. That the effects of *cable control* were unusually strong in this district can be seen from the behaviour of the second series of equipotential lines which were obtained with the cable along $AOXYB$. It will be observed that all signs of the original conductive zone have disappeared and that the new set of lines exhibits a pronounced fan-like distribution—such that each line, if produced, would cross the cable at right angles. It will also be noticed that where the cable remained in its original position, close to the point A, there was no change in the direction of the equipotential lines, but that a marked deviation occurs just beyond the point O, where the cable was deflected. From this behaviour it must be concluded that the distribution of the return current through the ground is controlled to a remarkable extent by the position of the cable[a], and that a path of relatively high current density actually follows the line of the cable even where sharp changes occur in its direction, as at the point O. It is understood that a somewhat similar condition is met with in the case of submarine telegraphy where the return current is said to be concentrated along the outside of the cable. In view of these abnormal circumstances it was considered that there was little hope of achieving results of a reliable nature by the equipotential line method, and the investigations were discontinued.

High Frequency Work. The high frequency method[b] was used at Moonta in the hope of gaining more precise information about indications given by the equipotential line survey. It was realized, however, that the chances of success were not very great owing to the extremely high ground conductivity.

Most of the work was carried out at a frequency of about 60,000 cycles per second, and tests were carried out over five small and indefinite equipotential indications at the southern end of the field. In two of these cases very weak indications were obtained, but in the others the results were negative. In the northern part of the field, where the equipotential method had given an extensive indication, the high frequency method also revealed fairly large conductive zones but, with little doubt, these were due to local anomalies in the ground conductivity.

Resistivities of Surface Formations. As the working conditions had proved even more abnormal than had been anticipated a short test was carried out with the Gish and Rooney resistivity apparatus. This revealed a highly conductive layer within a few feet of the surface, and apparently extending uniformly over the entire district. To investigate this feature a pit was sunk to a depth of 10 ft. and resistivity determinations were made from samples taken from it at 1 ft. intervals. From Fig. 76 it will be seen that a layer of clay having a resistivity of 271 ohm cm. occurs within 6 ft. of the surface and immediately overlies the felspar porphyry in which mineral

[a] This concentration of the alternating return current in the surface of the ground adjacent to the cable is known as the skin effect. See Russell, *Alternating Currents*, vol. I, chap. VII, University Press, Cambridge, 1914.

[b] See p. 69.

is to be sought. On geological evidence it can be assumed that this conductive formation extends as a thin sheet over the whole area and there is no doubt that it is the primary cause of the electrical anomalies that were observed. In the light of this information the analogy with a submarine cable is still greater since the ground at Moonta is shown to be highly conductive within a few feet of the surface on which the cable was lying.

Testing of Indications. At the conclusion of the field work in this district a report was sent to the Department of Mines for South Australia in which details were given of the several conductive zones that had been located. It was explained, however, that there were strong grounds for fearing that the majority, if not all, of these indications were due to salinity conditions rather than to the presence of copper mineralization. Nevertheless, three of the most distinctive indications were thought to be rather more favourable than the remainder and the Director of Mines agreed to put a diamond drill-hole down on each of them in the hope that their true character would be revealed. Each of these three holes penetrated the highly conductive surface formations before passing into the underlying porphyry. Hole No. 1 passed through 11 ft. of surface clay having a mean resistivity of 300 ohm cm. and a water sample from the standing ground water level, met with in this hole at 28 ft. from the surface, gave a value of 44 ohm cm. Hole No. 2 revealed a siliceous copper lode 30 in. wide, carrying 1·4 per cent. of copper in the form of chalcopyrite. This was cut at a depth of 190 ft. below the surface and within 18 ft. of the axis of the electrical indication. Hole No. 3 was quite barren of mineral.

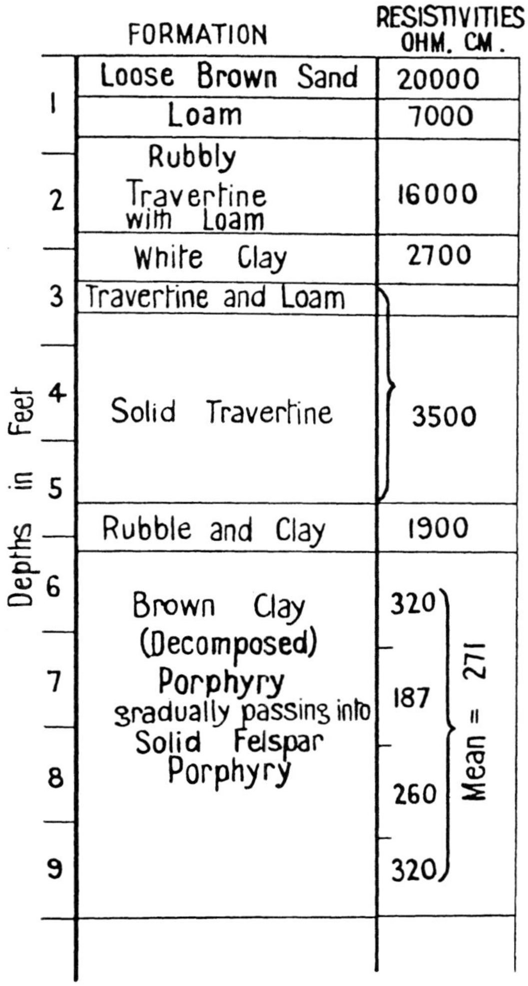

I·G·E·S

Fig. 76. Section of test pit at Moonta showing strata and resistivities

Conclusions. Although a low grade copper bearing lode was proved by Hole No. 2, within a few feet of the position indicated by the electrical survey, it cannot be said that this affords satisfactory evidence of the efficacy of electrical prospecting methods in this district. The investigations carried out at Moonta have shown that a satisfactory application of these methods is almost, if not entirely, impracticable in this locality. As already explained, a highly conductive sheet of clay covers the entire area to a depth of about 10 ft., and beneath this there is in many places a standing underground water level which is so saline as to be almost as good an electrical conductor as the copper ore which is to be located. The screening effect of such a conductive layer is so great that there can be little hope of discovering conductors that lie beneath it. The 10 ft. layer of clay alone would have such an adverse effect that, after passing through it, the primary field employed would have only about 35 per cent. of the intensity it had at the surface. At many places the thickness of clay increases to 20 ft., and under these conditions the intensity of the primary field would fall off to about 15 per cent. before passing into the underlying formations. It can well be understood that the response at the surface from deep-seated conductors would be so feeble as to be beyond the limits of detectability.

It should be mentioned that the frequencies employed at Moonta were too high for an area of such low ground resistivity. Frequencies of the order of 100 cycles per second, or the use of direct current methods, would have been more appropriate under such conditions, but even if suitable equipment had been available it is very doubtful if satisfactory results would have been obtained.

PORT LINCOLN, SOUTH AUSTRALIA (E 11)[a]

Occurrences of graphite are known in several localities in the Eyre Peninsula, along a belt of pre-cambrian schists and gneisses which extends over a length of 50 miles between Sleaford and Yaranyacka. The area examined by the I.G.E.S. is situated on this belt, within the Hundred of Uley, an undeveloped part of the Peninsula lying about 14 miles south-west from Port Lincoln. At the time of the investigations the ground in question was held under lease by an Adelaide company[b] which has developed one of the previously known deposits and has erected a flotation plant for the recovery of the graphite.

Geology and Topography. The pre-cambrian rocks are highly folded and, prior to regional metamorphism, they appear to have consisted of magnesian limestones, associated with arenaceous and argillaceous sediments. The view has been expressed[c] that the graphite occurrences are connected with the metamorphism of the calcareous schists, but there is also some evidence of pegmatite dykes which appear to have acted as foci of graphitization.

The country at Uley is of a mildly undulating character, mostly covered by low scrub, which did not present any serious obstruction to the geophysical survey work. The graphite bearing formations are concealed by a covering of aeolian calcareous sands, although in the area investigated by the I.G.E.S. they are largely hidden from view by a superficial sheet of travertine[d] and iron-stone.

[a] This account is based on a report by I. W. Morley, B.M.E., Assoc.Inst.M.M.

[b] To this company (Graphite, Ltd.) the Survey is indebted for many facilities.

[c] *Mining Reviews*, Nos. 26 (1917), 27 and 28 (1918), 31 (1919), and 45 (1926), South Australian Department of Mines, Adelaide.

[d] See footnote (a) on p. 110.

The true character and distribution of the graphite occurrences in this district are not known with any degree of certainty but, from the evidence provided by trenches and shafts, it would appear that graphitization is rather general over the entire area and is not confined to definite lode channels. Nevertheless, from an economic viewpoint the workable deposits are limited to ground carrying 8 per cent. or more of flake graphite. The graphitic zone already exposed in the open-cut workings has the appearance of a flat lying lenticular body of a rather irregular shape and with ill-defined walls. Both the ore and the surrounding country rock are so soft and decomposed that their original characters cannot be fully determined.

Objects of the Electrical Investigations. Since the graphitic bodies of economic size and grade are likely to have so sporadic an occurrence, the thorough prospecting of the district by ordinary methods would be a difficult matter; particularly so since the graphite bearing formations are largely concealed by an extensive sheet of overburden. It was hoped, therefore, that some electrical method would prove effective in locating the richer deposits.

Resistivity Determinations. A preliminary investigation of the local resistivity values showed that the conditions for electrical prospecting at Uley are reasonably favourable.

TABLE VI. *Mean Resistivity Values at Uley Graphite Area. Port Lincoln, S.A.*

Material		Resistivity ohm cm.
Graphite ore	Average grade	0·5
	Low grade	150·0
Country rock	Soft kaolinitic clay	20,000·0
Overburden	Red clay soil	2×10^8
	Black surface soil	10^6
Water	From local wells	1,000·0
	From mine workings (contaminated)	500·0

From the above table of resistivities it will be observed that the surface formations (overburden) are, fortunately, of high resistivity and that the country rock, although soft and clay-like, still affords a large resistivity ratio with respect to the average grade of ore. Under such circumstances it was thought that one of the surface potential methods might prove effective and, therefore, it was arranged that test surveys should be carried out, first by equipotential lines and afterwards by the A.C. potential ratio method. Port Lincoln was the last district in Australia to be examined by this section of the I.G.E.S. and circumstances did not permit more detailed investigations by inductive or other methods.

The Survey Area. The surface features and general lay-out of the survey work are shown in Fig. 77. The known ore-body, which has partly been exposed by the open-cut, appears to have a strike trend a few degrees east of north, but the structural features of the deposit are still rather obscure and its extensions beyond the limits of the present workings are not known with any degree of certainty. Although flake graphite has been encountered in shafts and trenches at a number of points within the area, there was no definite evidence of other commercial bodies at the time the electrical investigations were carried out.

For the purposes of the electrical survey a base line was laid out along a north-south bearing, past the western edge of the open-cut. It was then marked at 100 ft. intervals both to the north and south of the zero station shown in Fig. 77.

Equipotential Line Survey. The intention was to carry out a quick reconnaissance survey by this method, with a view to locating the principal conductive zones. The short linear electrodes (*AA* in Fig. 78) were placed about a mile apart in a north-south line passing through the known ore-body (open-cut), the cable connecting them being laid 800 ft. to the east of the base line in order to be clear of the mineral bearing ground.

The tracing of the equipotential lines over the surface was found to be a slow and difficult matter owing to the extreme dryness of the ground and exceptionally large out-of-phase conditions. Although the area was covered very thoroughly by lines starting from the base line at 100 ft. intervals, the results were quite inconclusive. No distinctive indication was obtained in the neighbourhood of the known ore-body, although an abnormal current distribution was observed at several points.

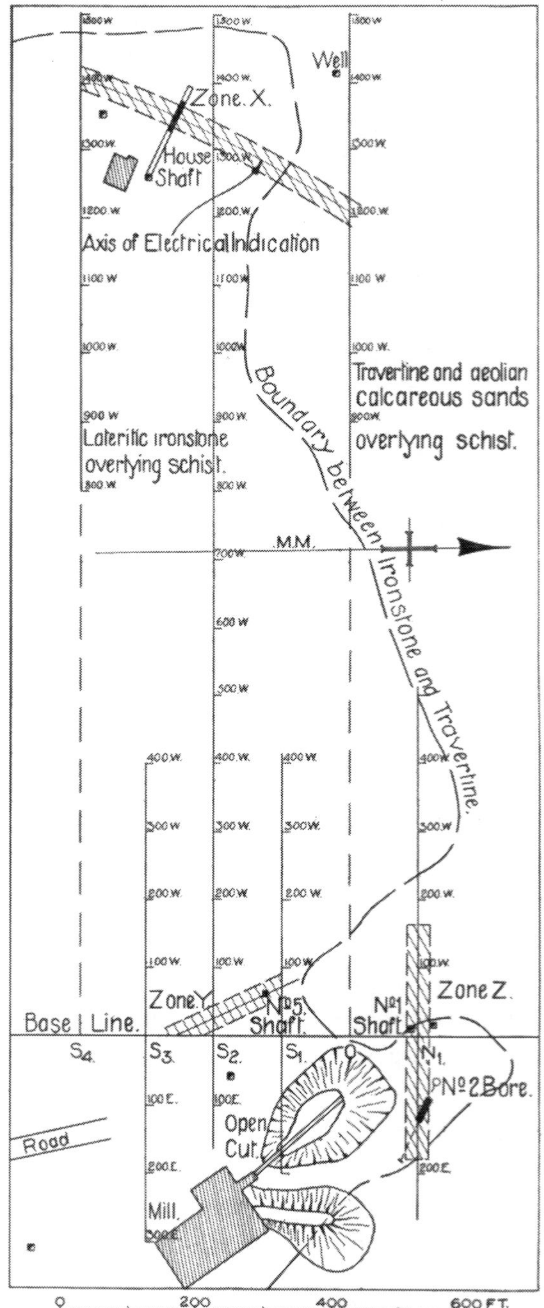

Fig. 77. Surface plan showing area of survey and conductive zones, Graphite Mine, Port Lincoln, S.A.

These unsatisfactory results were attributed to the ill-defined character of the graphite mineralization, the deposits having no distinctive boundaries but really consisting of local enrichments gradually grading off laterally into weak impregnations. The flat dip of the deposit and the rather low resistivity of the country rocks may also account to some extent for the absence of recognisable indications by the equipotential line method.

A.C. Potential Ratio Tests. Since marked out-of-phase conditions were observed during the course of the equipotential line survey, tests were carried out by the A.C. potential ratio method which is capable of dealing with such phase anomalies.

For the purpose of these investigations a series of parallel traverse lines was laid off to the east and west of the base line stations, N 1, O, S 1, S 2, etc., as shown in Figs. 77 and 78. These lines cross the survey area in a direction approximately at right angles to the general strike trend of the graphitic body exposed in the open-cut which, presumably, is indicative of the strike of any other bodies that occur within the area. Investigations of the potential and phase anomalies were then repeated along these traverse lines for each of the three separate electrode systems shown in Fig. 78; the object being to compare the results obtained and so determine the most suitable electrode lay-out for a survey of this kind. In the cases of these three tests the portable electrodes connected to the ratiometer were spaced along the traverse lines at 50 ft. intervals.

In Test A the electrodes *AA*, already laid out for the equipotential line survey, were employed. In Test B the electrodes *BB* were used, the electrical traverses being carried out westwards from the long linear electrode *B*. In the case of Test C, the same long linear electrode *B* was connected to a distant electrode *C*, lying 4000 ft. away to the west of the base line. For purposes of comparison the potential and phase curves obtained along Traverse S 2, with these three separate electrode systems, *A*, *B*, and *C*, are shown together in Fig. 79.

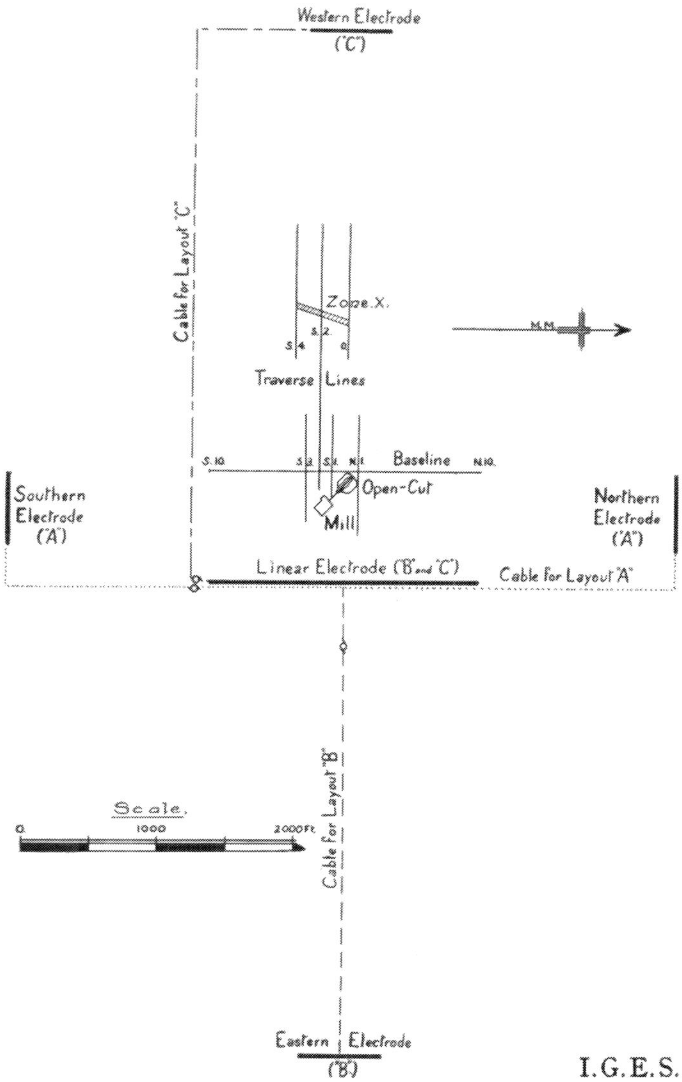

Fig. 78. Plan of electrode systems *A*, *B* and *C*, Graphite Mine, Port Lincoln, S.A.

The general resemblance between these three sets of curves is marked, even in the case of Test A, in which the direction of the current distribution through the ground was at right angles to that of Tests B and C. As was expected the anomalies revealed by Test A are not very distinctive since, in this case, there was a close degree of parallelism between the equipotential lines and the directions of the electrical traverses. Traverses by the A.C. potential ratio method should not, as a rule, be made under such conditions, since slight changes in the direction of the equipotentials, although of no practical significance, will often result in large values being recorded on the ratiometer. In the cases of Tests B and C, the current distributions are such that the traverse lines are approximately normal to the equipotential lines and, from Fig. 79, it will be seen that the potential variation curves *B* and *C* are very similar. In the former there is a rapid decline in current density with increase in distance westwards from the linear electrode, and this has the effect of exaggerating the potential anomalies at the eastern end of the traverse. In working between a pair of electrodes the normal potential gradient in the intervening ground is far more uniform, and for this reason it was anticipated that the electrode layout used in Test C would prove the most suitable for a traverse along line S 2.

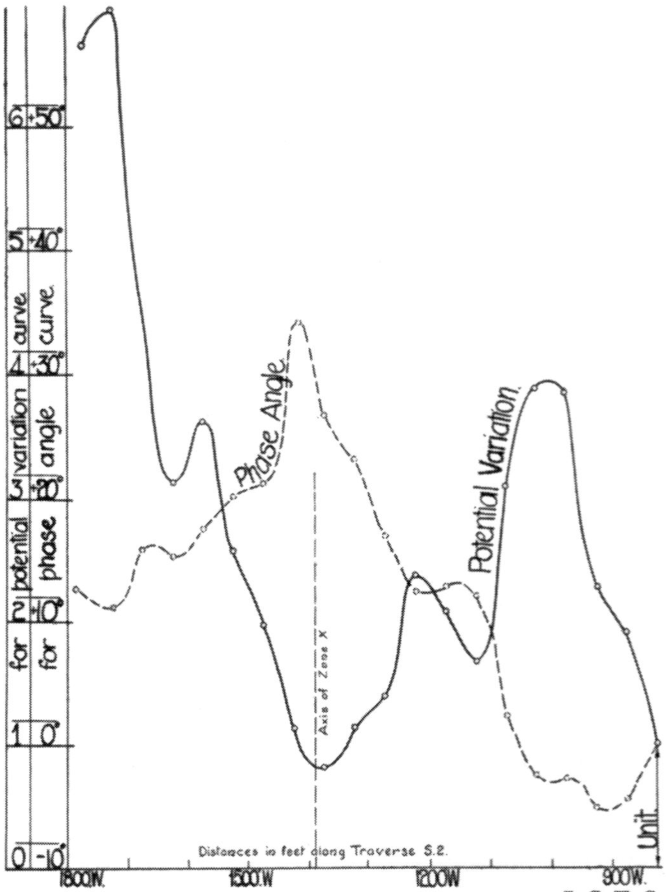

I.G.E.S.

Fig. 80. Potential and phase curves on Traverse S 4, Graphite Mine, Port Lincoln, S.A.

The potential profile *C* (Fig. 79) reveals two conductive zones, *X* and *Y*, the former being a large scale indication of characteristic appearance. As explained on p. 54 the potential peaks observed by this method on each side of a conductive body are due to the relatively high current density within the more resistant country rocks immediately bounding it. In Zone *Y* this feature is less apparent but the abrupt drop in the potential curve over the ground in the neighbourhood of the zero station (on base line) is quite significant. The anomaly revealed at this point by the phase angle curve is actually more distinctive than that which was observed in Zone *X* on this particular line of traverse.

In view of the magnitude of the potential indication recorded on Traverse Line S 2 (Zone *X*), two parallel traverses were run across the same piece of ground along Lines O and S 4, 200 ft. to the north and south of Line S 2 respectively (Fig. 77). In each case a large potential anomaly was recorded, of a character very similar to that obtained on Line S 2; the features shown by the phase angle curves being better defined however.

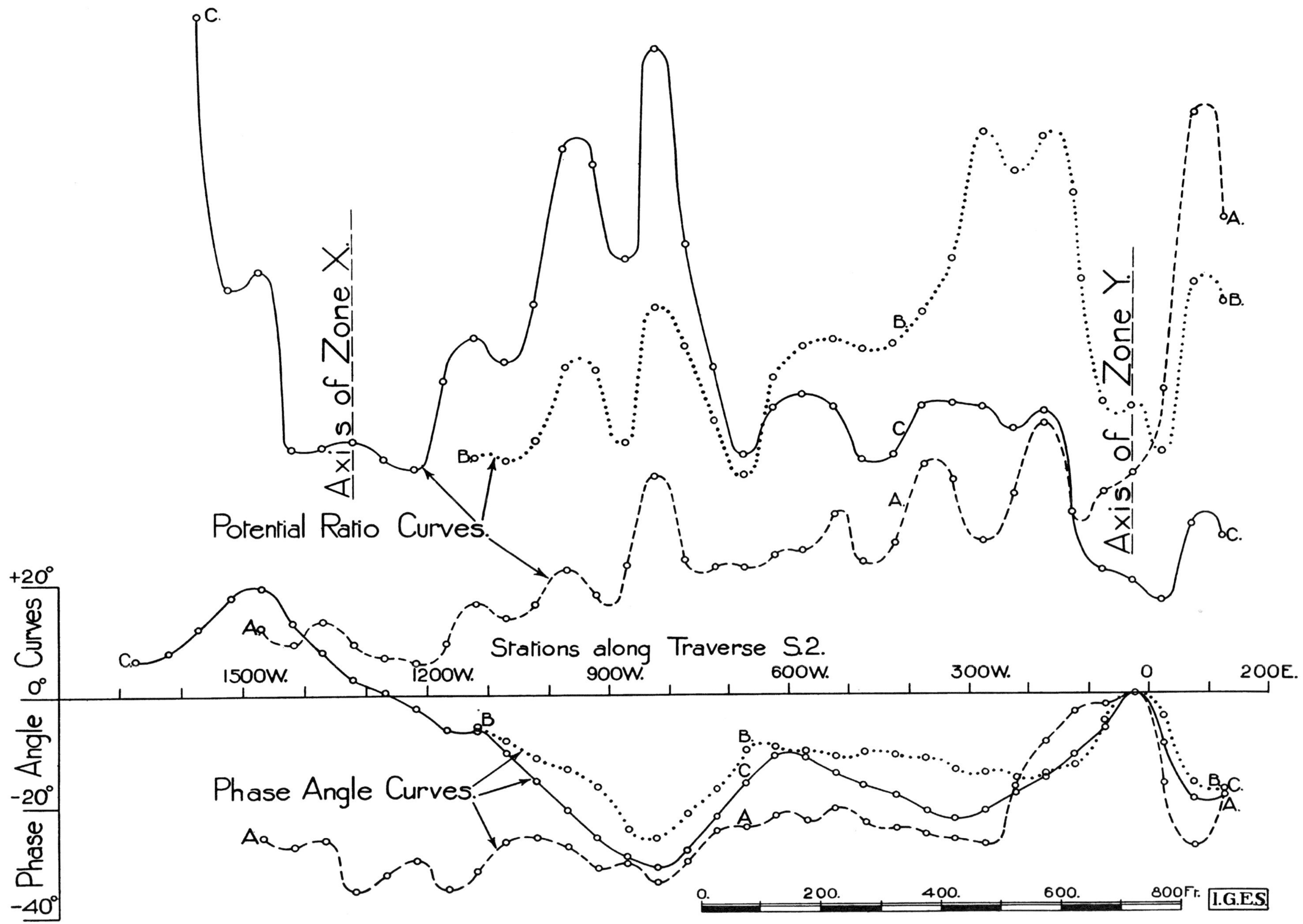

Fig. 79. Potential and phase curves on Traverse S 2, for electrode systems, *A*, *B*, and *C*, Graphite Mine, Port Lincoln, S.A.

From Fig. 80, which shows the results of Traverse S 4, it will be seen that the phase curve is almost an exact reflection of that plotted for the potential ratios.

The strike axis of the conductive zone was located by means of these three traverses and is marked in Fig. 77. Unfortunately, time did not permit the tracing out of its extensions to the north-east and south-west, but the evidence obtained on Traverses S 4 and O is suggestive of important continuations in both of these directions.

Zone *Y*[a] was also investigated further by parallel traverses along Lines S 3, S 1 and N 1, the results on S 1 being shown in Fig. 81. On Traverse N 1 a very extensive indication was recorded which reveals a highly conductive zone (*Z*) extending between stations 200 E. and 200 W.[b]. It is believed that this wide zone results from the convergence of Zone *Y* and the extension of the ore-body exposed in the open-cut, but owing to local disturbances caused by metal rails, etc., in the nearby workings a satisfactory investigation of this piece of ground could not be carried out.

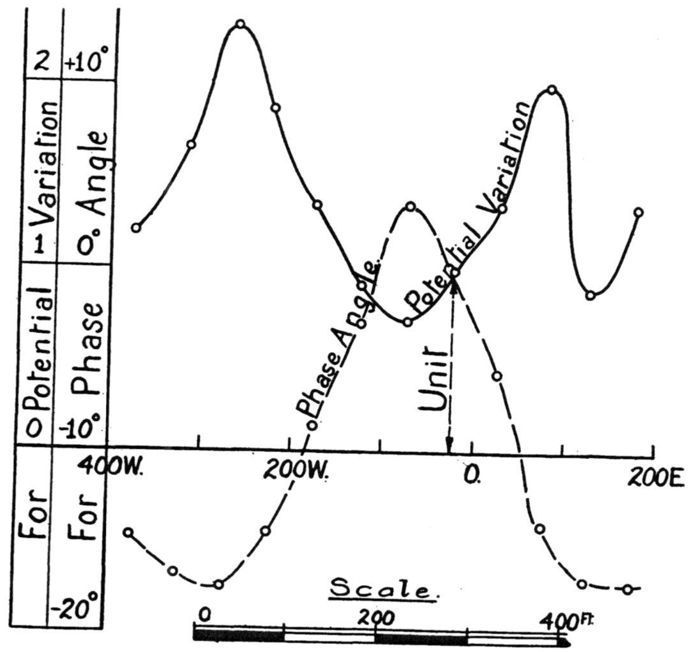

Fig. 81. Potential and phase curves on Traverse S 1 (Electrodes *BB*), Graphite Mine, Port Lincoln, S.A.

Testing of Results. The electrical indications obtained in the neighbourhood of the known ore-body were sufficiently convincing to show that the A.C. potential ratio method gives distinctive results over graphitic deposits occurring under the conditions that exist in this area. The fact that the conductive Zone *X* is better defined than Zones *Y* and *Z* suggested that priority should be given to the testing of the former and arrangements were made accordingly by the South Australian Department of Mines. Recommendations for proving Zone *Y* were also made but have not yet been carried out.

In the case of Zone *X* an old prospecting shaft (House Shaft), 40 ft. deep, was present within 100 ft. of the axis of the electrical indication and it was recommended that a cross-cut about 180 ft. in length should be driven from it in the direction shown in Fig. 77. This work was carried out under the direction of the South Australian Department of Mines, the results obtained being indicated by the following abstracts from a report prepared by the Deputy Government Geologist.

[a] Good flake graphite had been proved in No. 5 Shaft, prior to the electrical investigations.

[b] This portion of the traverse along Line N 1 passes close to the No. 1 Shaft and the No. 2 Bore-hole, in both of which flake graphite has been proved.

"The first 75 ft. were in decomposed gneissic material with sparse flakes of graphite of good quality, and then the lode was entered. The bands of clay and the richer graphitic zones show that the lode has a dip of 45° to 50° to the north-west. The lode which is 52 ft. thick is made up of clay with graphite, both of the flake and black amorphous types, together with hard magnesic nodules of secondary origin, similar to those which form a superficial crust just above the 37 ft. level in the main open-cut workings....It is not expected that the lode at this level will yield high returns, owing to the perishing of the flake graphite near the ground surface and the presence of magnesic accretions. On the analogy of the main workings there is very little doubt that good ore underlies the cross-cut....After traversing the deposit for 52 ft. the cross-cut passed into decomposed mica schist containing but little graphite and eventually it was stopped in this barren rock at a length of 153 ft.

...The results obtained by this cross-cut must be considered of great importance as they have established

(1) The existence of another large graphite deposit which, on the analogy of the main workings, is certain to be of better grade below the depth at which it was cross-cut.

(2) The value of the geophysical method employed."

In a subsequent communication the Director of Mines states that the average graphite content over the full width of the lode (52 ft.) is just over 5 per cent. He also explains that, owing to the presence of much secondary calcium and magnesium carbonates (kankar), the tenor of the lode is reduced, but that an improvement is confidently to be expected at a slightly greater depth. He concludes by saying "The results generally must be regarded as a complete proof of the suitability of the geophysical method employed for the exploration of this particular type of concealed graphite deposit."

SECTION 5. NORTHAMPTON DISTRICT, WESTERN AUSTRALIA (E 7)

The Northampton mineral field extends for about 60 miles on a north-south bearing between Galena, on the Murchison River, and the seaport of Geraldton. Its principal product has been lead, but copper ores have also been raised, the value of the combined outputs amounting to over £1,500,000. The field is of considerable historical interest, the first discovery of lead having been made in the year 1848 on the site of the Geraldine Lead Mine, near Galena. In recent years the mineral production has fallen greatly, and at the present time no large mines are being worked.

The annual rainfall is small, amounting to 15 in. at Northampton, and diminishing northwards to about 12 in. at Galena. The underground waters are distinctly saline; this feature being most marked in the northern portion of the field, as will be seen from the map (Fig. 82), on which their resistivities in ohm cm. are shown. It will be observed that the water from the Surprise Mine is far more saline than any other. This may be due to the fact that a considerble area around the mine has been cleared of bush, so that the rain water has had freer access to the salt stored up in the ground and has carried it down into the old mine workings.

Geology. For a detailed account of the geology of the district, and for descriptions of the mines, reference should be made to a valuable report by R. C. Wilson[a].

The country rock consists mainly of garnetiferous gneiss traversed by numerous basic dykes running in a north-east and south-west direction. The lodes occur at or near the junction of these dykes with the gneiss, usually in association with pegmatites. The majority of the lodes carry galena, sometimes accompanied by zinc-blende, with cerussite and more rarely, pyromorphite, in the oxidized zone. The copper lodes, which are less common, carry chalcopyrite and chalcocite, with malachite and azurite in the upper levels. The deepest mines in the district are Narra Tarra (550 ft.) and Baddera (442 ft.), both of which encountered impoverished zones at those depths, but the majority of the mines are considerably shallower. As a rule sulphide ore has been found to occur close to the surface; within 20 to 50 ft. in many instances.

Objects of Survey. It was originally intended that the work in the Northampton district should consist of routine surveys of the more promising localities; mainly with a view to locating new bodies along extensions of the known lines of lode. Although it was known beforehand that the underground waters in this district were rather highly saline it was hoped that this adverse factor would not entirely preclude the discovery of sulphide bodies lying close to the surface. It was also considered that, as saline waters are so prevalent in Australian mining fields, this would be a favourable opportunity for determining their effect on electrical prospecting operations. Actually two areas (Wheal Ellen and Baddera) were examined in the vicinity of Northampton; whilst, at the Galena end of the field, surveys were carried out at the Block 7, South Geraldine, and Mary Springs mines.

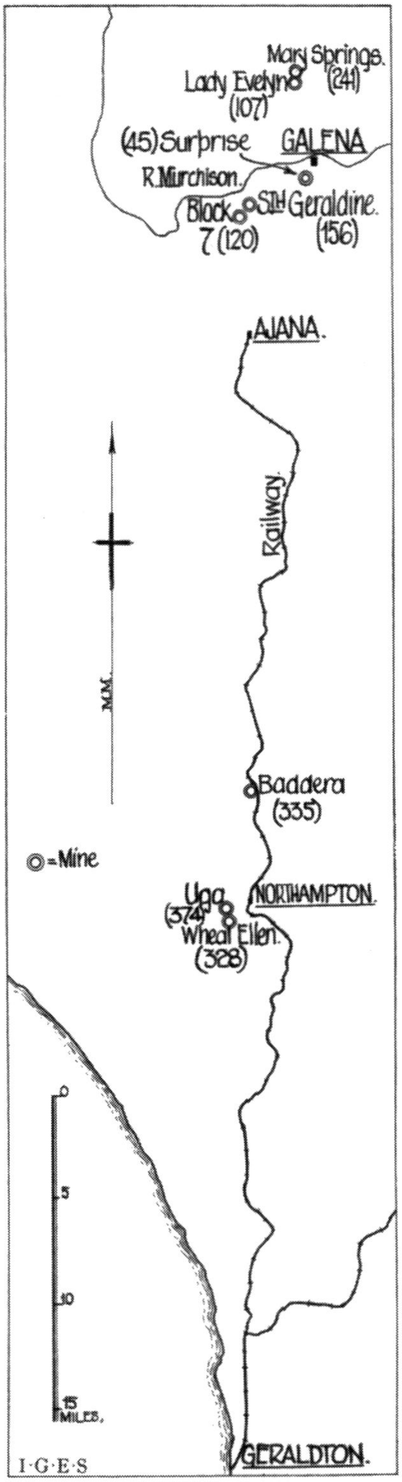

Fig. 82. Map of the Northampton district, W.A. The resistivities in ohm cm. of the underground waters are given in brackets beneath the name of each mine

[a] *The Northampton Mineral Field*, Mines Department, Western Australia, 1926.

Reconnaissance surveys were first made by the equipotential line method, but the experience gained during the earlier part of the work showed that this method could not be relied upon to locate the somewhat narrow type of lode upon which many of the smaller mines are situated. As this was attributed to high ground conductivity resulting from the presence of saline waters the more sensitive potential ratio method was then employed to some extent.

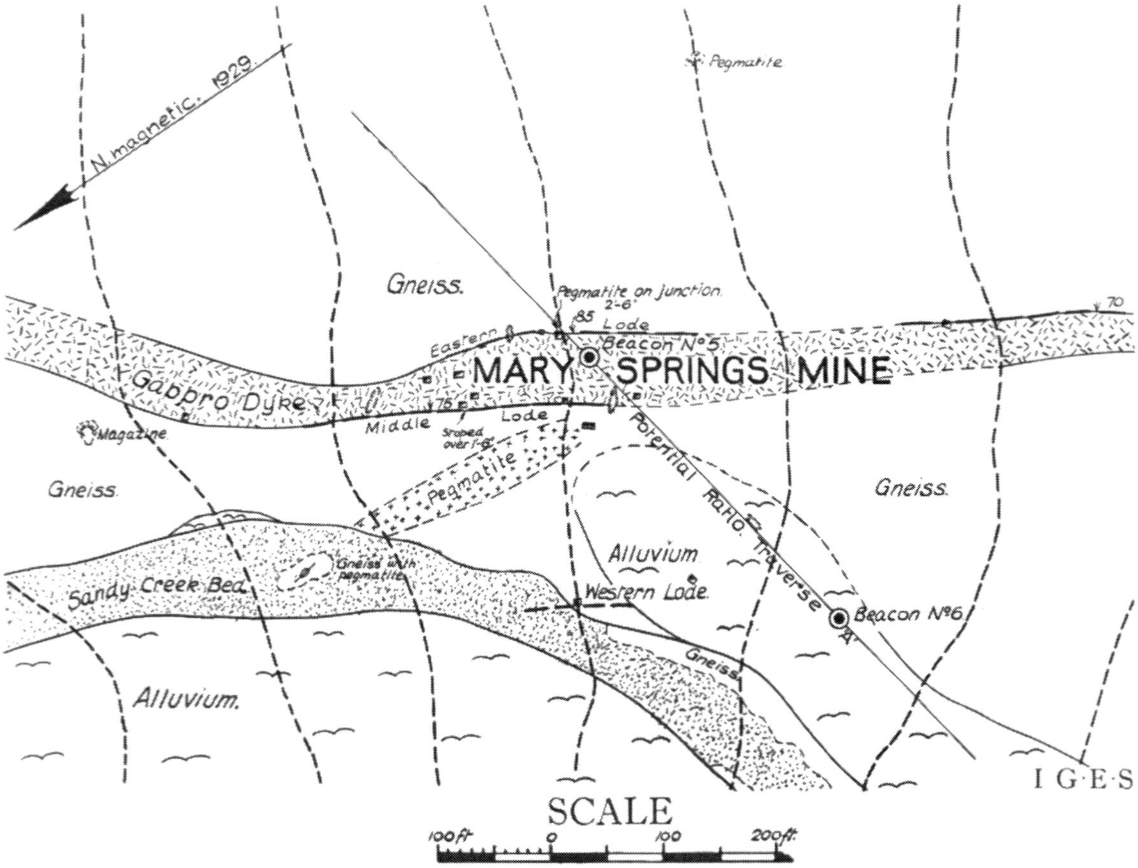

Fig. 83. Geological map showing equipotential lines, Mary Springs Mine, W.A.

Equipotential Line Surveys. Positive results were obtained only in the last area to be examined (Mary Springs and Lady Evelyn mines). At Wheal Ellen and Baddera there were no standing ore-bodies over which a test of the method could be carried out, but at Block 7 it was found that a 6 to 9 in. seam of galena, fairly free from oxidation and extending to within 15 ft. of the surface, gave no indication. As the resistivity of the underground water in this area is about 120 ohm cm. the conditions for the electrical operations were far from favourable.

At Mary Springs the feeble indication shown in Fig. 83 was obtained which, however, corresponds quite well with the positions of the known lodes that occur on each side of the gabbro dyke. In view of this encouraging result the survey was extended along the line of the dyke towards the Lady Evelyn Mine, which is situated 500 yds. south-west of Mary Springs. The lode formerly worked in this mine lies at the eastern contact of the same gabbro dyke, but although a seam of galena, 3 in. wide, is present at the 20 ft. level, no indication was obtained. On continuing the survey still farther to the south-west, the local but well-marked indication shown in Fig. 84 was recorded; the axis of which extends between Beacons Nos. 3 and 4, and coincides with the eastern boundary of the dyke. As no mining had ever been done here it was recommended that the indication should be tested by sinking a small shaft.

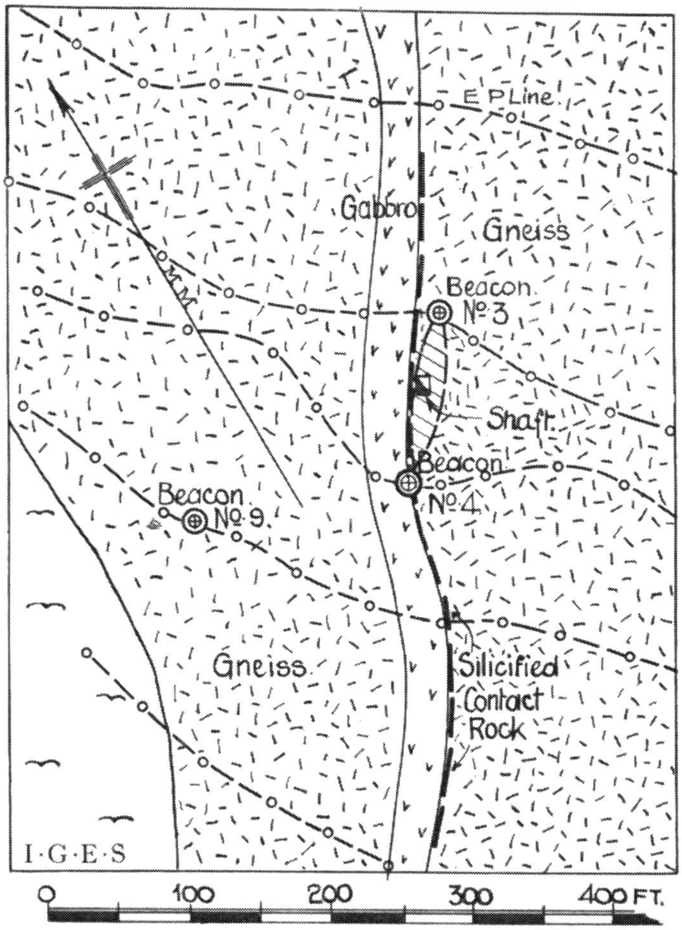

Fig. 84. Equipotential survey, near Lady Evelyn Mine, W.A.

A.C. Potential Ratio Method. This method was first applied in the Northampton district at the Block 7 and South Geraldine mines. On the former property a special electrode layout was employed in order to test the method over the known Eastern Lode[a], which the two equipotential lines shown in Fig. 85 had failed to locate. It must be understood that in this test by the A.C. potential ratio method the electrodes were re-arranged, so that the equipotential lines ran approximately parallel to the strike of the lodes and so that the ground to be examined lay about 1000 ft. beyond one of the electrodes and not between them as is usual with the equipotential line method.

The traverse, A, was made across the known lode at right angles to its strike, and observations were taken at stations placed 50 ft. apart. The results obtained are illustrated in Fig. 86, from which it will also be seen that the known lode dips in a north-westerly direction (i.e. away from the electrode) at an average angle of 75°. If the potential and phase curves are studied in conjunction with the accompanying geological section, it will be seen that there is a marked fall in potential over the Eastern Lode, and also over the western edge of the gabbro dyke. Notable changes in phase also occur at these points. Over the eastern contact of the dyke (i.e. the line of the Western Lode) there was no indication;

[a] Mineral formerly existed in the Western Lode also but has been worked out.

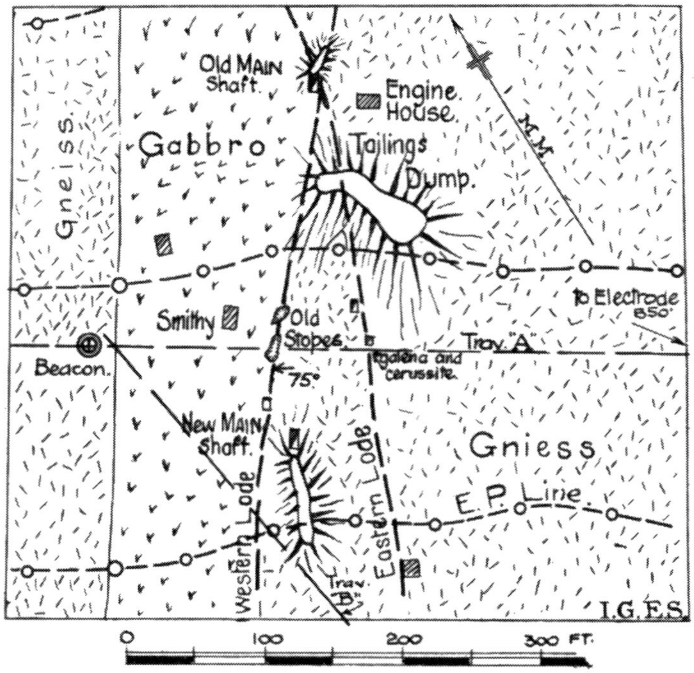

Fig. 85. Surface geological plan showing equipotential lines, Block 7 Mine, W.A.

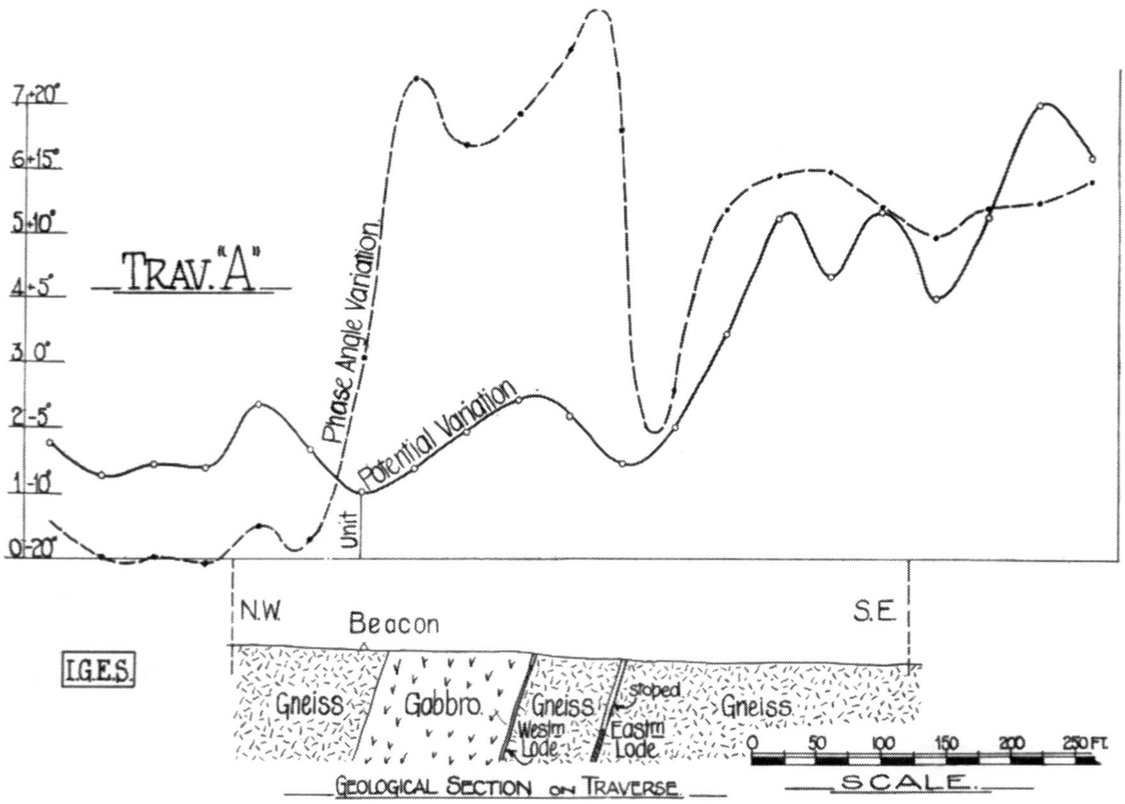

Fig. 86. Potential and phase curves along Traverse A, Block 7 Mine, W.A.

which is in accordance with recent developments underground, since this contact has been found to be almost barren at a depth of 110 ft., close to the line of traverse. Galena formerly existed closer to the surface in this lode but has all been worked out. The Eastern Lode, on the other hand, still carries galena, although it has been stoped out to a depth of 68 ft. This evidence suggested that the western contact of the gabbro dyke might be mineralized and trenching was recommended with a view to testing such a possibility.

At the South Geraldine Mine two traverses, C and D, 150 ft. apart, were made across the known line of the lode; one at each end of the mine workings. The two sets of profiles, which are shown in Fig. 87, reveal a marked similarity; the principal feature being the characteristic fall in potential and change of phase immediately above the conductive body. Traverse D crosses the line of lode beyond the limits of the mine workings, so that the existence of mineral at this point can only be inferred by comparison of the profiles with those obtained on Traverse C, which crosses a disused portion of the workings where the galena has only been stoped out down to a depth of 50 ft.

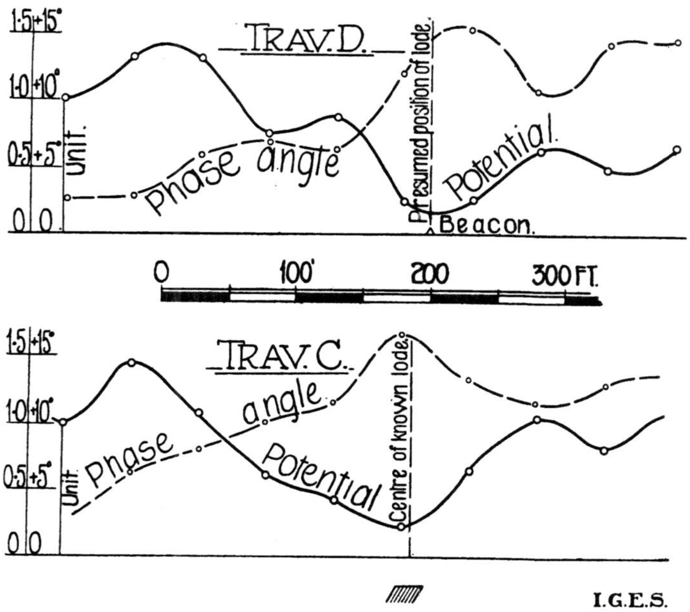

Fig. 87. Potential and phase curves, South Geraldine Mine, W.A.

At Mary Springs (Fig. 83) some potential ratio work was carried out for comparison with the equipotential line results. The curves plotted from the results of Traverse A are given in Fig. 88, and show a marked potential gradient and phase change as the axis of the equipotential line indication is approached.

Testing of Indications. At the request of the I.G.E.S. the following exploration work was carried out by the Western Australia Department of Mines.

Block 7. A trench was cut to expose the western contact of the gabbro dyke at the point marked by the beacon shown in Fig. 85. From Fig. 86 it will be observed that a marked fall of potential occurs at this point, accompanied by a change in phase of the same order as that recorded immediately above the known Eastern Lode.

The State Mining Engineer reported that the costean at this point exposed the western contact of the dyke in the immediate neighbourhood of the beacon erected by the I.G.E.S. The dyke was found to be very decomposed at and near its contact with the gneissic country rock. The costean was deepened to 7 or 8 ft., but no mineral was found.

South Geraldine Mine. The State Mining Engineer reported that a costean at the point marked by the I.G.E.S. on Traverse D (Fig. 87) showed that the beacon is situated at the junction of a pegmatite dyke with the gneiss that forms the country rock of the district. No mineral was exposed.

Lady Evelyn Mine. A shaft 50 ft. deep was sunk at a point midway between beacons Nos. 3 and 4 (Fig. 84), and two cross-cuts, each 25 ft. in length, were driven from its bottom in an east and west direction. The State Mining Engineer reported that the

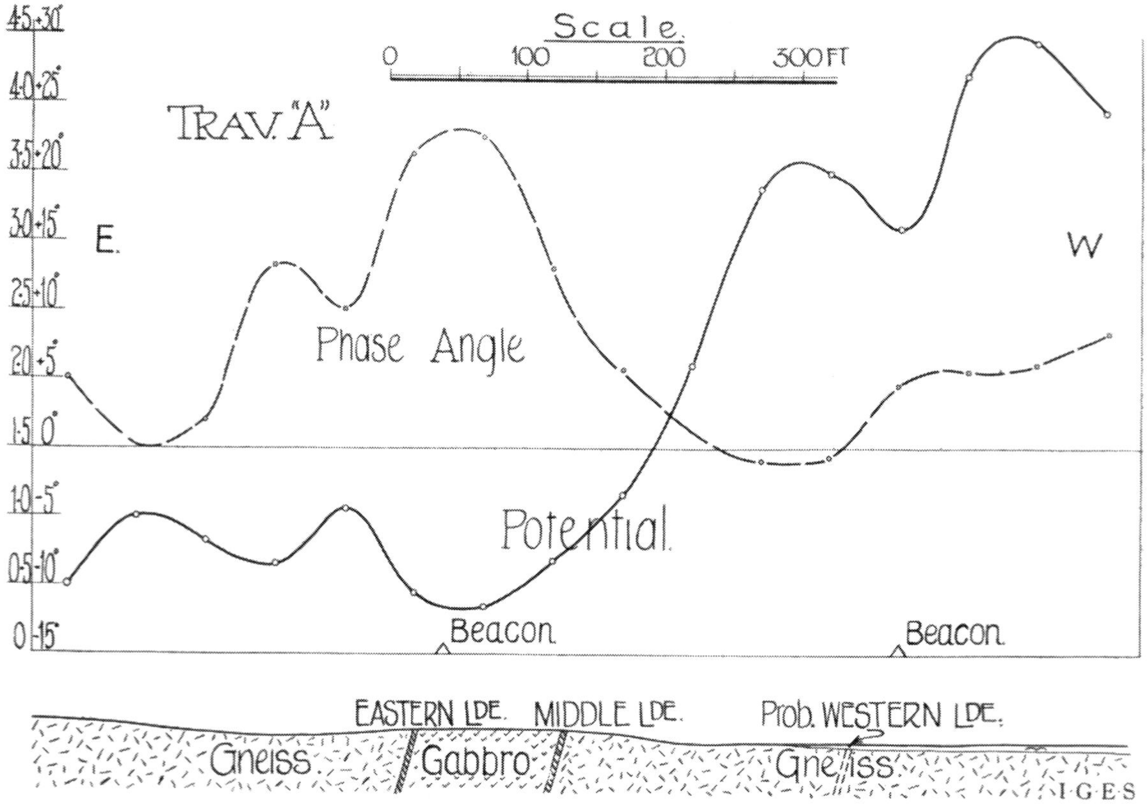

Fig. 88. Potential and phase curves, Mary Springs Mine, W.A.

shaft was sunk in soft decomposed basic dyke rock (i.e. gabbro) in which a little graphite was observed. He points out that the basic dykes in this district are usually hard and unweathered right to the surface. The decomposition of the rock in this instance he attributed to the proximity of a channel admitting of the percolation of water. This was confirmed by the fact that the shaft was making water, a sample of which was taken and sent to the Government Analyst for examination. This was found to be highly saline, carrying 796·6 grains of soluble salts per gallon, of which 654 grains were sodium chloride. It was reported that this saline water was making in the western cross-cut at the rate of 300 gallons per day. No mineral whatever was found either in the shaft or in the cross-cuts.

Conclusions. Although disappointing, in that no discoveries of mineral were made as a result of this work, the experience gained with regard to the adverse effects of saline underground waters was of definite value. Apart from this trouble, the Northampton district has all the features that are regarded as most favourable to electrical prospecting. Well-defined and continuous veins of clean galena occur close to the surface, and the country rock which, in most cases, is a garnetiferous gneiss, would be of high resistivity but for the fact that it is permeated by water of so saline a character.

Tests by the equipotential line method gave no response when conducted over known veins of mineral, and it was not until the A.C. potential ratio method was applied that any distinctive results were obtained. This method was also at a disadvantage, however, in that it was too sensitive to local variations in the surface conditions. In this district these are very pronounced; probably on account of the irregular distribution of saline material throughout the subsoil and to the fact that the ground salinity tends to be of a higher concentration along the contacts of the basic dykes where soft decomposition products will readily absorb the ground waters. There is little doubt that this was the cause of the strong indication recorded in Block 7 at the western contact of the gabbro dyke. The indication at the Lady Evelyn Mine can also be attributed quite definitely to the presence of the strongly saline water that was observed in the shaft and cross-cuts. This water had a resistivity of 60 ohm cm., and with little doubt had been impounded along the eastern contact of the gabbro dyke, which, owing to its state of decomposition, would be saturated with it.

SECTION 6. CHILLAGOE, QUEENSLAND (E 8)

The Chillagoe mineral field consists of a narrow belt of altered silurian sediments, extending in a general north-west and south-east direction and bounded both to the north and south by acid igneous intrusions (see Fig. 89). Limestones make up a large part of the sedimentary series, but conglomerates, sandstones, quartzites, and cherts also occur. The principal ore-bodies lie in the sedimentary rocks, usually close to the igneous contacts. In certain instances important bodies have been found at some distance from the contact but, in such cases, there is usually evidence pointing to the presence, at no great depth, of an offshoot from the main igneous intrusion, and there can be no doubt that the mineralization of the district is clearly related to the latter.

The ores raised in the district have been mainly those of copper and lead; the two classes of mineral sometimes occurring together as mixed ore, but in other cases as separate deposits. In the past, a large proportion of the output of this field has consisted of oxidized ores, but in most of the important mines the sulphide zones have been reached. The depth of oxidation varies greatly in different mines, from 60 to 300 ft., and it is impossible to give a general figure for the area as a whole. In some cases the oxidation has taken place in a very irregular manner and it is not unusual to find large bodies of oxidized ore well below the level at which sulphides were first encountered.

A characteristic feature of these deposits is the presence, within the lode formations, of large masses of ferruginous clay, much of which appears to be a residual material[a] produced by the solution of impure limestone. This clay, together with rock fragments enclosed by it, is known locally as lode formation or *casing rock*, and in some instances (e.g. Griffiths Mine) it has been found to persist to depths of over 200 ft. from the surface. Presumably, this residual clay may also occur in fissures in the limestone that are not mineralized but, generally speaking, local experience has proved it to be a good indicator of mineral, and most of the discoveries of rich secondary ores have been associated with clay material of this type.

The Chillagoe district was the last to be visited by the I.G.E.S. and the time available for its examination was short. The work carried out was of the nature of a quick reconnaissance survey, primarily in order to determine the suitability of the district to electrical methods of exploration, but also in the hope that indications of unknown ore-bodies might be obtained along certain lines of mineralization which had been proved to be of importance in the past. The Mount Redcap and the Griffiths areas were selected for examination since they lie well away from the railway line and other disturbing influences. They are situated respectively 4 miles north and 1 mile to the south of Mungana. The results obtained are described below.

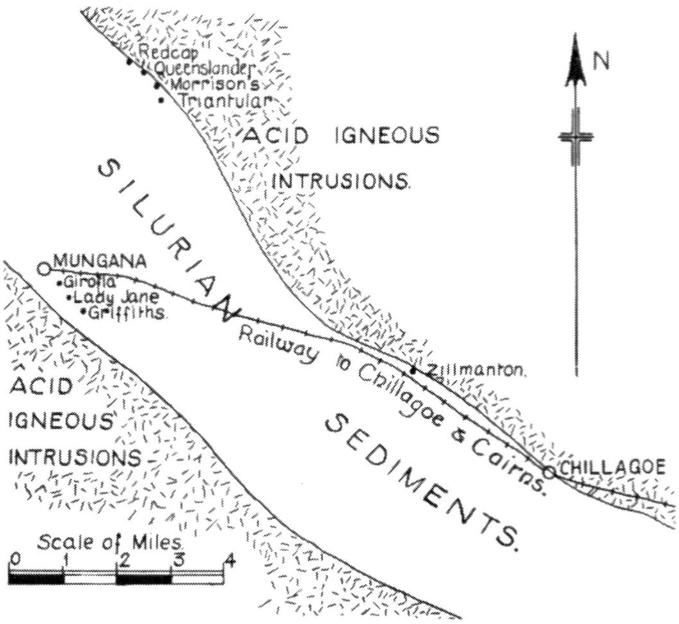

Fig. 89. Locality sketch map of Chillagoe district, Queensland

Mount Redcap Area.[b] The ground examined extends along the northern igneous contact which, in this part of the field, runs in a north-west and south-east direction. The intrusive rock is a porphyry (often decomposed and kaolinized near the contact) and the sedimentary rocks dip steeply away from it in a south-westerly direction. From Fig. 90, which shows the central portion of the surveyed ground, to the south-east of Mount Redcap, it will be seen that from north-east to south-west the following sequence of rock types is met with:

1. porphyry 2. quartzite 3. limestone 4. conglomerate

[a] Residual clays of this type are of common occurrence in limestone country in which solution of the rock is effected by meteoric waters circulating underground along fissures. The insoluble argillaceous impurities accumulate within the fissures in the form of clay. A similar material, often found in caverns, is known as *cave earth*.

[b] Morton, *Queensland Government Mining Journal*, vol. 27, pp. 426–430 (1926).

The sediments have undergone varying degrees of contact metamorphism; particularly the limestone, which has been altered, sometimes to a coarsely saccharoidal marble, and sometimes to garnet or garnet-wollastonite rock. Two principal lines of mineralization are present, the more important being along the limestone-quartzite contact, whilst the other lies at the limestone-conglomerate junction. The former line is marked by gossan outcrops throughout the greater length of the area investigated, and several important ore-bodies have been developed along it; namely, at Morrison's Main

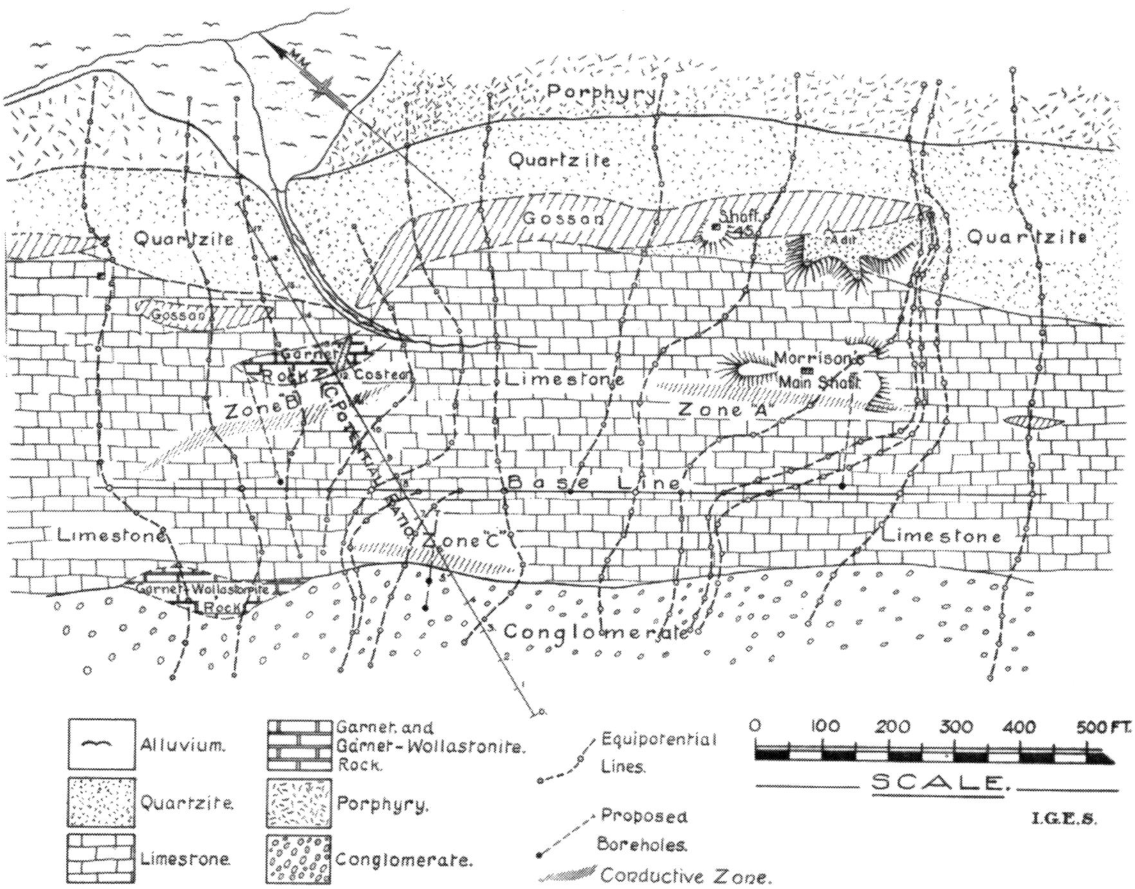

Fig. 90. Geological map and equipotential line survey. Morrison's section. Near Mt Redcap

Shaft, shown in Fig. 90, and at the Queenslander and Mount Redcap mines which lie to the north-west of it. No mining is at present being carried on in this part of the field.

At Morrison's workings sulphide ore was met with at from 100 to 150 ft. from the surface and was of a complex pyritic type carrying copper, although lead also occurred. From Morrison's Main Shaft the ore-body was worked to a depth of 250 ft., the principal stoping being confined to a strike length of 300 ft. Here the dip of the ore-body was about 45° to the south-west.

The Triantular, Boulder, and Victoria mines are situated along the second line of mineralization which, however, has so far proved to be of minor importance. The ore-bodies along this contact appear to be more pipe-like and less extensive laterally than those on the Redcap-Morrison line.

A notable feature of the Redcap area, from a geophysical viewpoint, is the occurrence of magnetite in association with some of the ore-bodies. This mineral has been observed at the Queenslander and Victoria mines and very likely occurs elsewhere, but its mode of occurrence appears to be sporadic and large continuous bodies probably do not exist.

Methods Employed. The preliminary survey was carried out by the equipotential line method; a strip of country just over a mile in length being covered, so as to include both of the lines of mineralization that extend between the Mount Redcap and the Triantular mines. Upon completion of this work traverses were made over the more interesting parts by the A.C. potential ratio method and tests were also carried out with the spontaneous polarization instruments. Electromagnetic methods were not employed since the necessary equipment was then in use in South Australia. The indications given by the electrical methods were finally tested with the Thalen-Tiberg magnetometer, in order to ascertain whether they were due to the presence of magnetite but, with the exception of a local disturbance near the Victoria shaft, no magnetic anomalies were detected.

Equipotential Line Survey. Four main indications were obtained along the Mount Redcap-Morrison's line of mineralization, in each case situated on the south-western or dip side of the gossan outcrops. Two of these conductive zones occur in the neighbourhood of Morrison's main shaft and are shown as Zones *A* and *B* in Fig. 90. Zone *A* may reasonably be attributed to the undisturbed portion of the sulphide body which was formerly worked from Morrison's shaft and which lies immediately beneath the axis of the electrical indication at a depth of 270 ft. from the surface. As no mining operations had ever been carried out in the vicinity of Zone *B* it was recommended that a diamond drill hole should be put down to test it.

On the second line of mineralization two indications were obtained, both of comparatively small size. One was centred near the Victoria shaft and could be correlated quite satisfactorily with the ore-body that had been developed there. The other, which is shown in Fig. 90 as Zone *C*, was situated close to Zone *B*, but at the limestone-conglomerate contact. This indication, although of limited extent, was very well defined and it was recommended that it should also be proved by diamond drilling.

A.C. Potential Ratio Method. Several traverses were made, using the same electrodes employed for the equipotential line lay-out, in order to investigate the character of the indications obtained by the preliminary survey. The position of the traverse line across Zones *B* and *C* is indicated in Fig. 90, and the results obtained are shown in Fig. 91. These curves reveal very clearly the marked fall in potential and change in phase angle that occur over the axis of Zone *C*; the agreement between the axis of the equipotential line indication and the peak in the phase angle curve being very close. This traverse was continued over Zone *B*, but no indication was recorded there, or could be expected in view of the disposition of the traverse relative to the equipotential lines, with which it is almost parallel for this part of its length (see Fig. 90).

Confirmatory Exploration. Diamond drill holes, in the positions shown in Fig. 90, were recommended in each of the three conductive Zones *A*, *B*, and *C*, that had been located electrically. They were to be drilled in northerly directions at inclinations of about 60° from the horizontal in order to prove the ground immediately beneath the axes of the indications, at depths of about 125 ft. in each case. The hole on Zone *C* is the only one in this area that has been completed so far and, unfortunately, it was not put down at the exact point recommended by the I.G.E.S. Although the site of the drill was moved only 40 ft. to the north-east of the selected point, the inclination was not altered in such a way as to allow for this change in position. In consequence the drill did not prove the ground at a depth of 125 ft., as intended, but did so at 65 ft., which, in all probability, is well above the sulphide zone. Nevertheless, the drilling results provide some important evidence, since a clay formation, 50 ft. in thickness, was intersected immediately beneath the axis of the electrical indication. The drillers' report states that lead ore of poor quality was encountered in this clay, which is considered to be practically identical in character with the ore-casing that occurs at the Redcap Mine. It certainly seems likely that a more substantial mineralization is to be expected at the depth originally specified.

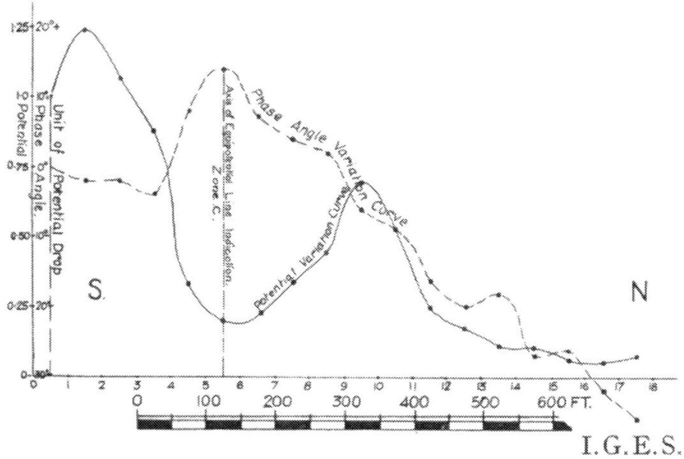

Fig. 91. Potential and phase curves over Zones *B* and *C*, Fig. 90

The Griffiths Area[a]. The ground investigated is situated on the southern border of the sedimentary belt, but at a greater distance from the igneous contact than was the case at Mount Redcap. The area is occupied mainly by limestone, which is frequently highly altered and garnetiferous, especially in the vicinity of the ore-bodies; a fact which points to the connection of the latter with the main intrusion. The limestone is associated with cherts and, more locally, with sandstones; the former being particularly prominent in the line of low hills which borders the limestone belt on the south. A general line of mineralization traverses the limestone in an approximate north-west and south-east direction (Fig. 89), and on it are situated some of the most important mines in the district, namely the Girofla, the Lady Jane, and the Griffiths mines. Of these, the

[a] Jensen, *Queensland Government Mining Journal*, vol. 21, pp. 150–152 (1920).

Girofla is the only one now in operation, and that on a much reduced scale. The ore-bodies discovered on this line have mostly been of large size and irregular shape, and usually have been of the nature of lenses and pipe-like bodies in the limestone. The Girofla and Lady Jane ore-bodies carry both lead and copper with silver values; but at the Griffiths Mine copper appears to have been the principal constituent of value. Both the Girofla and the Lady Jane mines have reached the sulphide zone at a depth of about 200 ft., the ore being of pyritic type, but in the Griffiths Mine area, which was examined electrically by the I.G.E.S., the deepest workings (200 ft.) were still in oxidized ore.

Objects of Survey. The ground to which particular attention was directed extends eastwards from the Lady Jane Mine to beyond the limits of the Griffiths workings, the area covered being about 1 mile long by a quarter of a mile wide. Some notable discoveries of oxidized ore have been made in this area, and there are reasons for believing that important bodies of sulphide ore remain to be found. As in the case of Mount Redcap the preliminary survey was made by the equipotential line method, and upon its completion confirmatory tests were carried out by A.C. potential ratio traverses.

Equipotential Line Survey. A conductive zone was located about 350 yds. south-east of the Lady Jane Mine, but the most striking indications were those obtained in the vicinity of the Magazine Face and Griffiths open-cuts, where the four indications shown on Fig. 92 as Zones *B*, *C*, *D*, and *E* were obtained. These indications all show a general west-north-west trend, and whilst Zone *B* is displaced to the north, the remaining three are suggestive of one line of mineralization. Very little mining has been done in the vicinity of Zone *B*, but it is quite reasonable to attribute Zone *C* to the Magazine Face ore-body. Zone *D*, adjoining the Griffiths open-cut, is on ground where a small amount of mining has been done in the past to a depth of 80 ft., in a body of ferruginous clay and ironstone, carrying sporadic occurrences of chalcocite. Zone *E*, which, perhaps, is the most striking of the four indications, is on unexplored ground. These last two indications, *D* and *E*, are distinctly promising in appearance and it is unfortunate that they should have been discovered at so late a stage in the Survey's work in Australia. A detailed examination by inductive methods might have yielded more precise information but the necessary apparatus was not available at the time and the survey party had to be withdrawn from the field.

A.C. Potential Ratio Method. Time did not permit special electrode layouts for each of the conductive zones, as would have been desirable, but a few traverses were made across the indications using the same electrode system that had been employed for the equipotential line survey. The results obtained on Traverse D are illustrated in Fig. 93, from which it will be seen that the features presented by the potential variation curve are in accordance with the disposition of the equipotential lines. The behaviour of the phase curve at the eastern end of the traverse has not been explained although a similar feature was observed at the western end of Traverse C.

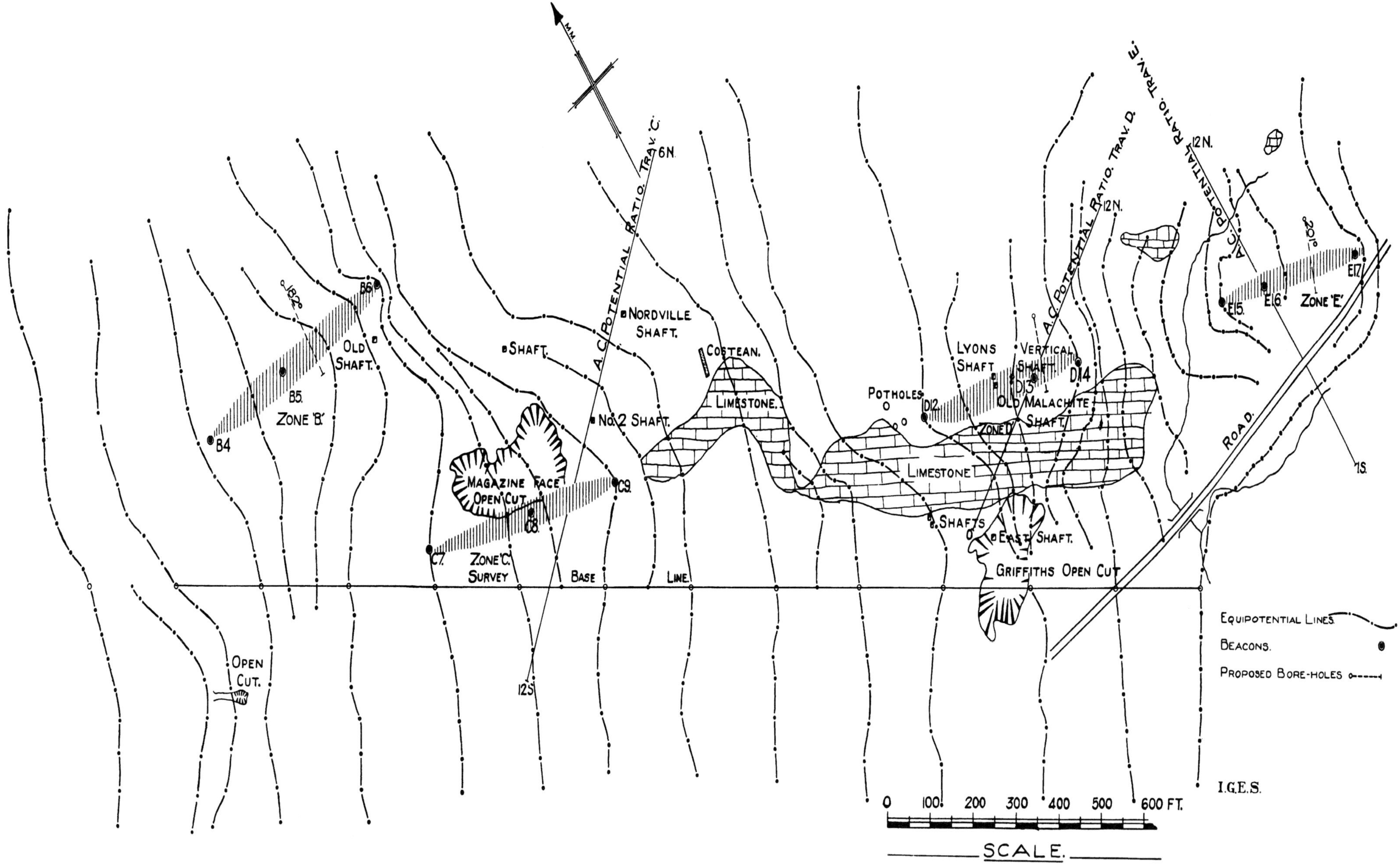

Fig. 92. Equipotential line survey of Griffiths area, Chillagoe district, Queensland

Confirmatory Exploration. Three diamond drill-holes have been put down in the Griffiths area by the Queensland Department of Mines, in order to test the conductive zones that had been located electrically. The full records of these drilling operations have not been received but the following brief accounts of the results obtained in Zones *D* and *E* can be given.

Zone D. The position of the bore-hole put down to prove this indication is shown in Fig. 92. The drill passed through decomposed limestone and soft clay for a distance of 142 ft., at which point it encountered hard limestone and was stopped. From the evidence of this hole and neighbouring shafts and pot-holes it appears that a ferruginous clay and ironstone formation extends along the entire length of Zone *D*. Although no mineral was met with in the bore-hole, it is reported that irregular occurrences of rich secondary copper ore were found some years ago when exploration work was being conducted at shallow levels from the nearby shafts.

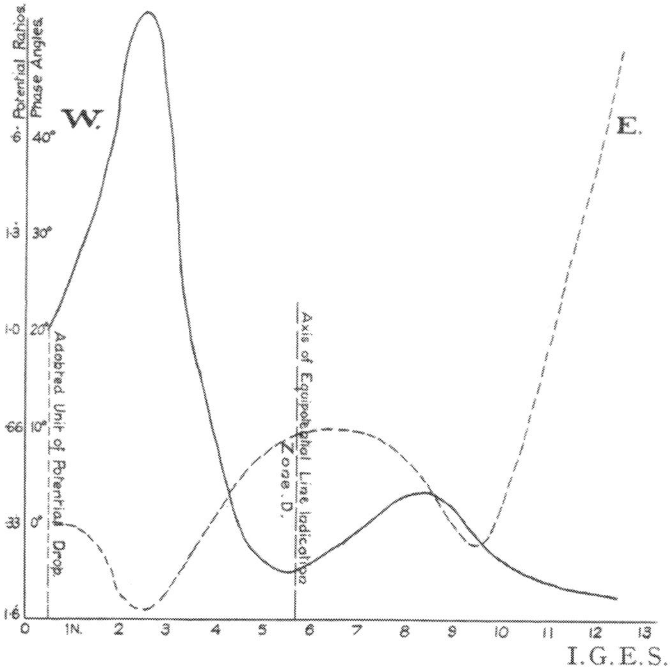

Fig. 93. Potential and phase curves over Zone *D*, Griffiths area, Chillagoe, Queensland

Zone E. The single bore-hole shown in Fig. 92 was drilled in a southerly direction at an angle of 50°, to prove the ground vertically beneath the axis of the equipotential line indication, at a depth of 100 ft. below the surface. From cabled information it appears that the first 70 ft. of the hole were in mixed limestone country carrying bands of calcite and ironstone. Ten feet of low grade copper ore, associated with clay seams, were then passed through at a point within 20 ft. of the axial plane (i.e. a vertical plane through Beacons E 16 and E 17) of the electrical indication. The hole was discontinued in hard limestone at a depth of 202 ft. In a report by the General Manager of the Chillagoe State Smelters it is stated that the low grade sulphides met with in this bore-hole will be of assistance in the local smelting operations.

Conclusions. Consideration of the results obtained by the electrical surveys in the Mount Redcap and the Griffiths areas, together with the information obtained by subsequent drilling operations, permits the following general conclusions.

(I) A number of sharply defined conductive zones are present in both of the areas examined. These zones conform closely in position and strike direction to some of the known lines of mineralization that traverse the district.

(II) The electrical methods employed proved quite effective in locating and defining the positions of these conductive zones.

(III) Although some of these zones may be directly attributed to the presence of sulphide ores (e.g. Zone *A* at Morrison's workings), it is probable that many of them owe their conductivity to the presence of clay-bodies of low electrical resistivity, lying in fissures within the limestone, or at the limestone contacts. Since these clay formations occur at shallow depths they will have a more profound effect on current distribution at the surface than will any sulphide ores that may lie at the base of the oxidized zone.

(IV) The value of electrical methods as a direct means of locating sulphide ores in the Chillagoe district has not been definitely established, but there can be little doubt that they can assist in discovering those mineral deposits that are characteristically associated with substantial quantities of clay or other lode filling of low resistivity. Although this is an indirect application and it seems likely that the limestone country may contain clay-filled fissures which have no mineral content, local experience over a period of many years has established the fact that these clay formations are of definite value as an indicator of mineral. This being the case, it would appear that electrical prospecting operations would prove of material assistance in the exploration of the Chillagoe field.

CHAPTER IV

GRAVIMETRIC METHODS

Section 1. PRINCIPLES AND APPLICATIONS

HISTORICAL[a]

The invention of the torsion balance has been ascribed to Mitchell and it was used by Coulomb in the investigation of magnetic and electrical forces. In 1798 Cavendish[b] used a torsion balance to determine the constant of gravitation[c] and hence, indirectly, the mass of the earth. A light rod, at the ends of which were two small metal spheres, was suspended by a fine fibre. Two heavy lead spheres were then brought up on opposite sides of the suspended system; one of them being close to each of these small spheres. The couple exerted by the gravitational attraction between the two pairs of spheres was balanced by the torque in the fibre. The torsion balance was elaborated by Boys[d] in 1895, who used a quartz fibre and evacuated the vessel containing the balance, and it was he who also was the first worker to place the attracted masses at different levels.

In 1896 Eötvös[e] devised a torsion balance to measure the rate of change in the value of the acceleration of gravity g at different points on the earth's surface and he was the first to adapt the instrument for use in the field. In his torsion balance he placed the two small masses at different levels, one of them being suspended from the end of the beam by a fine wire. This is the practice in many of the present-day instruments which are essentially the same as that described by Eötvös; even the dimensions being very little different from those used by him.

An important modification of the beam system was made by Schweydar in 1926[f] when he introduced the so-called "*Z* beam," in which one mass is supported above one end, and one suspended from the other end of the beam, by light aluminium rods. A radical change was the use of a circular beam[h] in the "gradiometer" designed by Shaw and Lancaster Jones[g] in 1927; this instrument being much more portable and more rapid in use than the ordinary torsion balance.

Eötvös[e] laid down the main principles which have since been followed in the investigation of geological structure by means of the torsion balance, and from 1901–12

[a] It is beyond the scope of the present work to deal with the history of pendulum observations for obtaining the value of the acceleration due to gravity.

[b] *Roy. Soc. Phil. Trans.*, pp. 469–526 (1798).

[c] Newton's law of gravitation can be expressed by $F = Gmm'/r^2$, where F is the attraction between two particles of mass m and m', and distance r apart. G, known as the universal constant of gravitation, has a value of $6{\cdot}66 \times 10^{-8}$ gm.$^{-1}$ cm.^{3} sec.$^{-2}$. The acceleration due to gravity g, which is equal to the gravitational attraction exerted by the earth on unit mass, is thus equal to $G\frac{M}{R^2}$, where M is the mass and R the radius of the earth.

[d] *Roy. Soc. Phil. Trans.*, vol. 186, pp. 1–72 (1895).

[e] *Ann. d. Physik*, vol. 59, pp. 354–400 (1896). *Verhandl. der* 15 *Allgem. Konferenz d. Intern. Erdmessung*, pp. 337 et seq. (1908).

[f] *Zeits. f. Geophysik*, vol. 2, p. 151 (1926).

[g] *Mining Magazine*, vol. 40, pp. 272–279 (1929).

[h] Brillouin, *Mém. de l'Acad. des Sciences*, Paris, 1908. British Patent Specification 272511, June 7, 1927.

he applied the method on certain parts of the Hungarian plain. Despite the full practical and theoretical details given by him, it was some time before the torsion balance was applied to problems of economic geology. Schweydar[a] in 1917 carried out the first torsion balance survey in Germany over a salt dome, part of the boundary of which was already known from bore-hole results.

In the Gulf Coast, U.S.A., torsion balance surveys for the location of salt domes were commenced in 1923 or 1924, and there the method has had its most striking successes. The first salt dome[b] was discovered in 1924, and since then several others have been found. At one time about 170 instruments, representing a capital of about a million dollars, were in use in this area.

THE EARTH'S GRAVITATIONAL FIELD

The gravitational attraction between two particles and the electric force between two point charges are both governed by an inverse square law of force. There is thus a physical quantity known as gravitational potential which is analogous to electrical potential. Consequently there are gravitational equipotential surfaces on the earth which are level surfaces (the surface of water at rest is an example), with respect to which the direction of gravity is everywhere perpendicular. In the case of a homogeneous sphere, or a sphere comprised of concentric layers of material of uniform density, the surfaces are spheres concentric with it. If two equipotential surfaces differing in potential by ΔU are at a distance Δz apart, then the intensity of the field is equal to the value of $\Delta U/\Delta z$. In the case of the earth's gravitational field this is the value of the acceleration of gravity g. The vertical, or direction along which the attraction of gravity is exerted at any point, is really a gravitational line of force, corresponding to electric or magnetic lines of force.

If the earth were a homogeneous sphere and if it did not rotate, the lines of the vertical would everywhere be perpendicular to its surface, and directed towards its centre; also the equipotential surfaces would be concentric spheres, and the value of gravity would be the same at all points on its surface. The state of affairs for a small portion of its surface would be that shown in Fig. 94 *a*, where the lines of the vertical are parallel and the traces of the equipotential surfaces are parallel to the surface and are equidistant.

Actually the earth is not a sphere but is a "spheroid of equilibrium"; the effect of which is to diminish gravity at the equator as compared to the poles and to make g increase with latitude. The rotation of the earth on its axis also lightens bodies at the equator, so this still further diminishes gravity there. The total effect of both is to decrease g at the equator to a value which is 0·53 per cent. less than the value at the poles. At points above the earth's surface gravity decreases proportionally to the height[c].

The acceleration of gravity has been determined absolutely by means of the pendulum at Potsdam, and at other points on the earth's surface it is usually found relative

[a] *Zeits. f. prakt. Geologie*, vol. 26, pp. 157–162 (1918).
[b] The Nash dome, Fort Bend County, Texas.
[c] The decrease is roughly 0·0003 cm. sec.$^{-2}$ per metre.

to Potsdam[a] by timing the swings of an invariable pendulum at both places. The error of modern relative determinations[b] is about ±0·002 cm. sec.$^{-2}$ or possibly a little less.

In many localities the measured value of g differs from that corresponding to the latitude and height above sea level. These differences between the observed and the normal value are termed gravity anomalies, and are due to the inhomogeneity of the earth's crust. Above a mass of rock of higher density than its surroundings the value of g is locally increased and a positive anomaly or "gravity high" occurs, and *vice versa.*

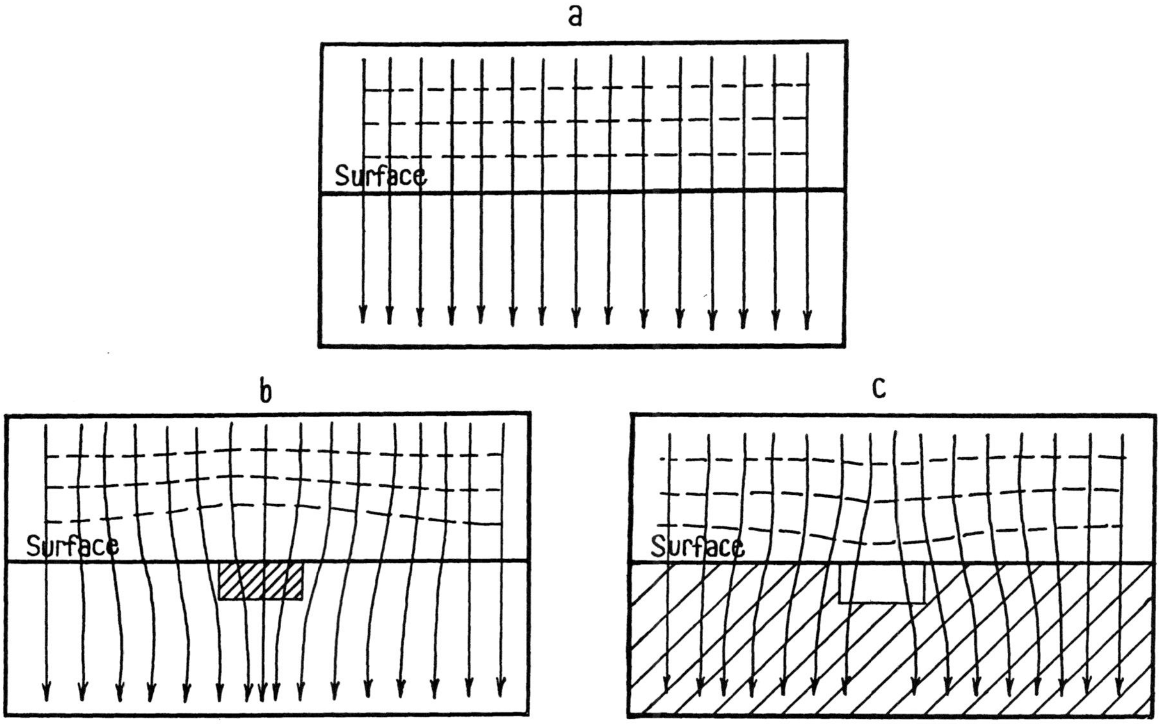

Fig. 94. Diagrammatic sketches showing gravitational field for a small section of the earth's crust, after Barton

a. With a homogeneous distribution of mass.
b. With a body denser than the surrounding country rock.
c. With a body lighter than the surrounding country rock.
- - - - Trace of equipotential surface.
⟶ Direction of vertical.

In Fig. 94 *b* and *c* these cases are shown diagrammatically. The lines of the vertical in *b* are bent in towards the denser body[c], and the equipotential surfaces are displaced by amounts which decrease with the distance from the body, so that immediately above it the equipotential surfaces will be closest together, i.e. the value of g will be increased, and there will thus be a positive anomaly.

[a] See Kaye and Laby, *Physical Constants*, 5th ed., p. 11, Longmans Green and Co., London, 1928.

[b] For a description of the use of modern pendulum apparatus, see, for example, Swick, *Modern Methods for measuring Gravity*, Government Printing Office, Washington, 1921.

[c] In the analogous magnetic case the magnetic lines of force are bent in towards the more magnetic substance.

Gradients. The rates of change of gravity with horizontal distance towards the north and east are termed the gravity gradients towards the north and east respectively. The horizontal gravity gradient is the resultant of these two components which we will term U_{xz} and U_{yz} respectively; so that the total gradient G is given by $G^2 = U^2_{xz} + U^2_{yz}$, and its azimuth α is given by $\tan \alpha = U_{yz}/U_{xz}$.

It is represented on plans by a vector pointing in the direction of the increase of gravity; see Fig. 95. For example, in Fig. 94 *b*, the value of *g* increases as the body is approached from either side. Thus the gradients point towards a body denser than its surroundings, and similarly in Fig. 94 *c* they point away from a lighter body.

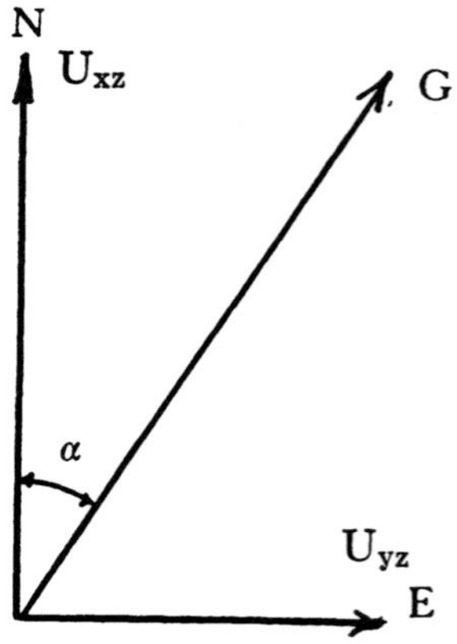

Fig. 95. Components of gradients

Where the anomalies are due to deep-seated and extensive rock-masses they extend over a large area and often can be measured by means of pendulums[a]. The gravity effects due to more localised features, such as the anticlines, dykes, etc., which are often dealt with in practice, are too small to be measured by means of pendulums. For instance, the anomaly due to a vertical dyke, 20 ft. wide, with its upper surface at a depth of 100 ft., and differing in density by 1 gm. cm.$^{-3}$ from its surroundings, would be of the order of 0·0002 cm. sec.$^{-2}$, or about 1/10th of the normal error in the determination of the value of g with a pendulum.

To measure the local departure of the equipotential surfaces from their normal position and direction we require a more delicate instrument than the pendulum. Eötvös made such measurements possible by his work on the torsion balance. The torsion balance and the gradiometer, which can be used to measure gravity gradients directly, depend for their action on very minute forces, and are effective although they are not required to measure the gradient with very great accuracy. The forces acting on a pendulum are very much larger, but as the pendulum measures the total value of gravity, its accuracy would have to exceed that of any known physical instrument for it to measure the small gravity gradients met in practice.

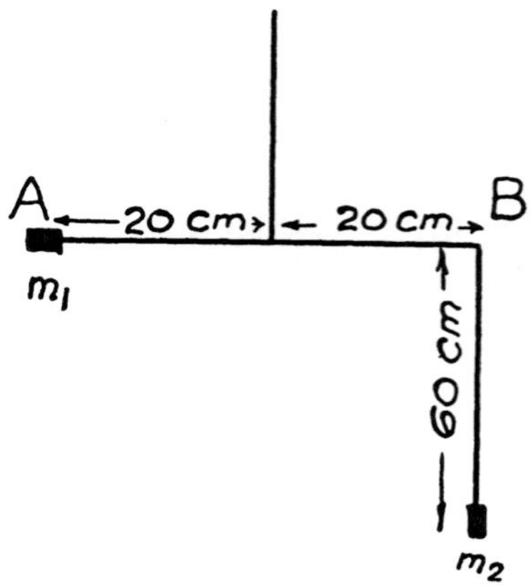

Fig. 96. Beam system of the Eötvös torsion balance

Eötvös Torsion Balance. The essential parts of the beam system of the Eötvös torsion balance are shown in Fig. 96. A fine wire usually of platinum-iridium alloy, and of a diameter of about 4×10^{-3} cm., carries the light aluminium beam AB, which is about 40 cm. in length. The masses m_1 and m_2 are each about 25 gm.; m_1 is at one end of the beam, and m_2 is suspended from the other by a fine wire about 60 cm. in length.

[a] An important case is that of a pendulum survey in the Harz Mountains, described by Galle, *Veröffentl. d. Kgl. Preuss. Geodät. Inst.*, vol. 61, Berlin, 1914. Pendulum surveys have also been of use in connection with the theory of isostasy; see Ambronn, *Elements of Geophysics*, p. 32, McGraw-Hill, New York, 1928.

Gradiometer. The beam system of the gradiometer is shown diagrammatically in Fig. 97. Two equal masses m_2 and m_3, and a light aluminium rod about 40 cm. in length carrying a third mass m_1, are mounted on equidistant spokes of the light aluminium wheel *ABC* of about 5 cm. radius; the combined mass of the rod and the mass m_1 being equal to that of m_2 or m_3. The torsion wire *OP* is ribbon-shaped and about 30 cm. in length. In both instruments a small light mirror is attached to the beam stem, so that by means of a suitable optical system the angle of twist of the wire can be measured.

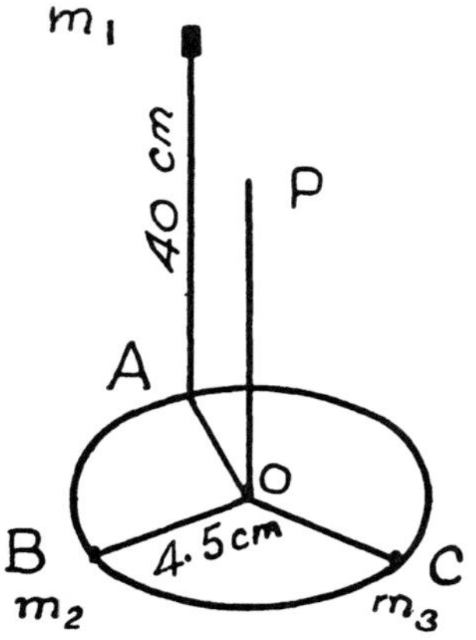

Fig. 97. Beam system of the gradiometer

Principle of Gradiometer or Torsion Balance. In order to understand the working of the gradiometer or the torsion balance system, let us consider Fig. 94 *b* and assume that the section line is in an east-west direction, and that the body extends with a uniform vertical cross-section for a long distance in the north and south directions, normal to the plane of the page. From considerations of symmetry it is clear that in this direction there is no change in the value of gravity due to the body.

Let the balance or the gradiometer[a] be set up on the western side of the body so that the upper mass m_1 is in the north. The equipotential surfaces S_1 and S_2 (Fig. 98), which pass through the masses m_1 and m_2 respectively, are not parallel but become closer together towards the east. The corresponding values of gravity g_1 and g_2 both have small horizontal components to the east, H_1 and H_2 respectively. The direction of g_2 is more inclined to the vertical than that of g_1, so that H_2 is greater than H_1; the magnitude of the difference $H_2 - H_1$ being proportional to the small angle between the directions of g_1 and g_2; i.e. to the angle between the perpendiculars through m_1 and m_2 to the two equipotential surfaces S_1 and S_2. This angle is a measure of the rate at which the two surfaces become closer to each other towards the east; i.e. to the rate at which the value of g increases towards the east. The horizontal component $H_2 - H_1$ causes a couple which rotates the balance arm in a

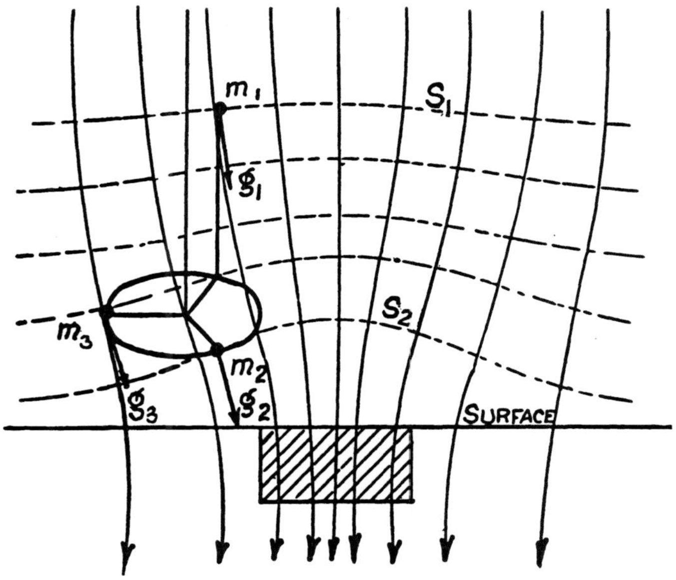

Fig. 98. Diagram to illustrate the principle of gradiometer
- - - - - Trace of equipotential surface.
——— Direction of vertical.

[a] For the sake of simplicity, and for analogy with the torsion balance, only one of the lower gradiometer masses is dealt with. What has been said about m_2 and g_2 applies with equal force to m_3 and g_3.

clockwise direction until the couple is balanced by the torque of the wire. The angle of twist of the wire is then proportional to $H_2 - H_1$, i.e. to the rate at which the value of g increases towards the east, i.e. to the gradient U_{yz}. If the beam were rotated through 180° so that m_1 were in the south, it would turn through an equal angle in an anticlockwise direction.

If the beam had been set up so that m_1 was in azimuth 90° or 270°, the horizontal components[a] (in this case H_2' and H_1') towards the east would exert no turning moment on it. This of course neglects changes of g in a north-south direction, and in this case it could be shown in exactly the same way that the deflection is proportional to the rate of change of g in the north-south direction, i.e. to the gradient U_{xz}.

The gradiometer and the torsion balance are highly sensitive instruments, and under favourable conditions a gradient can be measured with an accuracy of $\pm 1 \times 10^{-9}$ C.G.S. units. This is equal to the value of the gradient occurring where the change in g per cm. is 1×10^{-12} of its total value. This quantity is usually termed the Eötvös unit and written as 1 E, which is thus equal to 1×10^{-9} sec.$^{-2}$. In the remainder of the chapter the values of gradients in numerical formulae will always be expressed in terms of the Eötvös unit.

In the case of the vertical dyke mentioned on p. 138, a maximum gradient of about 13 E would be obtained, so that such a dyke could readily be located with a gradiometer or a torsion balance if its effects were not masked by those due to topography, etc.

In addition to having a high sensitivity the instruments must be adapted for use in the field, i.e. they must be readily transportable and usable under all weather conditions. The torsion wire and beam system must be very carefully protected from convection currents and from mechanical shocks. The former is done by making the instrument double- or triple-walled, with alternate layers of conducting and insulating materials. For use in the field it is provided with some type of protecting hut, the walls of which are made of poorly conducting material which protects it from temperature fluctuations and from the wind.

Equation for Gradiometer. The equation for the equilibrium position of the gradiometer with the upper weight in any azimuth α is given by[b]

$$N_\alpha = (n_\alpha - n) = \frac{2Dmhl}{\tau}(U_{yz}\cos\alpha - U_{xz}\sin\alpha),$$

where h cm. = the effective height of the upper mass above the beam, i.e. the height of the centre of gravity of the system composed of the rod and mass;

m gm. = the mass of this system;

l cm. = the radial distance of the centres of gravity of the masses from the axis of rotation of the beam;

τ dyne cm./radian = the torsion constant of the wire;

[a] With the gradiometer the horizontal components H_3' and H_2' exert equal and opposite turning moments.
[b] See Part II, p. 300.

D scale divisions = the distance of the scale from the mirror;

n_a = the scale reading when the beam is in azimuth α;

n = the scale reading corresponding to the undeflected position of the beam.

There are thus three unknown quantities, viz. two gradient components U_{xz} and U_{yz} and the unknown value of n, so that for a complete solution readings in three different azimuths are required. With one of the instruments used by the I.G.E.S. and with azimuths 0°, 120° and 240°

$$U_{xz} = 7{\cdot}3\,(N_{240}-N_{120}),$$
$$U_{yz} = 12{\cdot}7\,N_0.$$

In these $h = 36$ cm., $m = 19$ gm., $l = 4{\cdot}5$ cm., $D = 1100$ scale divisions and $\tau = 0{\cdot}09$ gm. cm.^{2} sec.$^{-2}$.

When set in a new azimuth the beam takes 25 minutes to come to rest, but if a coarser wire ($\tau = 0{\cdot}16$) were used, only 20 minutes would be required, the sensitivity being correspondingly less.

Torsion Balance. In the above discussion the effect of the warping of the equipotential surfaces, in the case of the torsion balance[a], has not been fully taken into account. This introduces further horizontal components which are also effective in turning the beam. Thus, in the equation for the equilibrium of the torsion balance two additional terms occur, which are related to the curvature of the equipotential surface and will be designated by U_Δ and U_{xy}. Thus five unknowns enter into the equation, so that for a complete solution readings in five different azimuths are required. The general practice in the construction of torsion balances is to have two similar beam systems supported side by side, one of them being at an angle of 180° to the other. In this case readings in three azimuths give three equations for each beam which are sufficient for the solution of the above five unknowns and the unknown zero reading n', of the undeflected position of the second beam.

For the values of the constants of the torsion balance used by the I.G.E.S. the gradients and curvature terms were given by

$$U_{xy} = 5{\cdot}8\,N_0 + 3{\cdot}8\,N_0',$$
$$U_\Delta = 6{\cdot}7\,(N_{240}-N_{120}) + 4{\cdot}4\,(N'_{240}-N'_{120}),$$
$$U_{xz} = 1{\cdot}99\,(N_{240}-N_{120}) - 1{\cdot}31\,(N'_{240}-N'_{120}),$$
$$U_{yz} = 3{\cdot}45\,N_0 - 2{\cdot}27\,N_0',$$

where the dashes refer to the second beam.

It will thus be seen that this instrument has about four times the sensitivity of the gradiometer but has the disadvantage that when set in a new azimuth the beam takes about 1 hour to come to rest.

Calculation of Gravity Effects[b]. It is possible to calculate the gradients due to any given size or shape of body, but the calculation is excessively laborious unless the body can be treated as extending to a very great distance in one direction and remaining of

[a] See Part II, p. 301. [b] See Part II, p. 302.

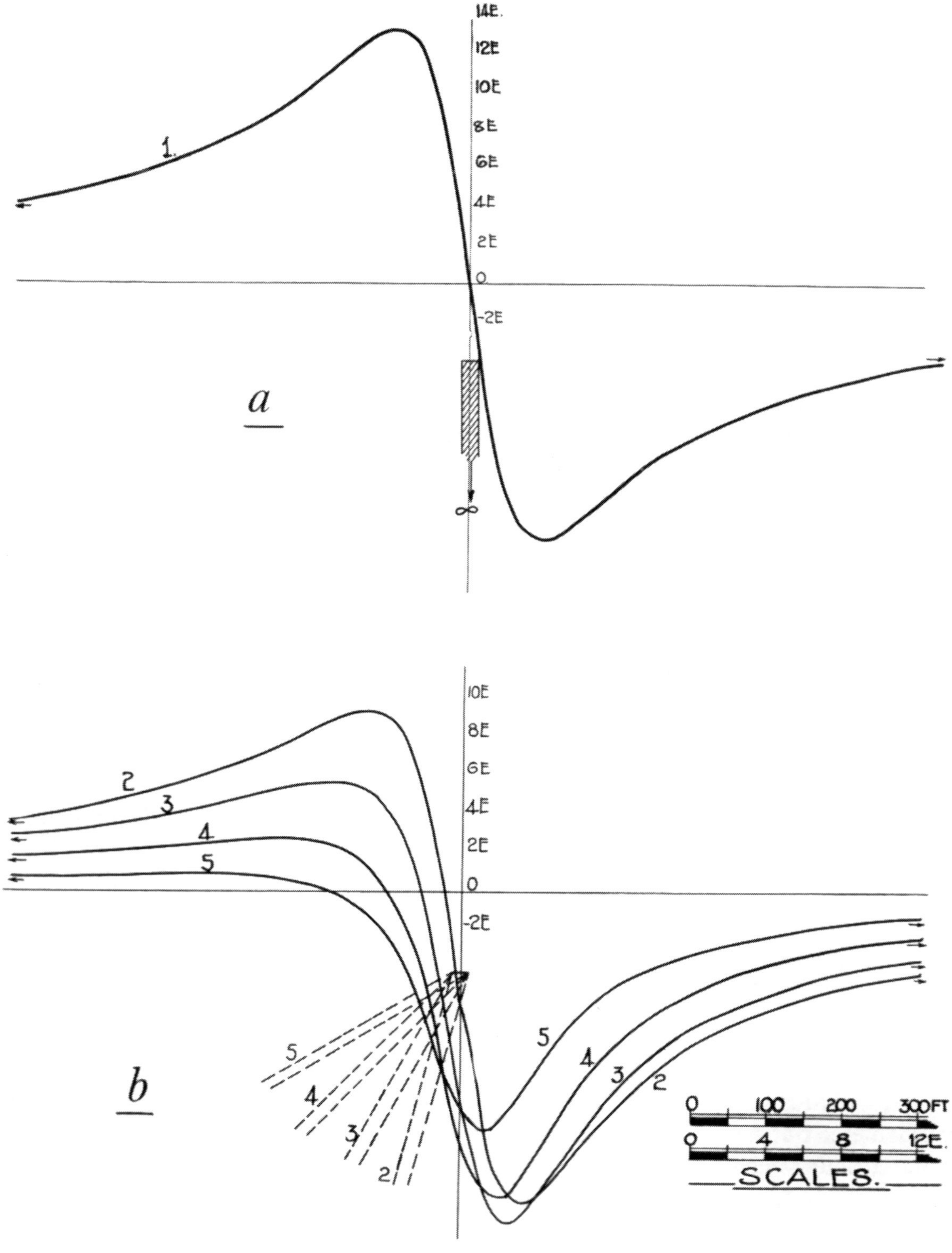

Fig. 99. *a*, Calculated gradients due to vertical dyke 20 ft. wide, at a depth of 100 ft. and having a density difference of unity. *b*, Case of a similar dyke dipping at various angles

constant cross-section. Such a body is termed a *two-dimensional* body, and various formulae have been derived giving the values of the gradients and curvature terms along a line perpendicular to its length for different types of cross-section.

The converse problem, of deducing the size and shape of a body from the values of the measured gradients, is not in general capable of a unique solution, but of the indefinite number of solutions there are usually only a limited number of probable ones when the geological information concerning the district in question is taken into account.

In Fig. 99 *a* are shown the results of the calculation for the case of a vertical dyke extending to a very great distance in a direction perpendicular to the plane of the paper. There is no component of gradient in this direction; the gradients all lying in the plane perpendicular to the line of strike. If this were in a north-south direction, then the gradient U_{xz} is always zero, and the graph shows the variation of U_{yz}. Above the centre of the dyke U_{yz} is zero, to the west it is positive and to the east negative. The effect of slope on the values of the gradients is shown in Fig. 99 *b*. In each case the upper surface of the dyke is in the same position; curve 1 is for the vertical dyke (90° dip), curves 2, 3, 4 and 5 are for westward dips of 75°, 60°, 45° and 30° respectively.

Fig. 100 shows the change in value of the gradient perpendicular to the direction of strike of a step fault[a]; this was calculated in connection with the results obtained at Gulgong[b].

In these diagrams the scale of length is immaterial; i.e. the maximum gradient due to a dyke 200 ft. wide and 1000 ft. below the surface is the same as shown in Fig. 99.

In the interpretation of the results the use of the gradients is much simpler than that of the curvature terms. The former are vectors, which point in the direction of the denser body, and in general have their maximum magnitude in the neighbourhood of the edge of the body. The latter are not vectors and, as they are functions of the curvature of surfaces in space, are difficult to visualise, and in the majority of cases they do not yield any information that could not be obtained from the gradients. In addition, the corrections due to topographical effects cannot be determined with the same degree of accuracy as they can for the gradients. Thus in rugged country there is so much uncertainty attached to the curvature values that they are of no practical value. For these reasons the I.G.E.S. did not deal with

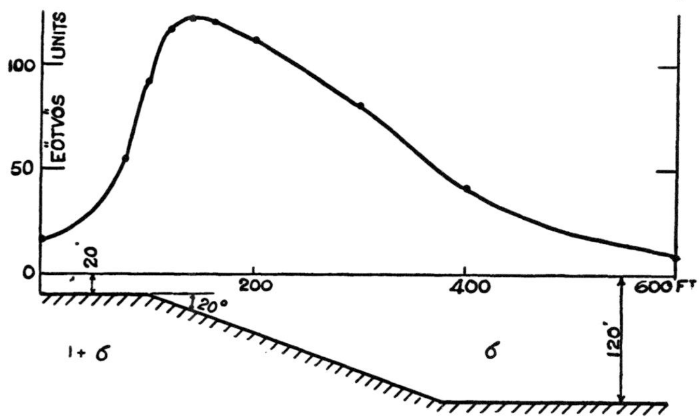

Fig. 100. Calculated gradients due to a step fault
The density difference is 1 gm. cm.$^{-3}$

[a] Shaw (*A.I.M.E. Tech. Pub.* No. 178, 1929) has calculated the gradient and curvature values for various slopes and ratios of the depth of overburden to the throw of the fault, expressing the results by means of curves. Unfortunately in the above publications they are shown on much too small a scale for convenient use.

[b] See p. 171.

them, and carried out most of its work with the gradiometer, which gives the values of the gradients but not of the curvature terms.

In the interpretation of the results of a gravity survey the following principles are observed:

(1) The gradients point towards the denser body.

(2) They usually have their maximum value near the edge of the body.

(3) From the gradients it is possible to draw "isogams" which are lines of equal gravity anomaly. These make the location of gravity "highs" or "lows" much clearer.

(4) In suitable cases the curve of the values of the gradients along a certain line will be found to agree with the curve calculated for a certain hypothetical structure. In practice, the depth and shape of the geological features are usually obtained as a result of such trial calculations, and it is here that experience of possible values and dimensions is of particular assistance in limiting the number of attempts that must be made.

Geological Applications of the Method. In the application of the gravimetric method to the elucidation of geological structure the main requisite is that a difference in density should exist between the material of which the feature concerned is composed, and the medium surrounding it. As a general rule, rock densities run in the following descending order—crystalline and volcanic rocks, older sedimentary rocks, recent sedimentary rocks and alluvium, black coal, brown coal. The following table gives the values of the densities of various materials:

TABLE VII. *Densities of Various Materials.*

Substance	Density	Substance	Density
Minerals		Sedimentary	
Barytes	4·4–4·7 gm. cm.$^{-3}$	Slates	2·6–2·9 gm. cm.$^{-3}$
Galena	7·5–7·8	Limestones	2·2–3·0
Haematite	4·5–4·9	Sandstones	1·8–2·8
Magnetite	4·9–5·4	Other materials	
Pyrites	4·9–5·2	Clay	1·8–2·6
Zinc blende	4·1	Gravel	1·8–2·0
Igneous rocks		Sand	1·4–2·0
Granite	2·5–3·05	Black coal	1·2–1·5
Basalt, etc.	2·7–3·3	Brown coal	1·1–1·4
		Oil	0·9

On account of the variation in the densities of these materials it is always necessary to use values that have been obtained for samples taken in the area under examination.

In certain cases the method may be used for the direct location and investigation of massive ore-bodies, such as haematite, magnetite, or sulphide ores; or of deposits of brown coal, or salt, but in general its use is indirect. Its application to metalliferous deposits is rather limited, as in the majority of cases the lodes are too small or occur in rugged or mountainous country where the method could only be applied with great difficulty or not at all. The gradients directly due to the presence of the precious metals would be quite inappreciable of course, since the latter are never found in nature in large enough concentrations.

The principal use of the method is in the location and investigation of geological structures, such as faults, anticlines, synclines, domes, etc., with which oil pools and mineral deposits are sometimes closely related. In the case of oilfield work the best known examples of its successful application have been in the Gulf Coast region, U.S.A., where the practice has been to search for the salt-domes with which the oil accumulations are associated. Many of these salt formations give rise to well defined gradients at the surface and in numerous instances their discovery, by means of the torsion balance, has led to the location of oil bearing ground.

To a great extent, gravimetric work has now been superseded by the seismic method, although it is still retained for certain purposes, such as the delineation of salt-domes and the more detailed investigation of structural features. It will be seen from some of the Australian field work, described later in this chapter, that there are particular types of problem which can be dealt with more effectively by the torsion balance, or gradiometer, than by the seismic or any other geophysical method. Under favourable circumstances, quantitative information as to the depth and dimensions of a concealed body, or structure, may be deduced from the gradients and, in the case of small scale problems, the gravimetric method is at a definite advantage in this respect.

It is difficult to express any definite views as to the depth limits to which the torsion balance will give information of practical value. It must be remembered that, in general, the gradient at the surface, due to a body, decreases roughly as the cube of its depth, but increases as the cube of its linear dimensions. That is to say, if the linear dimensions of the body increase proportionately to its depth from the surface, the effect will remain constant. Theoretically, therefore, there is no limit to the depth to which information can be obtained, provided that the body or feature concerned is of sufficiently large extent. In practice, however, the effective depth is limited by the fact that, in the neighbourhood of the body, and between it and the surface, unknown rock masses are usually present, which may also cause gradients at the surface. As these are undeterminable they will tend to mask the effects of a deep-seated body, and in many cases the interpretation of the results will either be uncertain or quite impracticable.

Before undertaking a torsion balance survey it is advisable to make preliminary calculations, based on such geological data as are available, in order to see whether the presumed mineral deposit, or geological feature, in question is likely to cause distinctive gradients at the surface.

SECTION 2. INSTRUMENTS USED BY SURVEY

Two types of instrument were used for gravimetric measurements; the Eötvös torsion balance and the gradiometer, both of which were manufactured by L. Oertling, Limited, London.

TORSION BALANCE[a]

The makers adopted practically the same dimensions as Eötvös[b] for the beam system. The beam is made of aluminium, and the upper weight, a flat block of electrolytic

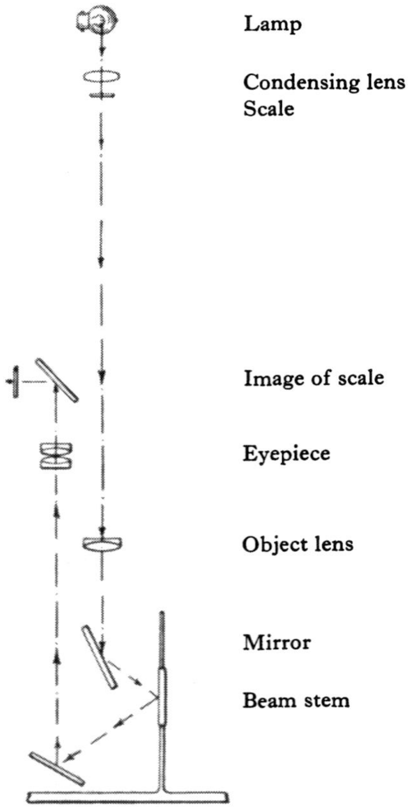

Fig. 101. Oertling torsion balance; optical system

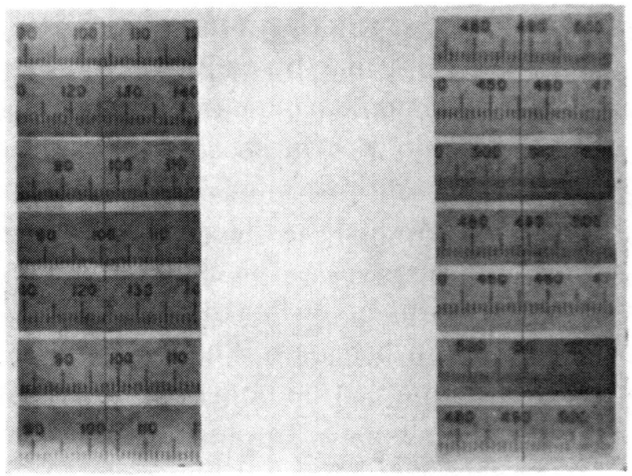

Fig. 102. Plate from Oertling torsion balance

copper, is proportioned so as to serve as a damping vane, while a cylinder of gold is used for the lower weight. The torsion wire is made of a platinum alloy and is ribbon-shaped. The damping can be varied by raising or lowering the movable platform beneath the beam. This platform also serves to clamp the beam for transport.

The deflection of the beam system is indicated by an optical system (Fig. 101) which can be arranged for visual or photographic recording.

Suitable provision is made for orientating and levelling the instrument, and the

[a] For full description, see the maker's handbook.

[b] *Verhandl. der 15 Allgem. Konferenz d. Intern. Erdmessung*, vol. 1, p. 337 (1908).

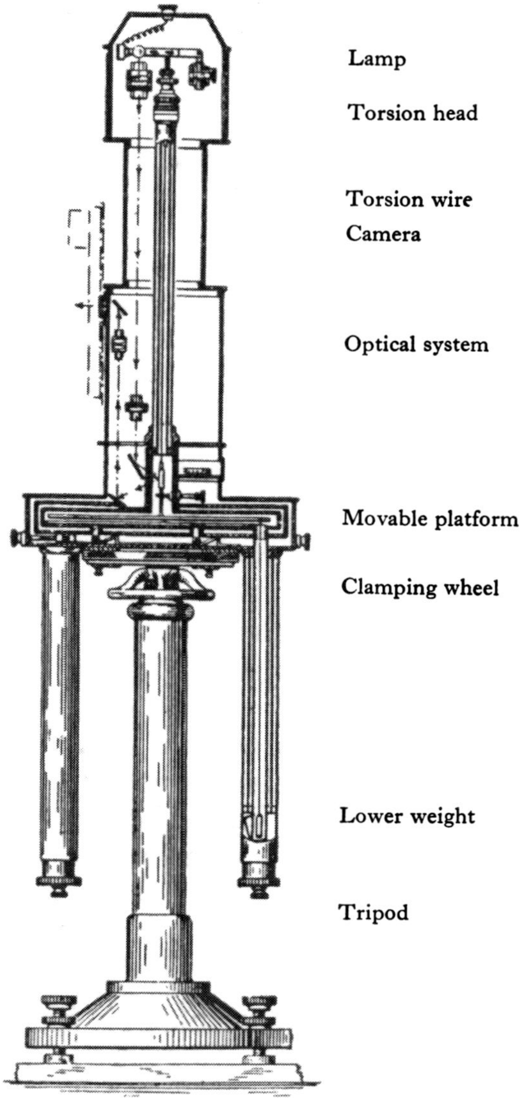

Fig. 103. Oertling torsion balance; sectional elevation

I.G.E.S.

Fig. 104. Torsion balance in the field showing metal hut

I.G.E.S.

Fig. 106. Transport of gradiometer

I.G.E.S.

Fig. 105. Gradiometer with outer cover removed

balance can be placed in the required azimuths either by hand or automatically. For photographic recording, seven exposures at hourly intervals are made by an automatic mechanism, which illuminates the scale, drops the plate, and finally turns the whole balance to the next azimuth. This series of operations can be started at any desired instant by suitably setting the master clock. Fig. 102 shows a plate recording the results obtained for a typical station in the field. Fig. 103 gives some idea of the thermal insulation of the balance, which consists chiefly of double or triple walls of aluminium separated by layers of ebonite. For further protection from temperature changes, and as a windshield, it is provided with a dome-shaped metal hut, shown in Fig. 104, consisting of triple walls of aluminium separated by kapok. In Australia this hut was enclosed inside a canvas hut lined with kapok and, where necessary, a tarpaulin was put over the outer hut, the whole being suitably stayed with wire guys.

For transport in the field, the balance can be firmly attached to the metal hut, and carried on a special trolley mounted on motor-car wheels.

GRADIOMETER

The beam system which has been described on p. 139 is much more compact than that of the torsion balance, and the overall height of the gradiometer is only about 4 ft. as compared with 6 ft. 3 in. for the torsion balance. This is largely due to the fact that the centre of gravity of the beam system of the gradiometer is above, and that of the torsion balance below, the beam. The torsion wires used were made by rolling a wire about 0·04 mm. diameter into a ribbon shape, thus diminishing its torsional rigidity without reducing its tensile strength. A damping vane, consisting of an annular ring of mica, is fastened to the underside of the beam and the damping can be varied by raising or lowering a platform under the beam by means of a graduated brass screw. By raising it to its limit the beam can be clamped securely against the roof of the beam box, thus removing the load from the torsion wire. Stops are provided to limit the swing of the beam.

Fig. 105 shows the gradiometer with its cover removed. The left-hand tube contains the upper weight which can be clamped or released by the cap at its upper end[a]; the middle tube contains the torsion wire, and the torsion head can be seen at the top; the oblong case to the right carries the optical system. Below these are the mirror chamber, beam box, and horizontal circle graduated in degrees and provided with sensitive levels and a trough compass.

The optical system is similar to that used in the torsion balance. Light from a small electric bulb is condensed by a cylindrical lens on to a scale, an image of which is reflected by a mirror attached to the beam and focused on to a ground glass screen for visual observation. No provision is made for photographic recording. Convection currents are diminished by using double and triple walls of brass and copper, with air as the non-conducting medium.

[a] See Shaw and Lancaster Jones, *Mining Magazine*, vol. 40, pp. 272–279 (1929).

The instrument is set up inside a canvas hut lined with kapok and large enough to permit the observer to move freely round the instrument. Whenever necessary the roof of the hut can be covered by a tarpaulin firmly pegged to the ground.

The instrument when detached from its tripod weighs only 65 lb. and can be carried by one man, for 100 yd., without undue distress (see Fig. 106).

SECTION 3. FIELD PROCEDURE

Testing of Instruments. It is necessary that the instruments to be used should be thoroughly tested in a suitable location, such as the basement of a building, before being taken into the field. They are first fully adjusted for use (in the manner described in the maker's directions) and, in the case of the torsion balance, the satisfactory operation of the automatic recording apparatus is checked. Consistent results should be obtained in the measurement of the gradient at any selected site in the basement, i.e. the probable error should not be more than one unit for the balance and two for the gradiometer. Owing to the non-uniform nature of the gravitational field in such a location, no purpose is served by comparing the results obtained on the same spot with different types of instrument, but with instruments which have the same dimensions these results should agree. The most satisfactory comparisons are made in the field, where the gradients may be taken as uniform over the space occupied by the largest instruments. The results are then compared for a gradient of appreciable magnitude, say over 30 *E*.

Personnel and Equipment. The I.G.E.S. field party, consisting of the leader, one observer, one surveyor, and two or three labourers, carried out photographic observations by night with one Oertling torsion balance and day observations with one gradiometer. Later on, day readings were made with two gradiometers only. This party, working the ordinary day shift, was quite sufficient under all circumstances, and under favourable conditions could have used a third instrument quite efficiently. A larger party would only be required if gradiometer observations were taken regularly at night as well as in the daytime. The following auxiliary equipment is required: for the torsion balance, an outer protecting metal cover, trolley for transport, and external portable hut; for the gradiometer, a similar smaller hut; spare torsion wires for both instruments. For field calibrations, a lead sphere about 30 lb. weight; for surveying, a light level with telescope focusing down to 5 ft. (such as Watts "Quickset"), a tacheometer theodolite (which in a flat country could be dispensed with) and the necessary staves, tapes, etc.

Planning the Survey. In planning the survey large-scale contoured maps are required but, in general, these are not available in Australia and in many other parts of the world. The ideal type of country for gravimetric work is a level open plain, and the method cannot be used in a rugged mountainous district. Satisfactory work can be done, however, in moderately rough terrain[a] and in heavily wooded country by selecting the sites for the instrument with sufficient care. It is here that good maps are required.

[a] At Lakes Entrance (Victoria) the slopes of some of the hill-sides were as much as 30°.

In general the stations should be set out along the ridges and spurs, and traverses on the slope of a hill should be avoided if possible. In such country the order of accuracy of the work is relatively low, so that the anomaly sought for must be such as to cause large gradients. When the terrain is rough the rate of work will be slow, and the cost of the survey will be correspondingly high.

Choice of Instruments. For use in Australia and other thinly populated countries, small, readily portable instruments of the gradiometer type are to be preferred to the full-sized torsion balance. They are more readily transported to the site of work, are more rapidly moved from one site to the next, and measure the gradient in a shorter time than does the torsion balance. Although they are less sensitive and do not give the curvature values it was actually found during the course of the work that the accuracy of the Oertling torsion balance was scarcely any greater than that of the gradiometer.

Observations taken at night are always more reliable than those made during the day and, in the experience of the I.G.E.S. and of McLintock and Phemister[a], the Oertling torsion balance, even with the extra protecting hut, is useless in the day time[b]. It is claimed by the makers of some of the other types of torsion balance that their instruments give satisfactory results even in the day time in the tropics. The I.G.E.S. party found that the gradiometer gives quite satisfactory day readings even with the high and irregular temperatures and intense sunlight prevailing in summer in the interior of Australia. The relative merits of photographic and of visual methods of observation have been the subject of discussion[c] but, for the Oertling torsion balance, photographic recording is to be preferred to visual reading, and every effort should be made to keep the automatic apparatus functioning properly. The party should therefore always be provided with a complete spare set of the clockwork used for this purpose. It is usually necessary to use the external hut, both for thermal protection and for protection from wind, so that unless this hut is very large visual observations cannot be made. Also, with a small party, on an extensive survey, it is very necessary from the point of view of the observer's efficiency that he does not have to stay up for the greater part of the night making observations.

In general, outside the tropics, the winter is the best time of the year for gravimetric work, as the diurnal changes of temperature are least during these months. For the same reason a dull day is better than one with clear skies.

[a] McLintock and Phemister, *A Gravitational Survey over the Swynnerton Dyke, Staffordshire*. Mem. Geological Survey. Summary of Progress of the Geological Survey for 1927. Part II, p. 1 (1928).

[b] This was found to be the case even on a dull winter day when the fluctuations of atmospheric temperature were slight, and much smaller than would occur on an ordinary clear night. The explanation may be that during the day the upper layers of air inside the hut are warmed by conduction through its top and thus are warmer than those in contact with the ground. These temperature differences cannot be equalised by convection currents inside the hut; the instrument is thus subject to a vertical temperature gradient. On the other hand, during the night the ground retains its heat longer than the air, so that the lower layers of air become warmed and rise. The temperature differences inside the hut are thus equalised by these convection currents.

[c] Pekar, *Zeits. f. Instrumentenk.*, vol. 42, pp. 173–178 (1922); vol. 43, pp. 187–195 (1923). Schweydar, *ibid.*, vol. 41, pp. 175–183 (1921); vol. 42, pp. 179–182 (1922); vol 43, pp. 307–311 (1923).

Layout of Stations. The manner in which the stations are laid out, and their distance apart, depend naturally on the type of problem under consideration. In many cases it is the location and investigation of a fault, syncline, or anticline, when the stations should be placed on a series of traverses perpendicular to the strike direction of the feature. If this is not known, it can sometimes be determined by a preliminary traverse. It is difficult to lay down a hard and fast rule as to the distance apart of stations, but as a general rule it should be of the order of the distance below the surface of the feature concerned, and should be considerably greater than this in the preliminary reconnaissance. Extra stations must then be placed in regions of rapidly changing gradient. For example, if a synclinal structure lying at a depth of 200 ft. were being sought, the stations would be say 400 ft. apart. When it was located, extra stations at 200 ft. and 100 ft. apart would then be used in order to get information as to its depth and slope. In other cases (salt-domes, etc.) the stations are best laid out on a chess board system. Thus in practice, the distance between stations will nearly always range from 50 ft. to 1000 ft.

The site for the station should be selected so that there are no marked irregularities in the ground, and no feature such as a large tree, ditch, etc. should lie within about 10 ft. of it. Then, for a radius of 3 ft., the station is cleared of scrub, grass, etc., and minor irregularities on the surface are smoothed out. Usually, but not always, it is necessary to drive pegs into the ground for the instrument to rest on.

Transport and Time Required. The transport of the Oertling torsion balance is very much facilitated by the method in which it is clamped to its metal cover and the whole carried on the trolley, as shown in Figs. 107 and 108. The gradiometer is readily moved by hand for distances up to 1000 ft. by two men, one taking the tripod and the other (Fig. 106) the upper part (weighing 65 lb.). This should be kept out of the direct rays of the sun; in hot weather it should be covered with a blanket or closely fitting felt cover. In bush country it is usually impossible to use vehicular transport for the huts; they are either carried whole by 3 or 4 men, or dismantled and each section carried separately[a]; see Fig. 109. The I.G.E.S. party found that the time taken for the complete move to, and setting up on a new site about 400 ft. away is normally 1–1½ hours for the torsion balance (including the preparation for photographic registration) and 11–15 minutes for the gradiometer.

I.G.E.S.

Fig. 107. Transport of Oertling torsion balance

[a] In this case one observer and one labourer are all that are required for work with the gradiometer. See p. 167.

The gradient, if there were no change in the "zero" of the instrument, could be derived from three readings of the gradiometer in the azimuths of 0°, 120° and 240°. Owing to this change, however, it is necessary to repeat some of these readings, so that in general five or six settings, at time intervals of 25 minutes, are required. Thus the total time for the station is usually 125+15 minutes, or 150+15 minutes. It will be seen that under favourable conditions three stations are readily done per shift of 8 hours. While the observer is making the readings the assistants are engaged in surveying and in the preparation of the site for the next station; they return in time to take part in the transport of the instrument and hut.

Provision must be made for the development of the plate from the torsion balance in the field during the daytime, as it is essential to see that the previous night's results are satisfactory before moving to a new site.

Each instrument should be calibrated from time to time during the course of the work, by obtaining the deflection when the lead sphere, mounted on a suitable stand, is placed in a definite position close to it.

I.G.E.S.

Fig. 108. Erection of torsion balance

Corrections to Observations. The gradient measured in the field is made up of various components:

(*a*) The gradient due to the non-sphericity and the rotation of the earth.

(*b*) The gradient due to the departure from the horizontal of the ground in the vicinity of the instrument.

(*c*) The gradient due to features such as embankments, cuttings, buildings, rivers, trees, etc.

(*d*) The gradient due to the distribution in the earth's crust of rocks of different specific gravity.

The curvature terms are also made up of four such components. The component (*d*) is the one which is required in applied geophysics.

(*a*) This is the so-called *normal* effect. Its value is constant for any one site and can be calculated from its latitude.

(*b*) This is divided arbitrarily into three classes: *terrain* (up to 100 ft. from the site), *topographical* (from 100 ft. to 1000 ft.), and *cartographical* (over 1000 ft.) effects. For their determination the value of the density of the surface material must be obtained. For the determination of terrain effects, levels, usually in eight azimuths at distances of 5, 10, 15, 20, 30, 50 and 100 ft. from the station, are required. In flat country, if gradients only are dealt with, it is usually sufficient to carry this only to

30 ft. Under favourable conditions these levels have been taken by the I.G.E.S. party (one surveyor and two men) for six stations in an hour. The calculation of the effects is very much facilitated by systematic working, and a party of three has done it for as many as 12 stations in an hour. For topographical effects, a plan of the country contoured at intervals of 5 ft. up to a distance of 1000 ft. from the station is required. In most cases this has to be prepared by the surveying staff of the party, and in broken or scrub-covered country it is a very laborious business. It should always be remembered when estimating the cost of a survey that it may take longer than all the rest of the work. In a flat locality these effects are negligible for the gradient; and in a moderately level one, in many cases the contouring need not be taken beyond 500 ft. A graphical method is used for the determination from this plan of the value of the effect, the computor requiring from 20 minutes to 2 hours per station. For cartographical effects a map of the district contoured at intervals of 50 ft. is necessary. As a rule these are only available in well-mapped countries, but for the gradients these distant effects may usually be neglected.

(*c*) This is the so-called *special* effect and must be calculated for each individual feature.

I.G.E.S.

Fig. 109. Transport of gradiometer hut

The value of the gradient (*d*), remaining after the components (*a*), (*b*), and (*c*), have been subtracted from the measured gradient, is usually termed the *residual* gradient, and it is this gradient which is plotted on the plan showing the results of the survey.

SECTION 4. SURVEYS IN AUSTRALIA

SURVEY AT GELLIONDALE, VICTORIA

Introduction. The Gelliondale brown coal field is in the south-east of Victoria, about 120 miles from Melbourne. The part over which the gravimetric survey was carried out (see locality plan, Fig. 110) was a roughly rectangular block about 4 square miles in extent, of very flat land lying about 20 ft. above sea level. The surroundings are also very flat, except for a few sand dunes half a mile to the south, about 100 ft. high, and a range of hills, 1000 ft. in height, which lies to the west and north-west, with its base about 2 miles away.

A large part of the area is open country with sword grass, tussocks, etc.; part of this is swampy, and in the winter and spring is covered with several inches, or even a foot, of water. The remainder is covered by very dense ti-tree scrub about 9 ft. in height, or by small eucalyptus and honeysuckle trees. The observation stations have been situated in all of these types of country.

The surface consists partly of sandy and muddy flats, and partly of wind blown sand, sandy loam, etc. Bore-hole records show that the brown coal varies in thickness in different parts of the area from a few feet up to 400 ft., and that it is overlain by various thicknesses (25 ft. to 100 ft.) of drifts[a], clays, and sand (see Figs. 114 and 115). Similar unconsolidated formations underlie the coal, but their thickness has not been determined at many points. The whole series, with the interstratified coal, is usually considered to be of miocene or pliocene age. It rests upon a jurassic bed-rock which in the area under consideration consists largely of a felspathic sandstone. The same formation outcrops in the range of hills to the north-west and west. Brackish water can be obtained nearly everywhere very close to the surface. The brown coal itself is very moist and contains over 60 per cent. of water. The mean values of the densities obtained for the coal, drift, and jurassic sandstone were 1·1, 1·9, and 2·4 gm. $cm.^{-3}$ respectively.

Field Operations. The field party consisted of 3 technical men and 3 labourers, and was equipped with one torsion balance, one gradiometer, and the auxiliary equipment given on p. 149. Two Schmidt magnetic field balances for vertical intensity were also included. At the commencement of the work it was found that the gradients had a predominant trend towards the north. The stations were set out accordingly along a series of parallel north and south traverses which were 400 ft. or 800 ft. apart. In general, stations were taken at intervals of 400 ft. along these lines, but in regions of large or rapidly changing gradients they were placed 200 ft., and sometimes 100 ft., apart.

The party was in the field for nearly 6 months. In the winter, night frosts were fairly common and during the day the shade temperature rose to between 60° and

[a] The term drift is used here, in accordance with local custom, to indicate soft and unconsolidated sands, clays, etc.

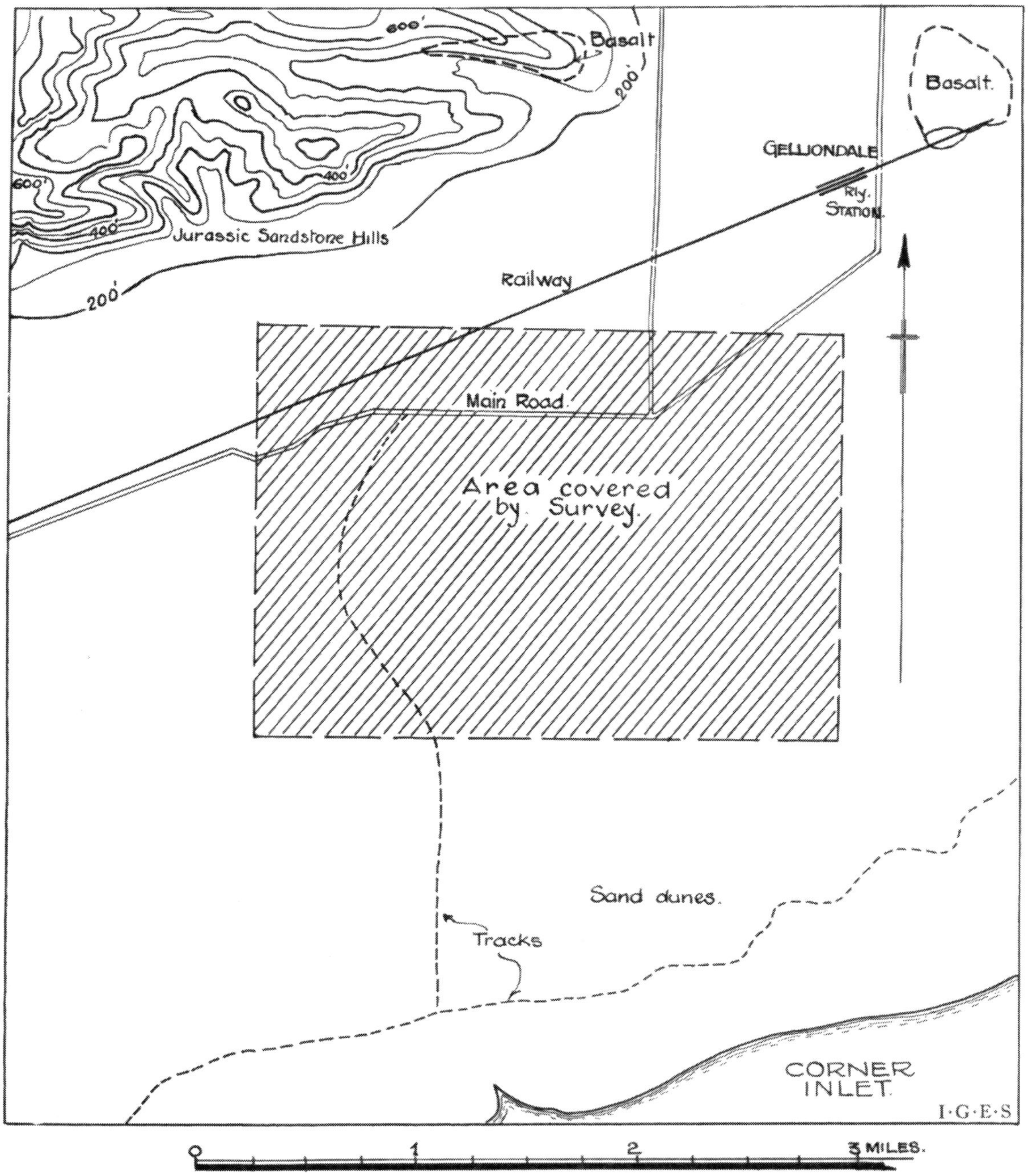

Fig. 110. Locality plan, Gelliondale, Victoria

70° F. In the spring the weather was very wet and windy[a], and in the summer the thermometer was scarcely ever over 90° F. The only interruptions in the work were for two days in October, when the ground was inaccessible owing to floods, and for a week at Christmas.

[a] During one stormy week-end the torsion balance in its hut was blown over. It had been pegged down with galvanized iron wire which snapped owing to the pressure of the wind on the sides of the hut.

Gradients were measured at 325 stations, i.e. an average of 13·5 per week, and measurements were made of the anomalies in magnetic vertical intensity over the whole area. In order to check the behaviour of the instruments, 23 of these gradients were measured with both instruments, while 15 were measured twice or oftener with the gradiometer, in some cases after the lapse of months, and 11 with the torsion balance.

Results. Owing to the flatness of the country topographical and cartographical effects for the gradients were negligible, while terrain effects in the majority of cases were less than 10 E, and in only one case over 20 E. The density of the terrain material was 1·2 gm. cm.$^{-3}$. The normal gradients were $-7·9$ and 1·1 E for U_{xz} and U_{yz} respectively. A special correction was made at one station for the effect due to a heap of road-metal.

The residual gradients are shown on the folding plan, Fig. 111[a]. At the stations where the torsion balance was used the value of the horizontal directing tendency[b] is also shown. The full lines represent the isogams, or lines of equal value of gravity anomaly, referred to some point in the field where the anomaly is arbitrarily taken as zero. The method of obtaining these lines is fully described in Part II, p. 304, and is analogous to the problem of drawing the contour lines over an area from a series of measurements of hill slopes[c]. Thus it is necessary to have a fairly detailed network of stations where the gradients are large and rapidly changing, while they may be taken farther apart in regions of small or slowly varying gradients.

The bores which had been put down by the Mines Department on the above area are marked by a circle and a number; a list of them is given on the opposite page. As the coal is less dense than the drift or the bed-rock so, in general, a gravity "low" will occur above it, and a gravity "high" above the other materials. The direction of the gradients will therefore be directed away from the brown coal area and towards the barren ground of higher density.

An inspection of the above plan shows two areas of gravity "high." The one in the south-west corner is very marked, and in this area the bores (Nos. 48, 73) show the presence of the bed-rock within 50 ft. to 100 ft. of the surface. It also appears from the plan that this elevation in the bed-rock extends farther to the north-east but at a gradually increasing depth. The evidence afforded by Bore No. 38 is in accordance with a pitch in this direction.

In the north-east corner, in the lesser gravity "high," one bore (No. 77) shows only 48 ft. of coal, and probably the bed-rock is not at a very great depth. This is not known definitely, however, as the bore had not been continued below the coal.

As a general rule, the largest gradients occur near the edge of a disturbing body. Thus in the south-west corner of the field there is a series of southerly gradients and

[a] For the sake of clearness several of the gradients in the north-eastern part are not shown.

[b] See Part II, p. 301.

[c] It is based on the following consideration. If the north component of the gradient at a point A is $a.E$, and at a point B at a distance r cm. to the north is $b.E$, where r is (in this case) of the order of 200 ft., then unless the gradient changes rapidly between A and B we can assume that the value of gravity at B is greater than that at A by $\frac{a+b}{2}.r.10^{-9}$ cm. sec.$^{-2}$.

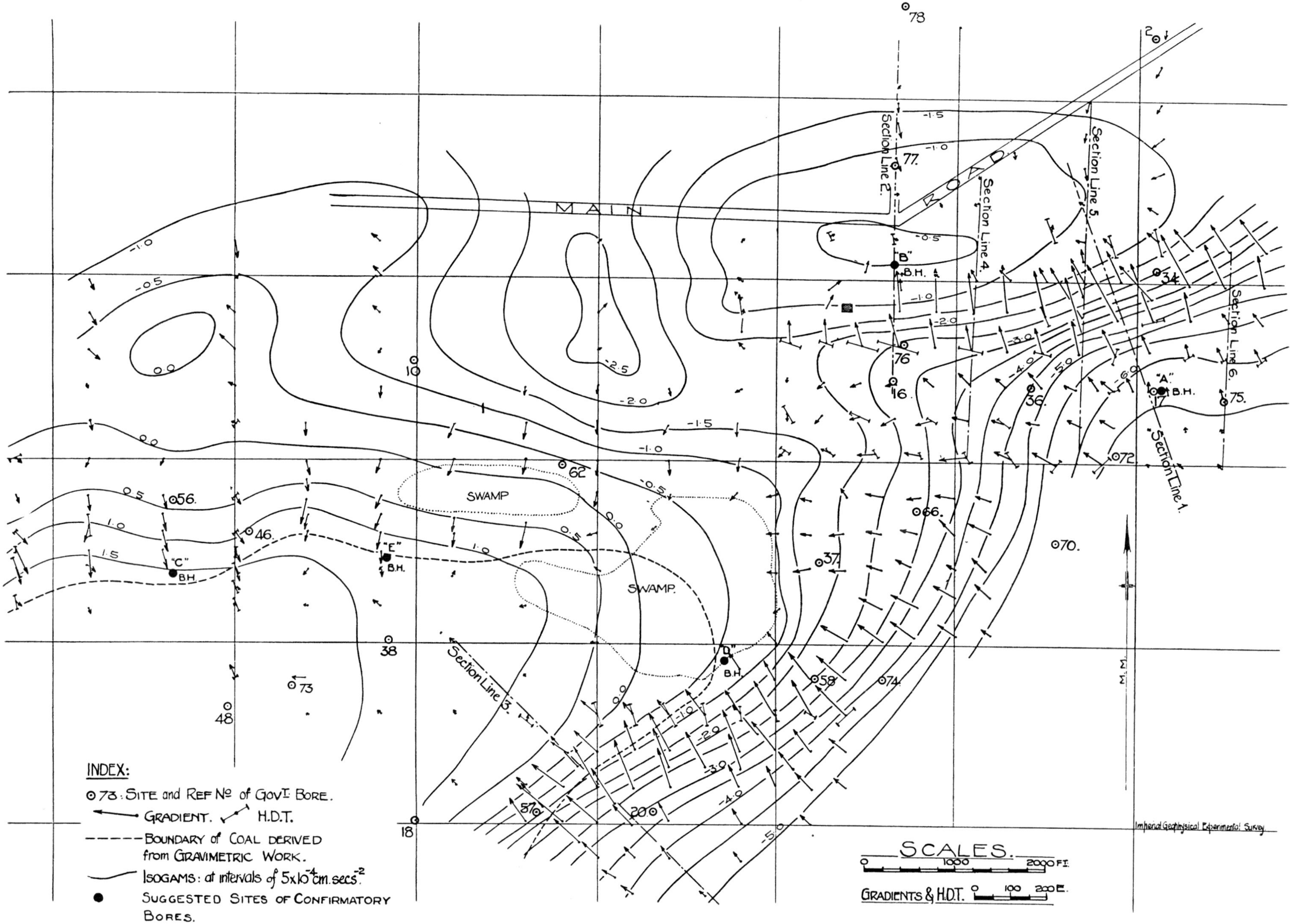

Fig. 111. Gravimetric survey in the brown coal field at Gelliondale, Victoria

TABLE VIII. *Bore Results at Gelliondale, Victoria*

Bore-hole No.	Brown coal: Struck at (ft.)	Brown coal: Thickness (ft.)	Bed-rock struck at (ft.)
2	30	158	—
10	34	250	—
16	56	Not determined	—
17	—	None	73 (reported as jurassic)
18	—	None	39
20	60	267	—
31	—	None	75
34	Not struck at 100	—	Not struck at 100
36	30	Not determined	—
37	60	Not determined	—
38	Not struck at 100	—	Not struck at 100
46	28	Not determined	—
48	—	None	73
56	30	100	209
57	—	None	61
58	64	Not determined	—
62	34	108	238
66	72	151	—
70	73	405	—
72	67	393	—
73	—	None	52
74	51	259	—
75	Not struck at 207	—	Not struck at 207
76	37	258	—
77	52	48	—
78	53	164	—

in the south-centre a series of north-westerly ones. Since boring has shown the presence of bed-rock near the surface in this part, it follows that the line joining the points where these gradients are largest is roughly the line marking the edge of the coal bed, i.e. the boundary between the coal and the heavier rock. Calculations show that in the south-west the largest gradients actually occur about 200 ft. north of the edge of the coal, and in the south-centre about 100 ft. south-east of the edge. The broken line in Fig. 111 shows the approximate position of the coal boundary obtained from the above considerations. The gradients remain southerly for some distance north of this line in the western portion of the area; indicating a thickening of the coal beds towards the north. This conclusion is confirmed by the data obtained from boring.

In the east-central part is a series of moderate gradients, all with a westerly tendency. This indicates a thickening of the coal bed towards the east, also in agreement with boring evidence, since Bore No. 70 has 405 ft. of coal, Bore No. 66, 151 ft., and Bore No. 64, not shown on the map, but lying to the west of No. 66, only 75 ft.

In the north-eastern corner the gradients to a very large extent are northerly. As one goes eastwards from Line 2, along a line running practically due east they are all large and change in direction, from almost due north to about 20° west of north at the eastern end of the line. In the neighbourhood of this east and west line one therefore expects a thinning out of the coal towards the north, due in all probability to a rise in the bed-rock towards the surface in the form of a buried ridge having a steep southerly slope. As one proceeds northwards in this area along Section Lines 2 and 5 the gradients diminish to zero and then become southerly and of moderate size. This shows that the northern slope of the ridge is more gentle and the coal thickens again. This point will be dealt with more fully below.

So far the gravimetric results have been in agreement with the data from boring, but in the case of Bore No. 17 it is impossible to reconcile them. At this bore jurassic bed-rock was said to have been struck at 73 ft. and the bore carried 1 ft. farther and then discontinued. At Bore No. 72, about 1000 ft. south-west of No .17, a total depth of 393 ft. of coal was proved. One would therefore expect very large north-easterly

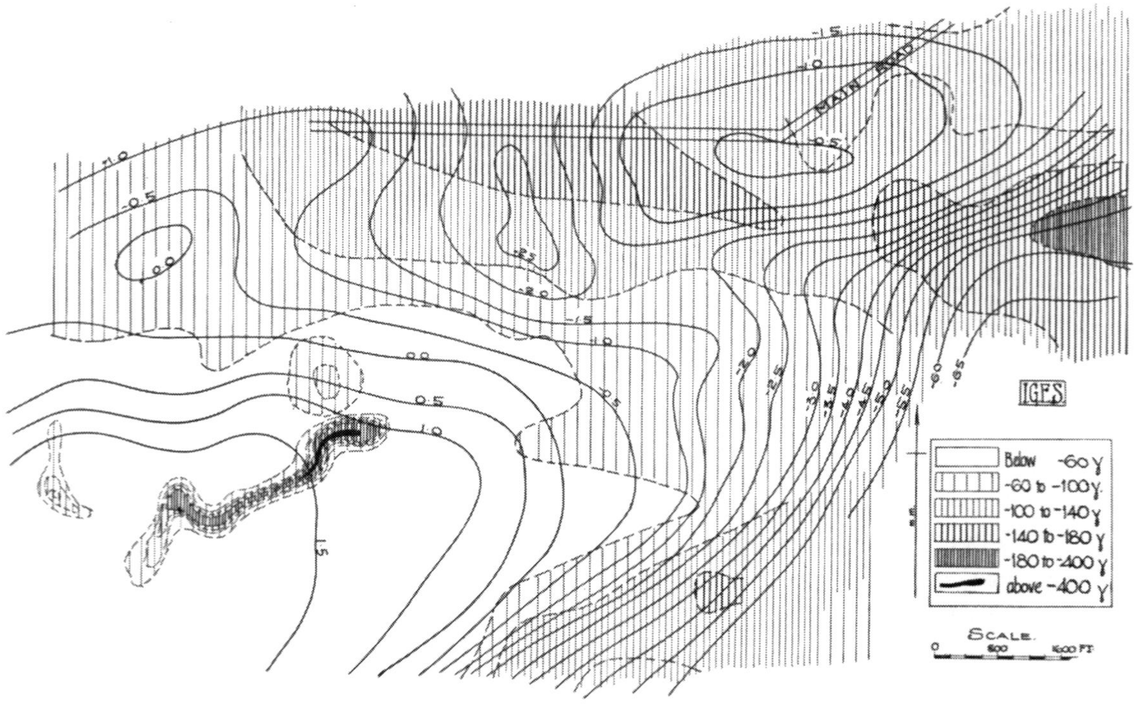

Fig. 112. Isogams and lines of equal anomaly in the vertical magnetic field at Gelliondale, Victoria

gradients between these two holes, but actually they are of moderate size and west to north-westerly in direction. At no other bore in the vicinity was bed-rock struck, though at No. 75 (800 ft. east of No. 17) no coal was met with at 207 ft., and at No. 34 (1300 ft. north of No. 17) no coal at 100 ft. It seems then that at Bore No. 17 the bed-rock was not actually reached, and that the rock encountered was probably an isolated boulder.

In the eastern part of the field the gradients are fairly small and in a westerly direction. There is no evidence therefore of any abrupt discontinuity in the coal bed in this area which would be indicated by large easterly gradients. It is presumed then that Bore No. 75 (no coal to 207 ft.) is still over coal, which is to be met with in considerable thickness but at a greater depth; and similarly at Bore No. 17, but with the coal coming nearer to the surface than at No. 75.

Comparison with Magnetic Survey[a] **over the same Area.** In Fig. 112 the values of the magnetic anomalies are shown by a series of hatchings; the closest hatching representing the areas of greatest negative anomaly in the magnitude of the vertical intensity. The full lines represent the isogams, which are the same as in Fig. 111. There is a rough correspondence between the two sets of anomalies; particularly in the south-west section of the area, where the gravitational and magnetic "highs" correspond respectively to the greater density and greater magnetic susceptibility[b] of the jurassic sandstone as compared with the sands, drift, and the coal. In the north-east corner there is also a rough correspondence; this can be seen more readily from Fig. 113, which shows both magnetic and gravitational anomalies along typical lines of section.

In the south-western portion there is a limited region of marked magnetic anomaly, the relation of which to the values of the gradients in that area is discussed on p. 187.

Calculations from the Results[c]. In Fig. 111 it will be seen that in certain parts of the field the isogams, for considerable distances, are roughly parallel. In these parts we can regard the gradients as being due to "two-dimensional" bodies which can, with sufficient accuracy, be taken as infinite in one direction which is parallel to that of the isogams. Then for sections in these areas, at right angles to the isogams, the values of the observed gradients can be compared with those calculated for various cross-sections of the two-dimensional bodies, i.e. of various possible distributions of coal and bed-rock which will conform with the data obtained from existing bore-holes lying on or close to the line concerned.

Considerable uncertainty exists on account of the lack of positive evidence as to the depth below the coal of the upper surface of the bed-rock. Practically all the existing bore-holes went through the coal and were carried some 50 feet or so into the underlying drift, but in most cases discontinued before bed-rock was reached.

In the calculations two sets of gradients were considered:

(1) Those due to the difference in density between the coal and the drift; the difference being taken as 0·75 or 0·8 gm. cm.$^{-3}$.

(2) Those due to the difference in density between the drift and the bed-rock; this difference being taken as 0·5.

The latter term was always the smaller, due to the greater depth of the bodies concerned, but it was by no means negligible.

The general procedure used was to assume a certain probable coal distribution, calculate the gradients due to it, plot the differences between the calculated curve of gradients and the observed one, and then to see if the resulting curve could be accounted for by any probable boundary between the bed-rock and the drift. If not, a new coal distribution was tried, and so on.

[a] See pp. 186–187.

[b] See table of susceptibilities on p. 177.

[c] See Part II, p. 302.

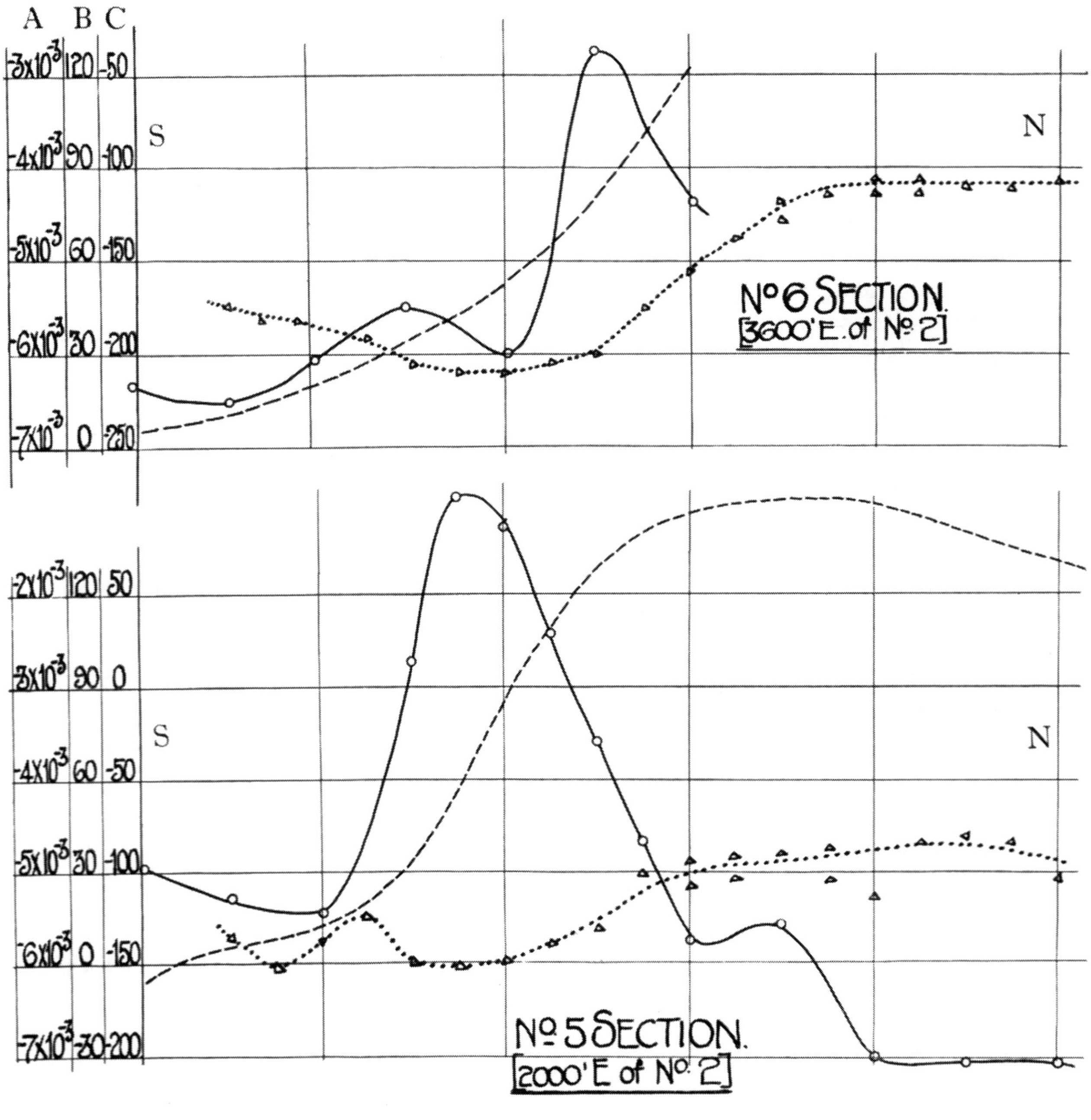

A = gravity anomaly (c.g.s. units)

B = gradient anomaly (Eötvös units)

C = magnetic anomaly (γ)

Fig. 113. Gravitational anomalies and anomalies in the

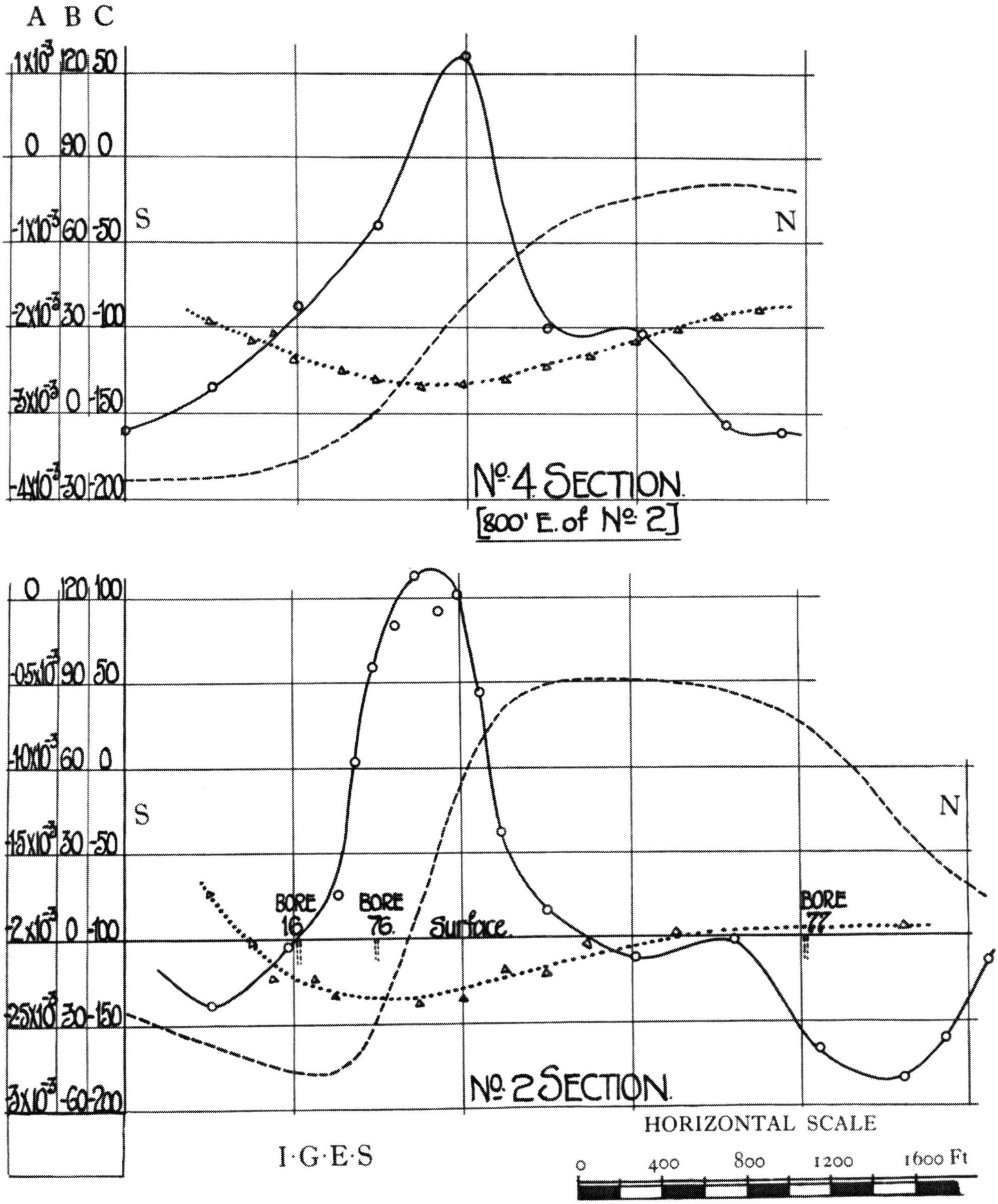

vertical magnetic field along various sections, Gelliondale, Victoria

It will be seen that a great variety of solutions could approximate to the observed gradients, even when the former are to a certain extent limited by evidence from boring. The solutions given below are by no means to be taken as unique, but rather to give some idea as to the type of distribution that could give the observed gradients. On this account it was not thought advisable to spend too much time on the calculations, which could be continued almost indefinitely. The curves actually calculated do not exactly fit the observed ones. They could readily be made to do so by assuming small irregularities in the top or bottom surfaces of the coal, but such assumptions would naturally be quite arbitrary and would be unjustified.

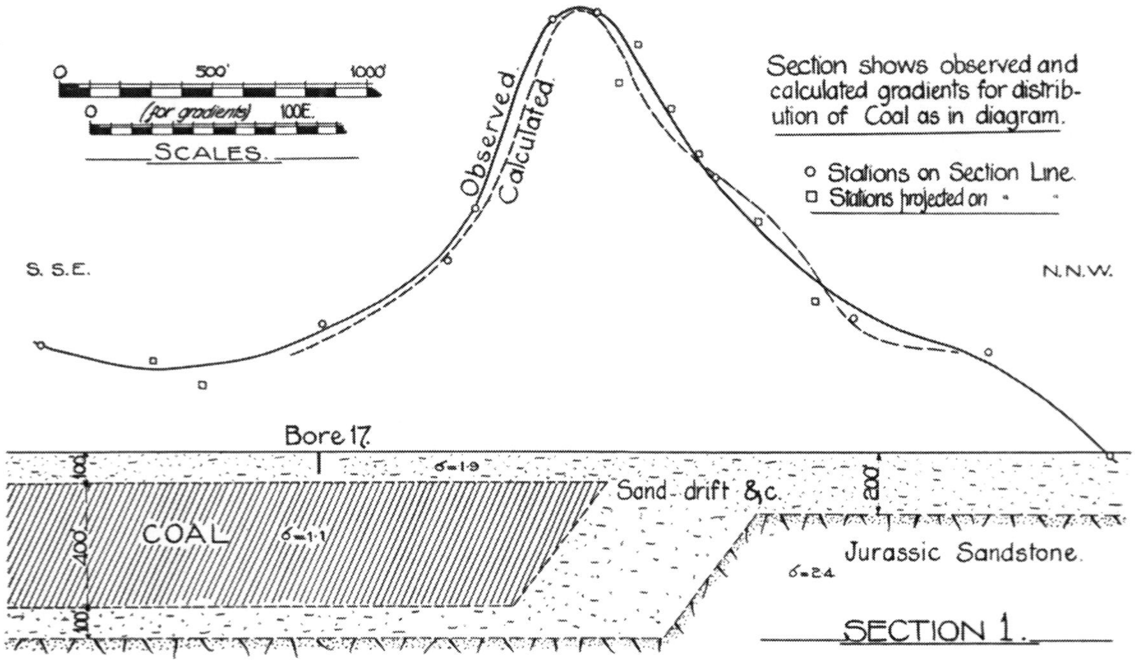

Fig. 114. Observed and calculated gradients along Section No. 1, Gelliondale, Victoria

The results[a] of the calculations[b] for three typical lines are shown in Figs. 114 and 115.

(*a*) *Section No.* 1, Fig. 114. This line passes through Bore No. 17 in a direction 20° west of north.

[a] Early in the course of the work marked discrepancies were observed in the gradient measured at the same station with the torsion balance and with the gradiometer. This will be dealt with in Part II, p. 317, where it will be shown that the two sets of results are brought into line by dividing those for the gradiometer by 1·36. The relative values of the gradients may be taken as correct within the limits of experimental error of the instruments, but the absolute magnitudes are uncertain, as even after a complete calibration of both instruments the cause of the discrepancy has not been fully determined. The qualitative deductions given on p. 157 are not affected, and even the calculations along the various section lines need not be seriously modified; an alteration in the magnitudes of the gradients could be compensated for by an alteration in the values of the density differences which probably are uncertain to within 0·1 or 0·2 gm. cm.$^{-3}$.

[b] The labour required was very much reduced by means of the graphical procedure given in Part II, p. 303.

In this figure the full line shows the curve connecting the components of the gradients along the line of section, measured at stations on or close to it. Now, since the isogams are roughly parallel over a considerable area, it follows that the gradients at all points along any given isogam are also roughly equal in magnitude and direction. Thus

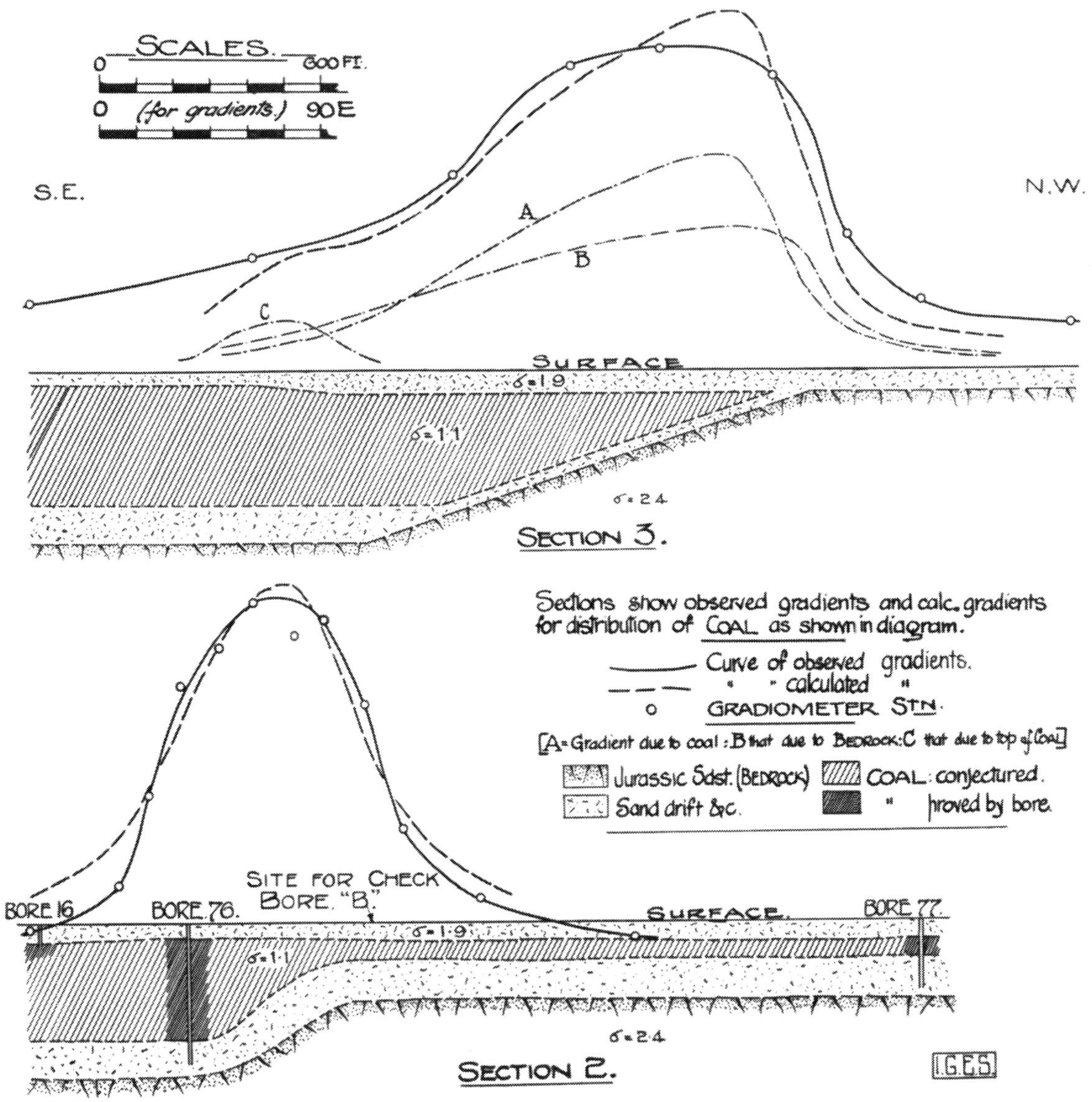

Fig. 115. Observed and calculated gradients along Sections Nos. 2 and 3, Gelliondale, Victoria

gradients at stations lying some distance off the line may be projected on to it. For example, if a gradient, whose components U_{xz} and U_{yz} are p and q respectively, be measured at a station S, whose co-ordinates referred to Bore-hole No. 17 as origin, with OX and OY in the directions of magnetic north and east respectively, are x and y, then the component of this gradient in the direction of the line is ($p \cos 20° - q \sin 20°$).

Now the isogam through S cuts this line at a point T, whose distance from the origin, measured along the line, is ($x \cos 20° - y \sin 20°$). At this point, therefore, the gradient is the same as at S. These components and distances are the ones plotted in the figure, where points representing stations close to the line are shown by circles, and those farther from it by squares. The fact that the latter fit very well on the curve through the former is justification for this process of projection.

The broken line is the curve of gradients calculated for the hypothetical section shown. It will be seen that this section entails a considerable thickness (400 ft.) of coal below the site of Bore No. 17; and that the coal remains of considerable thickness for about 800 ft. north of this, after which it thins out very rapidly.

(*b*) *Section No.* 2, Fig. 115. A fourth bore-hole, No. 78, lying on this line to the north of No. 77, is not shown on the figure. As before, the broken line shows the curve of gradients calculated for the assumed section shown, and the full line shows the gradients actually observed.

(*c*) *Section No.* 3, Fig. 115. This line is in a north-westerly direction. The graphical method was found to be unsuitable for the calculation of the gradients for the various sections assumed. Instead, the formula for an infinite horizontal block[a] was used for the calculation. The figure shows how the resultant calculated curve (broken) is obtained from the addition of three component curves A, B, and C.

Other lines of section could be considered; for instance, various north and south lines, in the west. The curves of gradients along these lines are typical of those for horizontal blocks having their southern faces with slopes of the order of 1 in 5, i.e. here the coal thins out in the form of a wedge with its upper surface horizontal. The calculations for these sections would be uncertain on account of the additional unknown factor introduced by the presence of the body causing the magnetic disturbances[b]. Probably the maximum gradients on these lines are about 200 ft. north of the boundary of the coal.

SUMMARY OF CONCLUSIONS

(i) The coal in the south-western portion of the field has been delimited.

(ii) Approximate positions, where rapid thickening or thinning out of the coal beds occur, have been found.

(iii) With the exception of one case (Bore No. 17) the gravimetric results are compatible with the data obtained from boring.

(iv) It is suggested that the records for Bore No. 17 may be incorrect.

Suggested Bore-holes. The following bore-holes are recommended for confirmation of these conclusions:

Bore A. Repeat Bore No. 17, in which jurassic bed-rock was recorded at 73 ft. Bed-rock at this depth is not in accordance with the gravimetric evidence and coal, about 300 ft. in thickness, should be met with at a depth of 100 ft.–150 ft.

[a] See Part II, p. 302.

[b] See p. 187.

Bore B. At about 900 ft.[a] north of Bore No. 16, and lying on Section Line No. 2. The purpose of this bore is to test the correctness of the section assumed in Fig. 115; the bore should show about 50 ft. of coal at about 50 ft. below the surface.

Bore C. At 800 ft. south of Bore No. 56.

Bore D. At 1000 ft. to the west and 200 ft. north of Bore No. 58.

The purpose of Bores *C* and *D* is to test the correctness of the location of the coal boundary, shown by the dotted line in Fig. 111. In both cases no coal, or coal of small thickness only, should be found.

Exploration Results. The execution of the four confirmatory bore-holes specified above was undertaken by the Geological Survey of Victoria after the close of the I.G.E.S. work in Australia. The following results were obtained:

Bore A. Sand and drift to 73 ft. Cemented sand, very hard from 73 to 75½ ft. Ligneous sand, 75½ to 79 ft. Cemented sand, 79 to 80 ft.

The cemented sand was tertiary, and definitely not the jurassic bed-rock. Owing to its hardness the bore, which was put down by means of a hand machine, could not be carried farther. It is quite evident that the previous log of No. 17 (see p. 157) is incorrect.

Bore B. This bore showed 67 ft. of brown coal, from 31 to 98 ft., thus confirming the results of the calculations along Section Line No. 2 (Figs. 111 and 115).

Bore C. There was no brown coal in this hole, and decomposed basalt was struck at a depth of 66 ft.

Bore D. It is understood that the site of this hole is within 50 ft. of the position indicated on Fig. 111. The bore showed brown coal from 47 to 83 ft., a thickness of 36 ft. only.

It will be observed that Bores *C* and *D* satisfactorily confirm the position of the assumed coal boundary shown in Fig. 111.

The results of Bore *E*, put down in order to obtain information of the cause of the magnetic anomaly mentioned on p. 159, are given in a footnote on p. 187. In this bore basalt was struck at 9 ft. from the surface, while in Bore *C*, where the magnetic anomaly was much smaller, it was met at 66 ft.

SURVEY NEAR LAKES ENTRANCE, VICTORIA

Introduction. For some years past bores put down in the search for oil near Lakes Entrance (about 200 miles east of Melbourne) have yielded favourable indications. There is, however, practically no information from the surface geology of the district as to the most suitable positions for the location of bore sites.

About 12 miles to the north of the township, just beyond the limit of Fig. 116, there is an outcrop of granite, which appears to be the southernmost of various outcrops which extend northwards for 50 miles or more towards the Dividing Range. At Lakes Entrance two of the three bores put down by the Mines Department bottomed on

[a] On Fig. 111 Bore *B* is erroneously shown 1300 ft. north of Bore No. 16.

granite at depths of about 1400 ft.; the third on ordovician slates at about the same depth. There are some isolated granite outcrops in the valley of the Mississippi Creek, about 7 miles north of Lakes Entrance. Thus it was thought that the granite might extend to the south of these in the form of a ridge at a depth below the surface of not more than 1000 ft., and that there were possibilities of oil accumulating on the flanks or crest of this ridge. Accordingly the I.G.E.S. made an attempt to locate it by gravimetric methods.

The surface consists of sandy loam, sands, clays, and gravels, overlying marls and soft limestone. The bores have shown that limestone is the predominating material down to the ordovician bed-rock. The density of a sample of the granite was 2·6, while values of 1·8 to 2·1 gm. cm.$^{-3}$ were obtained for samples of limestone taken from different depths[a].

The area investigated had a mean height of from 200 to 400 ft. above sea level but was very much dissected, in parts almost to sea level, by creek beds with steeply sloping sides. An idea of the broken nature of the country can be obtained from Fig. 238, p. 315, which shows the contours of the ground surrounding one of the stations on the first line of traverse.

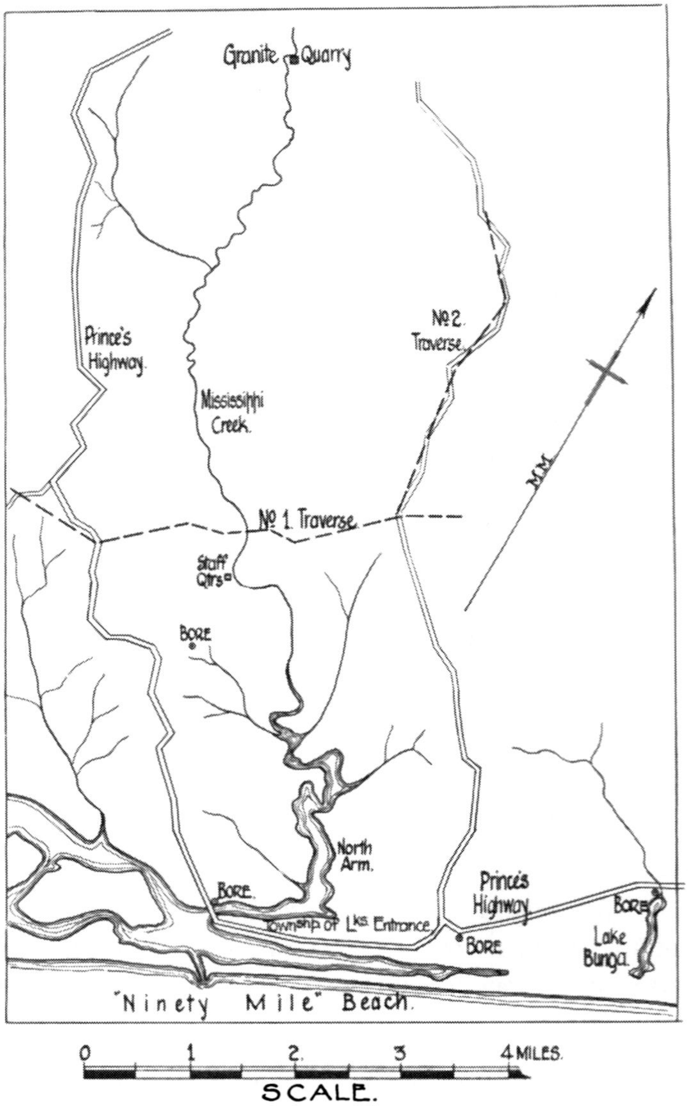

Fig. 116. Locality plan, Lakes Entrance, Victoria

Naturally these lines were laid out as far as possible where the topography was less severe.

In the valleys there was always a very thick growth of scrub and bracken. The ridges were much more open and carried a heavy growth of eucalyptus which, however, was reasonably free from undergrowth. A contoured plan of part of the locality on the

[a] These are rather low values for limestone. There is always the possibility that the limestone surrounding the granite formation is hard and has approximately the same density (2·6 gm. cm.$^{-3}$) as the latter, in which case gravimetric work could yield absolutely no information as to its structure.

scale of 4 in. to the mile, with contour lines at 25 ft. intervals, was available, and was found to be of very great use in selecting the lines of traverse.

Field Operations. The party, which used one gradiometer with an extra hut, consisted of two to three technical men and two labourers. About eight weeks[a] were required for the actual gravimetric work, gradients being measured at 63 stations. Owing to the small party, the rough nature of the country, and the distance of the outlying stations from headquarters[b], the normal rate of working was only two stations per day.

The surveyor, usually working with one labourer, spent about 14 weeks in contouring the region tacheometrically for about 1000 ft.[c] on each side of the lines of traverse. He required a further seven weeks for the office work and the preparation of the contoured plans. Since the surveying required so much more time than the actual gravity work, it rarely happened that the surveyor was near the gradiometer. Thus the most efficient mode of working was to have the observer and one labourer only with the gradiometer; the former took readings at one station, while the latter carried the other hut in sections from the previous station to the next one.

The locality plan, Fig. 116, shows the position of the two traverses that were made. Traverse No. 1 followed the direction of an old bush track which connects the two roads. It kept fairly close to the ridges, but naturally some fairly rough country was met with when crossing the creek. The total length of the line was about 5 miles; the 36 stations on it thus averaged some 700 or 800 ft. apart. Traverse No. 2, running in a northerly direction from the eastern end of No. 1, followed a bush road along the ridge to the east of the creek for some 4 miles, and had 27 gradiometer stations on it.

Corrections. The surveying for the terrain and topographical corrections was laborious, and in many cases it was necessary to use a theodolite for the terrain levelling. The highest value obtained for the terrain correction was about 50 *E*, and for the topographical correction 16 *E*.

Results. The residual values of the gradients along Traverse No. 1 are shown in Fig. 117. These are all small, except at some stations where the uncertainty of the result is greatest owing to the large values of the above corrections. Their general trend is away from the valley of the creek, which seems to suggest that they are largely due to these uncertainties, or else that the creek bed is in a locality where the limestone is softer and less dense than that in the surrounding district.

There is no evidence, therefore, of the existence of the granite ridge sought for, within the limits of the traverse made by the I.G.E.S. This means, of course, that there

[a] On several days no field work was carried out on account of rain. In general the weather was fine; frosts at night were very frequent, and in the day the thermometer usually reached 60° F. in the shade.

[b] The maximum distance was 5½ miles; the track was usually unfit for bicycles or motor vehicles, so that it was necessary to walk.

[c] When computing ther esults it was found that a survey of 500 ft. on each side of the line was sufficient for all except the central part of Traverse No. 1. The surveying done was, therefore, considerably more than was actually necessary.

is no ridge near enough to the surface, or with sides of sufficient slope, to cause an appreciable gradient at the surface[a].

Traverse No. 2 was then made in order to determine whether the granite to the south of the quarry is faulted, or whether it slopes downwards more or less uniformly to its known depth at Lakes Entrance. If the latter is the case, it drops about 1500 ft. in 8 miles at a slope of roughly 1 in 30, so that no appreciable gradients would be produced by it. In the former case, northerly gradients should be met with in the neighbourhood of the fault.

The residual gradients obtained along this line were all small and irregular in direction; the greatest value obtained for the northerly component U_{xz} was only 9 E, and there is thus no evidence of a fault with a roughly east-west direction of strike. The majority of the gradients have a negative value of U_{yz}, which again is small, being not over 14 E; this corresponds to gradients pointing in the direction of the known outcrops of granite in the valley. The results on this traverse thus indicate that the granite slopes downward from these known outcrops and that there is no well-marked fault. Obviously a traverse down the valley from the outcrops would give more information, but such a traverse would be scarcely feasible owing to the narrowness and the steep sides[b] of the valley.

Although the result of this survey is negative it is definite in that it shows that there are no appreciable gradients that can be attributed to a buried ridge of the type sought for. The work is interesting as being an example of a survey carried out under difficult conditions of topography.

SURVEY NEAR GULGONG, NEW SOUTH WALES

Introduction. This survey was made in an area where the seismic party had been working; an account of the geology and history of the locality being given on p. 224. Fig. 118 is a geological sketch plan of the district, and shows the location of the seismic and gravimetric traverse lines.

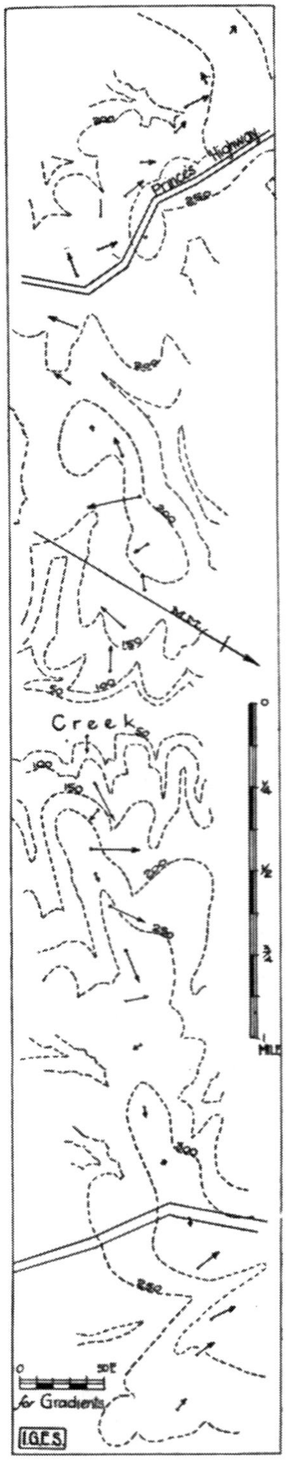

Fig. 117. Gradients along Traverse No. 1, Lakes Entrance, Victoria

[a] For example, a buried ridge rising to 1000 ft. above its surroundings, with its upper surface at a depth of 1000 ft., and with its sides sloping at an angle of 10° to the horizontal, would give gradients of the order of 20 E, directed towards the ridge, if the density difference between the granite and the limestone were 0·5 gm. cm.$^{-3}$. Along a traverse at right angles to the direction of the ridge gradients of this order would persist for over half a mile.

[b] In the topographical levelling slopes of up to 30° were met with when crossing the Mississippi Creek.

The results of the No. 1 Line of Bores put down in the search for the Home Rule Lead are shown in Fig. 170, p. 225. It was decided to make a gravimetric traverse along this line in order to compare the results with those of the seismic party and with those from boring. A previous calculation had shown that gradients up to 60 E were to be expected on the assumption that the density of the alluvium was 2·0, and of the slate and the granite 2·7 gm. cm.$^{-3}$. The latter value was obtained from samples,

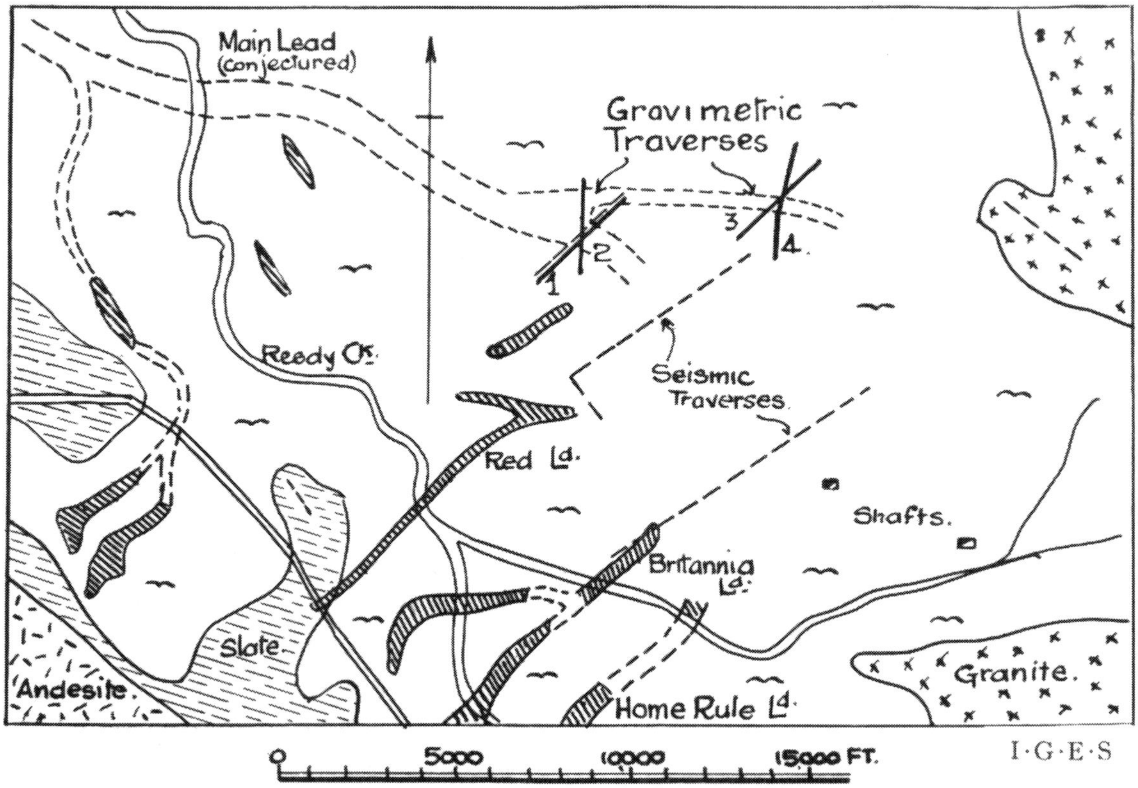

Fig. 118. Geological sketch map of Home Rule district, near Gulgong, showing location of seismic and gravimetric traverses

whilst actual determinations on the material near the surface gave values of 1·7 to 2·2 gm. cm.$^{-3}$. Two traverses farther to the east were made to try and locate any other lead coming in from that direction.

Field Operations. The party consisted of two to four technical men and two labourers, and was equipped with two gradiometers and two huts. The party was in the field for 7 weeks, and at the same time conducted a magnetic survey in an area about 10 miles away. One technical man and one labourer from the above party were usually engaged on this work during the day, but sometimes returned towards evening to assist in moving the gradiometers to new sites. On some occasions one observer only remained in the field and read one gradiometer on one station for a whole day in order to see how it behaved in hot sunny weather.

The weather was fine and sunny; the maximum shade temperature recorded was 98° F., and the maximum in the instrument hut was 88° F.; temperatures above 90° F. were not common. On two occasions the work was held up by flood waters in Reedy

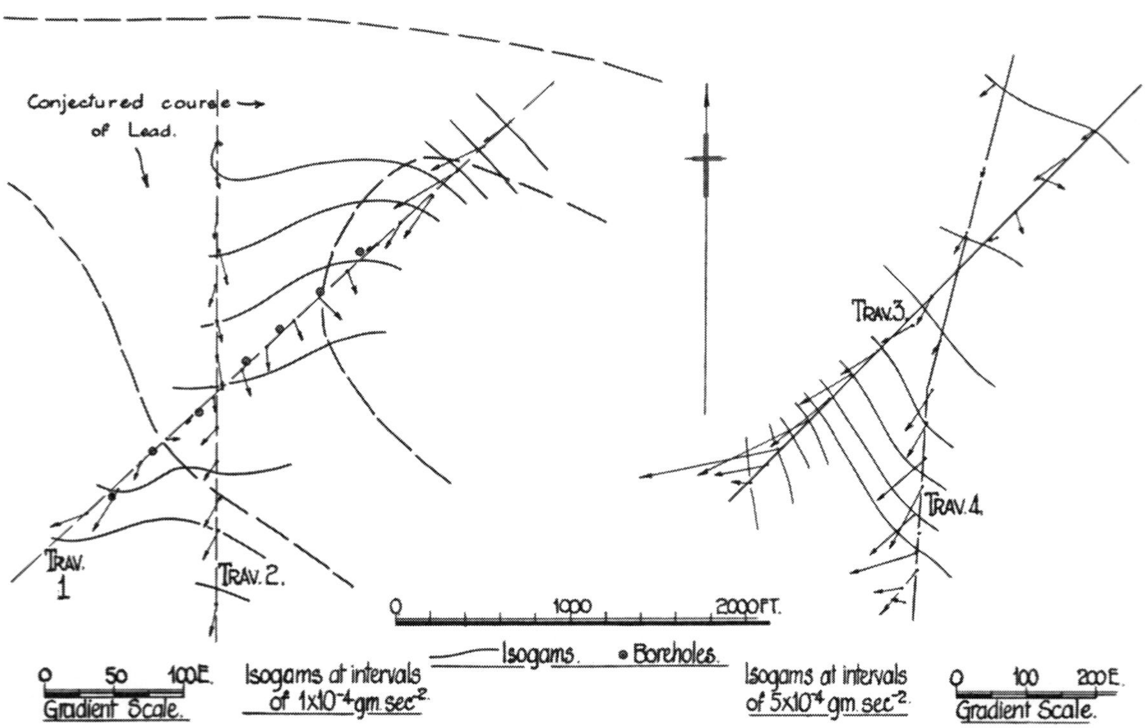

I.G.E.S.

Fig. 119. Gradients and isogams along Traverses Nos. 1 and 2, Gulgong, N.S.W.

Fig. 120. Gradients and isogams along Traverses Nos. 3 and 4, Gulgong, N.S.W.

Creek, which usually is quite dry for the greater part of the year, but becomes impassable for short periods following heavy rain in the hills where it has its source. The total number of stations was 58[a] and the average number per week thus 8·3.

Results. The gradients obtained along Traverses Nos. 1 and 2 are shown in Fig. 119 in which the course of the isogams is also indicated. The values of the gradients along Traverse No. 1 do not agree at all with those that would be expected for the simple cross-section shown in Fig. 170, which is derived from bore-hole records. The interpretation of the results does not appear to be easy, and quantitative deductions are not possible. They seem to indicate that the main lead runs in a direction rather west of north and that a branch lead, with its gutter a short distance farther along the line, enters from an

[a] All except two or three were done with the one gradiometer; the other did not give satisfactory results, probably owing to a defective torsion wire. See p. 172.

easterly direction. In Fig. 119 a possible arrangement of the lead system is sketched very roughly.

Fig. 120 shows the values of the gradients obtained, along the two traverses to the east, and the course of the isogams also is indicated. In this area there are no bores, but the bed-rock is probably granite owing to the proximity of the granite hill to the east. An examination of the gradient plan shows that the granite probably slopes downward towards the north-east and forms a valley, the bottom of which is somewhere near the intersection of the two lines. To the north-east of this point the gradients are small and irregular, though with a general south-westerly tendency. It appears, therefore, that the bottom of the valley is rather wide, or else the north-eastern side slopes upwards very gradually. If this valley is the tributary lead that is supposed to join the main lead near the No. 1 Line of Bores, its bottom cannot be at a greater depth than 260 ft. The components of the gradients along Traverse No. 3 have been plotted in Fig. 121 *a*, in which the curve shows the values of the gradients calculated for the geological structure indicated in cross-section. In this section the granite slopes from a depth of 10 ft. to one of 260 ft. The agreement between the measured and calculated values is fair; in general, the values of the observed gradients are too large. This is possibly related to the fact that the gradients on all four lines have a persistent south-westerly tendency which may probably be due to the effects of some more deep-seated structure being superimposed on those due to the lead systems.

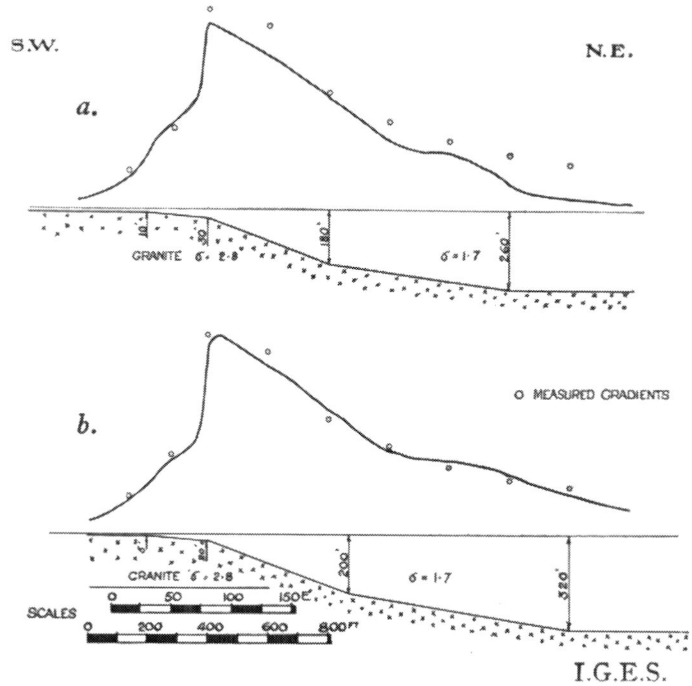

Fig. 121 *a*. Measured and calculated gradients along Traverse No. 3, Gulgong, N.S.W.
b. Recalculated for greater depths

The curve has been calculated for an extreme density difference of 1·1 gm. cm.$^{-3}$ (2·8 was taken for the granite, and 1·7 for the alluvium); thus, there was no justification for assuming a greater value, although the calculated curve could be raised and brought in better agreement with the measured values by doing so. Another solution is possible by assuming a greater depth. This has been done in the calculations of the curve shown in Fig. 121 *b*, where the granite has been assumed to slope from within 5 ft. of the surface to a depth of 320 ft.; the calculated curve and the measured values showing a remarkably good agreement. A source of uncertainty in these calculations is due to the fact that the feature with which one is dealing is not really two-dimensional; it is impossible, therefore, without further field work to discover the error resulting from the assumption that it is.

The most probable interpretation that can be put on these results is that Traverses Nos. 3 and 4 lie across a valley running roughly in a north-westerly direction, and that the floor of this valley is rather wide with no definite "gutter"; or that its north-eastern side slopes very gently. Its south-western edge may lie from 200 to 400 ft. to the south-west of the intersection of the two traverses at a depth of probably more than 250 ft.

This valley, which presumably forms a deep lead, would thus be a direct discovery of this gravimetric survey, if its presence is confirmed.

SECTION 5. CONCLUSIONS

(*a*) **Gradiometer.** The gradiometer, when once it has been set in proper adjustment, has proved itself quite satisfactory for use under Australian field conditions, and day readings taken with it in summer in latitude 32° S. and 150 miles inland were reliable. For instance, during the work at Gulgong, measurements of the gradients at four stations were repeated after an interval of one or more days. The average departure from the mean of the two values for each component was 0·8 *E*.

Comparisons were also made of the values of the gradient obtained with two gradiometers. Unfortunately the second one arrived towards the end of the survey, and in adjusting it the torsion wire was broken; the remaining wires did not behave in a satisfactory manner. The readings of the two instruments were compared on eight stations, and the average departure from the mean of the two values was 2·1 *E*[a].

The main drawback of the gradiometer is the difficulty and time required in making the preliminary adjustments before commencing a survey. Its circular beam requires very careful adjustment in order to swing freely, and in this direction there is room for considerable improvement in the design of the instrument[b].

The behaviour of the torsion wires supplied with the gradiometer has not been satisfactory. At the end of the work at Gelliondale a wire, which had been in use throughout the whole course of the work there, and which apparently was sound, was found to be defective and actually did not behave as an elastic body, i.e. had no definite torsion constant, when loaded with a weight equal to that of the gradiometer beam. Other wires were also found to be unsatisfactory[c]. On account of this behaviour of the torsion wires it was necessary to spend a considerable amount of time in the calibration and examination of the instruments.

[a] The only published example known to the Survey of the comparison of the values given by two instruments is that given by Numerov, *Zeits. f. Geophysik*, vol. 5, pp. 276–289 (1929), where the readings of two small-type Askania instruments were compared at 23 stations; the average departure from the mean of the two values was 1·8 *E*. This work was carried out in winter near Leningrad, where the fluctuation of temperature naturally would be much less than in the above case.

[b] It is understood that in a more recent design of the gradiometer allowance has been made for a very simple adjustment of the "balance" of the beam.

[c] An account of the behaviour of some of the wires is given in Part II, pp. 319–21.

(b) Torsion Balance. The Oertling torsion balance does not appear to be a satisfactory instrument for use under Australian climatic conditions. Its use was discontinued on account of the fact that it was slower in working and less readily transportable than the gradiometer, without having a correspondingly greater accuracy. It could not be used in the daytime, and even at night required an auxiliary protecting hut. The results at the 11 repeated stations showed that the average departure from the mean was 2·2 *E*. The values of the gradients obtained with this instrument were systematically lower than those with the gradiometer. Its behaviour is probably due to inadequate thermal protection of the beam system, so that convection currents have not been sufficiently reduced. Probably a more suitable insulating material than ebonite could be used; compressed granulated cork, for instance, which does not warp, is lighter, and a poorer conductor than ebonite. It is of the utmost importance[a] in the design of any torsion balance that these convection currents should be diminished. In the gradiometer the effects of convection currents are very much reduced by better design.

General Considerations. At present the chief drawback to the gravimetric method is its slowness and corresponding costliness. What is required is a readily portable instrument that will measure a gradient with greater speed. In the majority of cases the degree of accuracy that is claimed for the full-sized torsion balance, viz. 1 *E*, is greater than can be utilised, and in the opinion of the I.G.E.S. an accuracy of 5 *E* is sufficient for a large number of practical cases. For Australian conditions the large instrument is quite unsuitable, on account of the difficulty of transport to remote localities and over rough or thickly wooded country; also in view of its slow rate of working of one station per day. With the gradiometer, on the other hand, three stations per day can be completed working the usual 8-hour shift, and owing to its ready transportability the number of labourers required is less.

The type of problem to which the gravimetric method is suited has already been discussed. The practical value of the method depends of course on the economic consideration as to whether the information obtained is worth the necessarily high cost of a gravimetric survey.

In certain types of problem the gravimetric method has been largely replaced by the seismic method[b]. In the cases studied by the I.G.E.S. however, the former has some obvious advantages over the latter. For instance, at Gelliondale the gravimetric method delimits the coal itself, on account of the difference in its density from that of the surrounding formations, while the seismic method would most likely locate the bed-rock but would not be able to distinguish the coal from the drift. For small scale structural problems of this kind the gravimetric method is preferable, owing to the shortness of the paths for seismic waves and the corresponding difficulty of obtaining the velocities with sufficient accuracy.

[a] This has been stressed by Boys, *Roy. Soc. Phil. Trans.*, vol. 186, pp. 1 et seq. (1895).
[b] See *Geophysical Prospecting 1929*, p. 477, A.I.M.E., New York, 1929.

The gravimetric is the only geophysical method for which there is a complete and exact theory. Instrumental design is, therefore, the main avenue for improvement, and efforts should be specially directed to a reduction in the time and labour required in the measurement of a gradient. If it proves possible to devise an instrument which will measure the gradient of gravity in the vertical direction, with sufficient accuracy, the gravity variometer method would then have further applications to horizontal, or only slightly dipping, formations. With the torsion balance, or gradiometer, it is only the terminations or abrupt truncations of such formations that can be located and investigated in a satisfactory manner.

CHAPTER V

MAGNETIC METHODS

SECTION I. PRINCIPLES AND APPLICATIONS

The Magnetic Elements. The earth's magnetic field varies in direction and intensity from place to place on the earth's surface. It is directed in space, and hence needs three quantities for its complete specification. Those commonly used are its horizontal component H, the declination D, and the inclination I. As is shown in Fig. 122, D is the angle between the magnetic and geographical meridians and I is the angle, measured in the magnetic meridian, which the direction of the field makes with the horizontal[a]. The values of the quantities H, D, and I, at any place are termed its magnetic elements. The vertical component Z is sometimes used in place of one of these elements in specifying the earth's magnetic field. Its value is given by $Z = H \tan I$. The total intensity F is given by $F^2 = H^2+Z^2$. The declination for most places on the earth's surface is between $+20°$ and $-20°$; the inclination in England, Central Europe, and the greater part of the United States of America, is between 60° and 70°, and in Southern Australia between $-60°$ and $-70°$.

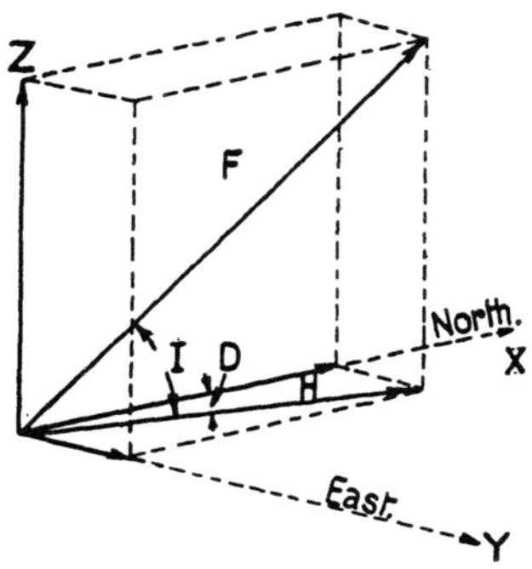

Fig. 122

The total intensity F is generally of the order of half a gauss, and the horizontal component H ranges up to 0·4 gauss; for Southern Australia the vertical component Z is about 0·5 gauss.

At any place the values of the magnetic elements vary with time; there is a very slow periodic alteration called the *secular variation* and a daily fluctuation called the *diurnal variation*. As an example of the former one may cite the value of the declination at London, which was $+11°$ in 1580, gradually decreased and passed through the value zero, attained a negative maximum of $-24°$ in 1815, and is now about $-13°$. The diurnal variation in D and I is of the order of a few minutes of arc, and in intensity usually from 10 to 50γ[b].

In addition there are irregular sudden disturbances known as magnetic storms in which the intensity may vary by a few hundred γ. These sudden changes are then followed by a gradual return to normal which may occupy some days.

[a] D is measured positive when the magnetic meridian is east of the geographical meridian; I is measured positive when the direction of the resultant field is below the horizontal.

[b] $1\gamma = 10^{-5}$ gauss.

Magnetic Anomalies. It is found that the value of the magnetic elements at a station may differ from the average values for the region surrounding the station; this difference being termed the magnetic anomaly at that station. It also requires three quantities for its complete specification.

These magnetic anomalies may extend over large areas (hundreds of square miles) or over limited areas; in the former case they may be regarded as being due to deep-seated causes, and in the latter, with which geophysical prospecting is concerned, to purely local phenomena.

There are three possible explanations of these anomalies:

(*a*) The earth's field inducing magnetism in the rock masses near the surface.

(*b*) The masses themselves may possess permanent magnetism.

(*c*) Magnetic effects due to local earth currents.

(*a*) The magnetic field within a body of magnetic material in the earth's field is greater than the earth's field by a factor μ which is termed the permeability of the substance. For most rocks and minerals μ is very little different from unity. Another quantity frequently used in describing the magnetic properties of a substance is the susceptibility κ which is connected with μ by the expression

$$\mu = 1+4\pi\kappa.$$

For highly magnetic materials such as metallic iron, magnetite, pyrrhotite, etc., the susceptibility κ is not a constant, but depends on the strength of the magnetic field which is used in making the measurements and on the previous magnetic history of the material.

The figures quoted by various writers[a] have in many cases been for fields some hundreds of times as intense as that of the earth. This partly accounts for the great range of values given for the same substance. Even then the values depend largely on the mode of occurrence, state of sub-division, etc., of the specimen. The matter has been discussed by Slichter[b], and by Stearn[a], and it appears that the susceptibility of rocks is mainly accounted for by their magnetite, and to a less extent, by their ilmenite content. Slichter[b] gives the following values for the susceptibilities of the more strongly magnetic minerals for a field strength approximately the same as that of the earth.

Material	Value of susceptibility $\times 10^6$
Magnetite	300,000 to 800,000
Pyrrhotite	28,000
Ilmenite	44,000
Specular haematite	4,000

[a] Ambronn, *Elements of Geophysics*, p. 94, McGraw-Hill, New York (1928). Haalck, *Die magnetischen Verfahren der angewandten Geophysik*, pp. 98–100, Borntraeger, Berlin (1927). Stearn, *Geophysical Prospecting 1929*, p. 327, A.I.M.E., New York (1929). Bartels, *Lehrbuch der Geophysik* (*Gutenberg*), p. 565, Borntraeger, Berlin (1928), etc.

[b] *Geophysical Prospecting 1929*, pp. 239–250, 342–343, A.I.M.E., New York (1929).

The figures for certain other rocks and minerals may be summarised as follows:

Material	Value of susceptibility × 10^6
Igneous rocks	200 to 7000
Non-ferrous minerals[a]	400 to −2
Sedimentary rocks, about	40
Clay, gravel, etc.	10 to 30
Limestone, dolomite, chalk	0 to 10
Coal	1
Rock salt[a]	−1
Graphite[a]	−8

The large range of values given for igneous rocks may probably be ascribed to their varying magnetite content.

If the rock mass has a greater permeability than its surroundings, the magnetic lines of force are bent in towards it, and become closer together within it. The state of affairs for such a body in the southern hemisphere is shown diagrammatically in Fig. 123. Above the upper end of the body the lines of force are closer together and thus the field is increased in intensity. The upper end where the lines of force leave the body will be a north-seeking pole[b], and the lower end a south-seeking pole. With an earth's field of high inclination, as shown in Fig. 123, the main increase will be in the vertical intensity Z. We thus say that there is a positive anomaly in Z above the upper end. The effect of the bending of the lines of force is to make them farther apart above the southern end where a corresponding but smaller negative anomaly occurs. The relative size and location of the anomalies depend, amongst other things, on the dip of the body. In Fig. 124 are shown the anomalies in vertical intensity obtained in north-south traverses[c] across sheets of iron with different angles of slope, and lying in an east-west direction.

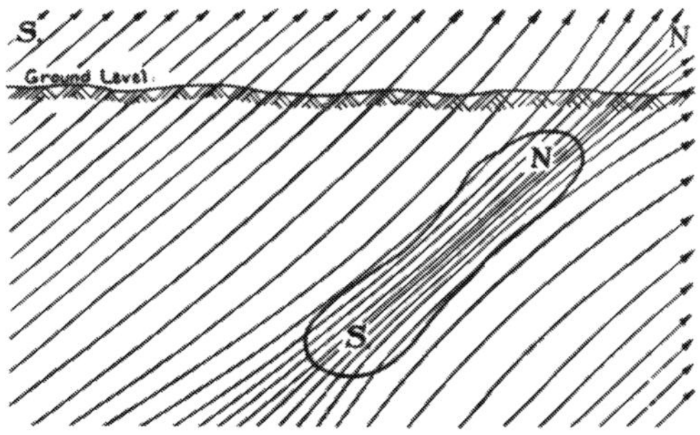

I.G.E.S.

Fig. 123. Diagram showing distortion of earth's field due to magnetic body

(*b*) A body of magnetic material could retain original permanent magnetism of its own, which would correspond in direction and intensity with that induced in it by the earth's field, provided the body had remained in its original position. If it has been subsequently displaced, then its magnetic effect would be different from the one it would have due to induction. Cases of this type apparently occur[d], although it is

[a] The negative sign denotes diamagnetic properties. The magnetic field inside a diamagnetic body placed in the earth's field is less than the earth's field.

[b] The north-seeking pole is usually called a north pole; the earth's north-magnetic pole, towards which a compass needle points, is thus of opposite polarity.

[c] Experimental work done by the I.G.E.S. at Gelliondale, Victoria.

[d] Heiland, Henderson, and Malkovsky, *U.S. Bureau of Mines Tech. Paper 439*, pp. 41–45 (1929). H. Reich, "Erdmagnetismus und glaziales Diluvium," *Jahrb. d. preuss. geol. Landesanst.*, p. 259 (1925).

difficult to see how such bodies could retain their magnetism when subject for long periods of time to the demagnetizing action of the earth's field. The matter has been fully discussed by Haalck[a], who considers that, with a better knowledge of all the conditions, the apparently abnormal negative anomalies which are attributed to this cause could be explained in terms of the magnetism induced by the earth's field. He concludes that the magnetism of rock-formations always corresponds to that due to the induction of the normal earth's field.

(*c*) Local earth currents will cause magnetic anomalies, but the currents that generally occur are usually so small that they do not cause anomalies of more than a few γ.

Haalck[b] divides magnetic anomalies into four classes:

Class	Range	Rocks
1	10,000 to 200,000γ	Large deposits of magnetite, usually of commercial importance
2	1,000 to 10,000γ	Large masses of volcanic and crystalline rock, rich in magnetite
3	100 to 1000γ	Large masses of the above, poor in magnetite
4	Up to 200γ	Sedimentary formations

Examples of the first class are the very large anomalies occurring at Kursk[c] in Russia, Kirunavaare[d] in Sweden, and at Caribou[e] in U.S.A. Cases of the other three classes have been studied by the I.G.E.S. field parties.

Calculated anomalies. The calculation of magnetic anomalies due to various rock masses is not a simple problem, and has been done for only a few cases, on

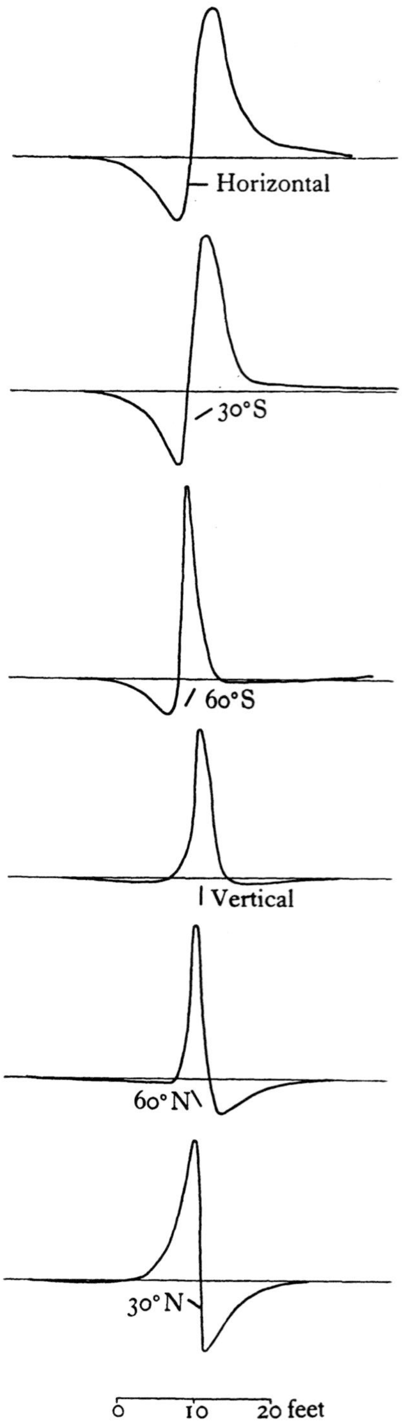

Fig. 124. Diagram showing the anomaly in the vertical magnetic field obtained in N.S. traverses across rectangular sheets of iron with long axes in an E.W. direction

[a] *Die magnetischen Verfahren der angewandten Geophysik*, pp. 41–45; *Gerlands Beitr. z. Geophysik*, vol. 22, pp. 385–389 (1929).

[b] *Die magnetischen Verfahren der angewandten Geophysik*, p. 108.

[c] Lasareff, "Die Kursker magnetische Anomalie," *Gerlands Beitr. z. Geophysik*, vol. 15, pp. 71–89, 97–102 (1926).

[d] von Carlheim-Gyllenskiöld, *A brief account of a magnetic survey of the iron ore field at Kirunavaare*, Stockholm (1910).

[e] Heiland, Henderson, and Malkovsky, *loc. cit.*

the assumptions that the magnetism is caused by the induction of the earth's field, that the mass consists of material of uniform susceptibility, and that it is uniformly magnetized. The matter has been dealt with by Haalck[a] who gives formulae for an ellipsoid of rotation, an infinite cylinder, and a sphere. The diagrams in Fig. 125, taken from Haalck, show the variation of the horizontal and vertical anomalies above these and some other bodies. The converse problem to the above, i.e. to calculate the size and shape of a magnetic mass from the measured anomalies, in general has not been solved, but as a rule it is fairly easy to make useful deductions by comparison of the observed anomalies with those due to ideal cases, such as those shown below.

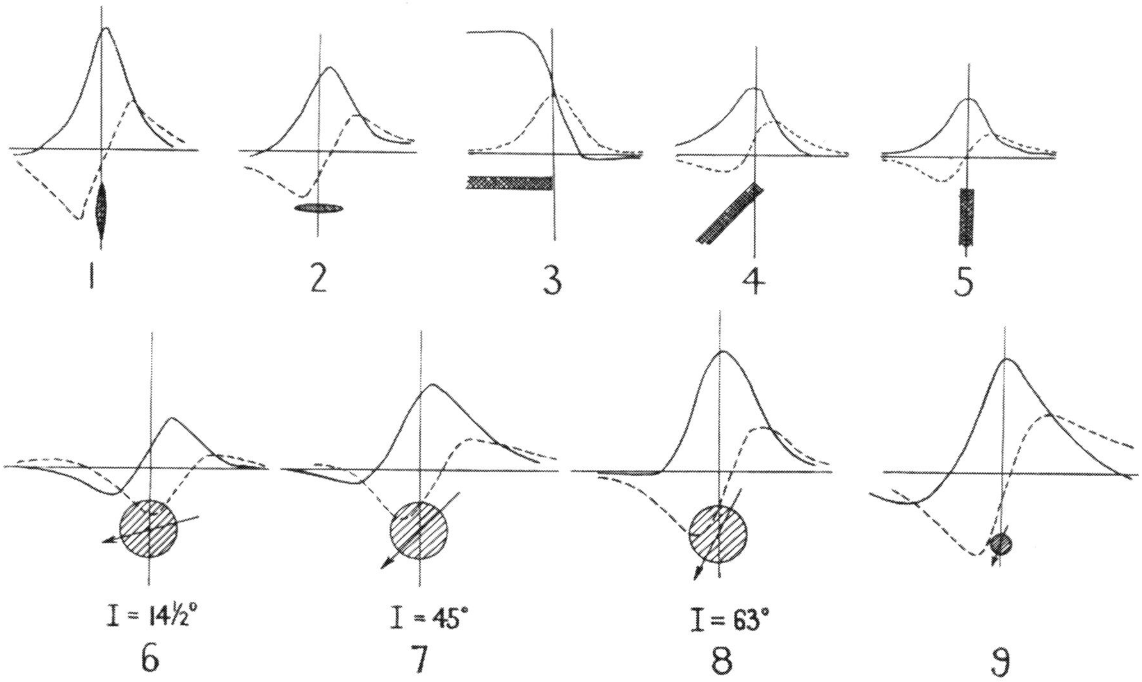

Fig. 125. Diagram showing calculated anomalies in vertical and horizontal intensity for traverses made over various magnetic bodies

It has already been pointed out that three quantities are needed for the complete specification of an anomaly. In many cases, however, very useful information can be obtained merely from determinations of the vertical anomalies, and if measurements of one component only are made the vertical is the one usually chosen. The interpretation of observed curves is easier for vertical than for horizontal intensity, because for the former the maximum is usually over the edge of the mass concerned (see Fig. 125). For instance, in the case of a vertical dyke (No. 5), the curve of vertical intensity has a well-marked maximum directly above the top of the dyke, while that for the horizontal intensity has a maximum and a minimum, which occur some distance to the sides of it. These remarks apply where the value of the inclination I is high, but as I decreases the horizontal anomalies become more important as will be seen from Nos. 6, 7, and 8, which

[a] *Die magnetischen Verfahren der angewandten Geophysik*, pp. 47–64.

represent the anomalies for different values of I. Thus in areas where I is low measurements of the horizontal anomalies may be more useful than those of the vertical ones.

Geological applications[a]. The most obvious economic application of the method is for the direct location of deposits of highly magnetic ores, such as magnetite and pyrrhotite, and for this purpose the method has long been used in Sweden[b]; in some cases with striking success. In these instances quite simple instruments can often be used. The sedimentary iron ores, such as spathic iron ore, bog iron ore, etc., and the majority of the non-ferrous minerals have relatively low susceptibilities and are not readily located directly by the magnetic method. The method, however, has many indirect applications, and frequently yields information of value in problems dealing with the extent and distribution of igneous rocks. In the case of sedimentary rocks the method is of less importance owing to their low susceptibility, but appreciable magnetic anomalies are often found in association with major geological features[c].

There are important applications in Australia in connection with the sub-basaltic tin and gold deep-leads, where the basalt flows are often concealed by alluvial deposits. The course of a lead can sometimes be traced by studying the magnetic anomaly due to the basalt that fills it[d]. In America the method has been of use in the investigation of alluvial deposits of gold and platinum, where concentrations of these minerals are sometimes associated with magnetite.

It is only in a very few cases that quantitative interpretations of the results of a magnetic survey can be made with any degree of certainty, owing to the large number of factors involved. The method can, however, be applied qualitatively in a very large variety of geological problems, and nearly always contributes some information of value towards their solution. The work can be carried out rapidly and cheaply and the instruments required do not need a highly trained specialist for their use. It is surprising that its importance as an auxiliary to geological survey work has received so little attention until within the last few years[e].

SECTION 2. INSTRUMENTS USED BY SURVEY

The I.G.E.S. used two Schmidt vertical magnetic balances manufactured by Askania Werke A.G. of Berlin, and also a Thalen-Tiberg magnetometer, made by Berg of Stockholm.

The Schmidt Vertical Balance. This instrument is illustrated in Fig. 126. The beam, Fig. 127, which rests on two quartz bearings fixed to the inside of the outer case, consists essentially of two parallel bar magnets of tungsten steel attached to a quartz knife edge, the direction of the knife edge being at right angles to the axis of the magnets. The

[a] An account of the application of the method in America is given by Heiland, *Colorado School of Mines Quarterly*, vol. 24, pp. 53–72 (1929).

[b] An interesting historical account is given by Lundberg, *Mining Magazine*, vol. 41, pp. 73–76 (1929).

[c] An example of this is afforded by the magnetic work at Gelliondale, Victoria, see p. 186.

[d] The survey of the gold deep-leads at Gulgong, N.S.W., described on p. 188, affords an example of this work.

[e] It is now being employed by the Geological Survey of Great Britain.

centre of gravity in this system is adjusted to be at a small distance below and to one side of the knife edge[a]. Suppose the knife edge to rest on a level surface and the magnets to oscillate in a vertical plane at right angles to the magnetic meridian. For this position the horizontal component has no effect on the magnets, but the vertical component will turn them about the horizontal axis. This couple is balanced by another couple due to the weight of the beam acting at its centre of gravity, which tends to turn the beam in the opposite direction. The position of the centre of gravity of the beam is adjusted by means of horizontal screws attached to it, so that the axis of the system is approximately horizontal when in use.

The sensitivity of the balance, which also depends on the position of the centre of gravity, can be varied by adjusting the vertical screw attached to the beam.

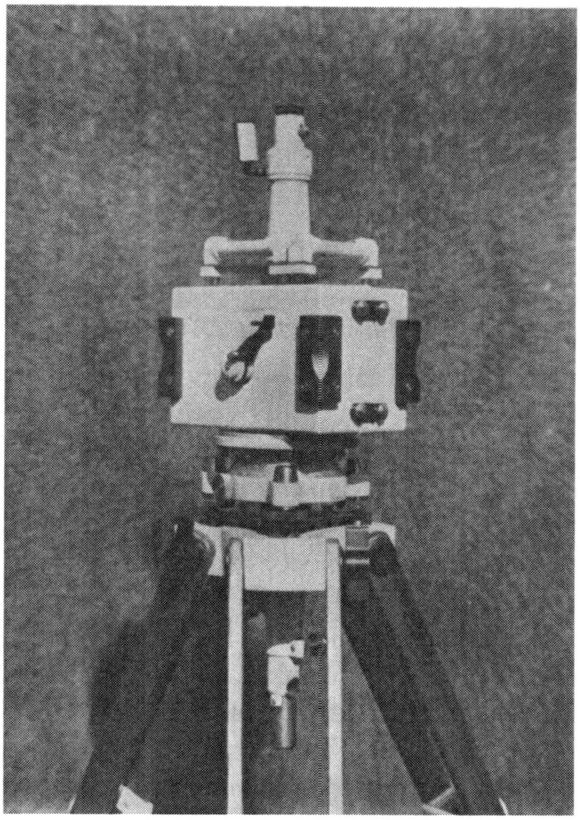

Fig. 126. The Schmidt vertical balance

The optical system is shown diagrammatically in Fig. 128. A fixed transparent scale is placed at the focal plane of an object lens, and is illuminated by light from the sky. The observer views this scale through the eyepiece, and also sees a parallel image of it formed by reflection at the mirror attached to the beam system. As the beam deflects this image is displaced with respect to the fixed scale. The difference between the scale readings at any two points is directly proportional to the difference in vertical intensity[b].

Fig. 127. Magnetic system of Schmidt vertical balance

Artificial cork separates an inner and outer case to protect the beam from rapid changes in temperature. Provision is made for levelling the instrument, for its rotation about a vertical axis through known angles, and for the attachment of auxiliary deflecting magnets for extending the range of the balance and for calibration purposes.

[a] For use in the southern hemisphere the centre of gravity lies on the north polar side of the knife edge.

[b] i.e. $s - s_0 = \frac{\Delta Z}{\epsilon}$, where ϵ is the scale value. See Part II, p. 323 for theory.

The Thalen-Tiberg Magnetometer. This instrument (Fig. 129) functions as a deflection magnetometer for measuring horizontal intensities, and as a magnetic balance for measuring vertical intensities. It consists essentially of a magnetometer M, which can be turned about a horizontal axis and placed in either a horizontal or vertical plane; the magnetic system being a magnetized needle carrying a sliding weight and pivoted at its centre between two hardened steel bearings to allow free movement in one plane. By altering the position of the sliding weight, the needle can be adjusted to balance horizontally in the field of survey. A fixed graduated horizontal arm A is provided for holding a deflecting magnet of known moment at any desired distance from the pivot. By placing this arm at right angles to the meridian and attaching the deflecting magnet, the needle is deflected to a position where the turning moment due to the magnet counterbalances that due to the earth's field. By observing the angle of deflection and knowing the moment of the magnet and its distance, the horizontal intensity of the earth's field can be calculated by the tangent formula[a]. Sights S S are attached to the fixed arm, and by placing them in the geographic meridian the compass needle gives the declination. The instrument can be accurately levelled and turned about a vertical axis through known angles and is supported on a light tripod.

By turning the instrument about its vertical axis until the magnet and needle are mutually perpendicular, the horizontal intensity can be calculated from the sine formula[b], by observing the deflection of the needle as it returns to the meridian after removing the magnet. A second detachable arm B is provided for modifying the sine method to make it more rapid. As both these methods determine the total horizontal field, the normal horizontal field must be subtracted vectorially in order to obtain the value of the anomaly.

For measuring anomalies in vertical intensity, the magnetometer M is placed in a vertical plane at right angles to the magnetic meridian, and is balanced to lie horizontally in the normal earth's field. In any other field the tangent of the deflection from the horizontal is proportional to the anomaly in vertical intensity. The principle of the method is exactly the same as for other balances; the deflection of the needle, however, is observed directly, as for an ordinary compass needle. The sensitivity, which of course is much less than that of the Schmidt balance, can be varied in the same way by altering the position of the centre of gravity with respect to the axis of suspension.

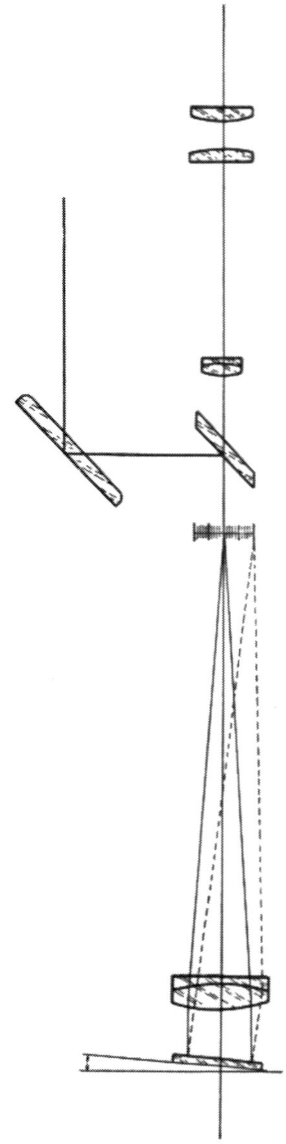

Fig. 128. Optical System of Schmidt vertical balance

[a] $H=\frac{2M}{r^3 \tan\theta}$ approx., see p. 324 for theory.

[b] $H=\frac{2M}{r^3 \sin\theta}$ approx., see p. 325 for theory.

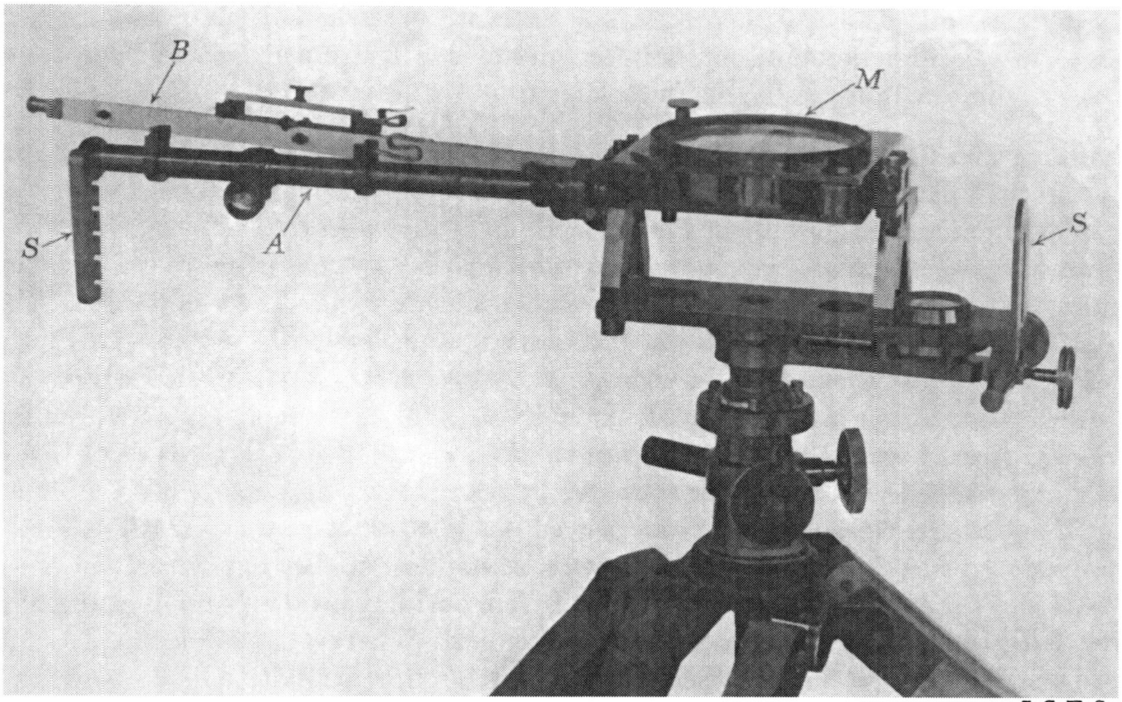

Fig. 129. Thalen-Tiberg magnetometer

SECTION 3. FIELD PROCEDURE

Nature of the Problem. Magnetic anomalies may be roughly classified into three groups, for each of which a slightly different method of attack should be adopted.

(1) Feeble anomalies up to about 200γ. All precautions should be taken to obtain the greatest possible accuracy in their measurement and corrections made for diurnal changes in magnetic intensity. These corrections range from about 10γ to 20γ.

(2) Anomalies from 200γ to 2000γ. The correction for diurnal variation can generally be neglected for this group.

(3) Anomalies greater than about 2000γ. For vertical fields of this magnitude it becomes inconvenient to use the Schmidt magnetic balance[a]. Deposits of magnetite and some other minerals are capable of producing magnetic anomalies much greater than 2000γ, and in these cases it is advisable to use a less sensitive type of instrument, such as the Thalen-Tiberg magnetometer.

SCHMIDT VERTICAL FIELD BALANCE

Personnel and Equipment. The I.G.E.S. party usually consisted of one observer and one assistant per instrument. When corrections for diurnal variation are necessary, another balance has to be set up at some suitable base and either read at intervals by an extra observer, or else it may be fitted with an attachment for photographic re-

[a] See, however, Part II, p. 326.

cording. In this case two instruments at least are necessary. If there is a magnetic observatory within a radius of, say 100 miles, one instrument is sufficient, as the necessary information can be obtained later from the observatory[a].

Planning the Survey. If available, a geological map of the district will prove of great assistance in planning the survey, especially when some information about the approximate position and magnitude of the magnetic body is available. As a rough guide, the distance between stations for preliminary work can be taken as equal to the estimated width of the body or to the thickness of the overburden; whichever is greater. When these particulars are not known the maximum spacing has to be determined by trial and error. When a considerable change in magnetic field takes place between two stations observations are made at closer intervals. For example, at Gulgong, the preliminary spacing was 200 ft., but this was reduced to $12\frac{1}{2}$ ft. at several places of rapidly changing magnetic intensity. The traverses in general were not less than 400 ft. apart.

The traverse lines may either be pegged out in advance so that the whole area of the proposed survey is covered by a network of stations, or they may be set out one at a time and the results of each line studied before deciding on the position of the next. The latter method is sometimes more economical. Wherever possible the I.G.E.S. used parallel traverse lines, either directed north-south, or east-west, or both, as it simplified the survey and drawing-office work. Where a magnetic survey is being carried out in conjunction with another geophysical method the same survey lines will often do for both.

Observations are made at successive stations along the traverse by levelling and orientating the balance, and taking six readings of the scale, while a note is made of the time and temperature. The manipulation of the Schmidt balance is fully described in the directions issued by the makers[b]. Even if every precaution is taken to avoid shocks to the instrument during its transport in the field, the centre of gravity of the beam system is liable to be displaced slightly, but the error thus caused[c] can be partly detected by repeating the first station on the completion of each traverse. The discrepancy thus obtained also includes errors due to diurnal variations in vertical intensity[d] and uncertainties in the temperature correction, but if small enough it can be distributed evenly among the various stations comprising the traverse; otherwise the traverse should be repeated as the error may have been introduced suddenly. Where the area of the proposed survey is small, a single base station is chosen and observations are

[a] Observations of diurnal variation made at Gelliondale gave values agreeing to within a few γ with those obtained from Toolangi Observatory, about 100 miles away.

[b] *Directions for the Use of the Vertical and Horizontal Field Balance*, p. 8, Askania Werke A.G. (1927). See also, Part II, p. 326, where several modifications which shorten the routine work considerably are discussed.

[c] A horizontal displacement of 10^{-6} cm. of the centre of gravity of the beam system causes an error of 1/5th scale division, i.e. 6γ if $\epsilon = 30\gamma$ per division. This can be calculated from the formula

$$\frac{S-S_0}{2f} = \frac{MZ-mga}{mgd}.$$

See Part II, p. 323.

[d] When they are not separately allowed for.

made on it before and after each traverse. For large areas, however, it is better to use several such bases and to make traverses between them.

Rate of Work. The speed at which the field measurements can be made depends upon the weather, type of country, the accuracy required, and the distance between stations. Where no great accuracy in reading is required as many as 13 stations, excluding repetitions on the base, have been measured in an hour. With 100 ft. spacing, perhaps the maximum possible rate with present methods is 16 stations an hour. When it is necessary to take six readings per station, which is the case for feeble anomalies such as those caused by sedimentary rocks, the rate is reduced to eight or ten stations per hour depending upon spacing, vegetation, and topography. As is explained in the makers' instructions, the observations require to be corrected for temperature, diurnal variation, and progressive changes in the instrument.

THALEN-TIBERG MAGNETOMETER

The use of an instrument of this type is restricted to the investigation of anomalies coming under group (3). The I.G.E.S. used a Thalen-Tiberg magnetometer for the measurement of both horizontal and vertical intensities. For the former determination, either the sine or tangent method may be used, the field procedure for each being similar. The magnetometer is set up with the magnetic needle in the meridian and a deflection is produced by means of an auxiliary magnet placed on the sine or tangent arm of the instrument, the amount of the deflection and the distance of the auxiliary magnet are recorded, and from them the horizontal intensity is computed. For the determination of vertical intensities the instrument is used as a vertical balance, the angle of inclination to the horizontal at each station being recorded. The magnetic needle is provided with a counterbalance, which must be so adjusted that the needle is horizontal when the instrument is in an undisturbed part of the earth's magnetic field. A simple calculation gives the vertical intensity from the angle of inclination[a].

The field procedure for measuring vertical intensities only differs in detail from that described for the Schmidt field balance. The observer can usually dispense with the services of an assistant, and base station readings are only taken in order to detect possible mistakes. In a region of large magnetic anomalies it will not be possible to use a prismatic compass for the preliminary topographical survey, and in rough unmapped country a theodolite would be required.

When using this magnetometer for measurements of vertical intensity only, the usual rate of progress was some ten or twelve stations per hour. As explained on p. 327 certain improvements in the design of the instrument should enable the field procedure to be simplified, with a corresponding increase in the speed of working.

[a] $\Delta Z = K \tan \theta$, see Part II, p. 324.

SECTION 4. MAGNETIC SURVEYS IN AUSTRALIA

MAGNETIC WORK AT GELLIONDALE[a], VICTORIA

The magnetic work in this field was subordinated to the gravimetric survey and was only carried out when an observer and an assistant could be spared from the latter work. Measurements were made on 32 days and as a rule only one instrument was used. The total number of stations was 470, and the usual rate of work for stations up to 400 ft. apart was six per hour. The work was carried out along a series of north-south traverses, in general making use of those already set out for the gravimetric stations. Finally, an east and west traverse was made in order to link up all the individual base stations. The field observations were corrected in the usual manner for temperature, diurnal variation, and change of zero. For the diurnal variations, figures obtained from the magnetic observatory at Toolangi, 100 miles distant from Gelliondale, were used.

In Fig. 130 are shown the lines of equal magnetic anomaly referred to the value of the vertical component at some arbitrary point in the area. In one part of the field the magnetic disturbances are very marked (up to 1000γ), and the iso-anomaly lines cannot be drawn at intervals of 20γ as they are on the rest of the plan, but are shown at larger intervals of 100γ, 300γ, etc. The values of

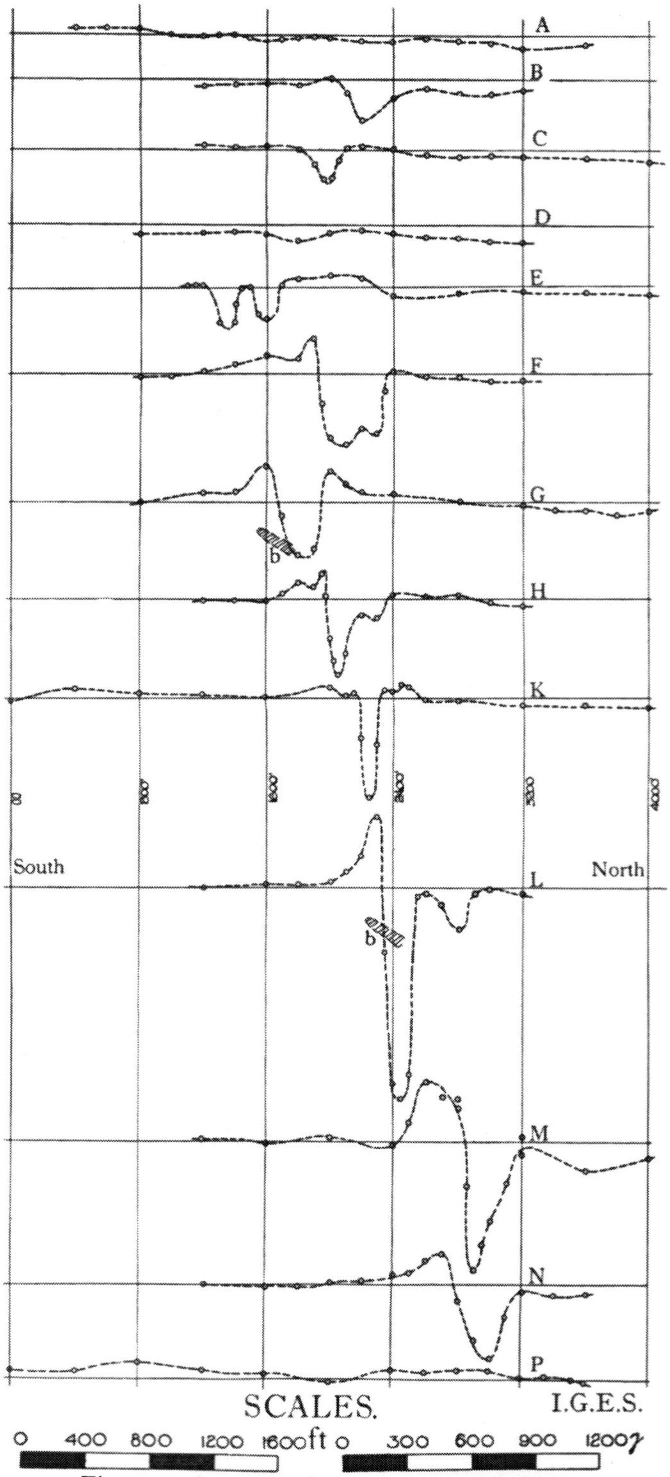

Fig. 131. Anomalies in vertical magnetic field on traverses *A–P*, Gelliondale, Victoria

[a] For geology, etc., see p. 154.

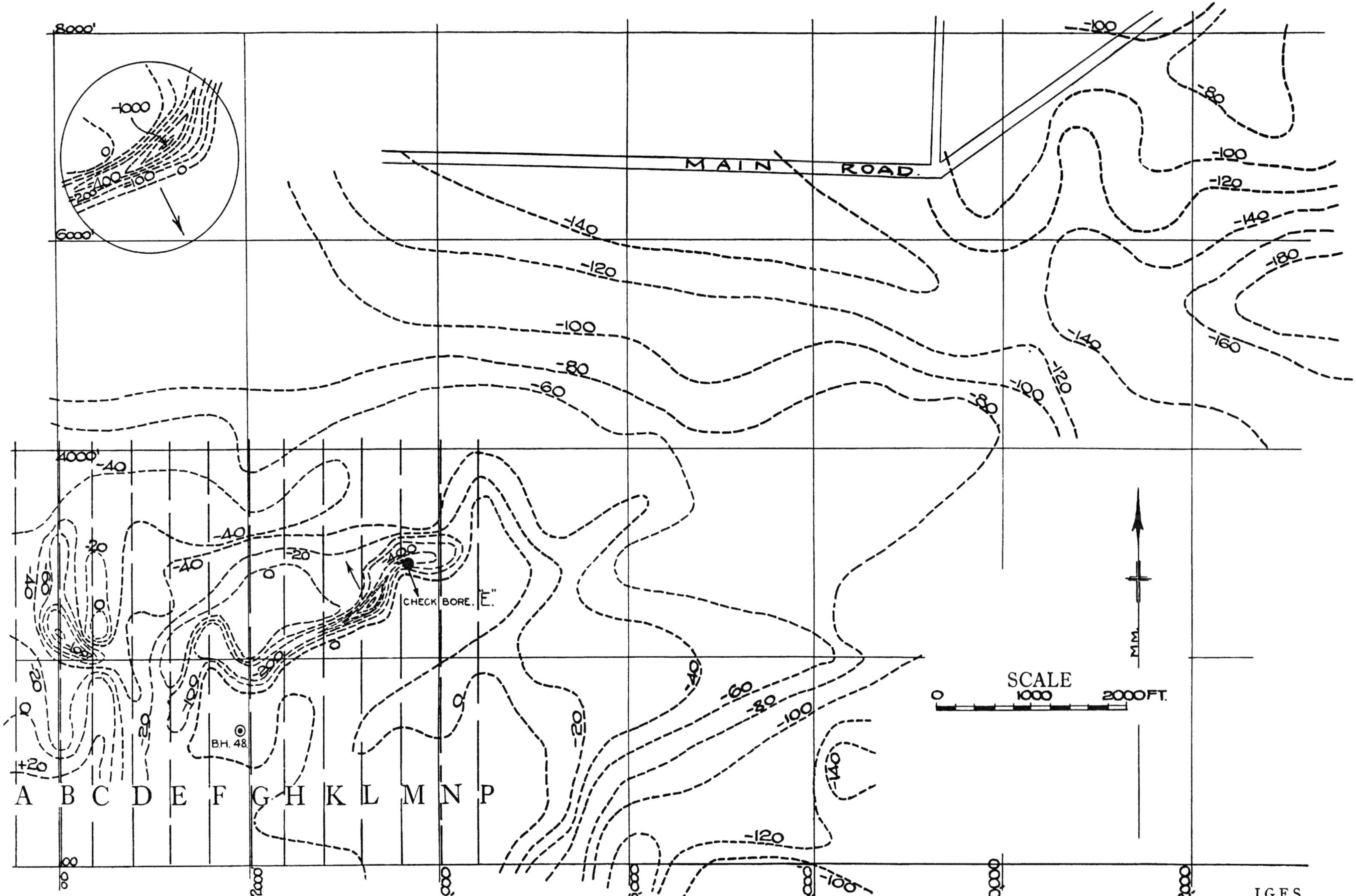

Fig. 130. Lines of equal magnetic anomaly in the vertical field, at intervals of 20 γ, at Gelliondale, Victoria

the anomaly are plotted along a series of traverses in Fig. 131. The curves are all similar except *E* and *F*, where as will be seen from Fig. 130 there is probably a break or some irregularity in the body which is causing the disturbance. Magnetic anomalies of the above magnitude cannot be ascribed to sedimentary rocks or to brown coal, but must be due to a much more strongly magnetic material.

As there are two outcrops of basalt in the vicinity, one 3 miles to the north and the other 4 miles to the north-east, it is not unreasonable to ascribe these anomalies to a basaltic intrusion, although there is no surface evidence of it in this locality. The most notable feature in these traverses is that in all cases the negative anomaly is greater than the positive one, which in traverses *B*, *C*, *E*, *H* and *K* is practically non-existent. In order to investigate these features experimentally, some north and south traverses were made over thin sheets of iron placed at different angles of slope. The results are shown in Fig. 124, where it will be seen that the curve for the sheet sloping at an angle of 30° to the north is closest to those for traverses *L* and *M* in Fig. 130. The suggestion is therefore made that the cause of the magnetic anomalies is a basaltic dyke[a] dipping to the north-west at an angle of about 30°, as indicated for example in traverses *G* and *L*. The specific gravity of basalt is considerably higher[b] than that of brown coal, drift, or the bed-rock, so that it is reasonable to expect that its presence would be indicated on the gradient plan. Such a dyke as is indicated above would cause appreciable north-westerly gradients for a distance of two or three hundred feet on each side of where it approached closest to the surface, but would cause no southerly gradient of any magnitude[c]. In the gradient plan, Fig. 111, there is one moderately large northerly gradient, 300 ft. to the north of Bore-hole No. 48, at a station where the maximum positive magnetic anomaly on traverse *G* was obtained. In addition the gradients on the westernmost line of stations in Fig. 111 all have an easterly tendency. This fits in with the small values of the magnetic anomalies on traverse *A*, which is probably due to the fact that west of traverse *B* the dyke becomes either non-existent or very narrow, or is at a relatively great depth.

The gradients measured in this area are not very numerous, since the bore-holes had shown that there was no coal. Nevertheless, the values of those that have been measured are not incompatible with the existence of a dyke.

The relation between the magnetic and gravimetric results in this area has already been discussed on p. 159.

[a] In order to verify this conclusion, a bore *E* was put down by the State Mines Department after the close of the Survey in 1930. It was located on line *M*, see Figs. 111, 112 and 130. In this bore, solid basalt was struck at 9 ft. from the surface and the bore was still in this rock when it was discontinued at a depth of 14 ft. (See p. 165.)

[b] The value is generally given as ranging from 2·7 to 3·3.

[c] See typical sections over sloping dykes, e.g. Fig. 99, p. 142, and Jung, *Zeits. f. Geophysik*, vol. 3, p. 264 (1927).

MAGNETIC SURVEY AT GULGONG, N.S.W.

Geology and nature of the problem. This survey was carried out over an area of about 1100 acres of flat, thinly wooded country, lying about 2 miles north of the town of Gulgong[a]. The history[b] of this field dates back to the year 1870, when the first important discoveries of gold were made. During the following seven years nearly half a million ounces of alluvial gold were won from a series of leads, of which the Black Lead was the most important producer. This lead is one of a group of tributaries belonging to an old river system, possibly of tertiary age, which now lies buried beneath a thick sheet of alluvium. Its position, together with the conjectured courses of the buried channels, are shown in Fig. 132. The heads of the Black Lead and Happy Valley Lead lie on high ground where the *gutters* actually extend to the surface and were worked without difficulty as *shallow leads*. On following them to the north the depths of overburden increased rapidly, and with the sinking of shafts and the adoption of underground methods of mining they became *deep leads* in the true sense of the term. When the workings reached the point now marked by the railway line (Fig. 132), an additional difficulty was encountered in the form of hard basalt, which covers and fills the whole deep lead system, and gradually increases in thickness in a northerly direction. This basalt was derived from a seat of volcanic activity lying away to the north of the area under consideration. The molten lava is supposed to have entered the Gulgong field by means of the Stubbo Lead, and to have welled back up the main lead and its tributaries for a mile or more, in addition to flowing westwards, down stream, for nearly 20 miles.

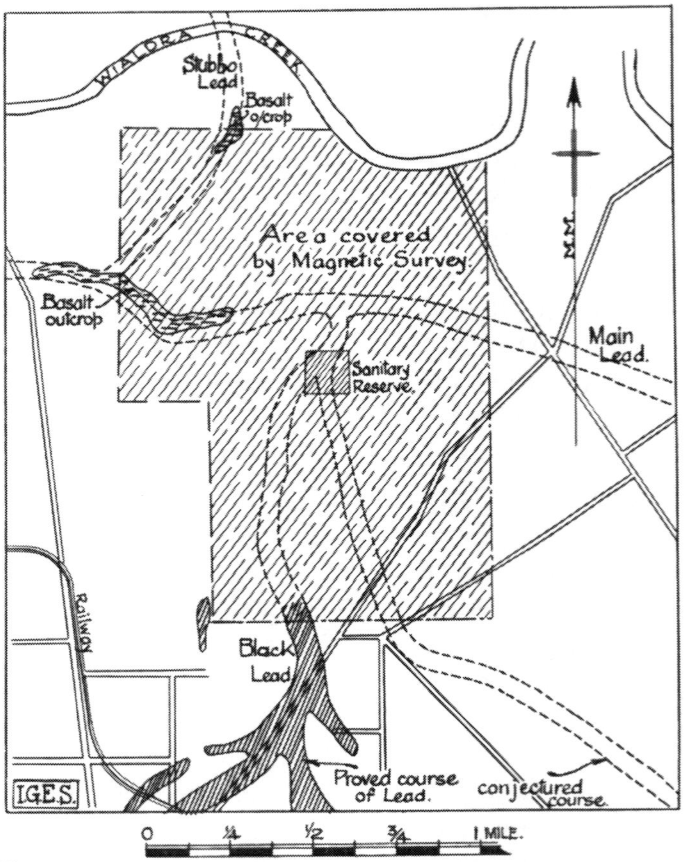

Fig. 132. Locality plan of magnetic survey at Gulgong, N.S.W.

[a] Other remarks concerning the geology of the Gulgong district are given on p. 224.

[b] These geological and historical details have been obtained from an unpublished report by L. J. Jones, Esq., Geological Surveyor to the Department of Mines, N.S.W., by kind permission of the Department.

In spite of the heavy costs of sinking through the basalt and of dealing with large volumes of water and running sand, the leads were followed and worked out to the point indicated in Fig. 132 by hatching. At and beyond this point difficulties were experienced in following the lead, and after several attempts to locate its position all work was abandoned, largely on account of the prohibitive cost of sinking exploration shafts through such difficult ground. Owing to the fact that good gold values were being obtained at this point, where work was abandoned, there have been several later enterprises, with a view to locating and testing the leads, beneath the basalt, in the unknown ground lying to the north. The two lines of bores and the shafts shown in Fig. 135 were put down with this object, but did not meet with any practical success. Excepting the outcrop areas shown in Fig. 135, the basalt flows are entirely concealed from view beneath later alluvial deposits. In the No. 1 line of bores the basalt was found at an average depth of 30 ft., and in the No. 2 line at 40 ft. The thickness of basalt was about 100 ft.

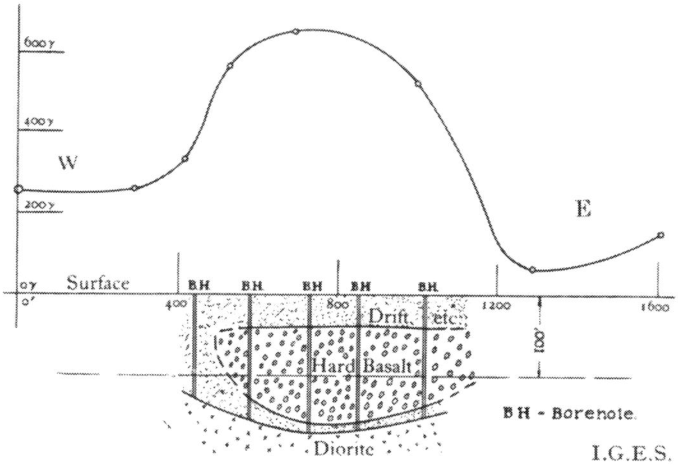

Fig. 133. Section showing analogy between the anomaly curve of vertical intensity and the thickness of basalt along traverse *C* (No. 2 line of bores), Gulgong, N.S.W.

Reason for Magnetic Survey. This section of the Gulgong Field was recommended by the Government Geologist of New South Wales as being one of considerable potential importance, in which any geophysical evidence as to the positions of the deep leads would be of material value. After a preliminary examination of the ground it was decided that a magnetic survey should be carried out. The magnetic susceptibility of basalt being higher than that of the bed-rock (granite and diorite) and alluvial formations, it was thought that the delineation of the basalt flows would be practicable. Since these are largely restricted to the courses of the buried channels, the general disposition of the deep lead system should be revealed in this way, and possibly the actual position of the gutters could be deduced in some instances from the changes in magnetic intensity along traverses made across the course of the basalt flows.

Personnel and rate of work. The magnetic observations were made with one Schmidt magnetometer and the whole of the work, including the setting out of the lines of traverse, was carried out by an observer and one assistant. The work was done concurrently with a gravimetric survey about 10 miles away, in which this personnel also took part. Altogether 23 days or parts of a day were required for the 430 stations; the average rate was ten[a] and the maximum rate sixteen stations per hour. The weather

[a] This is the mean for various observers; the two who did most of the work averaged 8·7 and 12·6 stations per hour respectively.

was calm and sunny; the temperature recorded inside the instrument being about 85° as a rule but with extreme values of 63° and 102°. The location of the traverses is shown in Fig. 135, where the positions of the magnetometer stations are denoted by dots[a].

Results. In Fig. 133, which shows the change in value of the vertical intensity along the line of traverse *C*, a very marked resemblance will be observed between the magnetic curve and the curve representing the lower surface of the basalt flow, as obtained from the No. 2 line of bores. It will be seen that the edge of the basalt corresponds roughly with the region of greatest rate of change in magnetic intensity, and that a positive anomaly occurs over the basalt flow. The boundaries of the basalt, shown in Fig. 135, were obtained from the magnetic curves along the various traverses by applying the above relationship as an empirical rule.

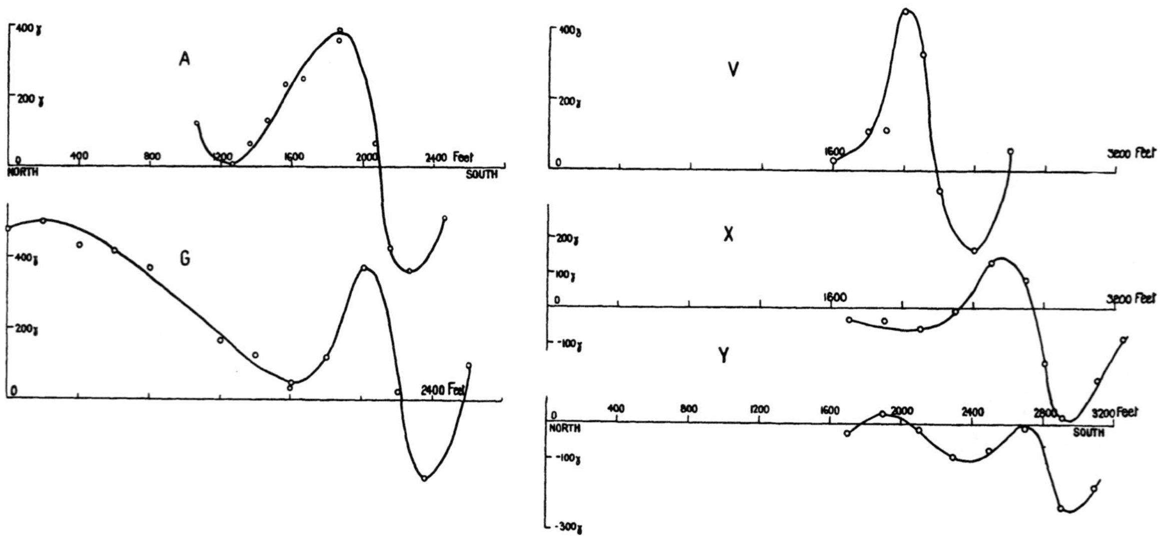

Fig. 134. Anomalies in vertical magnetic field along traverses *A*, *G*, *V*, *X*, *Y*, Gulgong, N.S.W.

Fig. 134 shows the profiles for the vertical intensity along the traverses *A*, *G*, *V*, *X* and *Y*. They are similar to the one obtained for a north-south traverse over a magnetic body[b], where the maximum positive anomaly is near the northern edge, and a negative anomaly occurs which has its maximum some distance south of the southern edge. The body indicated by the above profiles is fairly certain to be the basalt covering of a tributary deep lead, the course of which is readily determined from the profile and is shown in Fig. 135. This lead was previously unknown and thus is a direct discovery of this magnetic survey[c].

The two branch leads, which extend southwards from the sanitary reserve and are indicated in Figs. 132 and 135, have been followed for some distance by means of the lines *C*, *N*, *M*, *Q*, *R*, *T* and *Z*. On these lines the anomalies ranged up to 400γ, although on

[a] For the sake of clearness some of the dots where the stations fall closest together are not shown.
[b] See Fig. 124.
[c] Near where traverse *Y* crosses the road a well put down for water encountered basalt.

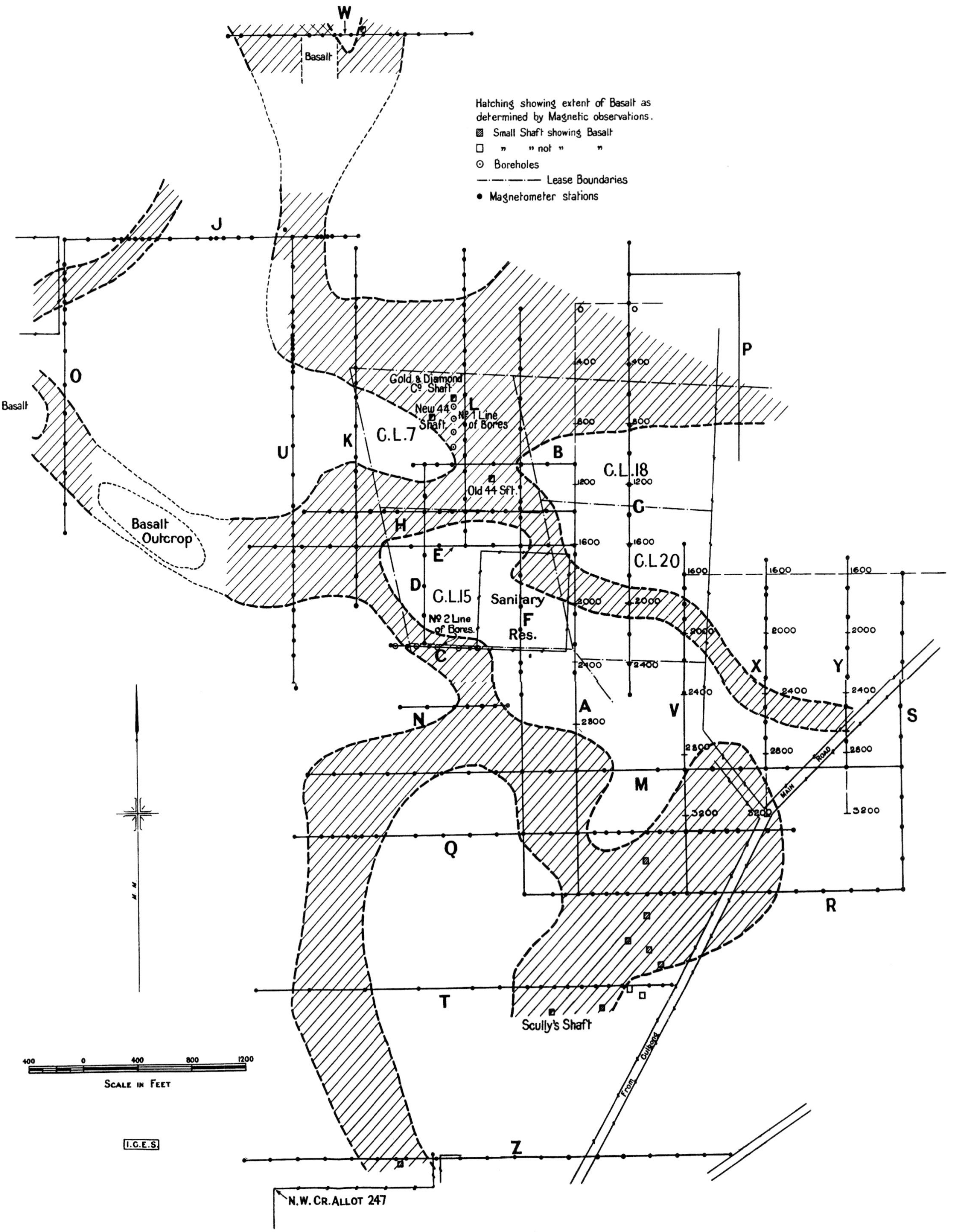

Fig. 135. Plan showing location of magnetic traverses and deduced boundaries of basalt flows, Gulgong, N.S.W.

N a maximum of 600γ was obtained. The position of the western boundary of the western basalt flow was rather indefinite on lines *M*, *Q* and *T*, and probably the flow extends farther to the west than is indicated in Fig. 135. The eastern branch appears to widen out into a rather indefinite area, and its boundaries could not be determined with any certainty from the results along line *Z*. Several prospectors' shafts had been put down in the area shown, and all those in the shaded area in Fig. 135 encountered basalt. The two shafts near the eastern end of line *T*, where no basalt was showing, are definitely outside the magnetically determined boundary.

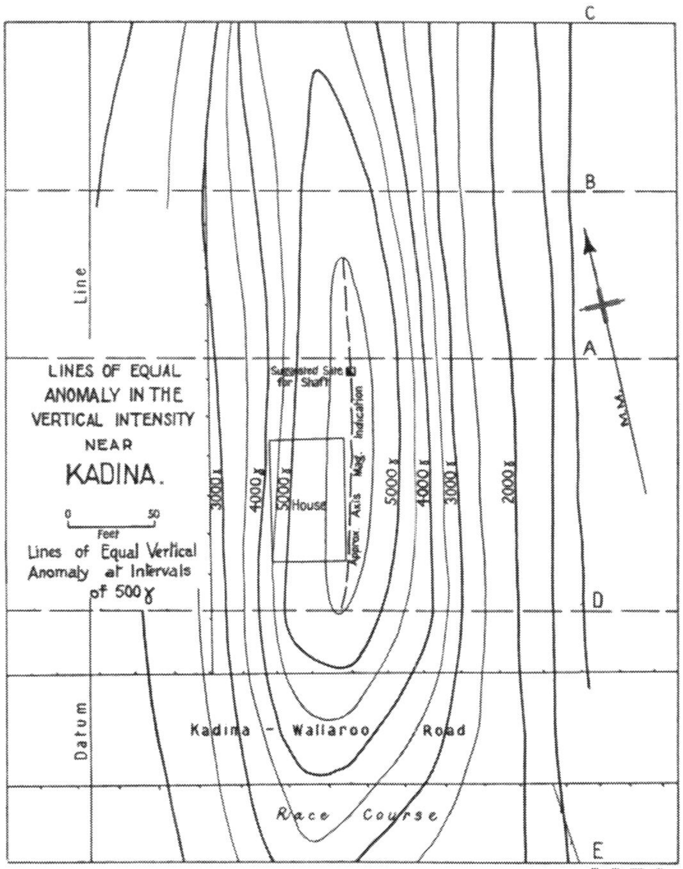

I.G.E.S.

Fig. 136. Lines of equal anomaly in the vertical magnetic field near Kadina, S.A.

In the northern part of the area the anomalies range up to 1500γ but the results are not so readily elucidated. A considerable portion of the area seems to be covered by a basalt sheet and there is no evidence that its northern boundary has been crossed, either on line *G*, or on line *P*, which extends a mile to the north and is only partly shown in Fig. 135. The boundaries of a flow are definitely obtained from the results on the southern portions of lines *K* and *U*, but on line *O* they are not so clear, although basalt outcrops on both sides of it. What appears to be a rather narrow flow crosses the northern end of *O* and the western end of *J*; it is probably too narrow to be the Stubbo Lead indicated in Fig. 132, but further work is needed to trace out its course. Clearly marked boundaries were obtained on line *W* and the eastern end of *J*, and from these a possible course of the Stubbo Lead is indicated in Fig. 135. Here again additional traverses are required, but on account of the termination of the I.G.E.S. field activities they could not be made. Even if they were made, the interpretation of the results in this part of the field would not be easy, and the positions of the gutters of the leads would be quite uncertain. The most important results are those obtained to the east of the Sanitary Reserve, where it is believed that a new deep lead has been discovered by this magnetic survey, also to the south-west of it where the course of one that was known to be very rich in its upper reaches has been traced for a considerable distance.

Conclusions. Although the area covered by the magnetic survey at Gulgong was comparatively small, there were sufficient bore-holes and shafts to afford a valuable check on the conclusions that were reached. From the results obtained in this field it appears that the magnetic method may have some most important applications in the sub-basaltic deep-lead areas of New South Wales, Victoria, and Tasmania. Where the basalt sheets are concealed from view by recent deposits, it should be possible in many cases to delineate them with considerable accuracy, and in special cases, such as the Black Lead at Gulgong, it may be possible to obtain evidence as to the approximate position of the gutters.

MAGNETIC SURVEY AT KADINA

Early in 1929 the Government Astronomer of South Australia (Mr G. F. Dodwell) located a series of magnetic anomalies in the Moonta-Wallaroo copper field, at a point some two miles east of the township of Kadina. After discussion with him an area of about four acres, comprising the region of greatest magnetic disturbance, was examined by the I.G.E.S. with the Thalen-Tiberg magnetometer.

Measurements of vertical intensity were made at stations 25 ft. apart, along a series of five parallel traverses *A*, *B*, *C*, *D* and *E*, and the lines of equal vertical intensity drawn at intervals of 500γ are shown in Fig. 136. The anomaly is of considerable magnitude, the maximum departure from the normal earth's field being about 5500γ, and positive in character. From the regular nature of the profiles (Fig. 137), it is probable that the disturbing magnetic body dips steeply or

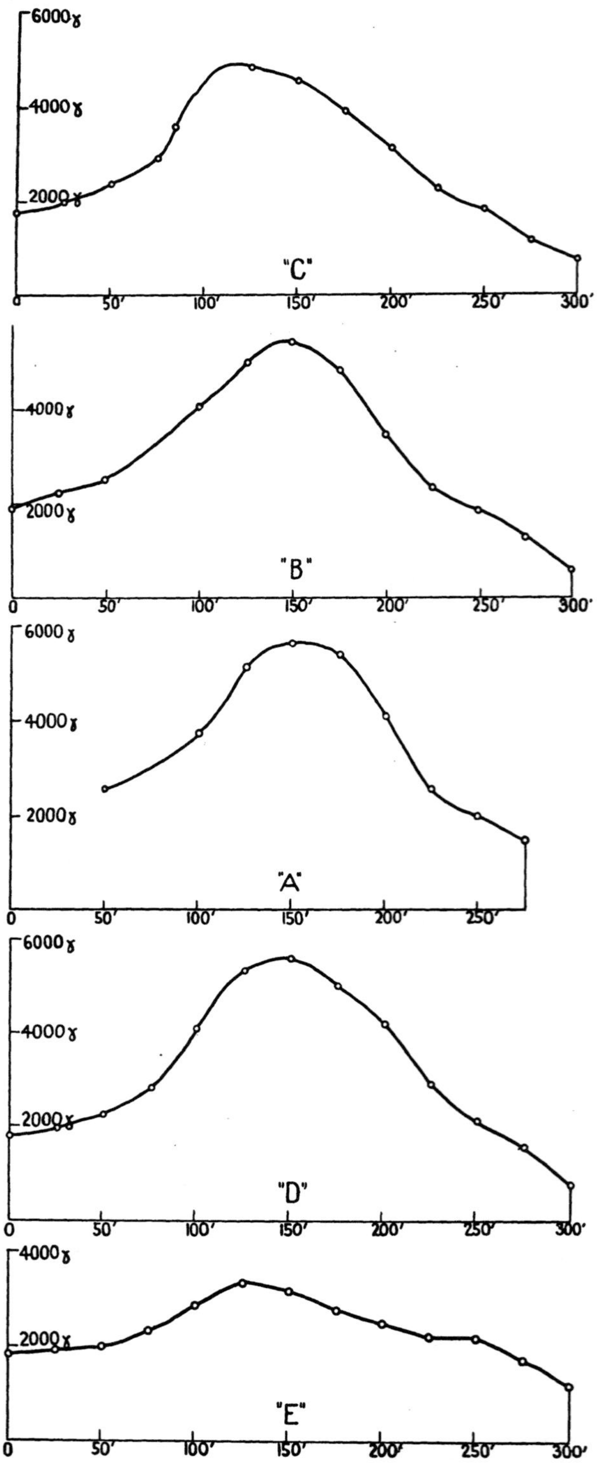

Fig. 137. Profiles of anomaly in the vertical magnetic field near Kadina, S.A.

is vertical, while the trend of the anomaly lines indicates that the strike of the body has a north-north-easterly bearing.

Since haematite and occasionally magnetite are known to occur in the district in direct association with the copper lodes, it was considered that the magnetic material causing this anomaly might possibly be associated with copper mineralization. In order to explore the ground a shaft was sunk to a depth of 60 ft. at the point where the anomaly was greatest. After passing through surface soil and a few feet of red clay a decomposed and highly ferruginous mica schist was encountered. At a depth of 60 ft., however, the schist contained so much magnetite that a small specimen caused a strong deflection in the needle of a pocket compass. The shaft was a small one and did not cut the full width of the ferruginous schist that has no doubt given rise to the magnetic anomaly. The Deputy Director of Mines for South Australia (Mr Lockhart Jack) was of the opinion that this schist is typical country rock, and that the magnetite and haematite it contains are its original constituents. It is unlikely, therefore, that there is even an indirect connection between this occurrence and the copper mineralization of the district.

MAGNETIC SURVEY AT RENISON BELL

This work has been described in ch. III as it was carried out in close conjunction with the electrical survey at Renison Bell.

CHAPTER VI

SEISMIC METHODS

SECTION I. PRINCIPLES AND APPLICATIONS

HISTORICAL

Relation to Seismology. The study of natural earthquakes has yielded much information about the composition and structure of the earth's crust. From the nature, and time of arrival at distant points, of the various types of elastic disturbances transmitted from the initiating shock, deductions have been made as to the position and character of the different layers through which the disturbances have travelled. In applied geophysics, similar methods are now used to obtain information about the subsurface structure, the initial shock being applied artificially—usually by a charge of high explosive. The effects are on a smaller scale than in natural earthquakes, and correspondingly greater detail is found for the structures investigated.

Development in Seismology. The beginnings of seismic prospecting are thus to be traced in the history of tectonic seismology. The first detectors were seismoscopes, indicating the arrival of shocks without recording them: for example, the settling of a heaped pile of fine sand, or the formation of ripples on the surface of a pool of mercury, was observed with a microscope. Milne, about 1880, constructed a horizontal seismometer[a], that is, an instrument for recording a horizontal component of the movement of the ground, and similar instruments were developed in Germany, Italy, and Austria. Seismology came into existence as a quantitative science with the introduction of such instruments, but theoretical interpretation of the records was lacking until Oldham[b], in 1900, identified in them the elastic waves whose existence had been demonstrated theoretically by mathematicians such as Poisson (1828), Stokes[c] (1845 and 1849), and Rayleigh[d] (1885). Important advances were made by Wiechert about 1900, especially the introduction of a satisfactory form of damping. Improvements in registration, damping, and magnification, continued until the War, 1914, and enabled ready separation of the component disturbances. Vertical seismometers were also constructed and developed. The theoretical work of Galitzin[e] and his development of an electrical seismometer are of outstanding importance.

[a] The work of Milne is outstanding in the establishment of seismometry, but he himself states: "From these short notes on horizontal pendulums it will be seen that these instruments...have within a period of sixty years been independently invented many times." Milne, *Seismology*, p. 63, Kegan Paul, Trench, Trübner, and Co., London (1898).

[b] *Roy. Soc. Phil. Trans.*, A, vol. 194, p. 135 (1900).

[c] Stokes, *Mathematical and Physical Papers*, vol. 1, p. 113, University Press, Cambridge (1880). *Loc. cit.*, vol. 2, p. 113, University Press, Cambridge (1883).

[d] *Lond. Math. Soc. Proc.*, vol. 17, p. 4 (1885–6). Rayleigh, *Scientific Papers*, vol. 2, p. 441, University Press, Cambridge (1900).

[e] Galitzin, *Vorlesungen über Seismometrie*, Teubner, Leipzig (1914).

Artificial Disturbances. Artificial shocks were observed by Mallet[a], 1860, in England, for the purpose of determining the velocity of earth waves from quarry blasts; and the Hell Gate blasts of 1876 and 1885 were observed by a chain of stations in America[b]. The tremors recorded from these artificial disturbances are of higher frequency than those from natural earthquakes, and seismometers suitable for their detection were developed by Mintrop and others; the former having patented a "tremormeter" in 1917. Working with Wiechert in Göttingen, Mintrop had for some years been recording artificial high-frequency tremors, but apparently with no specific interest in the possibilities of geophysical exploration. During the War, while the British and French armies developed methods for locating the firing point of a gun by recording the sound waves, the German experimenters studied the seismic disturbances produced by the recoil. This work gave a great stimulus to the development of applied seismology. Since the War some large artificial shocks have been recorded, such as the Oppau explosion of 1921.

Seismic Prospecting. The use of artificial disturbances for the examination of underground structure was suggested by Mallet and by Abbot. Afterwards Schmidt, in 1888, proposed the use of time-distance graphs for this purpose and later writers made similar suggestions. Fessenden attempted practical exploration in America in 1913, but, instead of an explosion and seismometers, he used the sonic sounder and a sonic receiver. After 1919 seismic prospecting passed beyond the experimental stage, inasmuch as various American field explorations were undertaken after that date, in addition to work in Germany. In 1923 Mintrop carried out seismic exploration work in Mexico for the Royal Dutch Shell Company, and similar investigations were commenced in the Gulf Coast districts. Since then various types of mechanical seismometer have been introduced, and the carbon microphone and an electromagnetic detector with valve amplification have also been employed. Seismic prospecting has been a commercial success in a variety of economic problems and, in the Gulf Coast area, some millions of dollars per year have been spent in applying it. In spite of this, little or no detailed information was published as to the methods employed the first technical accounts being given in 1928, e.g. by Barton[c] and by Heiland[d].

GENERAL THEORY

Sound Wave in a Fluid. Any elastic substance will transmit a disturbance as a single shock or a train of waves. Through a liquid or a gas the only disturbance transmitted is a simple oscillation of certain particles of the fluid in the direction of propagation of the disturbance. Such a wave is called a sound wave. It is known as a *longitudinal* wave, and its velocity depends upon the compressibility and the density of the fluid. In air, for example, the velocity is about 1100 ft. per second at ordinary temperatures.

[a] *British Assoc. Rep.*, p. 201 (1861).
[b] *Annual Report of the Chief of Engineers, U.S. Army*, Part 1, Appendix E, pp. 691–717 (1886).
[c] *Geophysical Prospecting 1929*, p. 572, A.I.M.E., New York (1929).
[d] *Ibid.*, p. 625, A.I.M.E., New York (1929).

Waves in a Solid. A solid, in addition to being compressible, has rigidity, and consequently more than one type of wave can be propagated through it. Thus there is added a *transverse* disturbance for which the oscillation is perpendicular to the direction of propagation of the disturbance. The longitudinal disturbance, or sound wave, is the one which is transmitted with greatest velocity, and consequently the one which will arrive first at a detecting device. For this reason it is the wave with which we are most concerned in prospecting by the seismic method. For a solid, the velocity of this disturbance depends upon the compressibility, rigidity, and density of the medium.

Because of the infinite variety of rocks and soils the velocity must be found afresh by the geophysicist for each medium with which he is concerned. Owing to the difficulty of exact determination of the elastic properties and density of a rock specimen, and to the improbability of a specimen being truly representative of the rock mass from which it is selected, earth transmission velocities must be determined directly on the ground, and cannot be calculated satisfactorily from elastic coefficients and densities.

Time-Distance Graphs. The velocity of the longitudinal wave varies from values less than 2000 ft. per second, for sandy soils, to values greater than 20,000 ft. per second, for granites. With ratios up to 10 : 1 between velocities in different media, the discriminating power of the seismic method may therefore be very great.

In seismic prospecting a charge is exploded upon the ground surface, or just beneath it, and the time of first arrival of the resulting disturbance is found at various distant points on the surface. The information to be derived as to the structure and character of rocks in the neighbourhood is studied conveniently by drawing a graph with distances from explosion point as abscissae and with times from moment of explosion as ordinates; plotting, for each observation point, the time of travel for the first tremor to reach the point, against the distance from the explosion. Such a graph is called a *time-distance graph*[a]. If the ground is uniform then the graph will consist of a straight line, passing through the origin. If the graph departs from a straight line then a lack of uniformity is indicated.

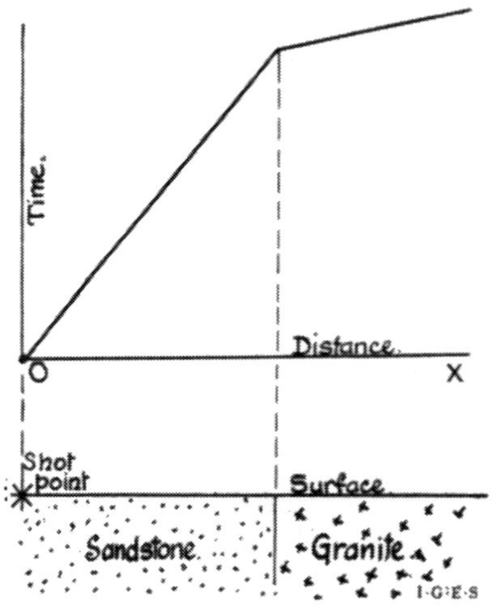

Fig. 138. Time-distance graph for surface discontinuity

For example, if the surface rock changes abruptly to one of[b] higher velocity, then the graph will, as in Fig. 138, bend sharply into a second straight line with diminished slope. The value of the velocity in each of the two types of rock may be derived directly from the graph.

[a] This will usually be referred to simply as "the graph." The firing of a charge in order to obtain such a graph will often be called a "shot." The procedure of exploding a charge at a point O, and obtaining a graph for points along a line OX, will be referred to as "shooting from O along OX," or by some similar phrase.

[b] Instead of speaking of a medium "in which the longitudinal wave travels with a certain velocity," it is convenient to speak of a medium "of such a velocity," and similarly to refer to "the velocity of the medium."

Reflection and Refraction. The graph may also be affected by structures which do not reach the surface. This is due to the effects of *reflection* and of *refraction*. These phenomena are familiar in the propagation of light. When a ray of light, *IO*, Fig. 139, is incident on an interface between two media, its energy is in general divided to follow two paths. A portion is reflected along a path, *OR*, back into the medium on the same side of the interface; the remainder is refracted along a path, *OS*, into the medium on the opposite side. If the second medium is of greater velocity, then the refracted ray will be bent away from the normal, towards the interface.

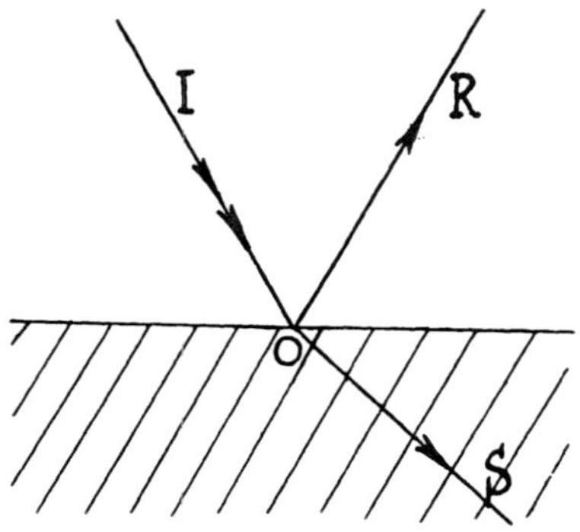

Fig. 139. Reflection and refraction of a wave

Critical Angle. If, when the velocity is greater in the second medium, the inclination of the incident ray to the interface is sufficiently decreased, the refracted ray will eventually lie in the interface itself, as in Fig. 140. This is known as the *critical case* of refraction, and the angle which the incident ray then makes with the normal to the interface is known as the *critical angle*. The critical angle depends upon the ratio of the velocities of propagation in the two media and, in this case, the incident energy lies almost entirely in the reflected ray, that given to the refracted ray being negligible. For incident rays less steeply inclined to the interface there is no refracted ray and the light is totally reflected.

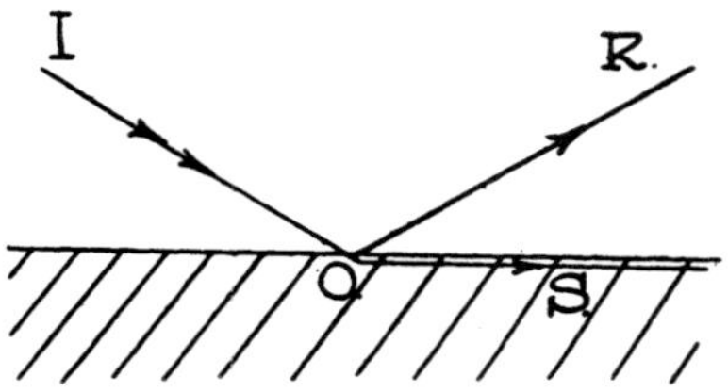

Fig. 140. Critical refraction. The lower medium is of greater velocity than the upper medium

Reflection of Elastic Waves. In a similar manner reflection and refraction occur in the propagation of sound, the "echo" being a simple case of reflection. They also take place in the propagation of elastic waves in solids, although the phenomena are more complex, owing to the different types of waves which can be transmitted through a solid. Thus, if a horizontal surface layer overlies a medium of different properties, some of the energy from a shot will be reflected from the interface back to any observation point, by a symmetrical path, *SRP*, Fig. 141. If the time of travel of the reflected wave is found, then the depth of the surface layer may readily be calculated, the velocity being known from the time of travel of the direct surface wave, *SP*. As the reflected wave will not arrive until after the surface wave, the graph of the latter will not be affected if it is drawn for the first-arrival[a] only.

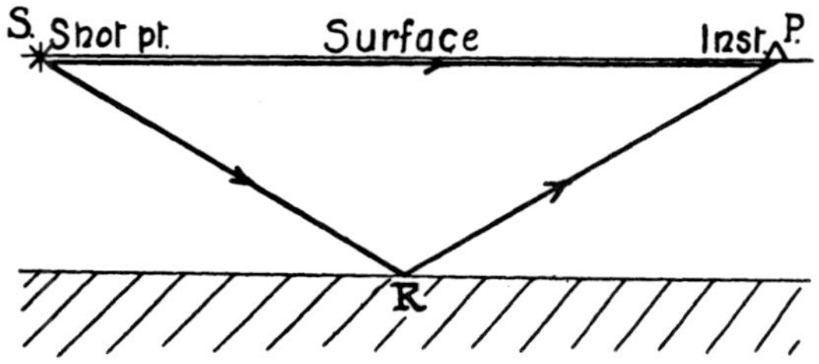

Fig. 141. Reflection from interface

[a] At an observation point a series of waves arrives from an explosion; the first of these waves to arrive is called the "first-arrival" at that point.

In practice the reflected wave will usually arrive while the detector is still excited by the direct surface wave[a] and, as it naturally carries less energy than the latter, the time of its arrival will be difficult to find on the record. Nevertheless determinations by methods employing reflection are sometimes made, when the structures are such that there are suitable intervals in the times of arrival.

Refraction of Elastic Waves. By means of refraction, however, a wave which has been down to a lower stratum may reach a point on the surface actually before the direct surface wave. Its time of arrival can then be determined more easily. This can only take place when the lower stratum is of greater velocity than the surface medium. Thus, if the velocity increases continuously with depth, a disturbance may follow a path which, by continued refraction, becomes continuously less inclined to the horizontal, as it passes into regions of successively higher velocities. After it has eventually become horizontal, the reverse effect takes place, and the path returns to the surface by a curve, *SRP*, Fig. 142, which, by refraction, becomes continuously more inclined to the horizontal, as it passes into regions of successively lower velocities[b]. The analogous effect in the propagation of light is familiar as the phenomenon of "mirage." If the detector is close to the shot, the refracted ray may reach it after the direct surface wave, because of the longer path. But at a sufficient distance the refracted wave will arrive first, because of the advantage of travelling, for the greatest part of the way, in the lower layers with higher velocity. Under these circumstances the apparent velocity for first-arrival would increase as the distance from the shot increased. The slope of the graph would diminish continually, as in Fig. 143.

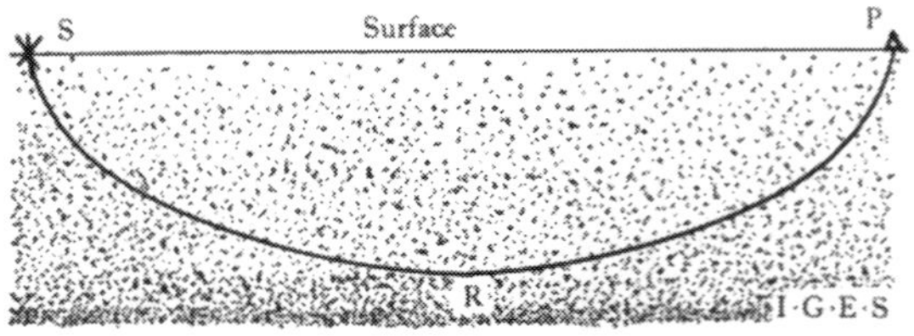

Fig. 142. The mirage effect for progressive increase of wave velocity with depth

Ambiguity of Interpretation. When such a graph is obtained, it may indicate the presence of higher velocities in the ground beneath the surface. The interpretation is however ambiguous; for a similar graph would result if the velocity increased continuously along the surface, i.e. if the variation were horizontal instead of vertical. This ambiguity could be resolved by shooting in the opposite direction over the same ground, the relative positions of the shot-point and observation station being reversed. For a vertical velocity variation the slope of the graph varies with the distance from the shot-point, so that the shape would be unchanged; for a horizontal variation, the slope of

[a] The reasons for this are (i) that the shock given to the ground by the explosion is not an instantaneous impulse; (ii) that an impulsive elastic disturbance will become diffused over a greater space, as it travels in the ground; (iii) that the seismometer, unless heavily damped, will take time to come to rest, even after a single impulsive shock. See example, p. 200.

[b] The longitudinal wave in the surface material then appears to arrive at the detector on a steeply inclined path, instead of horizontally. Schweydar and Reich (*Gerlands Beitr. z. Geophysik*, vol. 17, p. 121 (1927)) found the wave arriving vertically from below, at a distance of 103 metres.

the graph varies with the position on the ground surface, so that the curvature would be reversed, as in Fig. 144. This is illustrative of the fundamental ambiguity of interpretation of the time-distance graph, using a single shot; and of its method of resolution, using a second shot, with the opposite direction of observation.

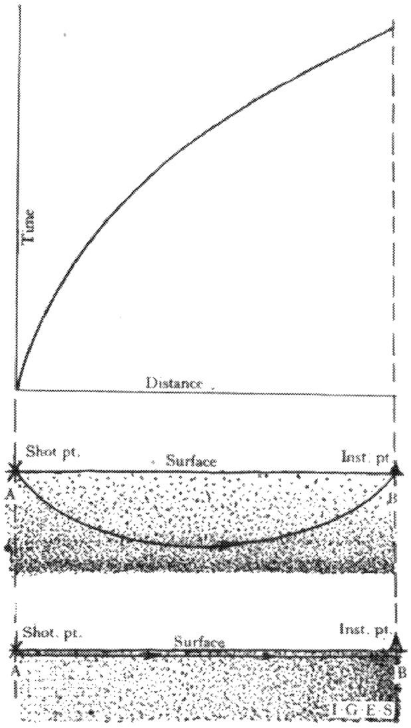

Fig. 143. Time-distance graph for two formations in which there is a progressive variation of wave velocity

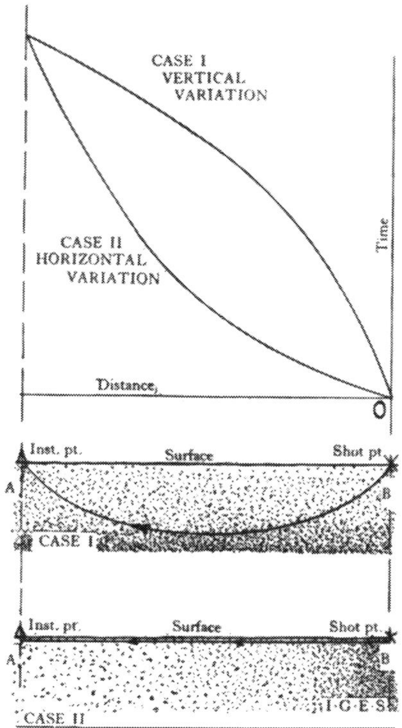

Fig. 144. Graphs for the same formations shown in Fig. 143 but with reversed directions of shooting

Mirage Effect from Single Discontinuity. The graph is modified in a similar fashion by refraction at a single interface between an upper layer of low velocity and an underlying formation of high velocity. At a sufficient distance from the shot-point, the shortest-time path which a wave could follow to a given detector point would penetrate to the bed-rock, follow the interface, and return to the surface beneath the detector, e.g. the path *SRQP*, Fig. 145. Such a path would not, of course, run vertically down to and up from the bed-rock, but would be at an inclination which would strike a balance between the increase in total length of path and the increase in average velocity. This inclination is at the critical angle for which, as described above, the incident ray is refracted to run along in the interface. Thus the path is

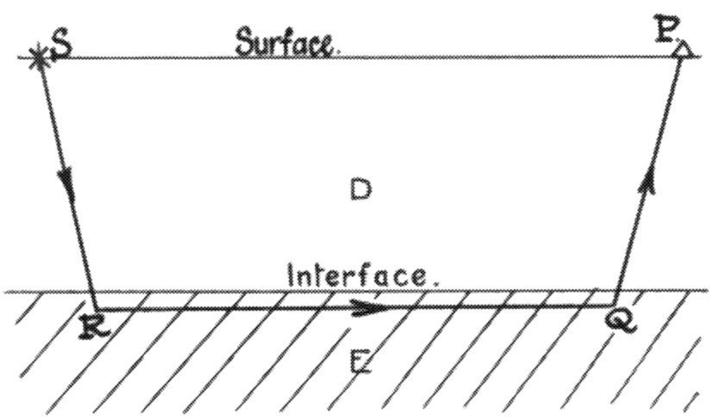

Fig. 145. The mirage effect, for an interface between formations *D* and *E*

theoretically possible[a], although it may be suspected that a negligible amount of energy would follow it. This is the case at great distances, and in the analogous case in the propagation of light, but the complexities of the propagation of elastic waves in solids modify the conditions, at short distances, as is discussed in Part II, p. 328. In practice quite an appreciable portion of energy is found to follow such a path, both from artificial shocks and from natural earthquakes.

Graph for Single Refraction. The effect described in the last paragraph is the basic phenomenon of the method of seismic prospecting prevailing at the present time. The time-distance graph of first-arrival (Fig. 146) now consists of *OX* and *XC*, parts of two straight lines, *OA* and *BC*. *OA* represents the surface wave and passes through the origin; from its slope[b] the velocity of the wave can be obtained. *BC* represents the refracted wave; it does not pass through the origin, and it intersects *OA* at *X*. From its slope the velocity of the wave in the lower layer can also be calculated.

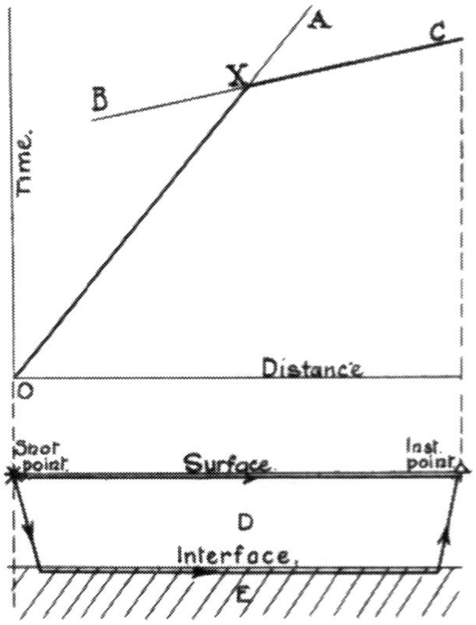

Fig. 146. Time-distance graph of Fig. 145

Such a graph, then, may show the presence of a bed-rock of higher velocity, may indicate its nature by virtue of the velocity, and may be used in calculating the depth of the interface. As usual, there is an ambiguity in interpretation for a single shot, for it will be noted that the graph of Fig. 146 is of similar form to that of Fig. 138, which corresponded to a vertical interface. This ambiguity could be resolved by shooting in the reverse direction, as described above[c].

If the bed-rock surface were not horizontal, but sloped away from the shot-point, then the return path from bed-rock to surface would increase in length continuously with distance, and the graph would not give a direct measure of bed-rock velocity. The apparent velocity given by "shooting down the slope" would be too small, that by "shooting up the slope" would be too large; the true velocity would lie between them.

Numerical Example of Paths and Times. A simple case of the longitudinal wave paths discussed above, and the calculation of times of travel, is illustrated in Fig. 147, where 200 ft. of argillaceous sand overlie a horizontal granite bed-rock. Path *A* is that of the refracted wave; path *B* that of the surface wave; path *C* that of the reflected wave. The path *A′*, *vertically* down into and up from the lower refracting medium, is

[a] The laws of refraction are, in fact, particular cases of a general physical law that propagation will follow a path of least time.

[b] The reciprocal of the slope, expressed in suitable units, is the velocity of the wave.

[c] There would in many cases be sufficient knowledge of the geology of the neighbourhood to eliminate one alternative as highly improbable.

substituted for a portion of path *A*. This is done quite frequently by field workers for simplicity of calculation[a].

Path *A*.	400 ft. at 3,000 ft./sec. = 0·133 sec.	}	0·196 sec.
	1,000 ft. at 16,000 ft./sec. = 0·063 sec.	}	
Path *B*.	1,000 ft. at 3,000 ft./sec.		0·333 sec.
Path *C*.	1,075 ft. at 3,000 ft./sec.		0·358 sec.

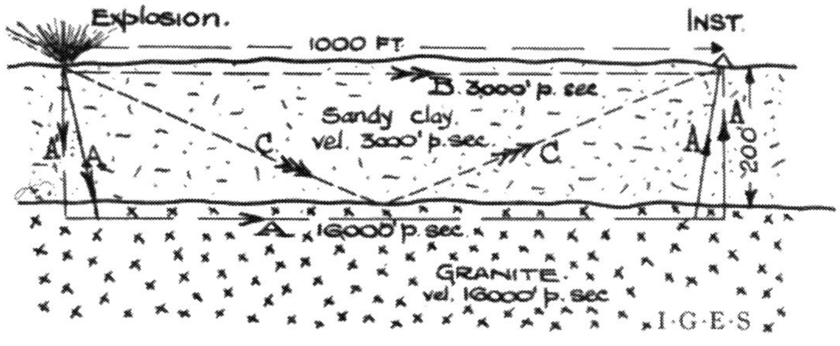

Fig. 147. Numerical example of refracted, reflected, and surface waves

The arrival of the disturbance along path *A* at 0·196 sec. instead of the expected 0·333 sec. after the initiation of the disturbance is easily detected, as it is the first arrival and the seismometer is at rest. The reflected wave *C*, at 0·358 sec., arrives while the instrument is still disturbed by the direct wave *B*, which arrived only 0·025 sec. earlier, or by other surface waves, or, in the case of most instruments, still oscillating from its first disturbance. Under such conditions the moment of arrival of the reflected wave *C* cannot as a rule be detected with any precision, and its very identification is usually unrealiable.

Complex Structures. It has been shown above that ambiguities may be resolved by shooting back from another point, towards the original shot-point. A shot from a second point, without reversing direction, would suffice theoretically, but is not usually as effective in practice. More complex structures than those which have been considered would yield a more complex graph. For example, refracted waves may also return from still lower interfaces. Theoretically it is only necessary to increase the number of shot-points and seismometer stations as the amount of detail requiring to be determined increases, and any sufficiently shallow structure may be solved, however complex. In practice the difficulties mentioned in the following paragraphs will arise.

Accuracy. The accuracy of determinations is limited by the following effects:

(1) Irregularities in the speed of the recording paper and of the time marker introduce errors in the time scale marked on the photographic record, and in the I.G.E.S. experiments these were of the order of 0·001 to 0·01 second.

[a] Although the assumption of such a path is apparently unsound theoretically, it is claimed by some workers that it is not only simple for calculation, but gives a more accurate result. Schweydar and Reich, *Gerlands Beitr. z. Geophysik*, vol. 17, p. 121 (1927); *Zeits. f. Geophysik*, vol. 2, p. 350 (1926).

(2) The velocity of the shock through the zone of pulverization[a] around the explosion point will be greater than the velocity of the disturbance in the ground away from the neighbourhood of the explosion point.

(3) There may be a lag in the recording, both of the moment of explosion and of the moment of arrival of the tremor; it is the difference of time between these events which is measured, so that only the difference between the two lags will introduce error.

(4) Inhomogeneities and irregularities of structure introduce errors which cannot always be detected or fully allowed for, and these may cause departures from the simple geometrical laws of propagation.

(5) There are reasons for believing that some of the assumptions fundamental to the conventional theory of seismic prospecting are not strictly true[b].

Taking into consideration the above effects, and the field experience of the I.G.E.S. and other prospecting parties, the order of accuracy claimed[c] under average conditions, both for velocity and depth determinations, is estimated at 10 per cent.

Applications of the Seismic Method. The seismic method is suited to the investigation of large scale geological structures of a simple character; in particular to conditions of extended horizontal stratification, for which the gravity balance method fails[d]. Under favourable circumstances the depths from the surface, the slopes (dips), and certain other features may be determined in cases of concealed structures, such as buried topographies, anticlines, synclines. In oil field work, seismic investigations have proved of value in locating key formations with which oil is associated; the special application to the discovery of salt domes, in the Gulf Coast region of Mexico, being the most sensational achievement in this connection. Apart from determinations of depth, and of details of the structure of a concealed formation, the method may also afford valuable evidence regarding its lithological character. For example, in a district comprised of wide expanses of slate and granite, covered by some hundreds of feet of alluvium, it may be possible to provide conclusive evidence (from the velocity values) as to which of the two types of rock is forming the bed-rock of the area under investigation. The most favourable structure for the successful application of the method is a low velocity material overlying one of high velocity. A material of high elasticity, such as an igneous rock, will have a greater velocity than an unconsolidated formation, such as alluvium, which may cover it. It is also the case that materials of low density, such as coal and salt, are characterized by high velocities, and when overlain by low velocity formations they may also come within the scope of seismic investigations. Although electrical methods are now being employed to an increasing extent in the investigation

[a] See p. 205.

[b] Pt. II, p. 328 et seq.

[c] Some writers claim greater accuracies, even to a fraction of 1 per cent. (see Ambronn, *Engineering and Mining Journ.*, vol. 128, p. 93 (1929)); but this does not appear to have been substantiated (see Barton, loc. cit., p. 180).

[d] Absolute determinations of gravity are effective in such cases, and improved pendulum apparatus is now employed for practical exploration work by several geophysical organizations.

of underground structures, it has not yet been shown that they can be used effectively when these lie at depths exceeding a thousand feet from the surface. In the case of seismic methods, however, convincing results have been obtained at depths of the order of two to three thousand feet[a].

Excitation of the Disturbance. Large shocks are readily obtained by the use of high explosives, which are now invariably used in seismic prospecting. Nevertheless the potential energy of the explosive is not all transmitted into the ground in seismic waves, the loss being much greater if the explosive is simply placed upon the surface of the ground than if it is buried at a depth sufficient to confine the effects beneath the surface.

Explosions. The term "explosive" is ordinarily restricted to solids or liquids which are "liable, on the application of heat or a blow to a small portion of the mass, to be converted in a very small interval of time into other more stable substances largely or entirely gaseous[b]." The gas is produced at a very high temperature and therefore exerts a high pressure. The explosive may be blasting or propellant, so that neighbouring substances are hurled away; or its violence may be confined to a very limited area, in which case it is said to be a "detonating" explosive. Explosives may be classified not only according to their *power*, i.e. the total available energy per unit mass, but also according to their *violence*, i.e. the rapidity with which the energy is made available. The latter varies as the rate at which the chemical transformation spreads through the mass of explosive, and this rate may be fast or slow, depending upon the manner in which it is initiated; the rate being greater if the initiating shock is large. Some substances will burn quietly when a flame is applied, but will explode violently when subjected to a shock by another explosive. The latter is then known as a "detonator"; and the travelling of the transformation through the explosive substance with explosive rapidity is known as "detonation," as opposed to "burning."

Types of Explosive. For blasting purposes, mixtures of potassium chlorate and certain combustible oils are sometimes used. These materials individually are not explosive, the potassium chlorate, usually in small bags, being dipped into the oil immediately before the explosive is required for use. A potassium chlorate explosive requires a detonator; preferably a high explosive, such as blasting gelatine, itself fired by a fulminate detonator.

Most high explosives are nitro-compounds. Blasting gelatine is a translucent jelly formed by the action of nitroglycerine on soluble guncotton. At normal temperatures it is exploded by a detonator, but when frozen it is unstable under a mechanical shock. Gelignite is of similar composition, but contains a proportion of inactive material, together with wood meal and potassium nitrate.

The detonator generally used for firing high explosives is a small metal cylinder containing an explosive of high specific pressure such as mercuric fulminate. The

[a] Rankine, *Nature*, vol. 123, pp. 684–686, 718–720 (1929).
[b] Marshall, *Explosives*, p. 414, Churchill, London (1915).

fulminate is ignited by heat, which is produced either by the flash from the end of a gunpowder fuse inserted in the cylinder, or, in the electrical detonator, by the fusing of a fine wire on the passage of an electric current.

The velocity of the wave of detonation, on which the "violence" of an explosive depends, is of the order of the velocity of sound in the explosive. It varies somewhat with the physical condition of the specimen, and with the violence of initiation of detonation. The following approximate values have been found for velocity of detonation in m./sec.: blasting gelatine (and other nitroglycerine derivatives), 7500 and 1500[a]; trinitrotoluene, 6500; guncotton, dry, 6000; guncotton, wet, 5500; mercuric fulminate[b], 4000.

Transference of Energy into the Ground. Very little quantitative detail is known about the initiation of a seismic disturbance in the ground. In the history of seismology only two full-scale earthquakes have had their point of origin at the surface of the earth, so that the magnitude of the initial shock can be estimated. These were the Pamir landslide, in 1911, in which about half a cubic mile of rock fell more than 1000 ft., and the Oppau explosion[c] of 4500 tons of high explosive on the surface of the ground, in 1921. In the former case, deducting the amount of energy dissipated as heat in the rock mass upon impact, the amount available for transmission into the ground has been calculated to be of the order 10^{14} ft. lb., this being 1/300th of the kinetic energy of the mass at the moment of striking the ground. From records of the earthquake it has been calculated that there was an amount of energy of the same order in the seismic disturbance produced[d].

I.G.E.S.

Fig. 148. Surface explosion of 30 lb. of lithyte at Tallong. This explosion formed a hole in the surface 6 in. deep. Compare the result of firing a buried charge, Fig. 163

Surface Explosions. The efficiency of the transference of energy was very much less in the case of the Oppau explosion[e]. The chemical energy of the explosive was 4×10^{12} ft. lb., and it has been calculated from the earthquake records that there was not more than 1/1000th of this amount of energy in the seismic disturbance produced.

[a] For nitroglycerine and its derivatives there are two distinct velocities, the slower corresponding to propagation through a liquid, and the faster to propagation through a solid.

[b] Mercuric fulminate at first burns slowly, with a velocity of about 10 metres per second; but detonation commences within about 1/3000th second.

[c] *Nature*, vol. 108, pp. 137, 278 (1921).

[d] Jeffreys, *Roy. Astron. Soc., M.N., Geophysical Supplement*, vol. 1, p. 22 (1923).

[e] Jeffreys, *The Earth*, 2nd ed., p. 109, University Press, Cambridge (1929).

As must always be the case with surface explosions, the greater part of the energy had been dissipated into the air, partly as sound. The action is equal in both directions, but the initial velocity in the air (greatly exceeding the velocity of sound) may be thousands of times greater than the velocity in the ground, and the energy is divided in similar proportions. To obtain the greatest downward blow from a surface explosion, detonation should start from a point at the top of the explosive. Moreover, from aerodynamic considerations, it appears that a more violent explosive will transmit into the ground a greater proportion of its energy than would a less violent explosive.

Zone of Pulverization. When a charge is exploded in a bomb calorimeter[a] the pressure of the gases on the walls reaches a maximum at a finite time after the moment of explosion. The ratio of the maximum pressure to the final pressure increases with the violence of the explosion. When a charge buried in the ground is exploded, the gas pressure produces a strain in the walls of the cavity in which the charge is placed, and this strain is transmitted to surrounding layers, and so out into the ground as elastic waves. During the explosion there is thus a flux of energy through the surface layers of the cavity. In practical cases the strain produced in the surface layers becomes greater than that which can be borne by the ground material without rupture. A zone of pulverization is formed, therefore, around the point of explosion. At the moment when this zone has reached its extreme size, the unbroken material on its walls is subjected to its limiting strain, the magnitude of which is dependent upon the nature of the material and not upon the size of the zone of pulverization.

Effect of Violence of Explosion. The energy passing through the walls of the zone of pulverization is carried out into the ground at a rate depending upon the nature of the ground material and upon the size of the zone, but it is supplied at a rate depending upon the violence of the explosive. A violent explosive may therefore produce a larger zone of pulverization than a larger charge of more powerful, but less violent, explosive. As the strain passes out into the ground in elastic waves, it diminishes in magnitude in direct proportion[b] to the distance from the point of explosion. Since the maximum strain is independent of the size of the zone, the maximum amplitude of the seismic disturbance at a distant point will be directly proportional to the radius of the zone, e.g. if the radius is 10 ft. the maximum amplitude at any distant point will have 10 times, and the maximum energy flux 100 times, the value which it would have if the radius of the zone were 1 ft.

In seismic prospecting it is more desirable that the disturbance shall arrive with the greatest possible amplitude than that it shall have the greatest possible energy content. The most efficient explosive is, therefore, the one of greatest violence, which will produce the largest zone of pulverization. Moreover, from a more violent explosive the energy will be concentrated in thinner shells, and will give a more simple form of shock to a seismometer.

[a] Marshall, *Explosives*, p. 358, Churchill, London (1915).

[b] For a spherical wave surface, neglecting absorption.

Microseisms. The ground surface is in a continual state of unrest. Small movements, of periods greater than, say, 4 seconds, are produced by the action of the sea on the coast and by other distant disturbances; movements of higher frequency are caused by wind among the trees and other local effects. Such fortuitous small disturbances are called *microseisms*. The disturbance arriving from an explosion has to be picked out from a background of microseisms; a difficult matter if it is so small as to be of comparable amplitude. There is a close analogy in radio-electricity, where a signal has to be detected against a background of atmospherics. In that case it is necessary that the signal intensity should be above the "noise level"; otherwise no advantage will be gained by amplification, since the background of noise would be amplified to a similar extent. The receiving set may, however, be made more sensitive to the signal by tuning to its frequency.

By analogy, one may speak of the "microseism level." Unless the intensity of the first-arrival is above the microseism level, an increase in the sensitiveness of the seismometer will not help to detect it. If the microseisms and the artificial tremors are of different frequencies, it may be possible to make the seismometer selective, so as to be more sensitive to the former than to the latter. Otherwise, the only remedy is to increase the charge of explosive, or, if the microseisms are of local origin and variable, to await more favourable conditions.

I.G.E.S.

Fig. 149. Cinematograph of explosion of buried charge. Charge, 30 lb. blasting gelatine, at depth 12 ft. in alluvium, Gulgong. The eight successive views cover a period of ½ second, from the moment at which some gas was blown out of the drill hole, to the moment when the ground surface was lifted to the maximum height (about 18 in.). After this the ground subsided, leaving an almost unbroken surface, as in Fig. 163

Absorption of Elastic Waves. The energy transmitted in an elastic disturbance is spread over larger areas as the disturbance spreads out from its point of origin; therefore the amplitude decreases with the distance. In addition to this, energy is absorbed[a] by the ground. This absorption takes place in two ways. Firstly, there is a loss by friction, whereby the energy is transformed into heat, and is dissipated in the ground. For this reason, loose sandy soils rapidly absorb the energy, whereas rocks transmit elastic disturbances with comparatively little loss.

Secondly, the energy is scattered, from the direction of propagation of the disturbance, by inhomogeneities and

[a] This effect is commonly spoken of as "damping" (of the amplitude) but it is preferable to call it "absorption," by analogy with light, X-rays, radio-electricity, etc.

discontinuities in the ground. For this reason a massive igneous rock transmits an elastic disturbance more efficiently than the majority of sedimentary rocks; just as the transmission of sound is more efficient when the density of the air is uniform, as, for example, over water at night. The scattering of elastic waves is greater for waves of higher frequencies. There is an analogy in the propagation of light; for example, the higher-frequency waves in the sun's rays are scattered by the earth's atmosphere to form the blue of the sky, while the lower-frequency waves are transmitted so that the sun appears of a yellow, or even red, colour. If a seismic disturbance, in the vicinity of its point of origin, includes waves of a larger range of frequency, those of higher frequency will be the most quickly absorbed by scattering, and the predominant or average frequency will decrease as the distance from the point of origin increases.

Frequency of Seismic Disturbances. The elastic disturbances generated in seismic prospecting, like those of earthquakes, are generally caused by an impulsive shock rather than by periodic forces. Near the point of origin the disturbance is therefore a pulse, and not a train of waves. At a sufficient distance from the point of origin, however, such a pulse becomes spread out into a train of waves[a]. It is known theoretically that, whatever the original form of pulse, a set of regular wave trains of different wave lengths may be found which combine to give such a form. In a perfectly uniform unbounded solid, elastic waves of all wave lengths would travel with the same velocity. But under the conditions which exist in the ground the velocity depends upon the wave length, and in consequence the pulse is transformed into a train of waves[b].

Very little is known about the form of elastic pulses in the ground produced, for example, by an explosion. A seismometer usually has one or more natural frequencies of vibration, and is especially sensitive to disturbances with frequencies of the same order as these. It follows that instead of a true record of such a disturbance it tends to show an oscillation at its own natural frequency. In recording a pulse of mixed frequencies, it is possible to determine a predominant frequency by varying the natural frequency of the seismometer until maximum sensitivity is obtained. The seismometers employed by the I.G.E.S. possessed natural frequencies of the orders 100, 30 and 3 per second[c]. A stamp on the ground excited the first of these frequencies in the seismometer; an explosion of blasting gelatine at a distance of a few hundred feet excited the second; at a few thousand feet, the second and third together. It is probable that the frequency recorded does not depend so much upon the method of excitation of the disturbance, as upon the distance from the point of origin. Since waves of higher frequency are most rapidly absorbed, the portion of the range of component frequencies at which there is the greatest concentration of energy will pass to lower frequencies as the

[a] In addition, when the pulse travels through a stratified formation it may set it vibrating as a whole in a periodic oscillation: thus the formations beneath Tokyo are believed to have a natural period of oscillation of 4·6 seconds. Knott, *The Physics of Earthquake Phenomena*, p. 208, Clarendon Press, Oxford (1908).

[b] This phenomenon is known in Physics as "dispersion."

[c] The Schweydar seismometer (see p. 214) itself possessed three natural frequencies comparable with these; the Western Electric microphone (see p. 212) possessed the second, and the electrical recording system used with it (see p. 213) possessed the first; the hot wire seismometer (see Pt. II, p. 346) the second and third.

disturbance travels onward through the ground. Thus the frequency recorded in the vicinity of a destructive earthquake may be of the order one per second; whilst at a distance of thousands of miles the period of one oscillation of the ground, in the main shock, may be several minutes. The waves recorded by the seismic section of the I.G.E.S. generally had frequencies of the order of 30 per second. It is commonly stated that waves from artificial explosions are of higher frequency than those occurring in natural earthquakes. This may be due merely to the proximity of the seismometer in the case of an artificial explosion[a].

SECTION 2. SEISMOMETERS

An instrument used to detect and measure a movement in the ground is called a seismometer[b]. Ideally it should be so designed that one point of the instrument remains stationary while its framework moves with the ground, the relative motion being recorded. The stationary point is invariably obtained by so supporting a heavy mass by a framework that the latter is free to move in one direction at least, relative to the mass. This mass is known as an "inertia mass."

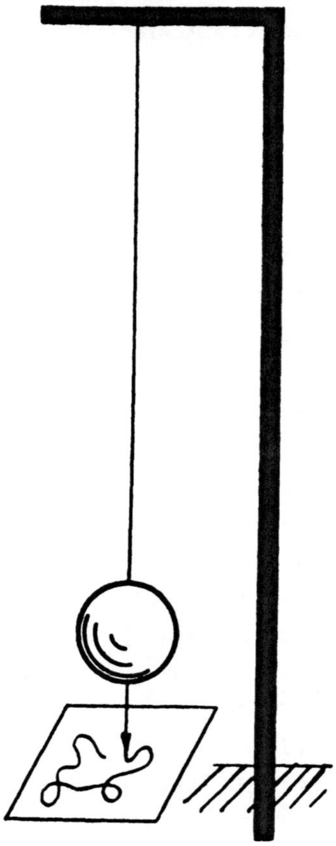

Fig. 150. The simple pendulum as a seismometer

The Simple Pendulum as a Seismometer. The simplest form of seismometer is a pendulum, in which a heavy bob is suspended by a long wire. If the point of support moves in a horizontal plane for a short period of time compared with the period of the pendulum, the bob will remain at rest; and the nature of the motion of the ground may be recorded by a pencil attached to the bob, as in Fig. 150. It is necessary that the ground movements should last only for a short period, for if they continue long enough the pendulum will be set in motion. The duration of the ground movements must be shorter than the period of the pendulum, for if they last for as long as its natural period the pendulum will be set into oscillation, giving a record which does not in its later phases represent the movement of the ground.

Further requirements of a Seismometer. There are a number of features in which this simple form of seismometer requires improvement.

(1) A pendulum would need to be about 3000 ft. long to have a period of 1 minute, and earthquakes lasting for more than a minute have to be recorded. The form of suspension must therefore be modified to give a possible instrument with a suitably low natural frequency.

[a] A natural earthquake rarely originates at a surface point, and even a seismometer directly above the point of origin is generally at a distance of some miles from it.

[b] Since the instrument usually gives a record, or "seismogram," it is sometimes called a "seismograph."

(2) Instead of the pendulum being free to move in any horizontal direction, it may be constrained to one, say the north-south direction; and may then record on a paper moving at right angles to this direction. A second pendulum would record the east-west movements.

(3) The movement may be magnified by means of a mechanical or an optical lever[a]; alternatively, electrical magnification[b] may be employed. Seismometers may be classified according as the recording system is mechanical (including optical) or electrical.

(4) As has been pointed out above, the pendulum will be set oscillating by every disturbance which lasts long enough, and it will continue long after the disturbance has ceased. A vane attached to the pendulum, dipping into a sufficiently viscous liquid, will damp out these natural oscillations; or electromagnetic damping may be used with advantage. In the absence of damping, the disturbance and the natural motion of the pendulum are combined in the record and cannot be separated satisfactorily. The seismometer may be so highly damped that the pendulum, when disturbed, returns to its position of equilibrium without passing beyond it, and therefore without oscillation. It is then said to be *aperiodic* and to be *critically damped* if it just fails to oscillate.

Simple Horizontal Seismometer. Modifications to meet these requirements have been embodied in the seismometer system shown in Fig. 151. The pendulum turns about an axis AB; the frame is therefore free to move in one direction only, relative to the inertia mass, so that only one component of the displacement is recorded[c]. The natural frequency is increased by bringing the inclination of the axis, AB, closer to the vertical. If it were vertical, then the period would be infinite, and the equilibrium would be neutral in type, which is not desirable.

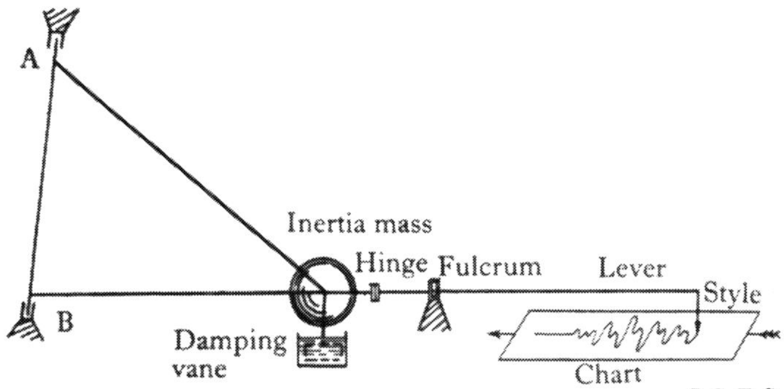

Fig. 151. Simple horizontal seismometer

A complete record of the motion of the ground[d] must include the vertical component of a displacement, in addition to the two horizontal components. For this purpose "vertical seismometers" are constructed, in which the relative motion of the inertia mass and the framework is in a vertical line only. The design of a compact vertical seismometer of low natural frequency is, however, more difficult than that of a horizontal seismometer.

[a] An example of an optical lever is given in the description of the Schweydar seismometer, p. 214.
[b] See examples, pp. 210–212.
[c] A rotation would also be recorded, but for seismic disturbances such movements are generally small.
[d] Excluding rotations.

The horizontal seismometer system shown in Fig. 151 is typical of those designed by Milne, Wiechert, and many others. Different types vary in the methods adopted for suspension, damping, and recording, but all seismometers, horizontal and vertical, have the same essential parts:

1. An inertia mass, suspended so as to have one or more "degrees of freedom," in a pendulum system of sufficiently low natural frequency.

2. A damping device, to suppress the natural oscillation of the pendulum.

3. A recording system, which usually includes magnification.

Measurement of Velocity and Acceleration. In the seismometer just described the recording is done mechanically or optically, and a record is obtained of the displacement of the ground particles in a known time. When a particle moves, its motion may equally well be determined by measuring its velocity or its acceleration. It so happens that if other means of recording are used the velocity or the acceleration is actually measured, as is shown in the following table:

Method of recording	Description	Measured
Electromagnetic	Coil of insulated wire moving relative to a magnetic field	Velocity
Carbon granule microphone	Carbon plates and granules as in a telephone microphone	Acceleration
Hot wire	Air set in motion by the seismic disturbance flows past an electrically heated wire, thereby cooling it	Velocity

Electrical Seismometers. In the electromagnetic seismometer[a] the relative motion of inertia mass and framework is translated into relative motion of a coil of wire and a magnetic field. The coil therefore cuts the lines of force of the field, so that a current arises in it which depends upon the rate at which the lines are cut, and therefore upon the relative velocity.

The carbon granule seismometer[b] utilises the change of resistance which occurs in carbon under pressure variations. The pressure is exerted by the relative motion of the inertia mass and the frame, and is proportional to the relative acceleration[c]. A constant voltage is applied to an electrical circuit in which the carbon granules are included. The current in the circuit undergoes a change which is proportional to the resistance change, and this in turn is proportional to the pressure change, and thus to the acceleration.

The hot-wire seismometer[d] first makes use of hydraulic[e] magnification, the inertia mass and the frame being arranged as piston and cylinder respectively. Their relative motion forces air through a small orifice (Fig. 152) containing a grid of fine platinum wire, which is normally kept at a dull red heat by an electric current.

[a] The type used by the I.G.E.S. is described in Pt II, p. 345.

[b] The type used by the I.G.E.S. is described on p. 212.

[c] Newton's "Second Law" of Motion.

[d] The type used by the I.G.E.S. is described in Pt. II, p. 346. This seismometer was developed by the I.G.E.S., who, so far as is known, have been the first to employ it in the field. Australian Commonwealth Provisional Patent no. 23,284 has been taken out by Major Edgar Booth.

[e] More strictly, aerodynamic.

The ratio of the velocity of air through the orifice to the relative velocity of piston and cylinder is inversely as the ratio a/A of the cross-sectional areas of the orifice and the cylinder. The platinum wire is cooled by the air flowing past it, and its resistance diminishes proportionately to the velocity of the air[a]. The current change, with constant voltage applied to the circuit, is therefore proportional to the velocity of the air current, and so to the velocity of the ground movement[b].

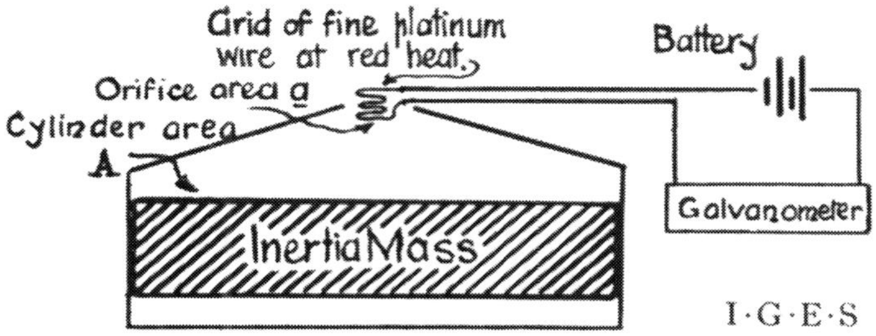

Fig. 152. Principle of the hot-wire seismometer. The hot-wire seismometer used by the I.G.E.S. is illustrated in Fig. 260

In each case the electrical circuit is connected with a galvanometer, either by shunt or Wheatstone bridge; or, to eliminate the possibility of deflections due to steady current, by transformer.

The change of current is measured by recording the motion of the galvanometer, which is proportional to the current. Thus a record is obtained; in the case of the carbon granule seismometer, of a component of the acceleration of the ground movement; and, in the case of the electromagnetic and hot-wire seismometers, of a component of the velocity of the ground movement. In each case further magnification may be obtained, if required, by means of valve amplification.

The Einthoven String Galvanometer. The requirements of a galvanometer for use in a seismometer system are high sensitivity, high natural frequency, and approximately critical damping. Moving-mirror galvanometers have been developed with frequencies as high as 10 per second, and it should be possible to mcdify them for recording the time of first-arrival. They would be unsuitable, however, for recording the remainder of the disturbance accurately.

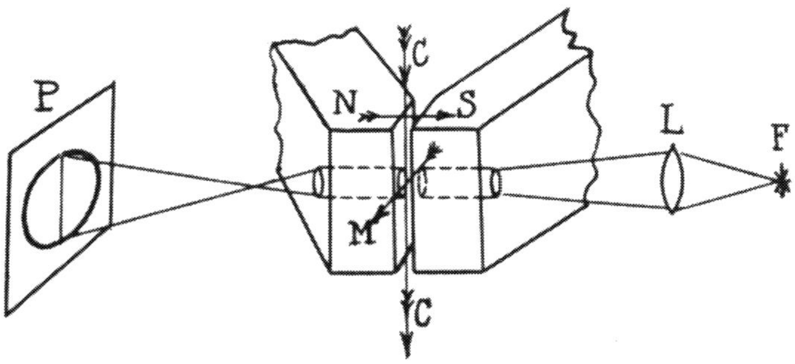

Fig. 153. Principle of the string galvanometer

M. Direction of motion of string. *C.C.* Conducting "string". *N.S.* Poles of electromagnet. *F, L.* Lamp filament and condensing lens. *P.* Projection screen.

The Einthoven string galvanometer is less sensitive and more cumbersome than these, but may be designed to have a very high natural string-frequency, and to be critically damped; it is especially suited for recording high-frequency disturbances. In this galvanometer the current passes through a fine wire, e.g. of platinum or plated

[a] Tucker and Paris, *Roy. Soc. Phil. Trans.*, A, vol. 221, p. 389 (1920–21).

[b] There are several other forms of electrical seismometers such as those depending on capacity variations and on the piezo-electric effect. The latter records accelerations.

glass, placed in the field of an electromagnet. The string is deflected in a direction at right angles to that of the current and to that of the magnetic field, as is shown in Fig. 153. A beam of light passes through a system of lenses in the manner shown in Fig. 154, and throws an image of the string on a moving strip of photographic paper or film. The beam is interrupted at regular intervals by the spokes of a wheel driven at constant speed, thus producing time marks on the record.

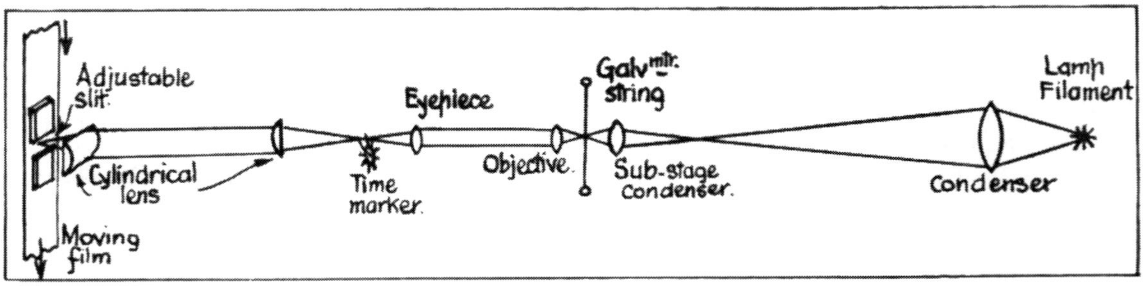

Fig. 154. Optical system for string galvanometer I.G.E.S.

Seismometers used by the I.G.E.S. Two types of seismometer were used in the I.G.E.S. experiments, namely, electrical and mechanical. The routine method developed employed a number of electrical seismometers connected to a central recording station. The electrical seismometers tested were: (1) several types of carbon granule microphone; (2) an electromagnetic seismometer; and (3) a hot-wire seismometer. The mechanical seismometer used was a Schweydar instrument made by the Askania Company.

Western Electric Microphone. Several types of carbon microphone were tested, but the field work was eventually carried out with the Western Electric Mining De-

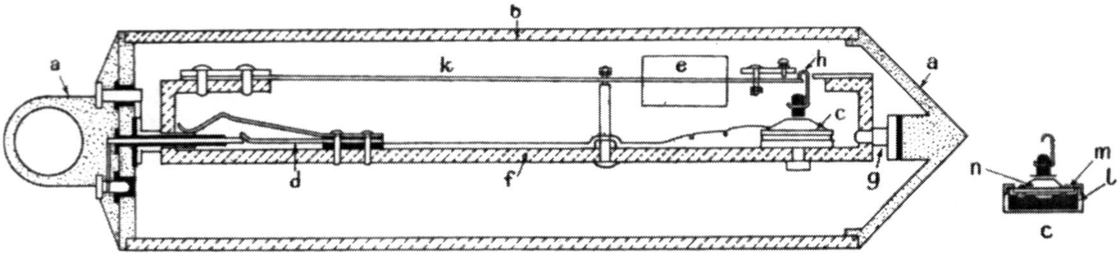

I·G·E·S

Fig. 155. Western Electric microphone

a. Aluminium cap	*e*. Inertia mass	*k*. Spring
b. Casing	*f*. Frame	*l*. Carbon
c. Capsule	*g*. Pivot	*m*. Mica
d. Lead to capsule	*h*. Pressure hook	*n*. Brass

tector which is illustrated in Fig. 155. This consists of a torpedo-shaped casing of brass and aluminium, 10 in. long and 3 in. in diameter, inside which a microphone system is pivoted in such a way that, when the instrument is placed with the long axis horizontal, the system rotates so that it will always be sensitive to both horizontal and vertical components of earth movements. The microphones employed had a natural frequency of about 30 per second, but the period can be altered by adjusting the position of the inertia mass, which is supported on a blade spring.

Electrical recording system. The recording apparatus used with the electrical equipment was a six-string Einthoven galvanometer, manufactured by the Cambridge

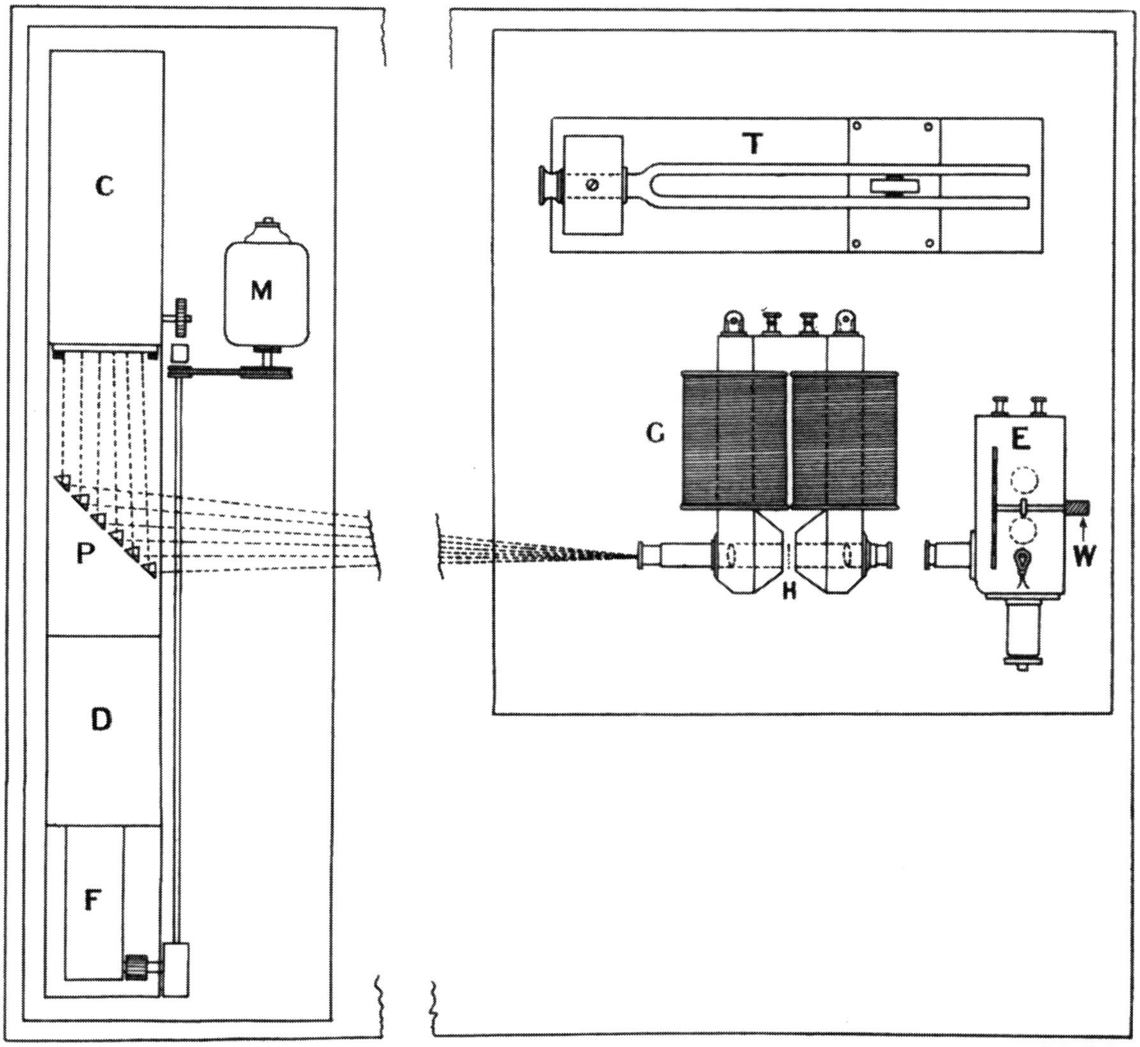

Fig. 156. Six-string galvanometer panel with automatic developing camera

G. Galvanometer. *H*. Harp of six strings. *E*. Electric lamp house. *W*. Time-marking wheel.
T. Tuning fork, governing speed of wheel. *P*. Prism bench. *C*. Camera. *M*. Motor.
D. Automatic developer. *F*. Automatic fixer.

Scientific Instrument Company, and obtained from the British War Office. By this system the records of six seismometers appear side by side on the one film. The time wheel, which was governed by a tuning fork, made 100 marks per second on the record, thus enabling times to be read to 1/1000th second. The time of explosion was marked by the movement of one string, which was connected by shunt or transformer with an electrical circuit, which included a fine wire placed in the explosive at the origin of detonation. The same string was also used for a seismometer, if the latter was at a sufficient distance from the explosion.

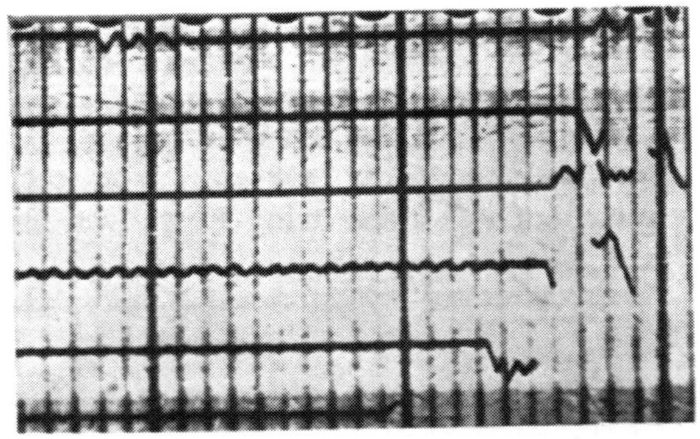

I.G.E.S.

Fig. 157. Typical microphone record. Strings are in sequence, Nos. 1 to 6, from top to bottom of the record. String No. 1 records the moment-of-explosion, in addition to the ground movement at the most distant seismometer

A typical record is shown in Fig. 157. The heavy cross lines are at intervals of 1/10th second, the lighter lines at intervals of 1/100th second. String No. 1 has been used to record the moment of explosion as well as the disturbance at the most distant seismometer. At the seismometers connected to strings Nos. 5 and 6 the surface disturbance was the first-arrival, while for strings Nos. 1 to 4 the refracted disturbance was the first-arrival. Except for string No. 4 the times of first-arrival may readily be picked out; but the seismometer connected to string No. 4 was evidently not in a condition of full sensitivity; the arrival of the refracted disturbance being indicated merely by a slight change in the small microseismic oscillations of the string, at about 0·013 second before the obvious "break." This "break," which might be mistaken for first-arrival, is in reality that of the surface disturbance.

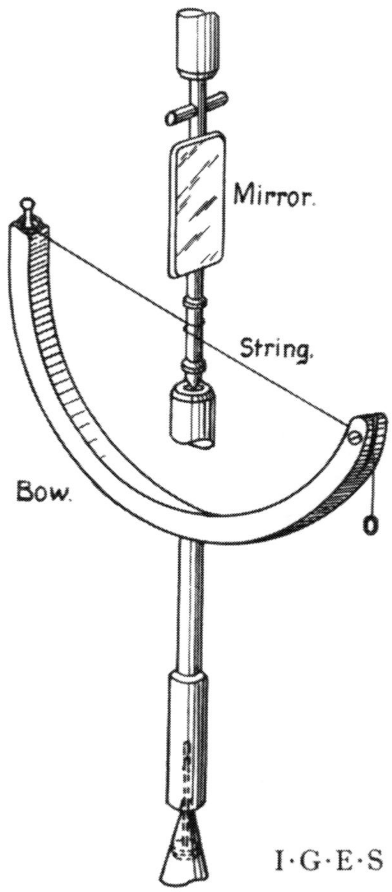

Fig. 158. Schweydar seismometer bow and string mechanism

The Schweydar Seismometer. The Schweydar mechanical form of seismometer is designed to be portable, and sensitive to the frequency of earth vibrations used in seismic prospecting. It has two inertia masses mounted on blade springs, responding to vertical and horizontal components of earth movement respectively. The movement of each mass relative to the frame of the seismometer is first magnified by a light conical lever,

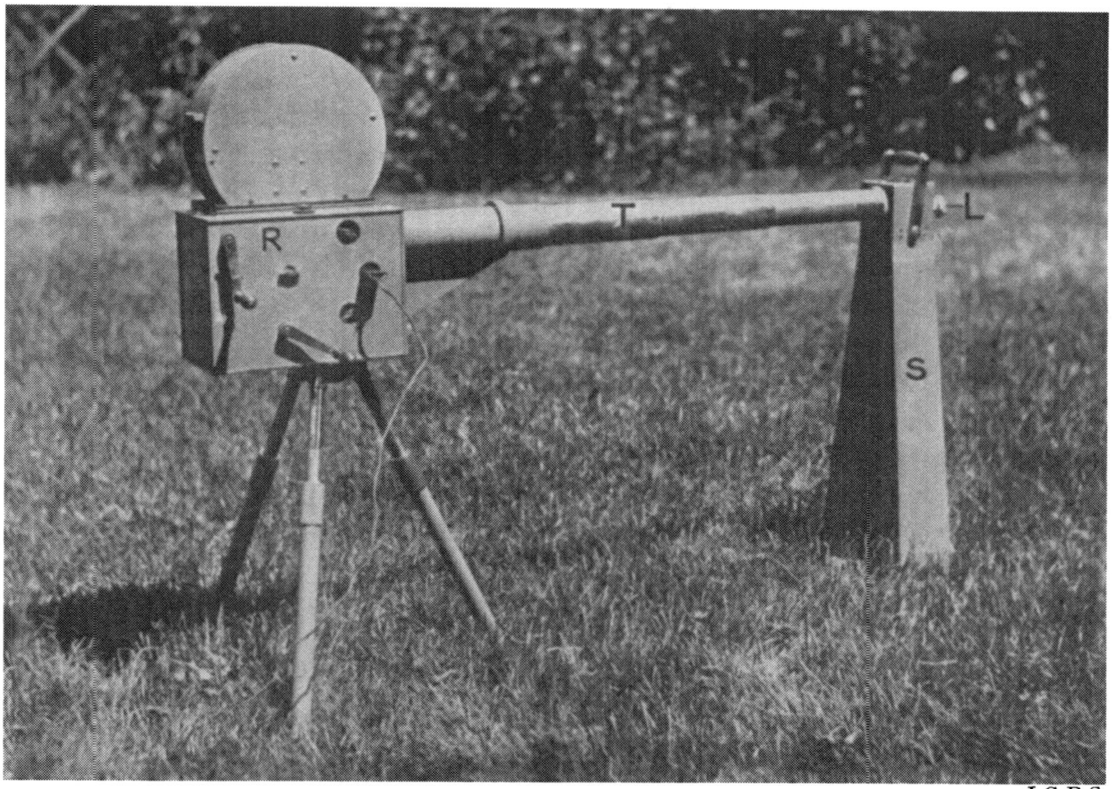

Fig. 159. Schweydar seismometer and camera
S. Seismometer. *T.* Telescopic tube. *R.* Recording camera.
L. Attachment for lamp for marking moment of explosion.

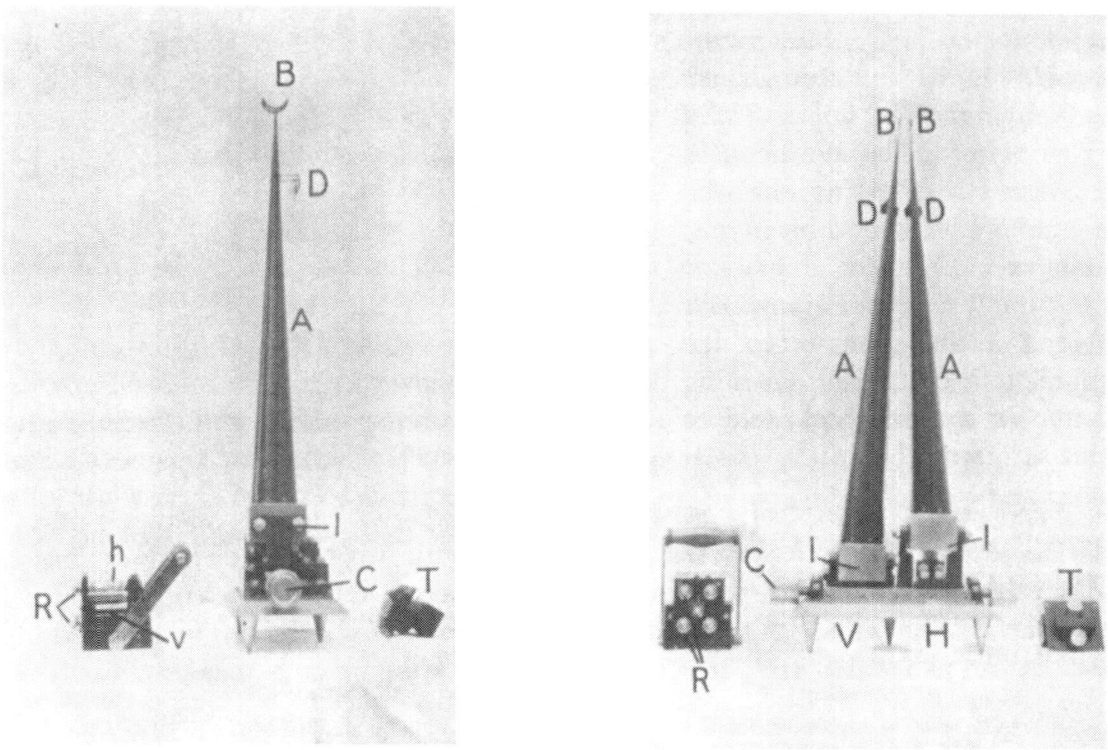

(*a*) (*b*)

Fig. 160. Schewydar seismometer dismantled
I.I. Inertia masses. *A.A.* Aluminium cones. *T.* Damping chambers. *B.B.* Bows. *R.* Adjusting screws. *V.* Vertical component system. *D.D.* Damping vanes. *H.* Horizontal component system. *C.C.* Clamping screws.

rigidly attached to the former. At the vertex of each cone is a bow carrying a string which passes around a spindle on which is mounted a light mirror, as in Fig. 158. Movement of the bow rotates the mirror, which reflects a beam of light to a focus a metre away on to a strip of photographic paper moved by clockwork, in a camera which is set up as a separate instrument. Apart from possible slipping between the string and the spindle, the record of an earth tremor is thus magnified mechanically and optically 14,000 fold. The motion of the cone is damped by means of a vane dipping into glycerine. The seismometer and its camera set up ready for use are shown in Fig. 159, the seismometer dismantled is shown in Fig. 160.

Time intervals are marked on the photographic paper by interrupting a beam of light, at intervals of 0·15 second, by a small pendulum. The moment of explosion was recorded by the I.G.E.S. by the extinction of an electric lamp which illuminated the photographic paper, and whose circuit included a fine wire placed in the explosive.

The Schweydar seismometer was used for comparison with electrical seismometers.

SECTION 3. FIELD PROCEDURE

Equipment and Personnel. The field party, in addition to the officer in charge and a temporary field assistant (changed from time to time), consisted of a youth trained as laboratory assistant and one or two miner-labourers. A 1-ton Chevrolet truck served for the transport of both equipment and personnel. The recording station was housed in a collapsible wooden hut, 10 ft. by 10 ft., later reduced to 7 ft. by 6 ft. The hut could be darkened for focussing the galvanometer strings by a blind over the windows. The accompanying photograph, Fig. 161, shows the larger hut in position in the Gulgong field. The recording equipment and the hut could be dismantled, as shown in Fig. 162, loaded on and unloaded from the truck, and re-established in working order, at the next station, within an hour.

I.G.E.S.

Fig. 161. The recording hut at Gulgong, N.S.W.

The Explosion. The explosive used for routine work was blasting gelatine[a]; lithyte[b] and a 50 per cent. gelignite[a] were also used for certain tests. The charge was usually buried at a depth sufficient to communicate as much of the energy to the ground as possible, and to avoid wasting it in the air or in throwing up large masses of earth. The

[a] See p. 203. [b] A potassium chlorate explosive, see p. 203.

efficient burying of charges had the additional advantage of avoiding damage to the surface by the formation of craters. Holes were drilled in the ground with jumping bars to a depth of 12 ft., and the bottom of the hole was enlarged by a preliminary "bulling" charge of 1/10th to 1 lb., to accommodate the main charge of explosive. At a depth of 12 ft. in hard soil a charge of 25 lb., well tamped, merely cracked the surface without forming a crater, as in Fig. 163. In soil too loose for drilling, and on rock, the explosive was buried in a shallow hole or was simply placed on the surface. Charges up to 35 lb. were used.

I.G.E.S.

Fig. 162. Dismantling the hut for transport to next station

The method of recording the moment of explosion, by the breaking of a fine wire placed in the explosive, was found to be satisfactory. Experiments in which explosions were brought successively closer to the seismometer showed that the lag in recording was not more than 1/500th second. On the other hand, the current in the circuit used for firing the detonator could not be employed, as it was found to involve a variable time lag, up to 1/50th second.

I.G.E.S.

Fig. 163. Ground surface unbroken after firing buried charge of 25 lb. blasting gelatine. The mouth of the drill-hole is at *D*

Disposition of Seismometers. For each explosion six microphones were usually employed, at stations established on a line passing through the explosion point. At least one of the stations was placed sufficiently close to the shot-point to receive the surface wave as the first-arrival, the remainder receiving refracted waves as first-arrival. For example, with a depth of 150 ft. of alluvium over a bed-rock of slate, distances of 300, 400, 600, 800, 1000, and 1200 ft. might be used; the maximum distance suitable for accurate recording under such conditions was about 2000 ft. The first one or two stations then give points on the graph for the determination of surface velocity[a]. Such a scheme of disposition is shown in Fig. 164, in relation to the corresponding time-distance graph. Where desirable, the shock from the bulling charge was also recorded. This was sufficiently strong for

[a] The arrival of the surface wave was frequently well marked on the record, even when the refracted wave had come first; corresponding points could then be added to the graph for the surface wave.

a determination of surface velocity within a few hundred feet, but not as a rule sufficient for recording the moment of arrival of refracted waves. The microphones were placed in narrow trenches about 6 in. deep, with their axes transverse to the line joining the seismometer point to the shot-point. Short lengths of fine wire were inserted between the leads and the microphone terminals in order to minimise the transmission of vibrations. The microphones were lightly buried, to damp out vibrations of these wires, and for protection from impacts due to sand carried by wind, to ants, etc. As the sensitivity of the carbon granule microphone is liable to large variations, each detector was tested after it had been placed in position[a].

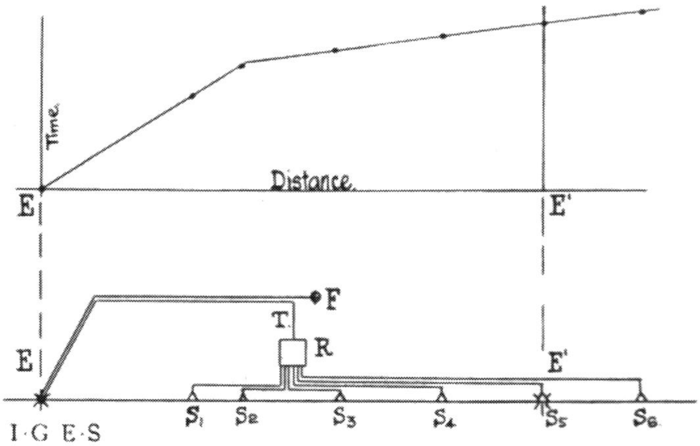

Fig. 164. Disposition of seismometers

E. Explosion point. *R*. Recording station. S_1, S_2, S_3, etc. Seismometer stations. *E'*. Point of succeeding explosion. *F*. Firing point. *T*. Moment-of-explosion lead.

Connection with Recording Station. To connect the microphones with the recording station, light insulated cable was used, laid upon the ground. The cable was divided into units, each consisting of a pair of parallel wires 300 ft. long, which could be joined in series, or to the terminals at the recording stations, by plug connectors. The charge was exploded electrically from a point within signalling distance of the recording station, the connecting cables being run at a sufficient distance from those leading to the microphones, so that the exploding current would not induce a current in the microphone circuits. The circuits for recording the time of explosion included an indicator lamp at the recording station; the cables for this circuit also had to be run at a sufficient distance from those leading to the microphones. The laying down and taking up of cables was found to occupy the greater portion of the total time required. The disposition of connecting cables is shown in the scheme illustrated in Fig. 164.

Disposition of Shots for Prospecting. As a general rule a series of shot-points was taken on a straight line, at intervals of 1000 ft., and seismometer points were taken at smaller intervals on the same line. A set, or network, of such lines might be designed to determine the geological structure of an area. Shooting was carried out in both directions[b] over the whole length of line. For this purpose two shots were fired at each

[a] For example, as a final test before the explosion, one member of the party walked past each microphone in turn down the whole length of line used, while the operator at the recording station watched the behaviour of the galvanometer string shadows on a screen. Testing by including headphones in the circuit is convenient, but it is difficult to distinguish between high sensitivity and low damping. Either of these methods is, however, a test of sensitivity to frequencies higher than those which predominate in the pulse received from the explosion.

[b] See pp. 198–200.

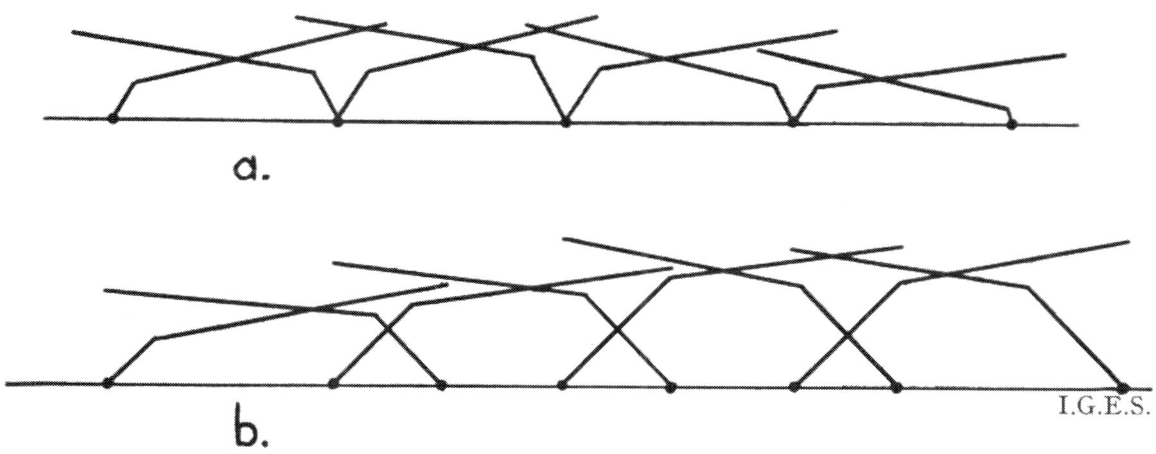

Fig. 165. Disposition of shots in prospecting along a straight line traverse, illustrated by the resulting time-distance graphs

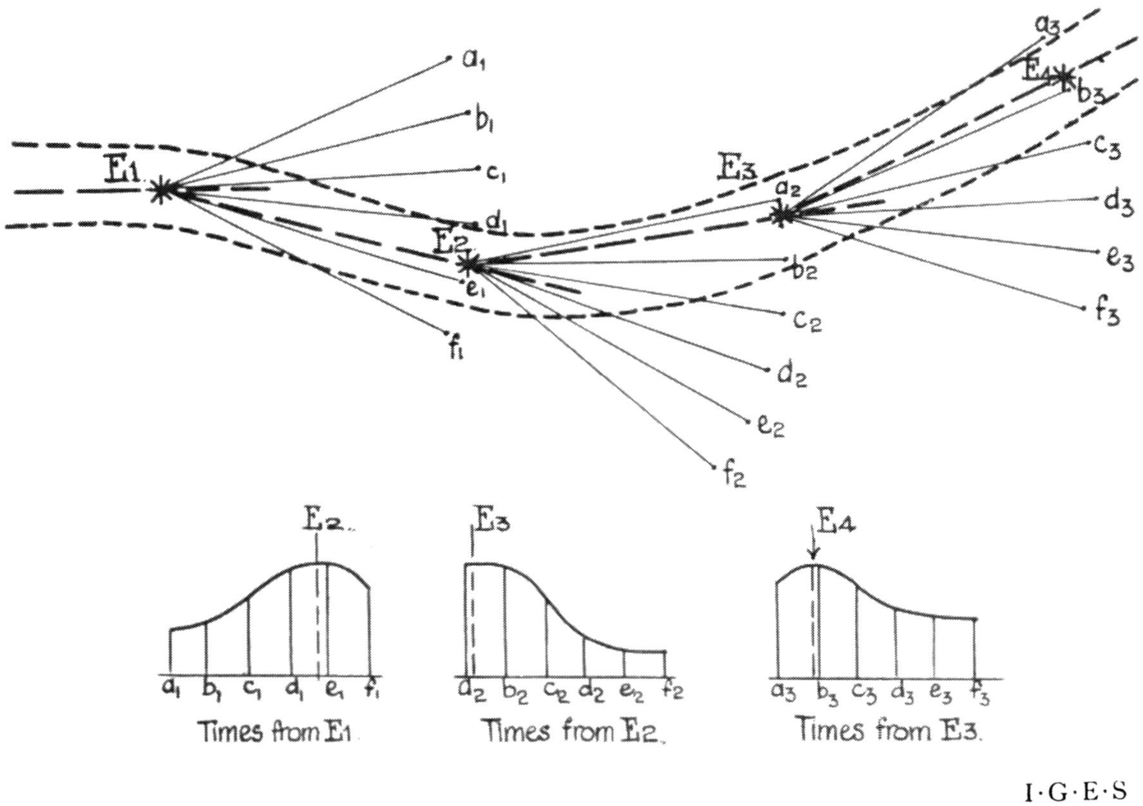

Fig. 166. Tracing course of deep lead by shooting in fans. Plotting time of first-arrival at the six seismometers, the centre of the lead is assumed to be below the point of maximum time and this point is taken for the next explosion point

shot-point. The seismometer points were spaced so as to extend up to, or beyond, the next shot-point, situated at a distance of 1000 ft.; or, where sufficiently strong waves could be obtained, seismometers were placed up to the second shot-point, situated 2000 ft. from the first explosion. This system would give a set of time-distance graphs such as is shown in Fig. 165 *a*, for which every portion of the line is covered by a graph of first-arrival of refracted waves travelling in each direction. For some depths of bed-rock it was found preferable to space shot-points at 500 ft. intervals, with one shot at each point, recording the waves travelling in alternate directions for alternate shot-points; this would give a set of graphs such as are shown in Fig. 165 *b*. When the rock was close to the surface, as in the determination of velocities in outcrops, seismometers were placed up to a total distance of 3000 ft.

An alternative method of exploration was to have the six seismometer points at the ends of six lines, of length 1000 to 2000 ft., radiating in a fan from the shot-point. The recording station was placed centrally amongst the seismometer stations. Such a method is used in preliminary prospecting to discover points of pronounced irregularity in otherwise regular sub-surface strata; for example, in the discovery of salt domes. It was used by the I.G.E.S. to trace out the course of a deep lead in the bed-rock surface. The vertex of the first fan of lines was taken at a point known to be over the centre of the lead; from the results of the observations a second point was selected as being over the course of the lead and was made the vertex of a second fan. Proceeding in this way, as illustrated in Fig. 166, the course could be followed fairly rapidly, if the structure was not too complex.

SECTION 4. FIELD TESTS IN AUSTRALIA

Choice of Areas for Field Work. As stated on p. 7, the investigations made by the Seismic Section of the I.G.E.S. were limited to tests of theory and methods for small-scale structures. For this purpose an area of simple structure was required, in which some definite quantitative geological details were available. The ideal conditions for these initial experiments would have been a homogeneous rock with a uniform cover of another homogeneous rock of lower velocity, extending over a considerable area. The Gulgong gold field, in New South Wales, was suggested by Dr Andrews, the Government Geologist of that State. Here the overburden is an alluvial sheet of sands, clays, and pebble beds, of varying depth, not exceeding 250 ft., resting directly on a bed-rock of granite and slate. Depths, however, were known from a line of bores and no more suitable conditions were obtainable within reasonable access. A period of 38 weeks was spent in this district, of which about 27 weeks were occupied in research upon instruments and upon wave propagation, and the remainder in trial runs prospecting for deep leads.

The Seismic Section also worked in the Tallong, New South Wales, district for a few weeks. This region, which was suggested by Assistant-Professor W. R. Browne, of the University of Sydney, included a large variety of geological materials and structures within a small range, thus affording opportunities for collecting data as to velocities in the different materials and as to types of graphs given by the different structures.

PRELIMINARY FIELD TESTS

Nature of Disturbances Propagated in Surface Formation. Fig. 167 shows three records given by the Schweydar seismometer at a distance of 800 ft. from a surface explosion, with a depth of alluvial overburden between 200 and 250 ft. They record the three components of vibration, x, in the direction of propagation, y, in the surface, perpendicular to the direction of propagation, z, perpendicular to the surface and to the direction of propagation. The small arrows indicate the times of arrival of waves which were also identified on records obtained at other distances from the point of explosion[a]. R, the refracted wave, appears in the z component only, and is thus a longitudinal wave coming from below, in accordance with the theory given above. A, B, and C, were shown, from the constancy of the corresponding velocity at varying distances, to be surface waves[b]. Although for the surface waves the x component records first, corresponding to a longitudinal wave travelling along the surface as assumed by the conventional theory, the motion quickly becomes more complex, involving both x and z components; there is, however, very little movement with the y component. S is the arrival of the sound wave, at which the record disappears; since not only does a far greater proportion of the energy go into the air, for a surface explosion, but also there is a smaller absorption in air than in alluvium.

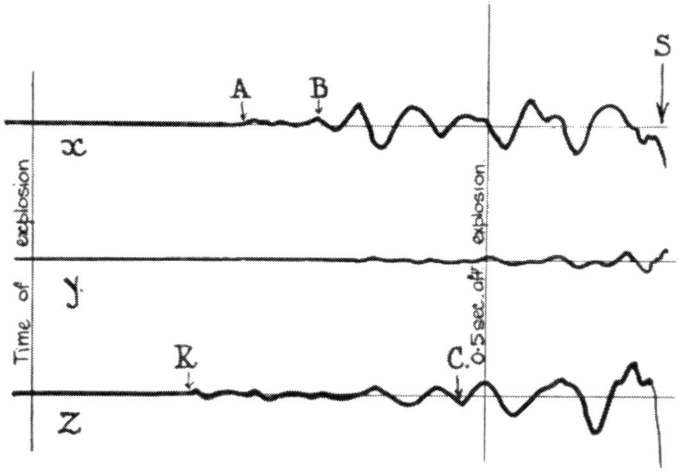

Fig. 167. Comparative records for three components x. Horizontal component in direction of propagation. y. Horizontal component at right angles to direction of propagation. z. Vertical component. Schweydar seismometer distant 800 ft. from surface explosion of 2 lb. blasting gelatine.

To compare the effect of a surface explosion with one in which the charge is efficiently buried, Fig. 168 shows records given by the Schweydar seismometer for such

[a] See Pt. II, p. 335.

[b] *Referee's comment.* In Fig. 168 the onset of the refracted disturbance apparently shows in the x component as well as in the z component, and also in Fig. 167 the onset A in x certainly appears to occur in z at the same time. This seems to conflict with the conclusion given later that the A onset has no vertical component and the R onset little horizontal component.

a pair of explosions, at a distance of 1200 ft., on a depth of alluvium between 200 and 250 ft. The upper records are of the horizontal, *x*, component; the lower of the vertical.

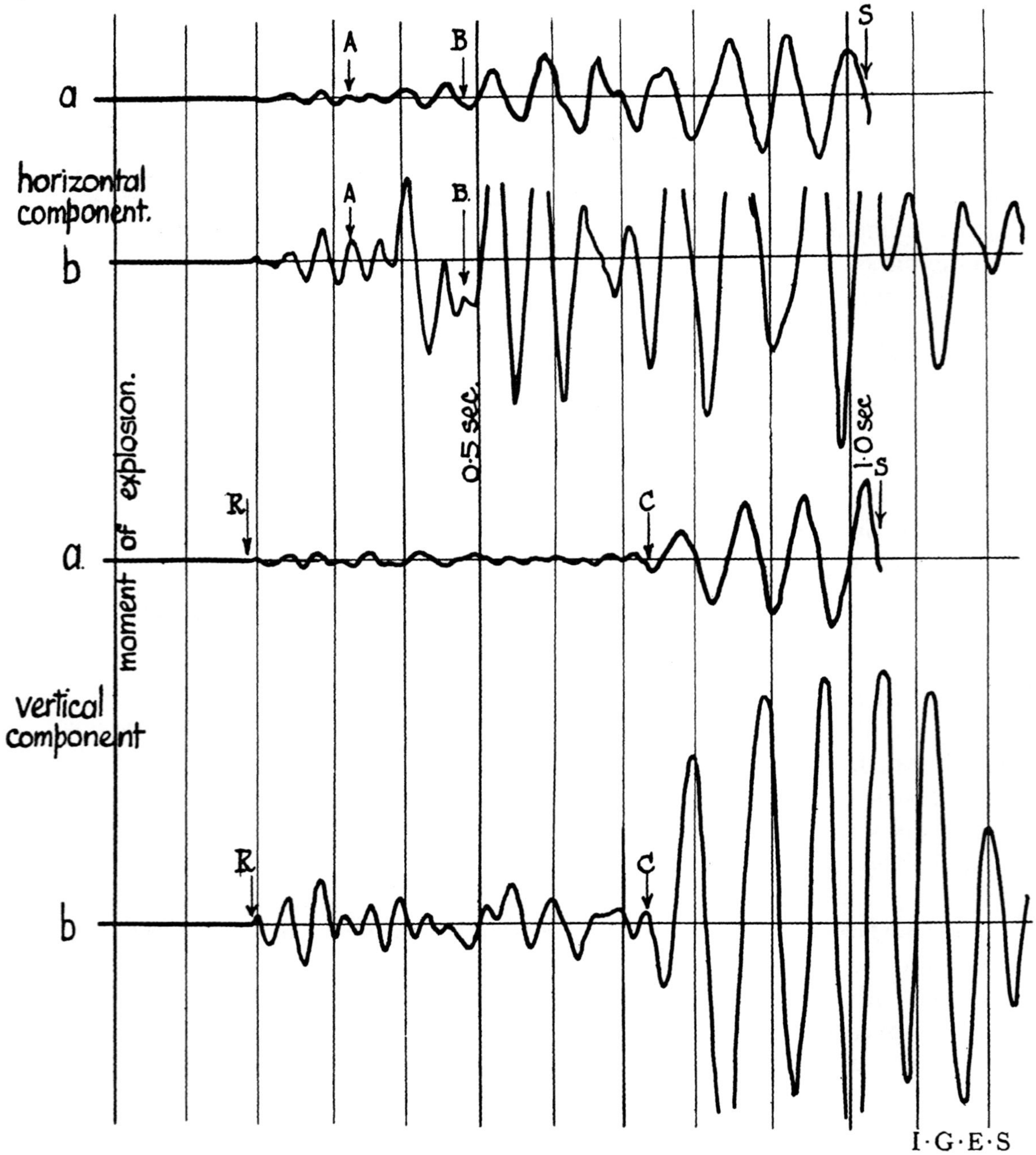

Fig. 168. Comparative records for surface and buried charges. Schweydar seismometer distant 1200 ft. from explosion

a. Surface explosion, 9 lb. blasting gelatine. *b*. Buried explosion, 12 lb. blasting gelatine.

Measurement of these records shows the buried charge—after allowing for its slightly greater size (12 lb. as against 9 lb.)—to be from 10 to 20 times as efficient in putting

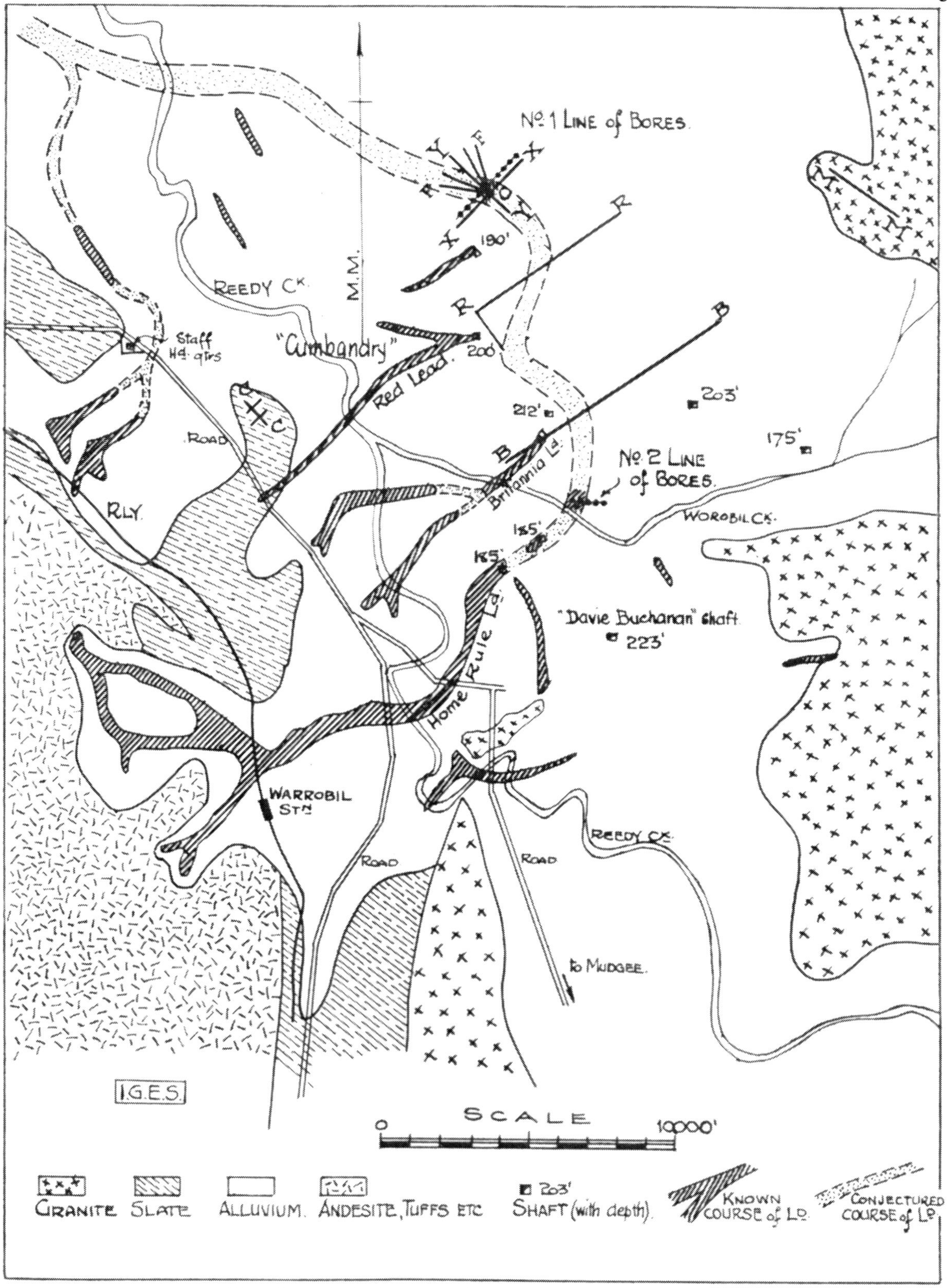

Fig. 169. Geological map of the Gulgong area, N.S.W.

OX, *OY*. Cross traverses at line of bores.
OF---F. "Fan" traverses at line of bores.
RR. "Read lead" traverse.
BB. "Britannia lead" traverse.
CC. "Cumbandry" traverses (for slate velocity).
MM. "Mountain" traverse (for granite velocity).

energy into the ground[a]. Although the two records differ in appearance, they agree in the times of arrival of waves at the points marked, the letters used again having the significance stated in the preceding paragraph. It will be noted that for the buried charge there is no sign of an air-borne disturbance[b].

FIELD TESTS AT GULGONG, N.S.W.

Geology[c]. The portion of the Gulgong gold field shown in Fig. 169 was suggested by the Government Geologist for New South Wales as providing some at least of the general characteristics required by the I.G.E.S. for its initial experiments in seismic methods. The presence of the two lines of bores giving definite structural information was so important a factor that the selection of the area was decided upon. The ground in question lies to the north of the village of Home Rule and 5 miles east of the Gulgong township, where the magnetic survey over sub-basaltic deep leads was subsequently carried out (p. 188). Those portions of the deep lead system which have been discovered and worked in the past (1871–94) are shown in Fig. 169. Although not so rich as the Black Lead and others lying near Gulgong town, some of them were of great importance—the Home Rule Lead itself averaging about 2 oz. of gold to the cubic yard over its southern end, where bed-rock lay at a depth of 100 ft.

From the general evidence of their levels it has long been concluded that they are tributaries to a main channel, approximately following the supposed course shown in Fig. 169, which must lie at a depth of over 200 ft. below the present land surface. Owing to the expense and difficulty of dealing with the water and running drift, the work in the tributary leads never extended as far as their junctions with the main channel; but the two lines of bore-holes put down by the Mines Department in 1901 and 1903, respectively, and the Davie Buchanan Shaft, afford conclusive evidence as to the existence of the latter.

By the No. 1 Line of Bores, which is 3000 ft. in length, the main channel was found just where anticipated, with a level clay-slate bottom about 700 ft. wide, lying beneath a cover of 250 ft. of heterogeneous alluvial material—mostly sands, clays, and gravels. It was deduced that on each side of the channel the bed-rock sloped rather steeply and, as will be seen from the cross-section in Fig. 170, the slate gives place to granite at the higher levels.

Most of the seismic investigations undertaken in this area by the I.G.E.S. were carried out in the neighbourhood of the No. 1 Line of Bores, where definite information regarding the position and depth of the main lead was available, and where the general

[a] This was in alluvium; if the charge were placed in a cavity in rock the efficiency should be much greater.

[b] A comparison of the records shows that the tremors appearing immediately before the shock of the sound wave are a real part of the sound wave, preliminary to the main shock, and not earth-borne waves of comparable velocity. It is probable that the main shock is due to the transmission to the ground of the vibration of the shelter in which the seismometer is housed, and that the earlier tremors represent the direct effect of the sound wave arrival upon the instrument itself.

[c] From information given in an unpublished report by L. J. Jones, Esq., Field Geologist, N.S.W. Dept of Mines, made available by permission of the New South Wales Government.

structure appeared to be simple. The two hundred or so feet of alluvium overlying the slate and granite bed-rock afforded satisfactory velocity ratios, and the general dimensions of the structure were considered suitable for the comparatively small scale in-

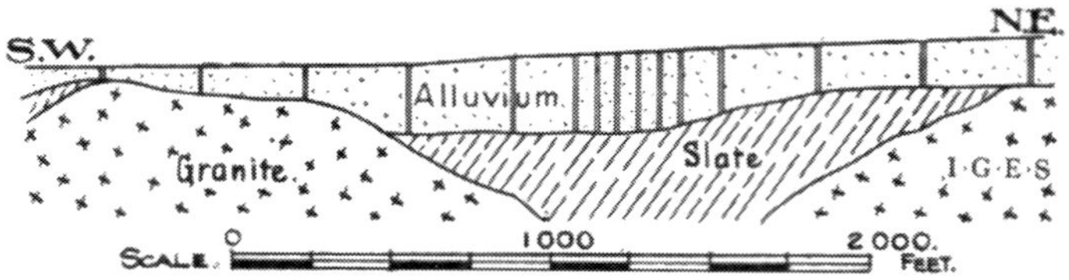

Fig. 170. Geological structure along line XX deduced from the No. 1 Line of Bores

vestigations that were to be carried out. Numerous outcrops of granite occur in the district, notably in the range of hills lying to the east, and some outcrops of slate; these afforded favourable opportunities for velocity determinations.

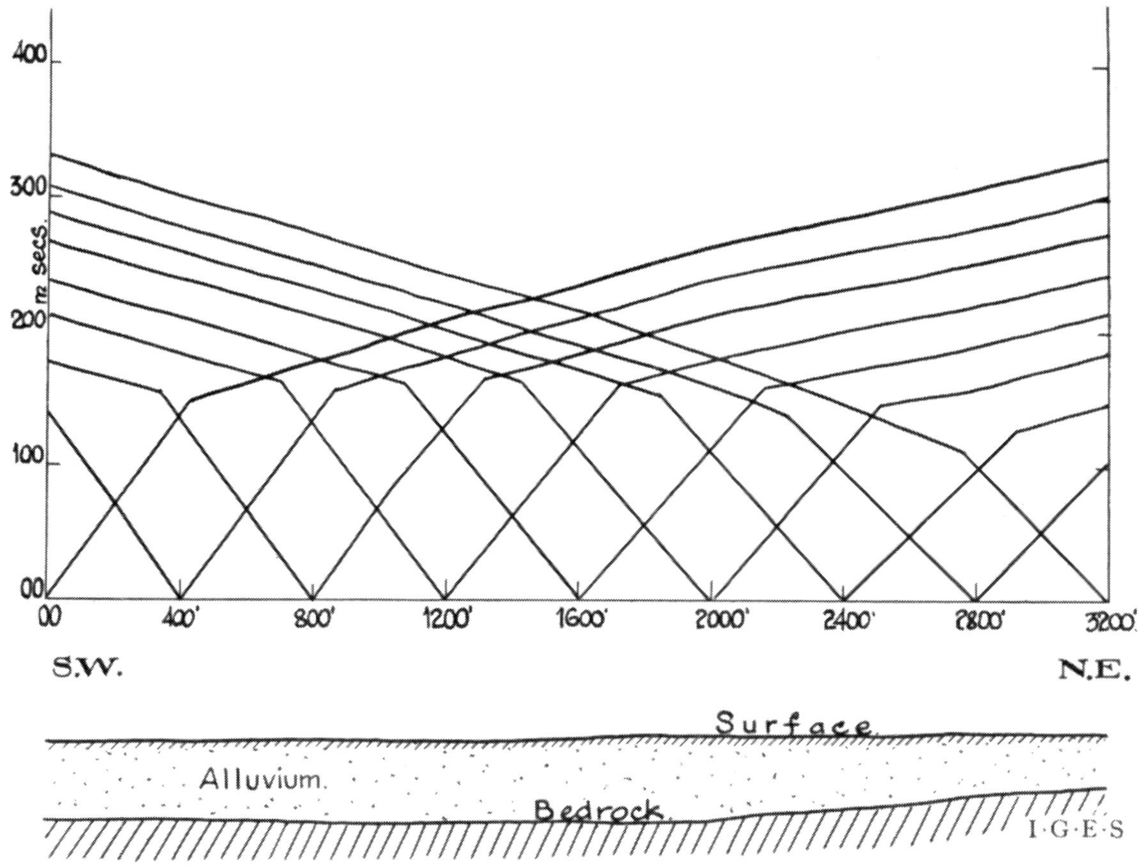

Fig. 171. Set of time-distance graphs on line XX and geological structure deduced from them

Check along the Line of Bores. The experiments to test the seismic method included a large number of observations taken over the known geological cross-section given by the No. 1 Line of Bores. A straight line was laid out to coincide as closely as practicable with the line of bores, which, however, was rather irregular in its alignment. More than 50 shots were fired on this line, at nine different points distributed over a length of 3200 ft., and over 300 seismometer records were obtained. The time-distance graphs for the nine shot-points are embodied in Fig. 171, the bed-rock profile calculated from them being shown beneath. The overburden velocity at the surface was found to vary between 2900 and 4200 ft. per second; the calculated velocity of the bed-rock lies between 13,500 and 14,500 ft. per second. The calculated depth of the channel agrees with that

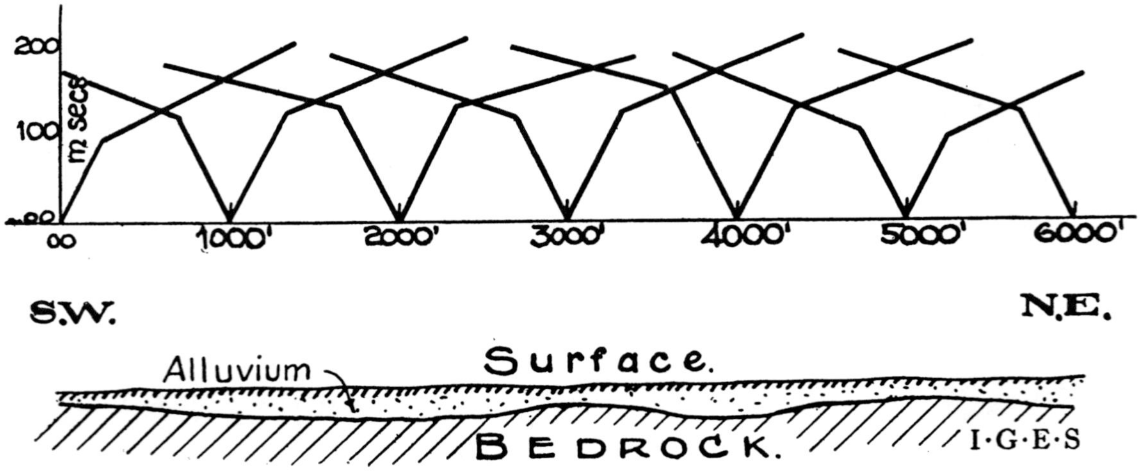

Fig. 172. Graphs and section along line *RR*

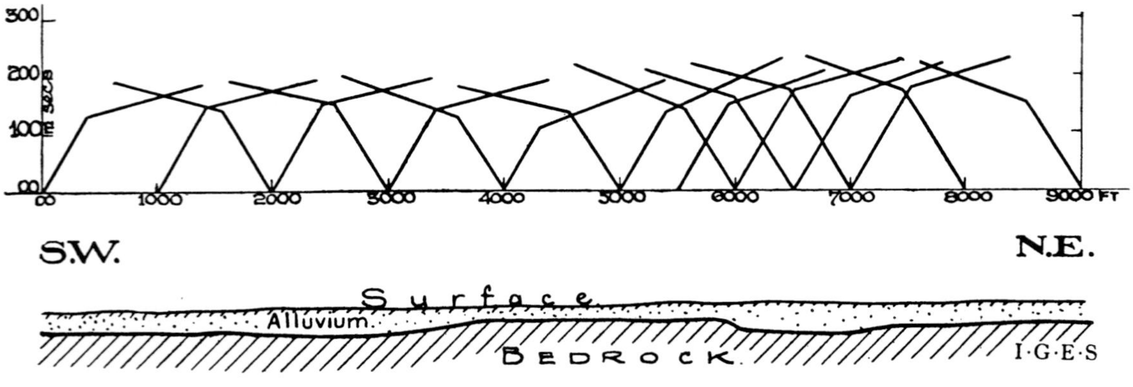

Fig. 173. Graphs and section along line *BB*

given by the bores, but the right-hand bank was not found to be as pronounced as in the cross-section deduced from the bores, and the left-hand bank was absent. These differences may be due largely to lack of coincidence of the two lines, especially as the later observations of the Gravimetric Section of the I.G.E.S. on the same site (see p. 170) showed that the structure is more complicated than had been deduced from the bores.

The number of observations is of course far greater than would be taken in prospecting, and the results have been examined in detail as a test of the refraction method. They give it a general support, but show that in such formations it is not as precise as

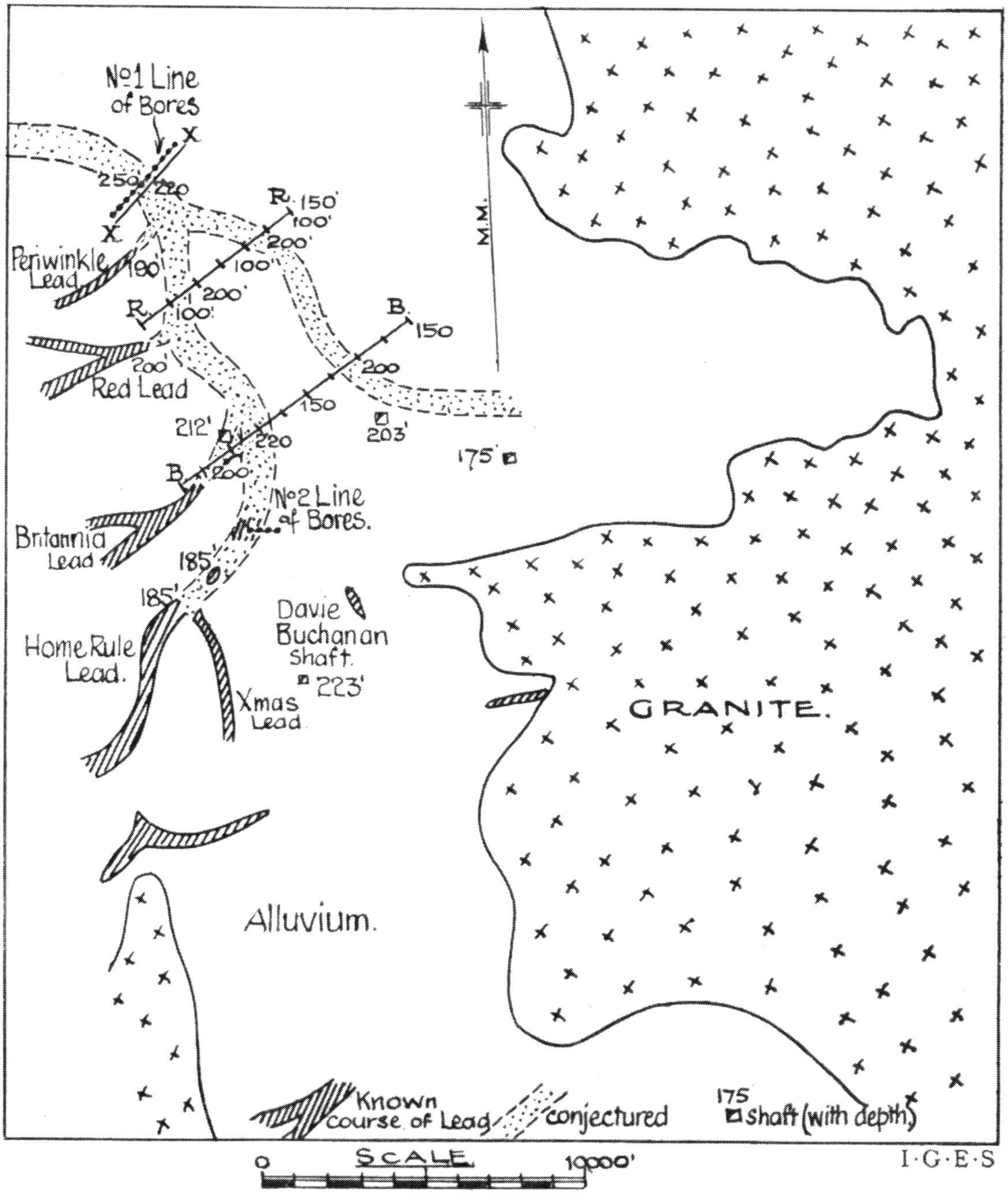

Fig. 174. Course of deep leads deduced from seismic observations

has at times been claimed[a]. It is very probable that both the depth to the bed-rock and the velocity in the bed-rock as calculated for a point, are in reality average values for a region extending for some distance around that point.

[a] See p. 202, footnote c.

Prospecting for Deep Leads. The method of prospecting by shooting in fans was tested in the vicinity of the line of bores, commencing with the "fan" of lines *OF - - F*, Fig. 169. Although it was successful in tracing the course of the lead, it was decided that the structures were not sufficiently pronounced, and probably too complex, for the advantages of the method to be fully realized. Next, the two lines *BB* 9000 ft. long, and *RR* 6000 ft. long, also shown in Fig. 169, were run across the supposed course of the lead, and at right angles to its supposed direction. The series of graphs obtained on these lines, and the bed-rock profiles calculated, are given in Figs. 172 and 173. The features shown by the profiles are not very marked, possibly due to the averaging-out effect, but it is evident that each line has crossed two depressions in the bed-rock surface.

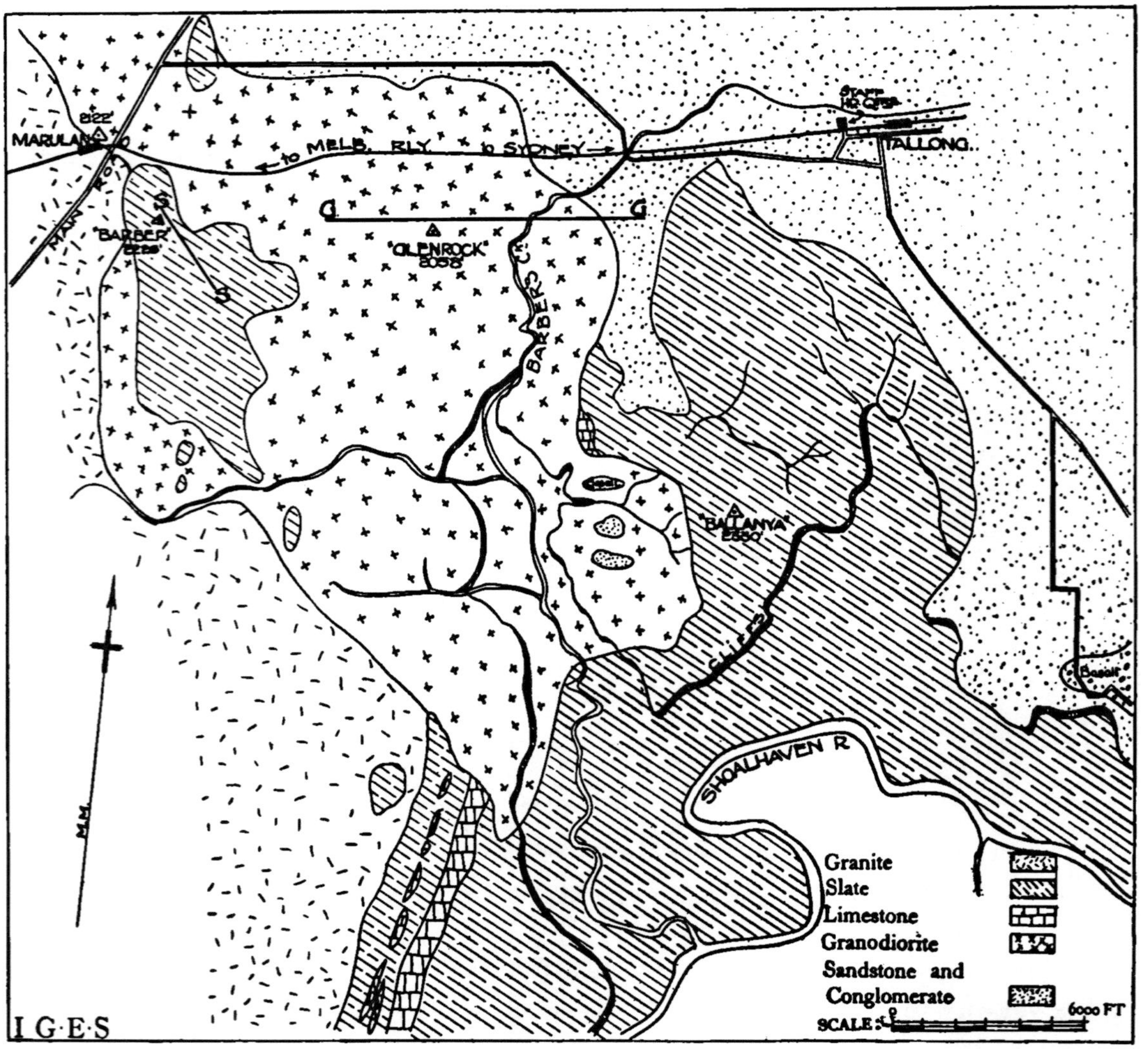

Fig. 175. Geological map of the Tallong area, N.S.W.

GG. "Glenrock" Traverse. *SS*. "Barber" ("Slate") Traverse.

The results of the three sets of observations suggest the courses sketched in Fig. 174 for two channels crossing the traverse lines *BB* and *RR*, and uniting close to the line of bores. It must, however, be emphasized that further work would be required before the existence of such channels could be fully proved.

FIELD TESTS AT TALLONG, N.S.W.

Geology. The Tallong district, of which a geological map is given in Fig. 175, is flanked on the west by a mass of granite, about 100 square miles in area. This gives place to the granodiorite which forms the centre of the Tallong area. A region of slate lies in the hills to the east of the granodiorite, and there is an isolated slate hill to the west. In permo-carboniferous times a level surface was eroded over the granite and granodiorite country, and sandstone and conglomerate formations of that age occur on the northern side of the area, forming the fringe of marine beds which extend over the coalfields of New South Wales. The region includes a narrow ridge of limestone, at least 1500 ft. high, and some caps of basalt.

Experiments carried out at Tallong. Velocities were determined for a number of the types of rock occurring in the district, and advantage was taken of the large mass of igneous rock to study wave propagation without complications arising from stratification, or from the proximity of rocks of different type. Boisterous weather conditions continued throughout the whole of the four or five weeks spent at Tallong, and were found to have a serious effect on seismic experiments. Except for a few isolated periods of short duration, the microseism level was above the intensity of the initial tremor from explosions of practicable size. The need of developing methods to cope with high-frequency microseisms was strongly indicated.

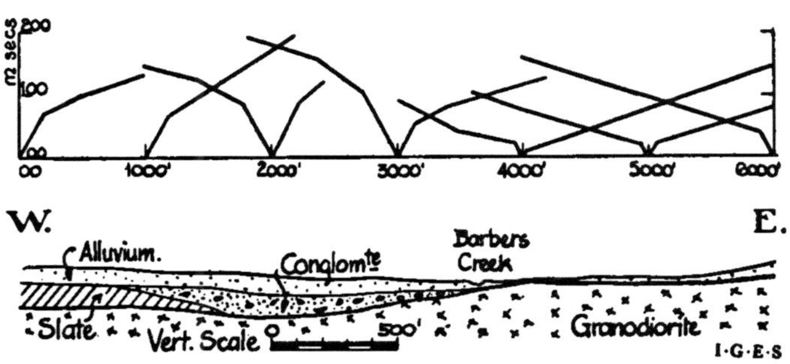

Fig. 176. Graphs and geological section along line *GG*. The section is derived from the graphs given above. Compare with the map previously deduced by geologists from outcrops only, Fig. 175

A seismic traverse *SS* was commenced on the slate hill to the west of the area shown in Fig. 175. Sufficient work was done to show that, on the line taken, the slate is standing on a granite base at a shallow depth. An east-west traverse *GG* was then run, from the slate, across a shallow tongue of sandstone conglomerate, to the granodiorite. The graphs for a length of 6000 ft. of this line, and the geological structure deduced from them, are shown in Fig. 176. It will be seen that the point of transition from the graphs over the granite to the steeper graphs over the boundary between the two rocks is not as far to the east as is indicated by the map, but lies closer to Barber's Creek. This may be accounted for by the fact that the geologists had to guess the actual position, since the boundary was hidden beneath an alluvial covering. Floodings of the creek prevented the I.G.E.S. party from effecting a more exact determination of the boundary. At the eastern end of the line the bed-rock was found, from the velocity appearing in the graph, to be granite beneath slate.

TABLE IX. *Comparative table of velocities*

The I.G.E.S. determinations were made at Tallong and at Gulgong, N.S.W.

Unit: Feet per second.

Material	I.G.E.S.	Salfeld*	Gutenberg†	Barton‡
Rubble or gravel	—	1,970–2,600	—	—
Wet loam	2,500	—	—	—
Cemented sand	2,800–3,200	—	—	—
Sandy clay	3,200–3,800	—	—	—
Cemented sandy clay	3,800–4,200	—	—	—
Water-saturated sand	—	—	4,600	—
Sand	—	—	4,600–8,400	—
Clay, clayey sandstone	—	5,900	—	—
Limestone, etc.	—	—	5,500–6,900	—
Sandstone	—	7,200–7,900	—	—
Sandstone conglomerate	8,000	—	—	—
Chalk (subsurface)	—	—	—	9,900–13,800
Hard clay-slate	10,500–11,500	10,500–12,400	—	—
Hornfels slates	11,500–14,500	—	—	—
Rock salt of salt domes	—	—	—	15,500–17,100
Granodiorite	15,000	—	—	—
Granite	18,500	—	25,300–30,200	—
Limestone, dolomite / Metamorphic rocks / Massive rocks	—	16,400–20,200	—	—

* *Vortrag auf dem Internationalen Geologenkongress*, Madrid, 1926.
† *Lehrbuch der Geophysik*, p. 608, Gebrüder Bornträger, Berlin, 1927.
‡ *Geophysical Prospecting 1929*, pp. 572–624, A.I.M.E., New York, 1929.

The range of velocities for each type of rock appearing in Table IX is indicative of the fact that formations which are grouped in the same geological classification may vary widely in the velocity with which they transmit elastic waves: for example, the

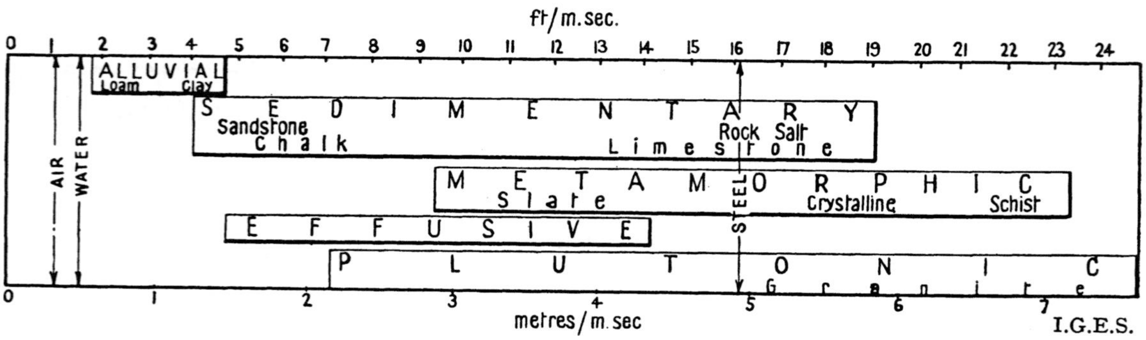

Fig. 177. Velocities of longitudinal waves

I.G.E.S. party found a 20 per cent. variation in velocity in a distance of 1000 ft. for an outcrop of slate. A general scheme for the range of longitudinal velocities found in the various types of geological formation is shown graphically in Fig. 177.

SECTION 5. CONCLUSIONS

Utility of Seismic Methods. The I.G.E.S. has gained sufficient experience with the equipment at its service to say that seismic methods are readily applicable to the examination of such simple structures (see p. 202) as extended stratifications, either horizontal or dipping, or large intrusions; or of buried topographies, provided they are not complex. Igneous intrusions, even when buried beneath a considerable depth of cover, are readily observed, provided that the velocities in the adjacent materials differ sufficiently, and that the intrusions are sufficiently extensive. It is essential, if an accurate determination of the nature, depth, and dip of the lower material is to be obtained, that the upper and lower materials should be homogeneous, or that inhomogeneities should be on a large scale and well-defined, so that the necessary factors may be taken into consideration in the calculations. In this connection it should be pointed out again that repeated observations showed marked variations in surface velocity in different parts of what was presumed to be the same material; so that any minor irregularities in the profile of buried topographies might be masked, or non-existent irregularities deduced, because of unknown variations of velocity in the surface formation. For this reason alone, it is felt that a claim for an order of accuracy higher than 10 per cent.[a] would have to be justified by the special conditions of the experiment—i.e. simplicity of structure, and marked homogeneity of the materials concerned.

In all cases investigated by the I.G.E.S. the lower material was that of the higher velocity, and the so-called *refraction* method was employed. With a seismometer design improved so as to simplify the record subsequent to the first "break," it is probable that *reflection* methods would be more widely used than at present; for reflection shooting, for example, smaller charges of explosive are required than for refraction shooting. The reflection method, however, does not indicate the nature of the lower material; and simple structures are desirable, if complex reflections are to be avoided.

Suggested Development of Apparatus. Defining "magnification" as the ratio of displacement on the photographic record to displacement of the ground, a magnification of 20,000 would be suitable for the normal working conditions experienced by the Seismic Section of the I.G.E.S. It would, however, be desirable to be able to employ higher magnifications in places where, or at times when, the microseism level was abnormally low. The calculated magnification of the Schweydar seismometer was 14,000, and that attained by the electrical system was comparable with this. Greatly increased magnification in the latter system could readily be obtained by the use of transformers of more suitable design than those employed, and by decreasing the string diameter and increasing the field strength of the galvanometer (which had been designed for military "sound-ranging" purposes); further, the optical magnification, which was only 30, could readily be increased by a factor of, say, 5.

[a] See pp. 201, 202.

It is unlikely that any method devised for increasing the sensitivity of the carbon-granule microphone itself would be of advantage, for it would probably be at the expense of steadiness and reliability. On the contrary, the experience of the I.G.E.S. leads to the conclusion that, using carbon-granule microphones, it would be advisable to have a microphone designed so as to be more highly damped and more stable, and correspondingly less sensitive; and to secure a compensating increase in magnification by changes in the recording system as suggested above.

The hot-wire seismometer shows considerable promise. It is not in any way unstable, and the sensitivity is readily varied from headquarters (by varying the current through the hot wire). It would be preferable to introduce greater air damping to make it aperiodic if an accurate record beyond first-arrival is desired. The sensitivity could be increased, if required, by preliminary mechanical magnification, the inertia mass then being separated from the diaphragm and hot wire. This, however, should not be necessary if the sensitivity of the recording system were suitably increased.

Development of Field Methods. Experience showed that, for general field work, it would be advisable to mount the recording station as a unit on a light motor truck, from which it could be lifted and carried if it should happen that the country were inaccessible to motor transport. This would be feasible with the apparatus as ultimately employed by the I.G.E.S.

The time occupied in the placing of seismometers and the laying of cables enormously exceeded the time spent on instrumental work, calculations, etc. It is recommended that the seismometers be supported on pointed legs which would penetrate the ground surface, and that the instruments be protected by heavy covers resting on the ground and completely enclosing them without touching. In countries where labour is expensive it is essential to have a wheeled vehicle bearing reels for the paying out or taking up of cable. For large scale work cable communications would of course be replaced by radio-electrical connection between explosion point and self-contained seismometer stations[a], each of which would include a single- or double-stringed galvanometer and recording system. Actually, one string could pick up the "break" of a radio note at the moment of explosion, and later record the arrival of the earth movements at the seismometer station.

Mechanical and Electrical Recording. Seismometers employing mechanical recording alone are not used in seismic prospecting, as the magnification procurable is not sufficient. A further stage of magnification by an optical lever system is generally introduced. An electrically recording system to give this second stage of magnification would possess many advantages; this might be done by the addition, at the end of the lever arm, of a light coil moving in a magnetic field or by the compounding of a hot-wire seismometer. From the experience of the I.G.E.S. electrical recording offers the following advantages over mechanical recording.

[a] *Referee's comment.* This of course is the general practice in large scale investigations by the seismic method, and the only remaining question concerns the relative merits of electrical and mechanical seismometers.

(*a*) One recording station is sufficient for a number of seismometers, especially with small-scale work. The seismometers may then be placed in position by unskilled assistants, whereas a skilled man is required to adjust a mechanical seismometer, even if not required to operate it. The records of several seismometers may, under such conditions, be obtained side by side on the same film, for ready comparison with each other, and with a common time-scale and common time-of-explosion mark. Moreover, the seismometers would be at a distance from disturbances caused by movements of the operator, whereas with mechanical recording the operator is stationed at the seismometer.

(*b*) The degree of magnification is readily variable and is at the immediate control of the operator, so that it may be chosen to suit the microseism level.

(*c*) Electrically recording seismometers may be designed as velocimeters or accelerometers[a], whereas sensitive mechanically recording seismometers record displacements. As far as the theory of the refracted disturbance has been investigated (by Jeffreys) it has been shown that the arrival is characterised by zero displacement and velocity, only the acceleration having a finite value. It would appear, on these grounds, that, for the accurate determination of the time of arrival, the accelerometer has a decided advantage over those instruments which employ velocity or displacement.

[a] See p. 210.

PART II

CHAPTER VII

ELECTRICAL METHODS

Section 1. SPONTANEOUS POLARIZATION METHOD

FIELD EQUIPMENT

As explained in Part 1 of this volume, this consists simply of a portable potentiometer mounted on a tripod, to which a pair of non-polarizing electrodes are connected by means of two insulated wire leads (Fig. 3, p. 18). Although simple in character the question of design is of some importance and, unquestionably, the construction of the non-polarizing electrodes is a matter which requires special attention.

The Potentiometer. In order to comply with field requirements the potentiometer and galvanometer must be incorporated into one compact field instrument of robust design which, when mounted on its tripod, should be capable of quick setting up without any special attention to levelling. Some simple device should also be provided for checking the value of the current in the potentiometer circuit, and for readjustment when necessary. For practical purposes a precision instrument is not required and the use of a standard cell is not necessary, excepting for a periodical check of the performance of the instrument which, of course, need not be carried out in the field. The potentiometer circuit illustrated in Fig. 178 has been employed by the survey parties in Australia and has proved quite satisfactory. The galvanometer *G* is a Weston portable instrument having a sixty division scale with central zero (30–0–30), each division being 1 mm. in width and approximately equivalent to 0·25 microampere[a]. It serves the double purpose of balancing the unknown ground potential which is applied across the terminals $T_1 T_2$ and also, by means of the shunt *S* and switch *Q*, for standardizing the current supplied to the potentiometer circuit by the small dry-cell *C*. The value of the shunt resistance *S* is such that, when the switch *Q* is in the position shown in Fig. 178 and the circuit is completed by the double contact key *K*, 1 milliampere will pass through the potentiometer wire *P* when the

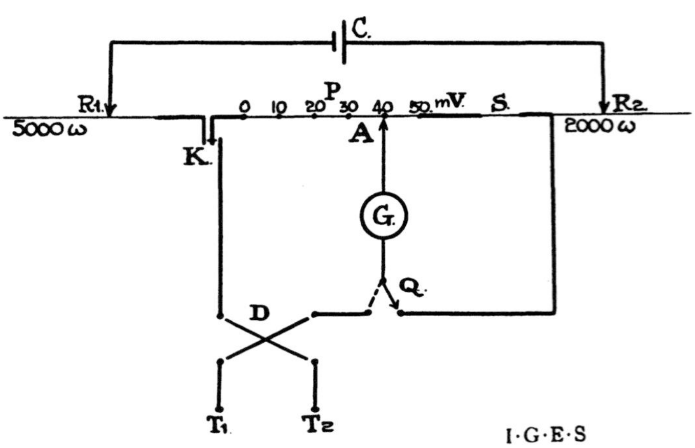

Fig. 178. Potentiometer circuit

[a] This instrument had a resistance of 154 ohms.

galvanometer needle is over the twentieth division of its scale. When the necessary adjustment has been made, by means of the variable resistances $R_1 R_2$, the position of the switch Q is altered, so that the potentials applied across $T_1 T_2$ may then be determined directly by adjustment of the contact A upon the slide-wire P. The spirally wound slide-wire P has a resistance of 50 ohms and consequently the instrument will have a range of 0–50 millivolts. Other ranges (0–25 and 0–100 millivolts) may be obtained when required by adjustment of the resistances $R_1 R_2$. Further points in connection with the instrument may be summarized as follows:

(i) In standardizing the instrument by means of the resistances $R_1 R_2$, the switch knob Q should be in the position indicated in Fig. 178 and the sliding contact arm A must be moved to the far end of the scale (reading 50 millivolts). In packing the instrument for transport the switch Q and contact arm A should also be placed in these positions, as the movements of the galvanometer coil will then be effectively damped. Even when moving from station to station in the field the switch Q should be moved over to the shunt circuit.

(ii) The dry cell C is of ordinary make about $2\frac{3}{4}'' \times 1\frac{1}{2}'' \times 1\frac{1}{2}''$ in size. As it must be replaced from time to time it should be easily accessible. The hinged door T, shown in Fig. 179 at the side of the instrument, will be convenient for this purpose.

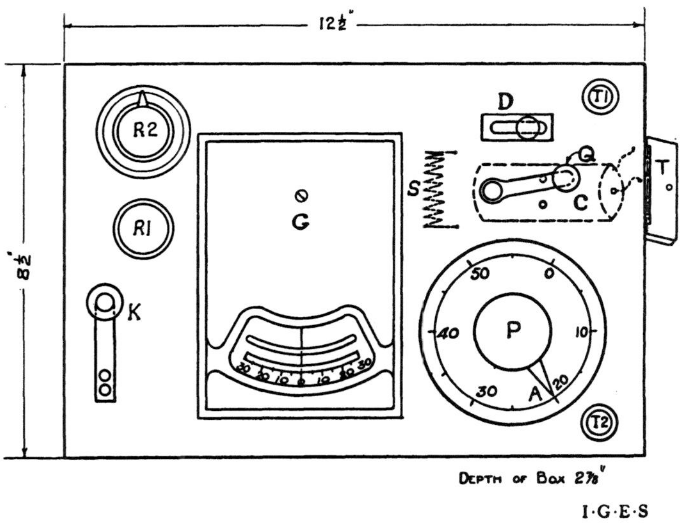

Fig. 179. Panel arrangement of potentiometer

(iii) The variable resistance R_2 should be provided with a pointer and blank ivorine dial, upon which the various settings may be marked in pencil and obliterated as required.

(iv) The base of the instrument should be provided with a quick-fitting device for attachment to the tripod head. The tripod legs should have a length of about $3\frac{1}{2}$ ft.

The dimensions and panel arrangement of the potentiometer used by the I.G.E.S.[a] are illustrated in Fig. 179. The instrument weighs $7\frac{1}{2}$ lb. and is provided with a removable cover and carrying handle, to assist in transport in the field. The size and weight could be further reduced by substituting a smaller type of galvanometer, but difficulty has been experienced in finding an instrument that combines the necessary robustness with a sensitivity which, certainly, should not be less than 1 microampere per millimetre of scale deflection. It will be observed that the component parts are all mounted on the panel, excepting the cell C which is held in position by brass clips.

[a] This instrument and the non-polarizing electrodes (Fig. 180, Type B) were designed originally by A. Broughton Edge for use in Cyprus.

Non-Polarizing Electrodes. The satisfactory operation of the equipment is largely dependant on the behaviour of the non-polarizing electrodes which require close attention, not only as regards their construction, but also during their use in the field. For general purposes they may be considered to be satisfactory if their internal polarization effects do not exceed 1 millivolt, and, if sufficient care is exercised, it is possible to maintain this low average value.

Several patterns of electrode have been made and tested by the I.G.E.S. and the general conclusions published by Schlumberger have been fully confirmed. The contact polarization effects produced by metal electrodes—even if gold plated or made of platinum—are usually of the order of one or two hundred millivolts and, moreover, are of a very irregular character. The same behaviour, but to a somewhat smaller extent, was observed in the case of electrodes made from block graphite, and it can only be concluded that direct contact with the ground by means of solid conductors is impracticable in the spontaneous polarization method.

These large polarization effects may be overcome by the use of electrolytes, and the form of electrode recommended by Schlumberger has been found quite satisfactory. This is shown in Fig. 180 (type *A*), and consists essentially of a flat-bottomed porous pot containing a saturated solution of copper sulphate into which a perforated tube of copper is inserted. In this type of electrode the long copper tube is also filled with the solution, together with crystals of copper sulphate, as it is claimed by Schlumberger that the large liquid contact surface is an important factor in maintaining the potential equilibrium of the system; since the concentration of the liquid film in contact with the copper will be more readily adjusted after the passage of current. Although this may be the case to some extent,

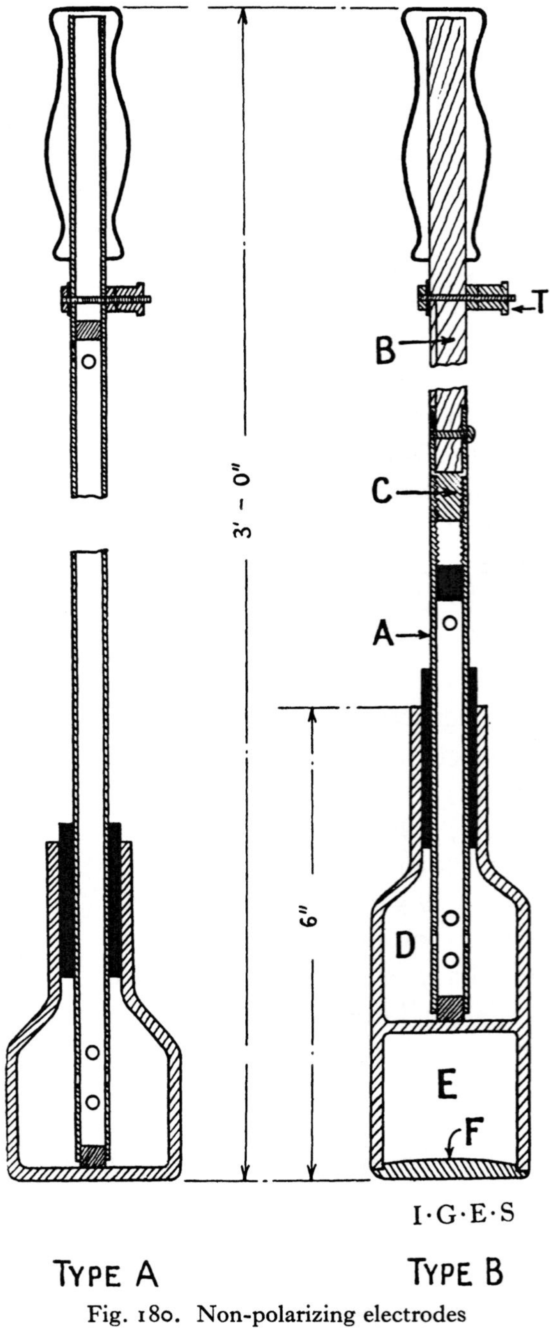

Fig. 180. Non-polarizing electrodes
Type *A*. After Schlumberger.
Type *B*. I.G.E.S. pattern.

it was found that electrodes prepared with short tubes (Fig. 180, type *B*) gave equally good results, and that they were more convenient to use. Not only may the electrode heads be unscrewed from the shafts for ease in packing, but a new pair may be substituted in the field in case of damage, or excessive polarization.

The construction of the electrodes requires close attention if internal polarization is to be reduced to a minimum and, from the details that are summarized below, it will be observed that a pair of electrodes should be identical in every possible respect—not only in design but more particularly with regard to the materials from which they are made.

Non-Polarizing Electrodes, type A. (i) The half-inch diameter copper tubes employed should be cut from the same length, in order to ensure that their chemical and physical characters are similar. A high grade of electrolytic copper is preferable, and on no account should copper tubes derived from different sources be used in the construction of a pair of electrodes.

(ii) The pots should have the same degree of porosity and should be made from the class of material that is commonly employed in Daniell and Leclanché cells. The walls of the pot should be $\frac{3}{16}$ in. in thickness (base $\frac{1}{4}$ in.) and, excepting the flat base, the outer surface should be covered with a uniform salt glaze. A satisfactory degree of porosity will be indicated if minute beads of liquid appear uniformly over the bottom within a minute or so of filling with solution.

(iii) Cork sleeves, which have been soaked in melted paraffin wax, are the most satisfactory means of securing the copper tubes into the necks of the porous pots. Rubber, unless unvulcanized and entirely free from sulphur, should not be employed, since a black film of copper sulphide is formed wherever it comes in contact with the metal. A small cork should be inserted into the lower perforated end of the copper tube to prevent the metal bearing directly on the base of the pot.

(iv) Before assembling the electrode a small quantity of copper sulphate, in the form of crystals, should be placed in the pot and within the tube. After inserting the latter, the electrode should be filled with a cold saturated solution of copper sulphate by means of the filling hole which lies immediately below the rubber handle and terminal. When the pot and tube are completely filled, the hole is sealed by means of a short length of black Para-rubber tubing, which is rolled down into position. Within a few minutes of closing the filling hole the base of the porous pot, which will have been sweating freely, will become dry and a minimum of solution will then be lost by evaporation. It will be necessary, however, to add a further supply of solution every seven or ten days, in order that a satisfactory level may be maintained.

(v) The copper sulphate employed should be the pure recrystallized substance and, if available, distilled water should be employed in preparing the solution. As a rule, however, any potable water of good quality will prove satisfactory.

(vi) When the electrodes have been completed and filled, their terminals should be connected together and the pots left standing overnight in a basin of concentrated copper sulphate solution. They may then be used in the field and should prove satisfactory. It is desirable that a number of these electrodes should be prepared at the same time and from the same materials. From a batch of four or six electrodes it is possible, as a rule, to select pairs which behave particularly well when used together.

(vii) In some cases it has been found that an improvement in the general stability of the electrodes may be obtained by plating the copper tubes. This may be carried out quite simply by standing the electrodes in a basin of copper sulphate with their terminals connected together. A piece of copper foil is then placed in the basin and is connected to the positive pole of a small dry cell; the negative connection being made to the joined electrode terminals. The cell should remain connected in this way for several hours and the electrodes should then be allowed to remain standing in the basin for a further few hours after the cell has been disconnected.

Non-Polarizing Electrode, type B. This pattern of electrode was devised with a view to increased stability and portability, and also in order to minimise the loss of solution by evaporation from the porous contact surface. It has proved satisfactory, even in areas where the surface soils presented most unfavourable conditions and the loss of liquid over a period of several months was negligible. Their constructional features are similar to those which have been described above for type *A*, but the following modifications should be noted:

(i) The copper tube *A* is reduced to 7 in. in length, and is attached to the wooden shaft *B* by means of a screw-thread fitting *C*. Thus the entire electrode head may be removed or replaced with the greatest ease. The terminal *T* is connected to the fitting *C* by an insulated wire, which is let into the surface of the shaft *B*.

(ii) The pot is double-chambered and, excepting the porous lid *F* which closes the lower chamber *E*, its outer surface is completely glazed. The upper chamber *D* and tube *A* are filled with a solution of copper sulphate and loose crystals, as in the type *A* electrode, but in this case 2 per cent. of pure gelatin are added.

(iii) The lower chamber *E* is filled with a hot solution of gelatin which is prepared by soaking the dry sheets in a cold concentrated solution of copper sulphate for 6 hours. The soft sheets are then melted in a beaker over a water-bath, with careful stirring until the solution is clear and free from bubbles. The hot material is then poured into the base of the pot, which should be filled to within an eighth of an inch of the lip. The lid *F* is then cemented into place, so that it will be in contact with the gelatin without any air space beneath it.

(iv) The electrodes prepared in this way require very little attention, and may be packed for transport without emptying or fear of leakage. As in type *A* they should be kept, when not in use, with their terminals connected together, and with their bases immersed in a solution of copper sulphate.

FIELD PROCEDURE

A general account of the application of the method has been given in Part I, but experience in the field has shown that the following details of procedure may also be followed with advantage:

(i) In dry country it is necessary to prepare special contact surfaces before the electrodes are applied to the ground. As a rule a shallow hole, a few inches in diameter and an inch in depth, will provide a satisfactory contact after it has been well moistened. The water employed for this purpose must be reasonably free from impurities. Stream or river water is satisfactory, provided that it is not brackish or contaminated with liquors derived from mine workings or treatment plants.

(ii) When taking observations along straight-line traverses the operator with the potentiometer should always occupy the forward stations, so that he may supervise the preparation of the contact-surfaces. This should not be left entirely to an unskilled assistant. On the other hand, this procedure is not practicable when determining the positions of equipotential lines, since the potentiometer mounted on a tripod is not sufficiently mobile. In this case the assistant should go ahead and make the trial contacts with his electrode, whilst the observer directs his movements by watching the instrument at the rear station.

(iii) The non-polarizing electrodes should bear distinguishing marks—preferably upon the sides of the porous pots (e.g. *A* and *B*)—and their behaviour should be checked at frequent intervals. It is always possible that internal polarization may be set up unexpectedly owing to some unforeseen cause, and if this should arise to a serious extent and remain undetected, then the potential observations will have little value, and may well lead to erroneous conclusions. To guard against this danger it is recommended that the potentials of the first five stations occupied at the commencement of a day's work should be checked by a second observation with reversed electrodes. The behaviour of the electrodes will then be revealed in the manner shown by Table X, which is a page taken from one of the Survey field books. Here it will be seen that the electrode polarization is rather high at the first three stations, but then drops to 0·25 millivolt; the *A* electrode being positive at stations 3/4 and negative at stations 4/5. This is quite satisfactory, but should any serious polarization persist, then a fresh pair of electrodes should be employed. As a rule, however, it will be found that after the first few stations have been completed the electrodes will be behaving normally. It will then be good practice to make the reverse observation at every fifth station until full confidence is established.

(iv) When not in use, the pair of electrodes should be left standing in a dish of concentrated copper sulphate solution with their terminals connected by a short length of wire. If this practice is followed after the day's work it will ensure that the electrodes will be in equilibrium when the survey is continued on the following morning. A wide-mouthed vessel containing clean sand saturated with copper sulphate may conveniently be used in place of the dish of free liquid.

TABLE X. *Spontaneous polarization*

Date, 12th October, 1928. *Locality*, Leadville, N.S.W.
Operator, J. C. F. *Traverse*, North from station 625.

All potentials in millivolts.

Station	*A* to north	*B* to north	Polarization electrode *A*	Ground potential forward station	Summed potentials
0					0
	+ 6·0	+ 9·0	−1·5	+ 7·5	
1					+ 7·5
	−22·5	−17·5	−2·5	−20·0	
2					−12·5
	− 6·0	− 4·5	−0·75	− 5·25	
3					−17·75
	−17·0	−17·5	+0·25	−17·25	
4					−35·0
	+18·5	+19·0	−0·25	+18·75	
5					−16·25
		+ 4·0			
6					−12·25
		+ 4·5			
7					− 7·75
		0·0			
8					− 7·75
		+ 2·0			
9					− 5·75
	+ 6·0	+ 4·0	+1·0	+ 5·0	
10					− 0·75

Page of field book—Spontaneous polarization method.

THEORETICAL CONSIDERATIONS[a]

It has already been pointed out that a qualitative analysis of spontaneous polarization results is generally sufficient to locate the position of the body. Under favourable circumstances, however, it is possible to make some estimate of the depth and dip of the body causing the indication. The case of a sphere of ore with a polarization following a cosine law has already been calculated by A. Petrowsky[b], but the application of such a system is small, and in the following the more general case of a sheet of ore has been considered.

Rod of Mineral. The general characteristics of a potential traverse across a sulphide body in a state of electro-chemical activity may be obtained from the case of a thin rod of ore. Let *AB* (Fig. 181) be the rod of ore contained in the plane of the paper which is at right angles to the ground surface *LM*. Let the direction of the intersection of these two planes be the axis of x, the axis of y being perpendicular to it and passing through O, the projection of A on the ground surface. It will be assumed that the potential at any point may be regarded as the resultant potential of a current $-I$ at A and $+I$ at B in

[a] By J. McGarva Bruckshaw, M.Sc.
[b] *Bulletin of the Institute of Practical Geophysics, Leningrad*, vol. 1, p. 87, 1925.

a semi-infinite medium of resistivity ρ ohm cm. Then the potential V at a point (xy) on the ground surface may be represented by

$$V = -\frac{\rho I}{2\pi}\left[\frac{1}{(x^2+y^2+h_1^2)^{\frac{1}{2}}} - \frac{1}{\{(x+a)^2+y^2+h_2^2\}^{\frac{1}{2}}}\right],$$

taking into account the images A', B' of A, B in the surface, where h_1, h_2 cm. are the distances of A and B below the surface, and a cm. the projection of AB on LM. It will be seen that the shape of the potential surface is independent of the values of the current and the resistance, and depends only on the depth and size of the body. Further, if two bodies of the same dimensions and depths are in the same state of chemical activity so that the polarization potential set up between their ends is the same, ρI is a constant, and the values of the surface potentials will be the same. Fig. 182 shows the profile for

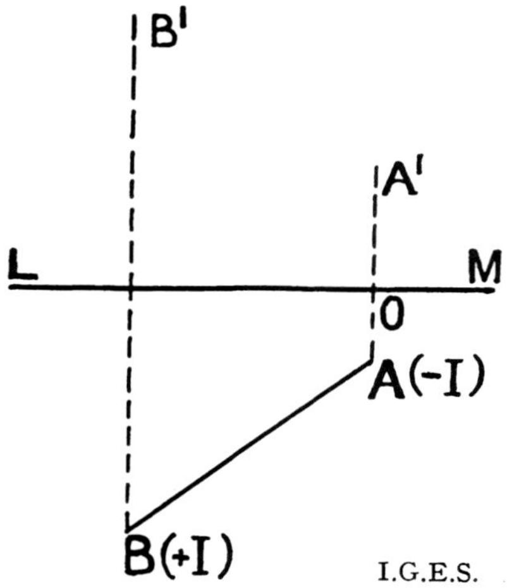

Fig. 181. Case of a rod of mineral

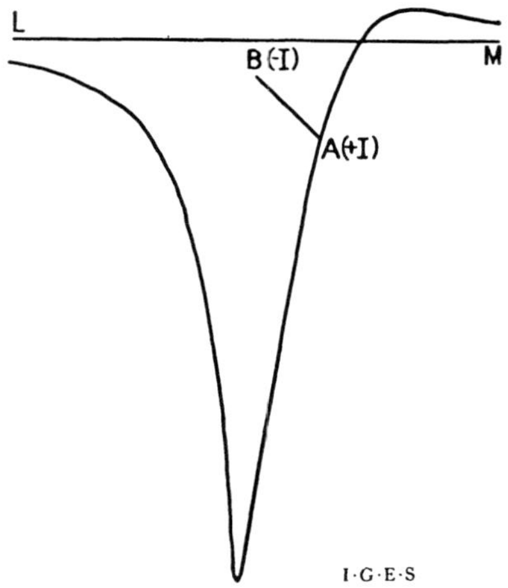

Fig. 182. Potential profile due to polarized rod of mineral

$y = 0$ due to a body in which $h_1 : h_2 : a = 1 : 3 : 2$, the potential being shown on an arbitrary scale as ordinates. The main features of the curve are that the potential gradient in the direction of the dip is greater than in the opposite direction, and the negative centre is not immediately over the upper end of the body, but is displaced away from the source of current. The displacement of the negative centre will depend on the position of the far end of the ore-body. When it is far removed compared with the depth of the upper end, or when the rod is perpendicular, there will be no displacement. The maximum displacement will occur when the length is very small so that it forms a dipole with its axis parallel to the surface, in which case the displacement is 0·707 of the depth of the doublet.

Sheet of Ore. Let $pqrs$ (Fig. 183) be a sheet of ore of negligible thickness dipping downwards from pq towards sr. The upper and lower edges pq, rs are parallel to the line Ox which lies in the ground surface. O is a point immediately above the centre of the upper edge and is taken as the origin of the rectangular co-ordinates $Ox . Oy$. It will be assumed that the current source and sink are uniformly distributed along *sr* and *pq* respectively and hence if I amps. is the total current and $2l$ cm. the length of the body, the potential at a point P (xy) on the surface due to an element dr of the upper edge and its image in the surface, is given by

$$dV_1 = -\frac{I\rho}{4\pi l}\cdot\frac{dr}{\sqrt{y^2+h_1^2+(r-x)^2}},$$

where h_1 cm. is the depth of PQ below the surface (r, O, h_1) the co-ordinates of the element dr and ρ ohm cm. the resistivity of the medium. Hence the total potential due to pq is

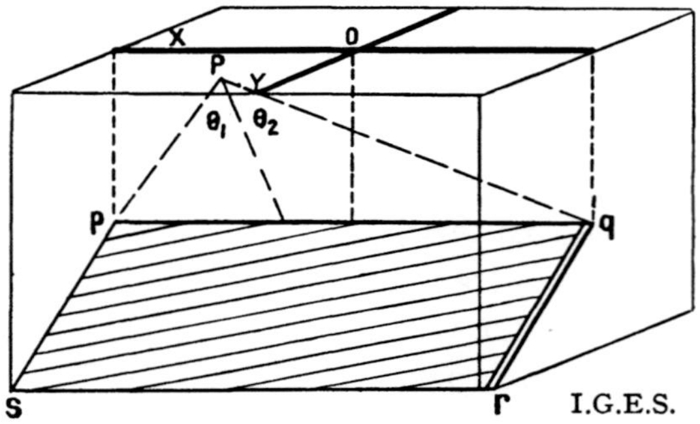

Fig. 183. Diagram for sloping sheet

$$V_1 = -\frac{I\rho}{4\pi l}\int_{-l}^{l}\frac{dr}{\sqrt{a^2+(r-x)^2}},$$

where $$a = PQ = \sqrt{h_1^2+y^2}.$$

Putting $(r-x) = a \tan\theta$ the equation assumes the form

$$V_1 = -\frac{I\rho}{4\pi l}\int_{\theta_1}^{\theta_2}\sec\theta\, d\theta,$$

where $$-\theta_1 = \tan^{-1}\left(\frac{l+x}{a}\right) \quad\text{and}\quad \theta_2 = \tan^{-1}\left(\frac{l-x}{a}\right).$$

Hence the potential due to the upper edge is

$$V_1 = -\frac{I\rho}{4\pi l}\left\{\log\tan\left(\frac{\pi}{4}+\frac{\theta_2}{2}\right)-\log\tan\left(\frac{\pi}{4}+\frac{\theta_1}{2}\right)\right\},$$

and for the lower edge

$$V_2 = \frac{I\rho}{4\pi l}\left\{\log\tan\left(\frac{\pi}{4}+\frac{\theta_2'}{2}\right)-\log\tan\left(\frac{\pi}{4}+\frac{\theta_1'}{2}\right)\right\},$$

where θ_1' and θ_2' are the corresponding angles for the lower edge.

The resultant potential at P is

$$V = V_1+V_2 = -\frac{I\rho}{4\pi l}\left\{\log\tan\left(\frac{\pi}{4}+\frac{\theta_2}{2}\right)-\log\tan\left(\frac{\pi}{4}+\frac{\theta_1}{2}\right)\right.$$
$$\left.-\log\tan\left(\frac{\pi}{4}+\frac{\theta_2'}{2}\right)+\log\tan\left(\frac{\pi}{4}+\frac{\theta_1'}{2}\right)\right\}.$$

The equipotential surfaces due to the body are shown in Fig. 184 where

$$h_1 : h_2 : l = 1 : 3 : 2,$$

the dip being 45°. It will be seen that the strike is given by the direction of the longer dimension of the equipotential curves, and the direction of dip by the shorter dimension, the body dipping from the negative centre towards the point where the potential gradient is steeper. This is shown more clearly by the potential profile (Fig. 185) along

AB. Other profiles along *CD*, *EF* show the same characteristics.

The main assumptions made in the above calculations of natural potentials are: (1) the current source and sink may be located in points or lines; (2) the country rock is homogeneous; (3) the distortion of the lines of flow, due to the size of the ore-body itself, may be neglected.

In the first, for sheets of ore, the assumption that the chemical activity may be confined to a line is justified, since it occurs along the region of contact between the ore and the zone of oxidation, and is sharply defined. The source of current, however, is not in a region of chemical activity, but is probably spread over a large surface area, but since this will be at a large distance from the surface compared with the distance of the upper pole of the body, the error introduced by the assumption will in general be small. When the body has large dimensions the actual distribution of current and potential in the surrounding rock may be represented by the source and sink together with their electrical images formed by the various boundaries of the body, and in general the system will be too complicated to allow an exact statement of the potential. If, however, the dimensions are small compared with the depth of the body, the distortion of the lines of flow and equipotential surfaces will be small, and the distribution of potential calculated on the above lines will be a close approximation to the actual state.

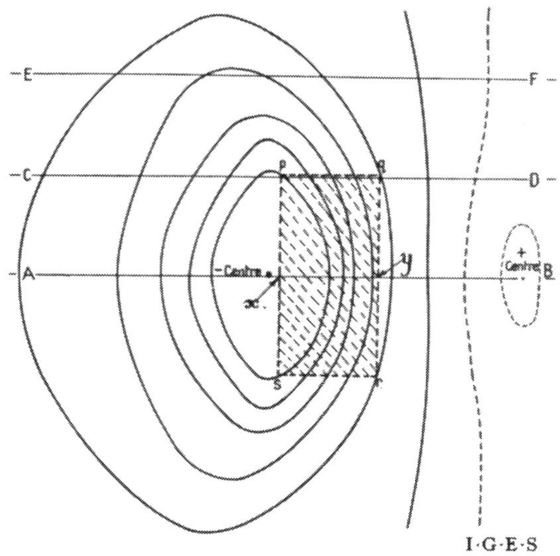

Fig. 184. Equipotential lines due to sheet of ore dipping towards the right

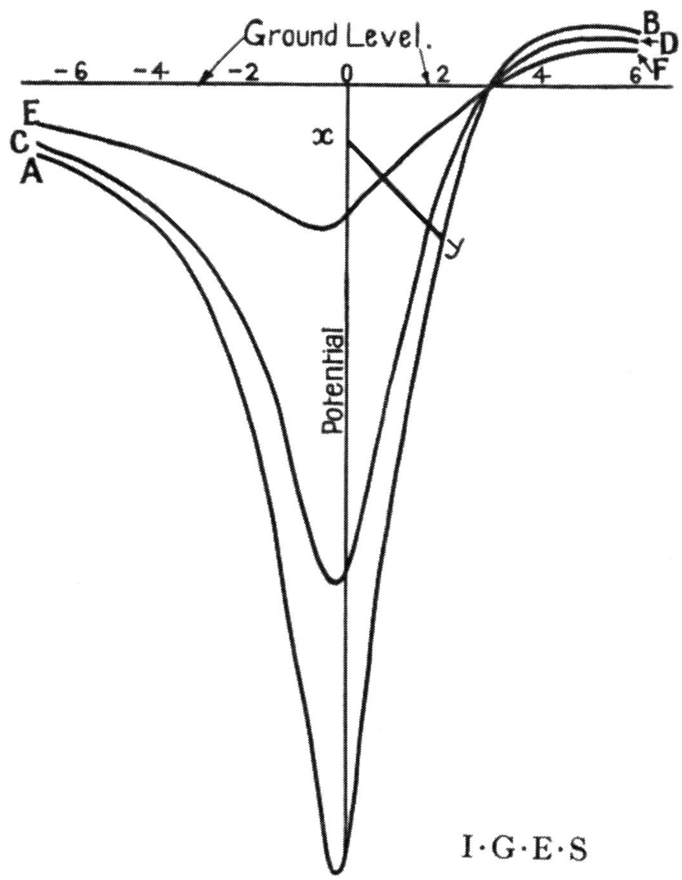

Fig. 185. Potential profiles across sheet of ore

SECTION 2. GROUND RESISTIVITY METHODS

Instrumental Details. The principal instruments used by the I.G.E.S. for this work were a milliammeter, a potentiometer, and a commutator. The milliammeter, a unipivot instrument by the Cambridge Instrument Company, had five ranges, 0–24, 0–120 and 0–1200 microamps, and 0–12 and 0–120 milliamps, and could be extended to 600 milliamps by an external shunt. Before using, the instrument was calibrated by a standard 1 ohm resistance and a potentiometer. A slide-wire potentiometer (Cambridge Instrument Company) with ranges 0–180 and 0–1800 millivolts, and a Tinsley portable mirror galvanometer, together with a 2-volt accumulator and a Western cadmium cell formed the potentiometer system.

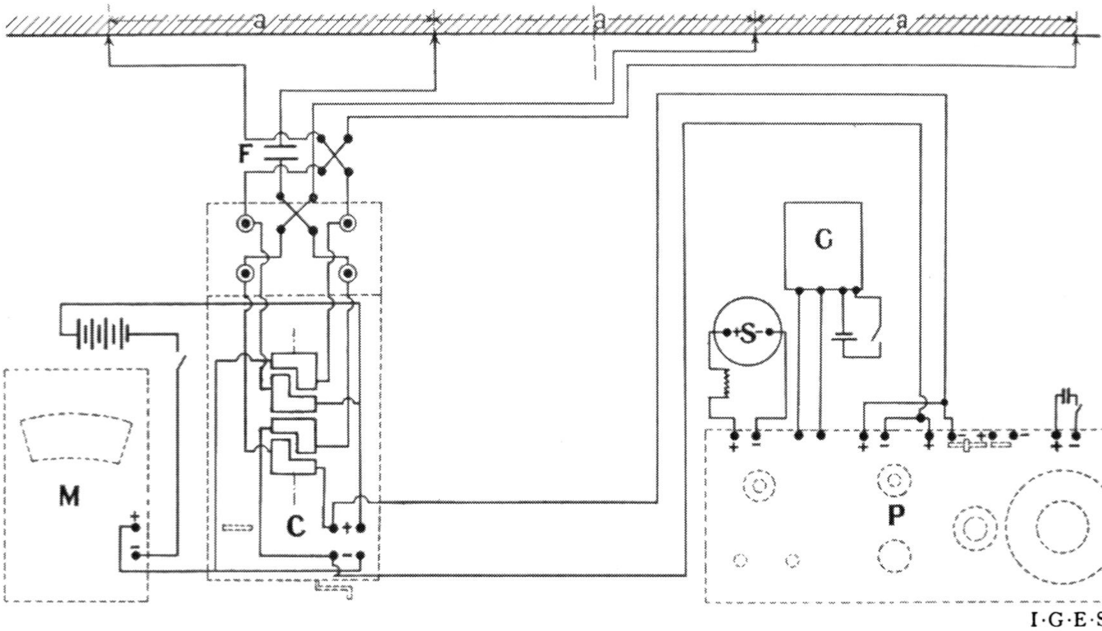

Fig. 186. Circuit for earth resistivity measurements

M. milliammeter. *C*. commutator. *S*. standard cell.
G. galvanometer. *P*. potentiometer. *F*. 20μ F. condenser.
a = electrode separation.

As too much time would have been taken to construct a commutator according to the designs kindly supplied by the Carnegie Institution of Washington, the pair of commutators of an Evershed and Vignole's Earth-tester were adapted to eliminate the effects of earth currents and polarization potentials. The essentials of such an instrument are robust construction, accessibility for cleaning, and a dust-proof case. Both commutators, which are mounted on the same shaft, have small gaps in the segments, and while the brushes are facing these the circuits are broken. The current registered by the ammeter is a time average of the current through the ground, while the potential measured corresponds to the actual current through the ground. Hence the measured resistance is greater than its actual value, but a correction factor may be determined by measuring a known resistance. As an improvement in design Gish and Rooney advocate a guard ring between the commutators to prevent leakage. This ring, together with an enclosing metal case connected to all neutral metal parts, is earthed. The circuit employed is shown in Fig. 186. It will be seen that a 20μ F. condenser has been placed in series with one of the potential leads. The action of the condenser is not known, but in its absence the galvanometer is unsteady. Large capacity wireless high-tension batteries were used to supply the current to the ground.

Sources of Error. It was found that within the limits of experimental error the values of E/I were independent of the current passing through the ground and, when induction effects were negligible, they were also independent of commutator speed. Other sources of error in the measurement of R arise from natural earth currents, polarization potentials, induction effects, leaks, errors in ammeter and potentiometer readings, contact resistances at the electrodes, finite size of the electrodes, and to small discrepancies in the position of the electrodes. Of these the first two are eliminated by the commutator, and by using different commutator speeds the induction effects may be corrected. Simple calculations show that the errors due to the discrepancies in the electrode positions and of the finite size of the electrodes may be neglected. It will be shown (see p. 249) that the effects of instrumental leakages are eliminated by taking observations with four combinations of the current and potential connections. Large contact resistances limit the accuracy with which the measurements can be made by decreasing the current and the sensitivity of the potentiometer circuit. The error made in reading the ammeter is negligible, but an error is introduced by the slight interval of time between the ammeter and the potentiometer readings, since the current falls continuously during the observation owing to polarization of the battery.

When the Megger is used for field measurements the contact resistances reduce the current through the potential coil; this results in a reduction of the potential applied to the instrument, and gives a measured resistance which is less than the actual resistance. If p_1 and p_2 ohms are the contact resistances at the inner electrodes and r ohms the internal resistance of the potential coil the readings must be multiplied by the factor $(r+p_1+p_2)/r$ to obtain the true values. The correction factor should be made as small as possible by keeping p_1 and p_2 small. This may be accomplished by watering the electrodes and knocking them well into the ground. In many cases it is desirable to have several stakes all connected together to reduce the resistance. Lee has carried out measurements both with the Megger and the Gish and Rooney apparatus, and finds that although the shapes of the curves obtained are the same, the absolute values differ; the Megger giving the lower values. A short traverse was carried out by the I.G.E.S. using both instruments, and a similar effect was noticed, but in this case the Megger gave the higher values.

Details of Field Procedure. The directions and intervals between the surveyed lines are matters of importance and will be determined by the geological knowledge available and by the special requirements of the survey. If it is the average depth of a discontinuity in a given area that is required, two mutually perpendicular lines over the area will be sufficient, while if some idea of the direction and magnitude of the dip is also to be obtained, a series of parallel lines will be necessary in addition. For most purposes the surface surveying can be carried out with sufficient accuracy with a prismatic compass, ranging rods, 100 ft. steel tape, etc., the lines being pegged at intervals of 100 ft.

The cable used by the I.G.E.S. was 300 ft. lengths of 3/036 insulated wire having a resistance of 8 ohms per 1000 yd. The leads to the inner electrodes were usually run out along the traverses but the current leads were carried parallel to it some 50 ft. away, as by this means the effects of mutual induction may be reduced. In most cases more than one length of cable will be required and it was found that if the cables were laid in position first the work was quicker. The connections between sections of cable were made by means of barrel connectors protected by rubber tubing. The connections between the cable and the electrodes were made by short lengths of wire terminating in two battery clips, one of which was placed on the electrode and the other on the free end of the cable. This quick change device proved quite satisfactory.

Electrode Contacts. Initially the electrodes used were made from drill steel, but later this material was replaced by T section steel which gave much better contacts. In damp country efficient contacts are obtained without difficulty by driving the electrodes into the ground, but in dry areas they must be watered, preferably with a dilute salt solution. A more efficient method is to make the holes for the electrodes and to fill them with water some 12 hours before they are to be used, and to re-water again when the electrodes are inserted into them. Even this may leave much to be desired, for when using this method at Gulgong, the current obtained with 90 volts applied to the electrodes was only 10 milliamps, while after a shower of rain with the same electrode layout (the electrodes were actually in the same holes) the current had risen to 70 milliamps. The increase in current is not due to a surface leak, but to an effectively increased contact area round the electrode. In dry sandy districts the contacts may be temporarily improved by watering just before the measurement. At Watheroo[a], Western Australia, where the surface is covered by dry sand, Gish and Rooney used a series of electrodes joined together to obtain a low contact resistance.

Field Measurements. A reversing switch is connected between the instrument and each pair of electrode leads, so that, if the two positions of the current switch are called D and R, and those of the potential switch d and r, readings can be taken for the four combinations Dd, Dr, Rr, Rd (see Table XI). As explained below, this results in the elimination of the effects of induction, and instrumental leaks from the current to the potentiometer circuit.

In areas of low resistivity it was found that when the effective resistance fell below a certain value, induction between the leads played an important part. To correct this effect it was necessary to rotate the commutator at a known rate, the rate of rotation

[a] *Terrest. Magn.* vol. 32, p. 49 (1927).

TABLE XI. *Imperial Geophysical Experimental Survey. Resistivity Survey*

District, Murrayville, Victoria. *Operators*, S. H. S. and J. McG. B.
Computer, I. W. M. *Date*, 14th August, 1929.

Commutator speed = 18 revs./sec.

Potential electrode separation	Applied voltage	Potential (millivolts) and Current (milliamps)				Mean potential and current	Resistance in ohms	Mean resistance	Resistivity in ohm cm.
		Dd	*Dr*	*Rr*	*Rd*				
230 ft.	22½	3·37 75·4	3·40 75·4	3·39 75·2	3·39 75·3	3·39 mv. 75·3 ma.	·0454	·0453	1600
	45	6·70 148·2	6·74 148·2	6·72 148·2	6·71 148·2	6·72 mv. 148·2 ma.	·0454		
	67½	9·98 218·0	10·09 218·0	10·00 218·0	9·99 218·0	10·02 mv. 218·0 ma.	·0451		
	45	6·67 148·2	6·71 148·2	6·71 148·2	6·70 148·2	6·70 mv. 148·2 ma.	·0452		
220 ft.	90	10·79 232·0	10·93 231·0	10·94 230·5	10·70 230·0	10·84 mv. 231·0 ma.	·0470	·0472	1600
	67½	8·00 170·8	8·13 170·8	8·14 170·3	7·99 170·8	8·07 mv. 170·8 ma.	·0474		
	45	5·35 115·4	5·45 114·4	5·46 114·4	5·40 115·4	5·42 mv. 115·0 ma.	·0470		
	67½	8·05 170·8	8·15 170·8	8·14 170·8	7·97 170·8	8·08 mv. 170·8 ma.	·0474		
210 ft.	90	9·64 198·5	9·70 197·5	9·74 197·0	9·57 197·0	9·66 mv. 197·5 ma.	·0490	·0490	1580
	67½	7·15 147·0	7·21 147·0	7·24 147·0	7·13 147·0	7·18 mv. 147·0 ma.	·0488		
	45	4·75 98·9	4·80 98·6	4·83 98·8	4·75 98·9	4·78 mv. 98·8 ma.	·0490		
	22½	2·42 49·8	2·45 49·8	2·45 49·8	2·42 49·8	2·44 mv. 49·8 ma.	·0490		

being controlled by a metronome. The resistances given in Table XI are not corrected for commutator factor, but the final column gives the correct value of the resistivity. Inspection of the table shows that the potential values in the two columns *Dd* and *Rd* are consistently lower than in the remaining two columns. In many cases the effect was much more pronounced than in the case shown. There are two possible causes for the difference: one the leaks between the cables and in the instruments, and the other the induction effects. The leaks create a potential always in the same direction, these aid or oppose the potential difference set up between the inner electrodes and if the leakage is sufficiently large the potentiometer readings may be negative. For the case of the induction effect if R ohms is the resistance, and L henries the self inductance of the ground included in the measurement, and M henries the mutual inductance between the current and potential leads, the effective resistance R' is given by

$$R' = \sqrt{R^2+(L\pm M)^2\,\omega^2},$$

where ω is the angular velocity of the commutator[a].

[a] This formula is not applicable to the case of the "Megger" Earth Tester on account of phase changes which occur.

For one pair of readings the term in brackets is $(L+M)$ and for the other $(L-M)$. There is no question of phase change, for, since the resistances in the current and potential circuits are large the currents in them are in phase with applied voltage, so in spite of the existence of a phase difference between the current and the measured voltage across the potentiometer terminals, since the potential is measured by balancing the current, the phase difference cannot be observed. That this explanation is sufficient to account for the discrepancies, in cases where leaks are small, is shown by the following figures, obtained in the Victorian Mallee, and is borne out in all cases where induction corrections have been applied.

From the values at two commutator speeds, two values of R and one value of L and M may be determined, and for any other two a similar set may be obtained. The following table shows a set of results:

	R (from $\sqrt{R^2+(L+M)^2\,\omega^2}$)	R (from $\sqrt{R^2+(L-M)^2\,\omega^2}$)	L	M
			Microhenries	
	ohms	ohms		
Using values at 18 and 27 r.p.sec.	·00325	·00320	1·8	20·2
Using values at 18 and 36 r.p.sec.	·00329	·00325	1·7	19·3
Mean	·00325		1·8	19·8

In the case where leaks are absent the differences obtained may be used as an indication of the presence of induction effects. For small separations, even in areas of low resistivity, the induction effects in the field are always negligible, but increase with increasing separation until finally they become the predominating factor.

THEORETICAL CONSIDERATIONS

The Resistivity Formula. Wenner[a] is responsible for the following calculation for the determination of the resistivity formula, but it may be derived by other means[b]. It has been shown[c] that problems of potential distribution and current flow in continuous media are analogous to the distribution of potential and lines of force in electrostatic problems, the electric charges in electrostatics being replaced by current sources and sinks and the dielectric constants by conductivities. Thus a point charge E in a medium of dielectric constant k produces a potential, at a distance r, of $\frac{E}{4\pi rK}$ using Heavyside units, while a current source I in a medium of conductivity σ gives a potential $\frac{I}{4\pi r\sigma}$. In the case of a point source of current I near the plane interface between two semi-

[a] *Bureau of Standards, Bull.* vol. 12, p. 469 (1916).

[b] Eve and Keys, *Applied Geophysics*, p. 240, University Press, Cambridge, 1929.

[c] Jeans, *Mathematical Theory of Magnetism and Electricity*, 2nd ed., p. 346, University Press, Cambridge, 1911.

infinite media, it can be shown[a] that the potential at any point in the medium containing the source is the same as the potential in a medium of infinite extent in all directions, and of conductivity σ_1, due to a source I and a source $\frac{\sigma_1-\sigma_2}{\sigma_1+\sigma_2}I$ at the mirror image of the source I in the interface, where σ_1 and σ_2 are the conductivities.

The potential in the second medium may be regarded as the potential due to a source $\frac{2\sigma_1}{\sigma_1+\sigma_2}I$ at the position of the source also in a medium of σ_1. If the corresponding resistivities are ρ_1, ρ_2 we have the potential in the medium containing the source as due to sources I and $\frac{\rho_2-\rho_1}{\rho_2+\rho_1}I$ and the potential in the second medium as due to $\frac{2\rho_2}{\rho_2+\rho_1}I$.

Let XY (Fig. 187) be the trace of the ground surface, which is perpendicular to the paper, and A the current source I and B the corresponding sink $-I$. The potential difference between P and Q is required.

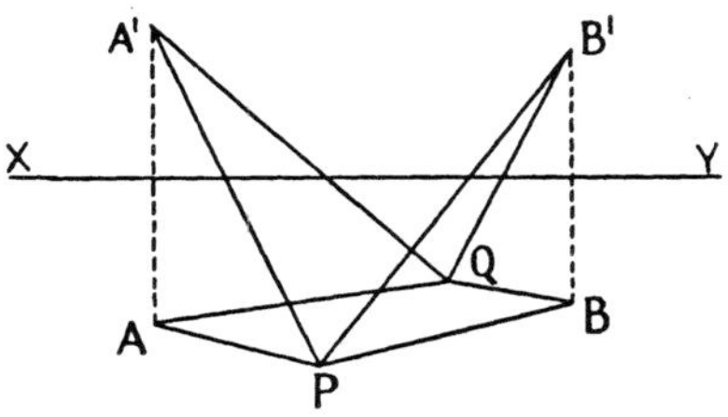

Fig. 187. Diagram for Wenner's formula

Let ρ be the resistivity of the lower medium, the resistivity of the upper being infinite. Then the magnitudes of the images A', B' of A and B are $+I$ and $-I$ respectively. This system of current sources obviously satisfies the necessary condition that no current flows across the surface XY, since XY is a surface of symmetry.

Potential difference between P and Q due to source I at $A = \frac{\rho I}{4\pi}\left(\frac{1}{AP}-\frac{1}{AQ}\right)$.

Similarly for $\quad I$ at A' potential difference $= \frac{\rho I}{4\pi}\left(\frac{1}{A'P}-\frac{1}{A'Q}\right)$,

$\quad -I$ at B potential difference $= -\frac{\rho I}{4\pi}\left(\frac{1}{BP}-\frac{1}{BQ}\right)$,

and $\quad -I$ at B' potential difference $= -\frac{\rho I}{4\pi}\left(\frac{1}{B'P}-\frac{1}{B'Q}\right)$.

Hence the total potential between P and Q is given by

$$V=\frac{\rho I}{4\pi}\left(\frac{1}{AP}+\frac{1}{A'P}+\frac{1}{BQ}+\frac{1}{B'Q}-\frac{1}{AQ}-\frac{1}{A'Q}-\frac{1}{BP}-\frac{1}{B'P}\right),$$

giving the general equation for the resistance between any two points in the semi-infinite medium.

In some cases resistivity determinations have been carried out with the electrode contacts at equal intervals in a straight line, all at a depth b below the surface. For this arrangement of electrodes

$$AP = BQ = a;\quad AQ = BP = 2a;$$

$$A'P = B'Q = \sqrt{a^2+4b^2} \text{ and } A'Q = B'P = \sqrt{4a^2+4b^2},$$

[a] Jeans, *Mathematical Theory of Magnetism and Electricity*, 2nd ed., p. 200, University Press, Cambridge, 1911.

and the equation becomes

$$V = \frac{\rho I}{4\pi}\left(\frac{1}{a} + \frac{2}{\sqrt{a^2+4b^2}} - \frac{1}{\sqrt{a^2+b^2}}\right)$$

$$= \frac{I\rho}{4\pi a}\left(1 + \frac{2}{\sqrt{1+4\lambda^2}} - \frac{1}{\sqrt{1+\lambda^2}}\right)$$

where $\lambda = b/a$.

When λ is large, that is the four points *ABPQ* are far removed from the surface compared with their separation

$$V = \rho I/4\pi a,$$

the equation for an infinite homogeneous medium, while for $\lambda = 0$, for four points lying on a straight line in the surface

$$V = \frac{\rho I}{2\pi a},$$

or

$$\rho = 2\pi a\, V/I = 2\pi a R,$$

the relation used in resistivity work.

In certain simple cases it is possible to determine the shape of the resistivity-electrode separation curves. Although the cases treated must of necessity represent ideal conditions which never obtain in the field, they tend to show that the present theoretical basis of the work is far from satisfactory, probably due to other factors entering into the problem which, at the moment, are not appreciated. The two cases treated below are for a simple vertical fault and for a uniform overburden. The results obtained from the fault tend to show that it may be possible to determine the boundaries of a hidden structure by resistivity measurements.

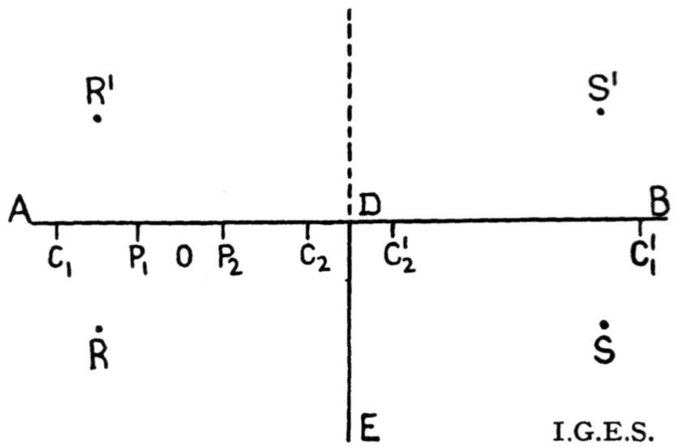

Fig. 188. Diagram for vertical fault *AB*. Ground surface. *DE*. Fault plane.

Vertical Fault. Let *AB* (Fig. 188) be the surface of the ground, and *DE* the fault plane dividing the region into two parts of resistivities ρ_1 and ρ_2. Both the plane of the fault and the ground surface are assumed perpendicular to the plane of the paper. If the centre point of the four electrode system is kept fixed at *O*, and the separations are varied, the resistivity curve may be divided into three parts, the first when all the electrodes are in the same medium, the second when one electrode only has crossed the fault line, and the third when two electrodes are in each medium.

The solution again involves the method of images. A source *I* of current at any point *R* in the medium of resistivity ρ_1, will give the same potential in the medium as a source *I* at *R* and *R'*, the mirror image of *R* in *AB*, together with sources of magnitude $\frac{\rho_2-\rho_1}{\rho_2+\rho_1} I$ at the images of *R* and *R'* in *DE*, the medium now being considered of infinite extent. For the second medium, sources at R and R' of magnitude $\frac{2\rho_2}{\rho_2+\rho_1} I$ are necessary. In the limiting case when R is on the ground surface the image of R in AB coincides with R.

All Four Electrodes in One Medium. This corresponds to the range of $a<\frac{2d}{3}$, if a is the electrode separation and d the distance from the centre of the electrode system to the fault. The system of a source I at C_1 and a sink $-I$ at C_2 in the system shown in Fig. 188 is equivalent to $2I$ at C_1 and $-2I$ at C_2, together with $2\frac{\rho_2-\rho_1}{\rho_2+\rho_1} I$ at C_1', and $-2\frac{\rho_2-\rho_1}{\rho_2+\rho_1} I$ at C_2' where $C_1D = DC_1'$ and $C_2D = DC_2'$ in an infinite medium of resistivity ρ_1.

Potential drop across P_1P_2 due to $2I$ at C_1 and $-2I$ at $C_2 = I\rho_1/2\pi a$.

Potential drop due to

$$2I\frac{\rho_2-\rho_1}{\rho_2+\rho_1} \text{ at } C_1' = \frac{I\rho_1 k}{2\pi}\left(\frac{1}{2d+2a}-\frac{1}{2d+a}\right).$$

Potential drop due to

$$-2I\frac{\rho_2-\rho_1}{\rho_2+\rho_1} \text{ at } C_2' = \frac{I\rho_1 k}{2\pi}\left(\frac{1}{2d-a}-\frac{1}{2d-2a}\right),$$

when

$$k = \frac{\rho_2-\rho_1}{\rho_2+\rho_1}.$$

Hence the total potential between the two may be written

$$V = \frac{I\rho_1}{2\pi a}\left\{1+4k\lambda\left(\frac{1}{4\lambda^2-4}-\frac{1}{4\lambda^2-1}\right)\right\},$$

where $\lambda = d/a$.

Comparing this with $V = \frac{I\rho'}{2\pi a}$ where ρ' = effective resistivity

$$\frac{\rho'}{\rho_1} = 1+4k\lambda\left(\frac{1}{4\lambda^2-4}-\frac{1}{4\lambda^2-1}\right) \qquad \text{......(1).}$$

One Current Electrode in Second Medium, $2d/3<a<2d$. In this case the potential difference between P_1 and P_2 due to I at C_1 is the same as before, but the sink $-I$ at C_2 must be regarded as due to $-2\frac{2\rho_1}{\rho_2+\rho_1} I$ in an infinite medium of resistivity ρ_2, giving a potential difference of

$$-2\frac{2\rho_1}{\rho_2+\rho_1}\frac{I\rho_2}{4\pi}\left(\frac{1}{2a}-\frac{1}{a}\right) = \frac{(1-k)I\rho_1}{4\pi a}.$$

Hence

$$V = \frac{I\rho_1}{4\pi a}\left\{2+k+2k\left(\frac{1}{2\lambda+2}-\frac{2}{2\lambda+1}\right)\right\}$$

and the effective resistivity is given by

$$\frac{\rho'}{\rho_1} = 1+\frac{k}{2}\left(1+\frac{1}{\lambda+1}-\frac{2}{2\lambda+1}\right) \qquad \text{......(2).}$$

Two Electrodes in each Medium, $a>2d$. The potential at the point P_1 may be obtained from the following system of images:

$$2I \text{ at } C_1 \text{ in an infinite medium of resistivity } \rho_1,$$

$$2\frac{\rho_2-\rho_1}{\rho_2+\rho_1} I \text{ at } C_1' \text{ in an infinite medium of resistivity } \rho_1,$$

and

$$-2\frac{2\rho_1}{\rho_2+\rho_1} I \text{ at } C_2 \text{ in a medium of resistivity } \rho_2.$$

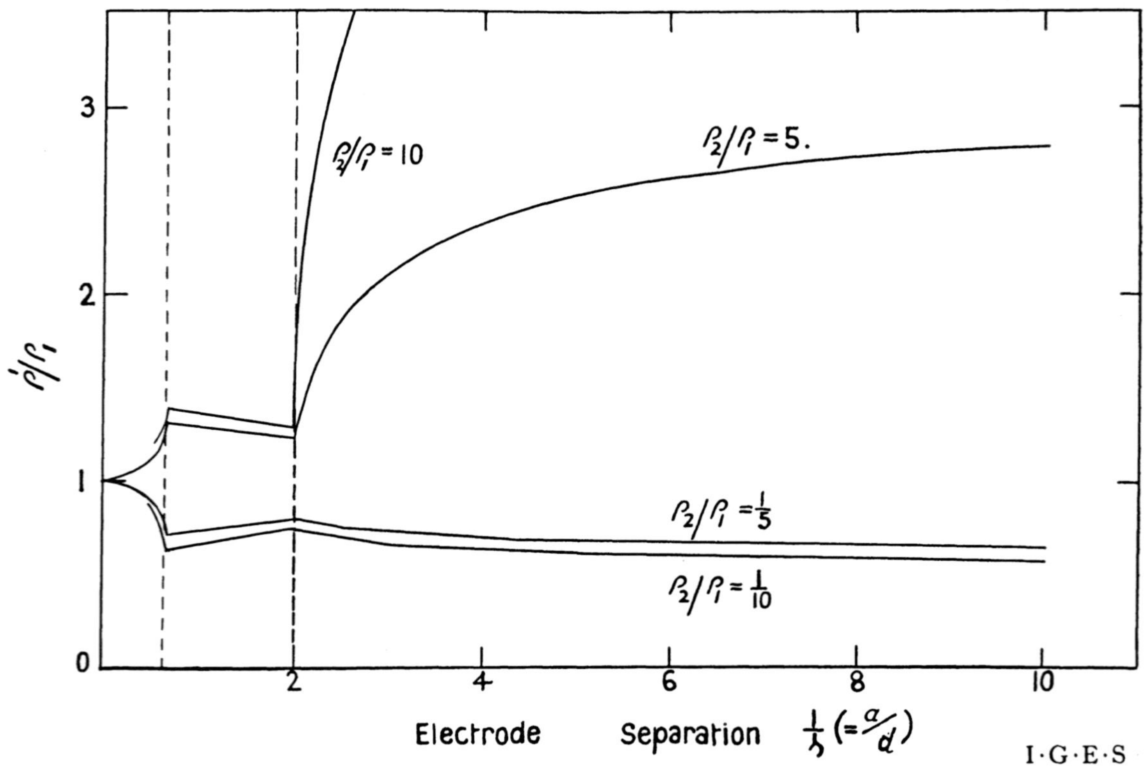

Fig. 189. Resistivity curve for vertical fault for various ratios of the resistivities on each side of the fault plane

A similar set of images will give the potential at P_2. With this distribution it can be shown that the potential difference is given by

$$V = \frac{I\rho_1}{2\pi a(1-k)}\left(1+k^2-k\frac{k+\lambda}{1-\lambda^2}\right),$$

whence

$$\frac{\rho'}{\rho_1} = \frac{1}{1-k}\left(1+k^2-k\frac{k+\lambda}{1-\lambda^2}\right) \qquad \text{......(3).}$$

From equation (1) for small separations, i.e. for large values of λ

$$\rho' = \rho_1,$$

while for large separations using equation (3)

$$\rho' = \frac{\rho_1}{1-k} = \tfrac{1}{2}(\rho_1+\rho_2).$$

Further, by differentiating (1) and (2) at $\lambda = \frac{3}{2}$ it can be shown that the slopes of the curve are discontinuous at this point, and a second discontinuity occurs at $\lambda = \frac{1}{2}$. The changes in slope are very marked, the slope of (2) being negative throughout the range $\frac{3}{2}>\lambda>\frac{1}{2}$, while the slopes of (1) and (3) are positive if $\rho_2>\rho_1$. Fig. 189 shows the resistivity curves for various ratios of the two resistivities, ρ'/ρ_1 being plotted against $1/\lambda$ (a/d). A complete solution of the problem would be possible from an experimental curve, the discontinuities giving the position of the fault and the resistivities being obtained from the effective values at small and large electrode separations. This method could be applied to the location of boundaries of various structures even when covered by an overburden, so long as the distance d is large compared with the thickness of the overburden. In the case of the central point of the electrode system being practically over the fault, i.e. d small compared with the thickness, the characteristic shape of the curve would be masked and the effect would be somewhat similar to that of a uniform overburden on a medium whose resistivity was the mean of the specific resistances on each side of the fault.

It seems peculiar in the second part of the curve, that as the electrodes move into an area of higher resistivity, the effective resistivity should fall. This section corresponds to one current electrode moving into an area of higher resistivity; the potential drop between the electrode and the edge of the fault being greater than normal, with a corresponding decrease in the potential difference between the inner electrodes implying a smaller effective resistivity.

Uniform Overburden[a]. In Fig. 190 let C_1C_2 be the surface of the ground and AB the interface between two media, the upper medium having a resistivity ρ_1 and the lower ρ_2. It can be shown by standard methods[b] that the potential at points in the overburden due to a source I at C_1 is the same as a source $2I$ at C_1 together with a series of sources $2k^n I$ at $\pm 2nd$ in a medium of infinite extent and of resistivity ρ_1, where d is the thickness of the overburden, $k = (\rho_2-\rho_1)/\rho_2+\rho_1)$ and all distances are measured from C_1 along a line passing through C_1 and perpendicular to the ground

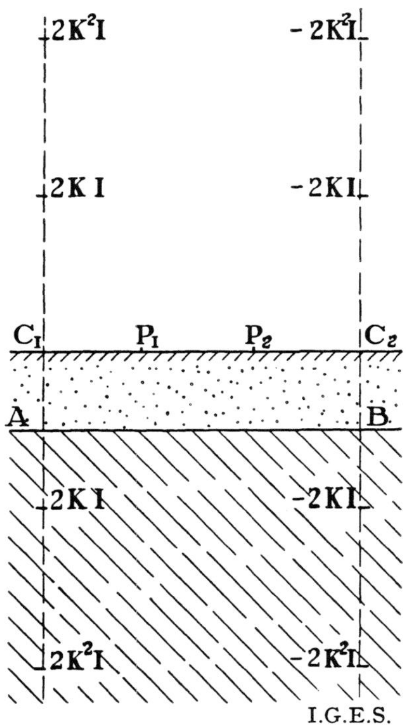

Fig. 190. Diagram for uniform overburden

C_1C_2. Ground surface.
AB. Interface.

[a] It now appears that this calculation has been done independently by J. N. Hummel, *Zeits. f. Geophysik*, vol. 5, p. 89 (1929).

[b] Maxwell, *Electricity and Magnetism*, 3rd ed., vol. 1, p. 435, Clarendon Press, Oxford, 1904.

surface. Hence the potential difference between P_1 and P_2 due to a source I at C_1

$$= \frac{2I\rho_1}{4\pi}\left(\frac{1}{a}-\frac{1}{2a}\right)+2\sum_1^\infty \frac{2k^n I\rho_1}{4\pi}\left\{\frac{1}{(a^2+4n^2d^2)^{\frac{1}{2}}}-\frac{1}{(4a^2+4n^2d^2)^{\frac{1}{2}}}\right\},$$

and a similar expression will give the potential due to $-I$ at C_2.

The total potential difference is given by

$$V = \frac{I\rho_1}{2\pi}\left[\frac{1}{a}+4\sum_1^\infty k^n\left\{\frac{1}{(a^2+4n^2d^2)^{\frac{1}{2}}}-\frac{1}{(4a^2+4n^2d^2)^{\frac{1}{2}}}\right\}\right]$$

$$= \frac{I\rho_1}{2\pi a}\left[1+4\sum_1^\infty k^n\left\{\frac{1}{(1+4n^2\lambda^2)^{\frac{1}{2}}}-\frac{1}{(4+4n^2\lambda^2)^{\frac{1}{2}}}\right\}\right],$$

corresponding to an effective resistivity ρ' where

$$\frac{\rho'}{\rho_1} = 1+4\sum_1^\infty k^n\left\{\frac{1}{(1+4n^2\lambda^2)^{\frac{1}{2}}}-\frac{1}{(4+4n^2\lambda^2)^{\frac{1}{2}}}\right\} \qquad \ldots\ldots(4),$$

and $\lambda = d/a$.

Now, $\left\{\frac{1}{(1+4n^2\lambda^2)^{\frac{1}{2}}}-\frac{1}{(4+4n^2\lambda^2)^{\frac{1}{2}}}\right\}$ is always less than one and for all practical values $|k| = \left|\frac{\rho_2-\rho_1}{\rho_2+\rho_1}\right|$ is less than one since ρ is positive and so the series is convergent.

For small values of a, i.e. λ large

$$\rho' = \rho_1,$$

and for large values of a

$$\rho' = \rho_2.$$

Hence for small separations the resistivity obtained is appropriate to the overburden and for large values the resistivity of the lower medium is measured.

For large values of n

$$k^n\left\{\frac{1}{(1+4n^2\lambda^2)^{\frac{1}{2}}}-\frac{1}{(4+4n^2\lambda^2)^{\frac{1}{2}}}\right\} = k^n\frac{1}{2n\lambda}\left\{\left(1-\frac{1}{2}\frac{1}{4n^2\lambda^2}+\ldots\right)-\left(1-\frac{1}{2}\frac{1}{n^2\lambda^2}+\ldots\right)\right\}$$

$$= k^n\frac{3}{16n^3\lambda^3},$$

neglecting higher powers than $\left(\frac{1}{n}\right)^2$. It can further be shown that if $n>\frac{1}{\lambda}$

$$k^n\frac{3}{16n^3\lambda^3}>k^n\left\{\frac{1}{(1+4n^2\lambda^2)^{\frac{1}{2}}}-\frac{1}{(4+4n^2\lambda^2)^{\frac{1}{2}}}\right\},$$

and since $\Sigma\frac{1}{n^3}$ is a standard series this may be used to determine the accuracy with which numerical calculations can be made. The greatest error is obtained when $k = 1$ for the series then converges slowly, and for this case the estimated errors are shown in the following table.

TABLE XII

λ	No. of terms	ρ'/ρ_1	Error
4	2	1·01	·001
2	5	1·09	·002
1	8	1·50	·001
$\frac{1}{2}$	14	2·77	·005
$\frac{1}{4}$	20	5·49	·05

A series of curves for various cases is shown in Fig. 191; series (*b*) illustrating curves for a lower medium of greater resistivity, and series (*a*) for lower resistivities. The resistivities are plotted in terms of the resistivity of the overburden, and the

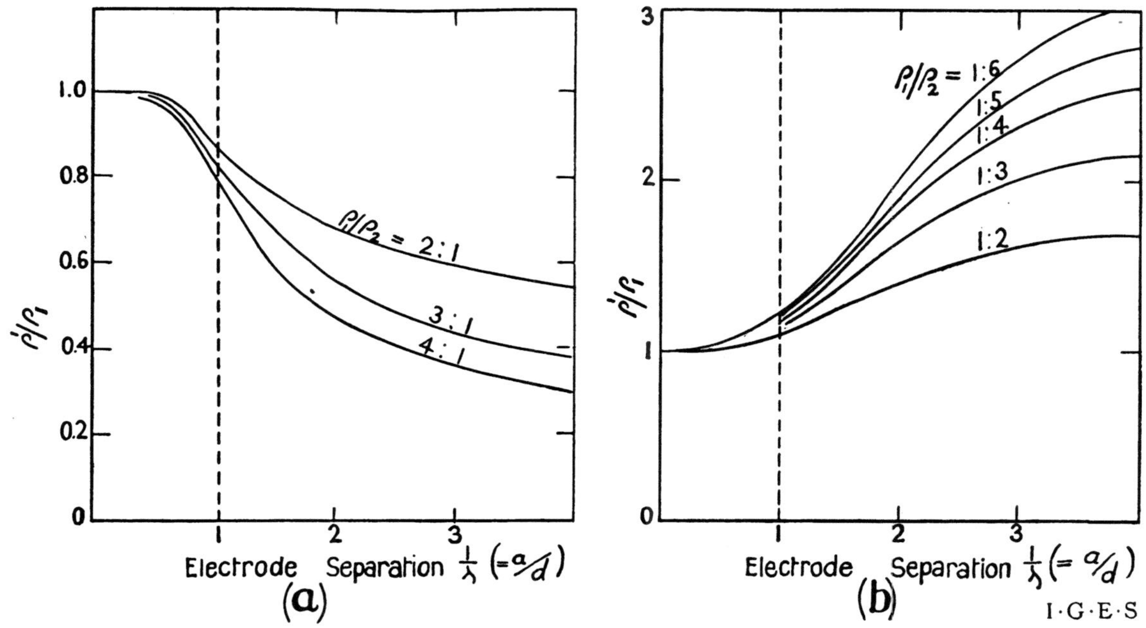

Fig. 191. Resistivity curves for uniform overburden for various ratios of the two resistivities
(*a*) Overburden of greater resistivity. (*b*) Lower medium of greater resistivity.

separation in terms of the depth of the overburden. Gish and Rooney analyse their experimental results by assuming that the depth to the discontinuity is equal to the separation corresponding to the greatest curvature or turning point, and it will be seen that this point corresponds approximately to a separation equal to about half the thickness of the overburden.

A further advance in the theoretical treatment has been made by Hummel[a] who has investigated the form of the resistivity curve for "four strata" problems. The system consists of two superimposed homogeneous strata of constant thickness resting on a semi-infinite homogeneous medium. The fourth medium is the air which acts as a

[a] *Zeits. f. Geophysik*, vol. 5, p. 228 (1929). See also *Mining Magazine*, London, June, July and Sept. 1930.

medium of infinite resistivity. If h and h_1 are the thicknesses of the strata, ρ, ρ_1, ρ_2 the resistivities, then the apparent specific resistance ρ' is given by

$$\frac{\rho'}{\rho} = 1+4\sum_0^\infty k^{n+1}\left\{\frac{1}{\sqrt{1+\left[2(n+1)\frac{h}{a}\right]^2}}-\frac{1}{\sqrt{4+\left[2(n+1)\frac{h}{a}\right]^2}}\right\}$$

$$+4(1-k^2)k_1\sum_0^\infty(n+1)k^n\left\{\frac{1}{\sqrt{1+\left[2\frac{h_1}{a}+2(n+1)\frac{h}{a}\right]^2}}-\frac{1}{\sqrt{4+\left[2\frac{h_1}{a}+2(n+1)\frac{h}{a}\right]^2}}\right\}$$

$$+4(1-k^2)k_1^2\sum_0^\infty\left[\frac{n(n+1)}{2}(1-k^2)-(n+1)k^2\right]k^{n-1}$$

$$\left\{\frac{1}{\sqrt{1+\left[4\frac{h_1}{a}+2(n+1)\frac{h}{a}\right]^2}}-\frac{1}{\sqrt{4+\left[4\frac{h_1}{a}+2(n+1)\frac{h}{a}\right]^2}}\right\}$$

$$+4(1-k^2)k_1^3\sum_0^\infty\left[\sum_{m=0}^{n}\frac{(m-1)m}{2}(1-k^2)^2-n(n+1)(1-k^2)k^2+(n+1)k^4\right]k^{n-2}$$

$$\left\{\frac{1}{\sqrt{1+\left[6\frac{h_1}{a}+2(n+1)\frac{h}{a}\right]^2}}-\frac{1}{\sqrt{4+\left[6\frac{h_1}{a}+2(n+1)\frac{h}{a}\right]^2}}\right\}+\ldots \qquad \ldots\ldots(5),$$

where $k_1 = (\rho_2-\rho_1)/(\rho_2+\rho_1)$ and $k = (\rho_1-\rho)/(\rho_1+\rho)$.

If the simple case of $h_1 = h$ and $k_1 = 1$, i.e. the two strata are of equal thickness and the lowest medium is a non-conductor the expression simplifies to

$$\frac{\rho'}{\rho} = 1+4k\sum_0^\infty\left\{\frac{1}{\sqrt{1+\left[2(2n+1)\frac{h}{a}\right]^2}}-\frac{1}{\sqrt{4+\left[2(2n+1)\frac{h}{a}\right]^2}}\right\}$$

$$+4\sum_0^\infty\left\{\frac{1}{\sqrt{1+\left[4(n+1)\frac{h}{a}\right]^2}}-\frac{1}{\sqrt{4+\left[4(n+1)\frac{h}{a}\right]^2}}\right\} \qquad \ldots\ldots(6),$$

and for large separations it becomes

$$\rho' = \frac{2}{\frac{1}{\rho}+\frac{1}{\rho_1}}\,2u\log 2 \qquad \ldots\ldots(7),$$

where $u = a/h$.

The types of curve obtained for such structures are shown in Fig. 192, which has been extracted from the original paper by Hummel.

The various broken lines passing through the origin are the asymptotes to the curves given by equation (7). The three other examples (Figs. 192 *b*, *c*, and *d*) are for more general cases of different thicknesses of the upper strata, for various ratios of the specific resistances. In each case the details are given in the subscript. The broken curves are for the corresponding simple cases of a uniform overburden on an infinite medium and they enable an approximation to the final resistivity curve to be obtained. This approximation may be extended to a system containing a larger number of layers.

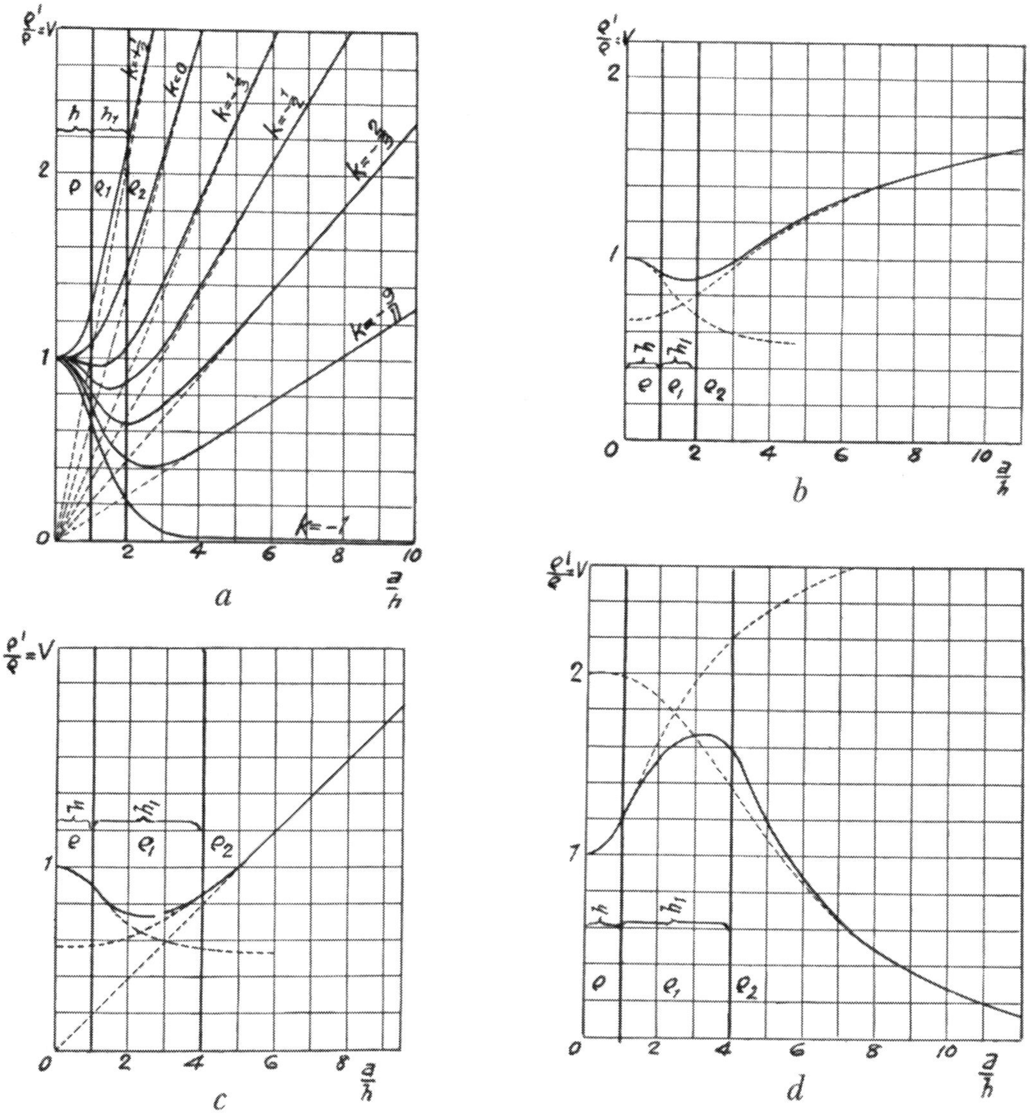

Fig. 192. Resistivity curves due to three parallel strata (after Hummel)

a, Case of $h_1 = h$, $\rho_2 = \infty$ for various values of $\rho : \rho_1$.
b, Case of $h_1 = h$ and $\rho : \rho_1 : \rho_2 = 1 : 0{\cdot}5 : 2$.
c, Case of $h_1 = 3h$ and $\rho : \rho_1 : \rho_2 = 1 : 0{\cdot}5 : \infty$.
d, Case of $h_1 = 3h$ and $\rho : \rho_1 : \rho_2 = 1 : 3 : 0$.

Although these curves approach closely to the type of curve obtained in practice there appears no simple method of interpreting them in terms of strata thicknesses and resistivities. It will be seen that the turning points on these curves bear no simple relation to the lower interface, and a theoretical basis for the method of interpretation used by Gish and Rooney is still required.

SECTION 3. EQUIPOTENTIAL LINE METHOD

FIELD PROCEDURE

Layout of Survey. Although the general principles of the equipotential method have been explained on pp. 43–50, the more detailed information given here may prove of assistance to those intending to use the method who are unacquainted with the field procedure. In practice it usually happens that the area to be examined is limited, on geological grounds, to some definite strip of country within which the presence of ore-bodies is suspected. Such a stretch of ground may well be several miles in length but, as a rule, it will be more restricted in width—perhaps to a single line of contact or, in other cases, to a more widespread zone of mineralization up to a quarter or half a mile across. In most cases the probable strike bearing of any concealed deposits will

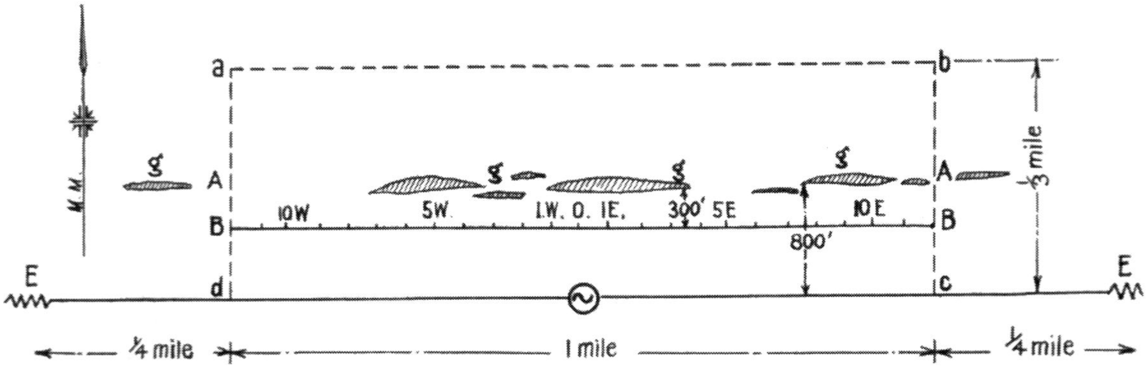

I·G·E·S

Fig. 193. Layout of equipotential line survey

AA. Axis of mineralized zone, indicated by gossan outcrops *g*, *g*.... *BB*. Survey base line.
EE. Earthed electrodes at ends of cable line. *abcd*. Area of equipotential line survey.

have been established on geological evidence. A typical instance is illustrated in diagrammatic form in Fig. 193, in which an east-west line of strike can be presumed from the general trend of the gossan outcrops, along the line *AA*. Under such circumstances the rectangle *abcd* would be convenient as the first area to be examined—since a mile is about the maximum length for effective and economical working. In preparing the layout for this survey the following are the principal points to be observed:

(i) A survey base line *BB*, marked by pegs at 200 ft. intervals, should be laid out parallel to the axis *AA* of the mineralized zone, and at a distance of about 300 ft. from it. This displacement of the base line is very desirable, as it is inconvenient to have large potential and phase anomalies at the points from which the equipotential lines are started. On the other hand it should not be too distant as it will be required in connection with intensive examinations by other methods, should any important indications be located by the equipotential reconnaissance survey. The centre peg on the line is taken as the zero point and pegs lying to the east and west of it are marked in the manner shown.

(ii) An insulated cable connecting the two electrodes, *EE*, should be run out parallel to the base line, but at a distance of about 800 ft. from the axis of the mineralized ground to be examined. This lateral displacement of the cable is desirable since some distortion of the equipotential surfaces usually occurs in the neighbourhood of the cable. In some cases this would be sufficiently marked to confuse the interpretation of indications which may be due to mineral bodies.

(iii) The two electrodes *EE* should be placed about 1000 ft. beyond the eastern and western extremities of the area to be surveyed, in order that the equipotential surfaces may penetrate to a sufficient depth and that their traces at the surface may cross the ground under examination as arcs of such great radii that they approximate to a series of parallel straight lines.

Locating Equipotential Lines. Where circumstances permit, it is desirable for reasons of economy in time and in operating costs that the tracing of the equipotential lines should be carried out by at least two parties concurrently; one taking the east and the other the west half sections as shown in Fig. 193. Providing the country is open and the cutting of undergrowth is not required, each party can be limited to two workers, as shown in Fig. 194; the operator with headphones and carrying the amplifier on his back, and his assistant following behind with the rear rapier[a], etc. In thick bush country one or more extra men will be required, armed with axes or slash hooks, for clearing a passage ahead of the operator. Under such circumstances, however, the equipotential line method becomes rather slow and tedious owing to the meandering nature of the lines and, as a rule, it will be better to adopt some other system, such as the A.C. potential ratio method, by which potential and phase changes can be observed along previously prepared straight line traverses. The procedure illustrated by Fig. 194 is quite simple and assuming that the party is to trace an equipotential line northwards from station 5 E on the base line shown in Fig. 193, the following steps would be taken:

I.G.E.S.

Fig. 194. Locating equipotential points

(i) The assistant is provided with a supply of blank card tickets, about 2 in. square, which he marks in soft lead pencil (not ink or indelible pencil) with the station numbers

[a] The slender pointed steel rods, or rapiers, employed for making contact with the ground are shown in Fig. 197.

5 E, 5 EN1, 5 EN2, etc., up to the anticipated number of stations on the line. He also obtains a bundle of thin reeds, or bamboo canes, about 2 ft. 6 in. in length, sufficient in number to mark each station.

(ii) The assistant embeds the point of the rear rapier into the ground at station 5 E and also places a cane in position there. The top of the cane is split with a knife and the corresponding ticket is inserted.

(iii) Meanwhile the operator has moved forward in a northerly direction to the full extent of the connecting wire (about 60 ft.) and makes one or two trial contacts with his rapier in order to locate the direction of the equipotential point, or rather the position of minimum sound in the headphones, which is the nearest approach to it. By matching the stronger notes at points on each side of the minimum it is possible, in most cases, to locate the latter with an accuracy varying from a few inches to several feet. Having found this point the operator marks its position on the ground and moves forward.

(iv) The assistant follows behind the operator and places the rear rapier at the point first located by the latter. While a second point is being located he puts a cane in position and inserts the ticket marked 5 EN1.

(v) The line is continued in a similar manner until it is estimated from the number of stations located that the limit of the survey area has been reached. The party may then return to the base line and commence a new line or possibly it will first survey the line just completed, following it back station to station with compass and tape.

(vi) In dry country the operator must moisten the ground at each point of contact. This may be accomplished by the self-watering rapier described on p. 265.

Surveying Equipotential Lines. In open country this may be done by tacheometer, but as a rule a $3\frac{1}{2}$-in. prismatic compass on a tripod and a 100-ft. tape will prove more satisfactory. Forward and back bearings at alternate stations are sufficient for reconnaissance work, although special precautions must be taken in areas in which magnetic anomalies are suspected. The plotting of the equipotential lines should be carried out on a scale of 1 in. to 100 ft., and the principal surface features such as shafts, open-cuts, railway lines, and pipe lines, should be shown on the plan. When electrical indications of any importance are discovered their positions should be marked permanently on the ground by numbered beacons. This should be done *immediately*, and before the temporary canes (with tickets) which mark the equipotential points have been removed or destroyed. This is necessary, as the positions for the beacons must be located with direct reference to the neighbouring equipotential point stations, which actually mark the true position of the indication. Owing to the rough methods of surveying employed it would be unsafe to fix the beacons later, at points measured off from the plan, and set off from the base line.

FIELD EQUIPMENT

Cable Laying. As a rule the cable may be laid directly on the ground—elevating it on poles or passing it through pipes or under brushwood mats where roads or paths have to be crossed. With practice this work can be carried out with considerable speed if the direction has been marked out beforehand, by blazing trees, etc., and provided that suitable cable reels, connectors, etc., are used. The following descriptions are based on the experience of the I.G.E.S.

(i) *Cable.* Multiple strand is better than single strand cable. That used by the Survey was 3/·036, rubber covered, double braided and vulcanized. It is usually sold in coils of 100 yd., which are convenient for handling in the field.

(ii) *Cable Reels.* These were specially designed for lightness and convenience in operation. The reel is cast in two pieces from a strong aluminium alloy, and to save weight the sides are perforated as far as is consistent with strength (see Fig. 195). The detachable side is easily released so that a 100-yd. coil of cable may be placed directly in position on the tapered barrel, or withdrawn from it during winding-in operations.

I.G.E.S.

Fig. 195. Design of cable reel

(iii) *Cable Connections.* Experience has shown that the individual 100-yd. lengths of cable should not be connected together too securely as, if cattle or other animals should become entangled in it, as they frequently do, it is better that a break should occur at the nearest connection than that a quarter of a mile or so of cable should be dragged out of position. Brass barrel connectors fitted as shown in Fig. 196 are quite sufficient and, when properly tightened with the fingers, they will withstand a steady pull of about 12 lb. In wet country they should be covered by a sleeve of thin-walled rubber tubing in the manner indicated. Under moderately dry conditions ordinary black friction or "outdoor" electrical tape will usually give sufficient protection.

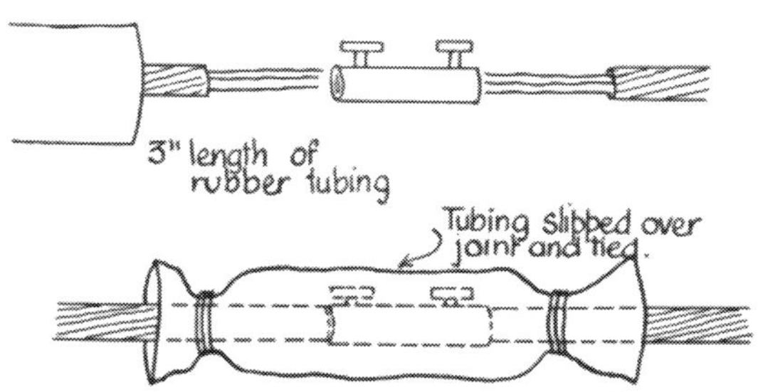

I.G.E.S.

Fig. 196. Method of cable connection

(iv) *Breaks in Cable.* When a mile or so of cable is laid across country, and directly on the ground, breaks in the line are not infrequent occurrences. As all work immediately comes to a standstill when this happens it is a matter of importance that a minimum of time should be spent in locating the trouble. This is facilitated by using a small testing set, consisting of an iron earth peg, a pocket milliammeter, and a small dry cell. With this apparatus the cable-earth circuit can be tested, and the break determined as being on one or the other side of the test point.

Electrodes. Passing current conductively into the ground is not always a simple matter and in very dry countries considerable difficulties may be experienced—which, however, can usually be overcome if sufficient trouble is taken. The general character and relative merits of linear and multi-point electrodes have been discussed on pp. 46–49, but there are a number of minor practical points which should also be mentioned. They can be enumerated as follows:

(i) Bare stranded (7/·028) copper wire is the most suitable material, whether linear or point electrodes are employed, but stranded galvanized iron wire is quite effective and from the point of view of cost is to be preferred.

(ii) A convenient type of peg for securing the electrode wire to the ground is made from a 12-in. length of ½-in. mild steel rod, one end of which is pointed, and the other bent over acutely at a point 2½ in. from its extremity. In order to improve the electrical contact a little solder, or spelter, is run into the crutch of this bend, which bears directly on the wire. These pegs are hammered into the ground so that the wire is pinned down securely into the soil.

(iii) Whenever possible a multi-point electrode should be placed in ground where the surface is moist. Swamps, dams, and stream beds are favourable for this purpose, and since the electrodes are at a distance of at least 1000 ft. from the area to be examined, any irregularity in their shape will have little effect. Numerous devices may be employed in order to reduce the contact resistances so that a satisfactory working current (2 to 3 amps at 150 volts) may be obtained. Increasing the area of the electrode and the number of contacts (metal spikes) is the usual procedure, but in cases of extreme dryness of the ground it may be necessary to have recourse to special methods. For example, the short metal spikes may be replaced by long iron rods (or lengths of wire) lowered into holes which have been jumped down through the dry overburden to a depth of 10 ft. or so. These holes may then be filled with salt solution which is replenished from time to time until each hole, and the salt saturated ground surrounding it, constitutes a large conducting column. In some instances in Australia effective electrodes have been made by lowering sheets of galvanized iron down into wells, and on other occasions the cable from the generator has been connected directly on to the underground water and compressed air pipe systems belonging to some small mine.

The Flying Circuit. The portable apparatus with which the equipotential lines are traced consists of the following components:

The Rapiers. These implements have been variously called secondary electrodes, prodders, etc. by different writers, but the design employed by the I.G.E.S. so closely resembles a fencing weapon that they have been called rapiers. They consist of $3\frac{1}{2}$ ft. lengths of $\frac{1}{4}$ in. hexagon steel with a hardened taper point at one end, and a hardwood handle and brass terminal at the other (Fig. 197 *a*). Being light and well balanced they are easy to use and in dry ground the point can be forced well down into the surface. In dry country, however, it is impossible to make an effective electrical côntact with these rapiers unless the ground is moistened with water. Although very little water is required for this purpose the operation is tedious and greatly retards the speed of work, even if an extra labourer is specially employed for the purpose. In order to reduce this inconvenience as far as possible, and so that extra labour may be dispensed with, the self-watering rapier illustrated in Fig. 197 *b* was developed by the I.G.E.S., and has proved most successful.

As the figure is drawn to scale and is self-explanatory, it is only necessary to draw attention to the following special details:

(1) Stainless steel is necessary, both for the barrel and the point, but the ball valve may be made of brass or bronze.

(2) The rubber bulb will perish in the course of time, but may be replaced in a moment by lubricating the outside of the barrel with glycerine and sliding a new one into position.

(3) There is no fixed terminal as in the ordinary steel rapier (type *a*). In this case an effective contact is made with a small battery clip which grips the barrel of the rapier just below the rubber bulb.

(4) The rubber tube shown at the upper end of the handle leads to a small metal tank holding half a gallon of water,

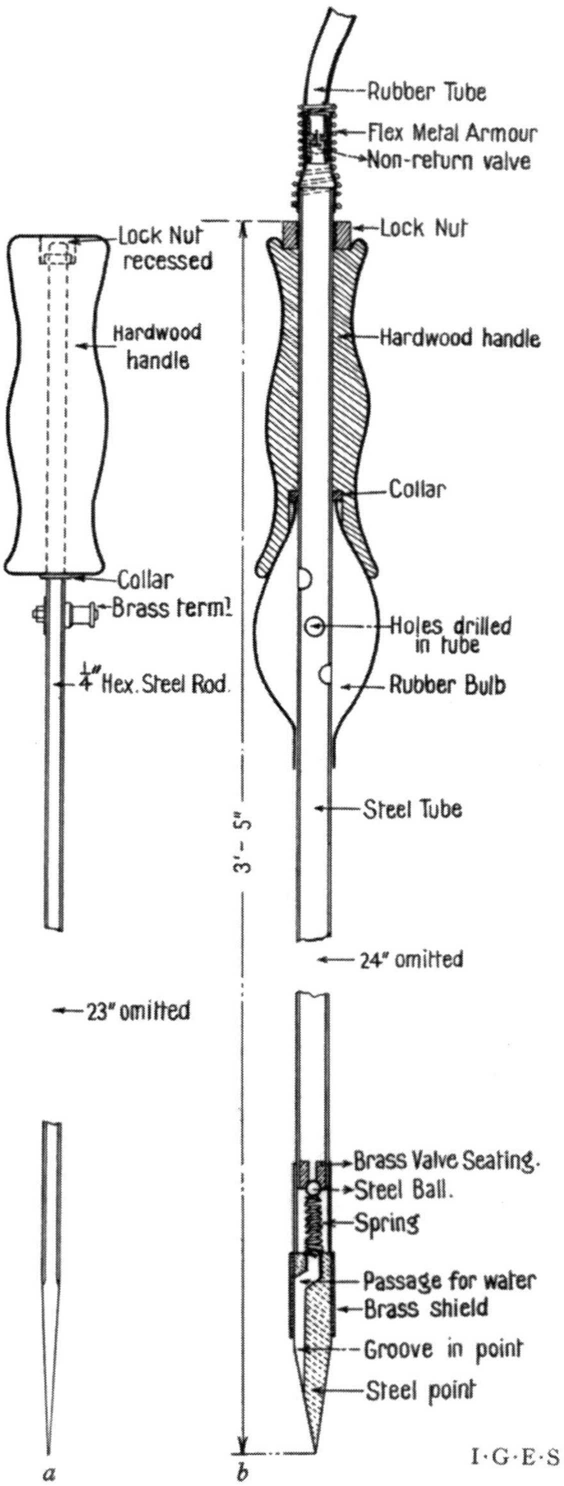

Fig. 197. Rapiers for equipotential line surveys

which is carried on the back of the operator on top of the amplifier. A reserve tank is carried by the assistant and this is substituted for the operator's tank when the latter becomes empty.

(5). The water used must be reasonably free from dirt and grit, otherwise the ball valve may fail to close properly. To guard against trouble from this cause a wire gauze filter is fitted to the mouth of the tank.

In using this rapier, the steel point is pressed into the surface of the ground for half an inch or so, and the rubber bulb squeezed at the same time. In this way a jet of water is forced down the point and into the broken ground. In loose soil or sand, the water outlet may occasionally become choked but this can usually be remedied by sharp pressure on the bulb. For the hard dry surfaces that are so common in Australia this self-watering device is most effective.

The Amplifier. In designing the amplifier used for equipotential line reconnaissance surveys, the reduction of weight to an absolute minimum was the most important consideration, since in countries, such as Australia, where the cost of an extra labourer would not be justified, the whole of the apparatus must be carried on the back of the operator. Although a tuned amplifier is necessary for several of the electrical methods, it is not essential in equipotential line work where the presence of harmonics makes very little difference to the accuracy of the determinations. For this reason tuned transformers were dispensed with and advantage was taken of the lightness of a resistance-capacity coupling.

The circuit diagram of the amplifier used by the I.G.E.S. is shown in Fig. 198[a]. It makes use of the principle developed by von Ardenne and Heinert and described by Colebrook[b]. In this amplifier the anode-coupling resistances are of the order of megohms instead of the usual values of 50,000–100,000 ohms, and Colebrook shows that under these conditions practically the full amplification factor of the valve can be realized with the use of moderate anode voltages. In addition it is desirable to use the valves at a considerably lower filament current than is required for their normal operation, and very little distortion will occur, provided the input is not too large; although this is not a matter of particular importance in the case in question.

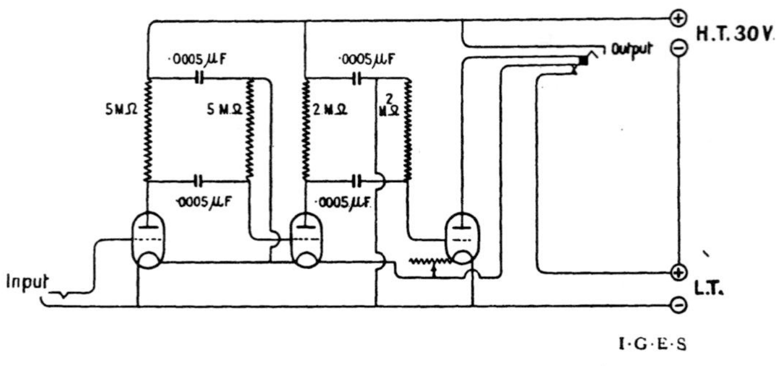

Fig. 198. Circuit diagram resistance-capacity coupled amplifier

All of these factors combine to produce an amplifier that is well adapted to those

[a] This amplifier has been used for electrical prospecting purposes in Rhodesia and elsewhere since 1926. For the following description of the circuit the Survey is indebted to R. O. Cherry, M.Sc., of the Laboratory of Natural Philosophy, Melbourne University.

[b] *Experimental Wireless and Wireless Engineer*, vol. 4, p. 195, London, 1927.

field purposes for which a resonance amplifier is not essential. The advantages of the arrangement are:

(i) Greatly reduced filament current consumption, which permits the use of small and light low tension cells.

(ii) Long life of high tension batteries. With this design of amplifier a small capacity 36-volt battery will last for several months, even when in constant daily use.

(iii) Extended life of the filaments. Actually, some of the valves (Osram, D.E. 3*b*) have been in constant use during the last four years without sign of failing, although tests have shown that the electron emission has decreased materially.

(iv) Great saving of weight by comparison with transformer-coupled amplifiers.

I.G.E.S.

Fig. 199. Method of carrying amplifier

I.G.E.S.

Fig. 200. A.C. generator

The components and batteries are enclosed in a welded sheet ($\frac{1}{16}$ in.) aluminium case, the valves being mounted in spring sockets, in order to minimise the effects of the bumping that is inevitable when working in the field. The case itself is protected by an outer cover of leather or webbing, in which holes are cut to accommodate the input and telephone plugs, and the central filament rheostat. From Fig. 198 it will be seen that the plate of the last valve of the amplifier is connected directly to the telephone jack and the H.T.+, and the insertion of the telephone plug also serves to switch on the filament current. The overall dimensions are $12\frac{1}{2} \times 8\frac{3}{4} \times 4$ in. and the total weight 11 lb. The general appearance and method of carrying the amplifier is shown in Fig. 199, from which it will be seen that a tubular metal carrier, with shoulder straps, is used for supporting the instrument on the back. These carriers (Bergen Meis pattern) have been found to be less fatiguing than any of the other devices that have been tried.

The A.C. Generator. The type of machine used by the I.G.E.S. is illustrated in Fig. 200. It was used both for equipotential line work and also for those electromagnetic methods which require current at a frequency of about 500 cycles per second. It consists of a 350 watt alternator directly coupled to a small four-stroke petrol engine which is run at 1800 R.P.M. The total weight of the machine is about 200 lb. and owing to the compactness of its design it has proved very suitable for field purposes. It is understood that it was made by the Bosch Magneto Co., during the War period, for use with field radio-telegraphy equipment, but, beyond the fact that it has proved very suitable for geo-electrical prospecting work, little is known about it.

The machine will supply 540 cycle alternating current (4·5 amps at 80 volts) and, by turning a switch, direct current with a 1080 cycle ripple can be obtained (16 amps at 50 volts). By using a condenser of suitable capacity an alternating current supply at the latter frequency is also available.

The following points may be mentioned regarding the alternator itself[a]:

(1) It is a four pole shunt wound self-exciting machine which, in a two pole form, has been described by C. P. Steinmetz[b], as the "Eickemeyer Inductor Alternator."

(2) The direct current ripple results from the pulsation in reluctance of the main armature flux, which is due to the alternate passage of the armature teeth across the slots in the pole faces, and across the pole teeth.

(3) Oscillograms (with the generator separately excited and run at a slow speed) of the A.C. potential wave form at 600 R.P.M. showed evidence of a 3rd harmonic, and the D.C. ripple a 5th harmonic, of the order of 10 per cent. of the fundamental.

SECTION 4. A.C. POTENTIAL RATIO METHOD[c]

The principles of this system have been briefly explained on pp. 50–54, but the theory of the method and details of apparatus and field procedure require further consideration.

Theory. Assuming that an alternating current is passing through the ground shown in Fig. 201 by means of a distant pair of earthed electrodes, and that the contact B is intermediate in potential with regard to A and C, it is possible by a simple adjustment of the capacities and resistances in the ratio arms to obtain a condition of balance in which silence will be observed in the headphones. Then, if Z_1 and Z_2 represent the impedances in the two arms

$$V_2 = \frac{Z_2}{Z_1} . V_1 .$$

The theory of this operation may be expressed as follows:

If a current i flows through a circuit of resistance R and reactance X, the potential across it is given in magnitude by

$$V = \sqrt{R^2+X^2} . i,$$

and in phase (relative to the current) by

$$\tan \theta = X/R.$$

[a] The thanks of the Survey are due to Associate Professor E. B. Brown of the Electrical Engineering Laboratory, University of Melbourne, who was good enough to examine one of the machines and provide this information.

[b] *Theory and Calculation of Electrical Apparatus*, p. 280, McGraw Hill, New York, 1917.

[c] For special applications see p. 352.

The two expressions may be incorporated in the vector equation

$$V = (R+jX)\,i \quad \text{or} \quad V/i = Z = R+jX,$$

which can be represented graphically, as in Fig. 202, where a line OA, equal in magnitude to R, is constructed in the direction OY. If from A a perpendicular AB is erected equal in length to X, then OB is the potential across the circuit for unit current and is

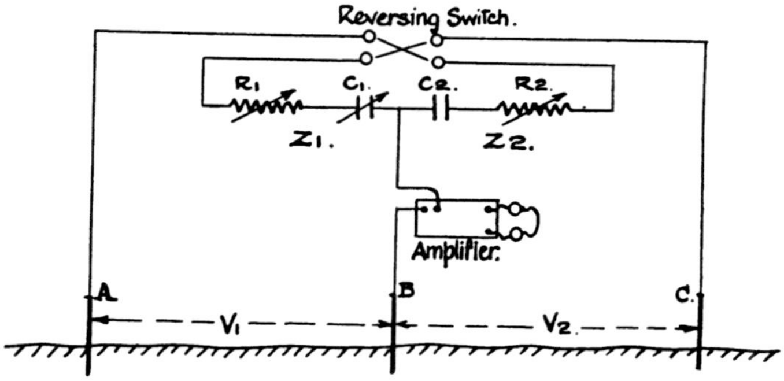

I·G·E·S

Fig. 201. Principles of A.C. potential ratio method

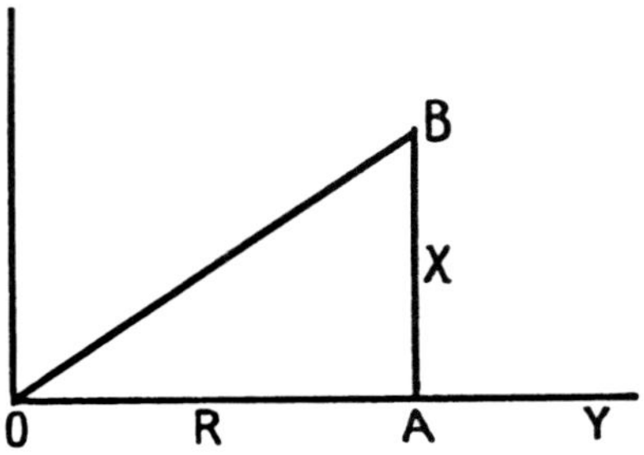

Fig. 202. Graphical representation of vector equation $\frac{V}{2} = (R+jX)$

known as the impedance. The angle AOB gives the phase angle between the current and the potential, for as far as numerical values are concerned we have from the diagram

$$OB = \sqrt{R^2+X^2} \quad \text{and} \quad \tan AOB = X/R.$$

In the ratio arms shown in Fig. 31 the current will be the same when the bridge is balanced, so the potential drops V_1 and V_2 will be directly proportional to the total impedances Z_1 and Z_2, as already stated, and

$$V_2 = \frac{R_2+jX_2}{R_1+jX_1}\cdot V_1,$$

when $R_1 X_1$ and $R_2 X_2$ are the resistances and reactances in the two arms.

This is shown vectorially in Fig. 203 according to the convention explained above, but for convenience the reactances are plotted as abscissae and the resistances as ordinates. In this case

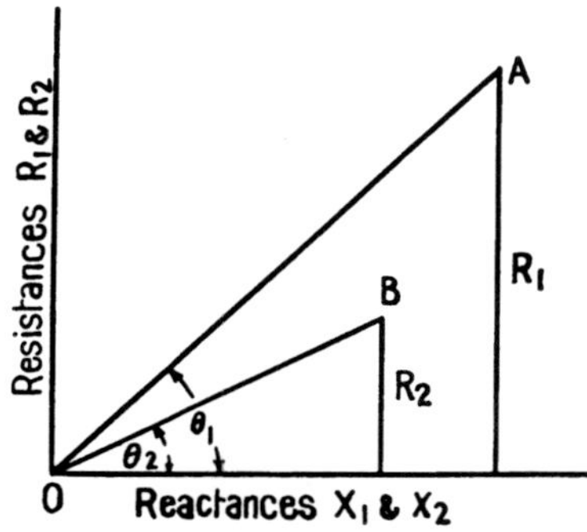

Fig. 203. Vector diagram, A.C. potential ratio method

$$V_2 = \frac{OB}{OA}\cdot V_1,$$

and V_2 lags in phase behind V_1 by an angle $\theta_2-\theta_1$. From Fig. 201 it will be seen that the reactances in the arms of the bridge, as used in the field, are introduced by condensers, the reactance of a condenser of capacity C being $\frac{1}{2\pi fC}$.

Apparatus. The series ratio arm instrument is illustrated below in Fig. 204, from which it will be seen that each arm contains a similar combination of non-inductively wound resistances giving a range of 0–300,000 ohms. A fixed condenser of 0·002μF is connected in series with the resistance in one arm, whilst in the other a parallel arrangement of a fixed and a variable condenser gives a range of 0·006μF. The reversing switch S is intended to be used for obtaining a second determination of the potential ratio. By taking two observations, not only is increased accuracy obtained by using their mean, but the behaviour of the instrument is checked, and any errors in adjustment or computation are revealed.

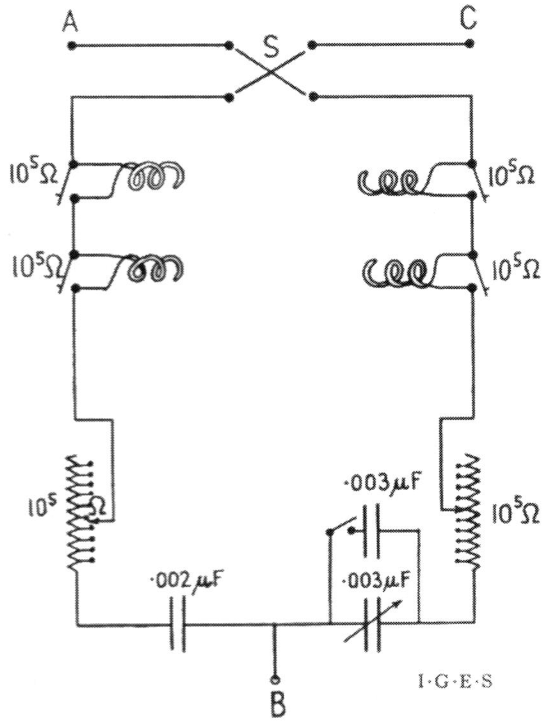

Fig. 204. Series ratio arm bridge

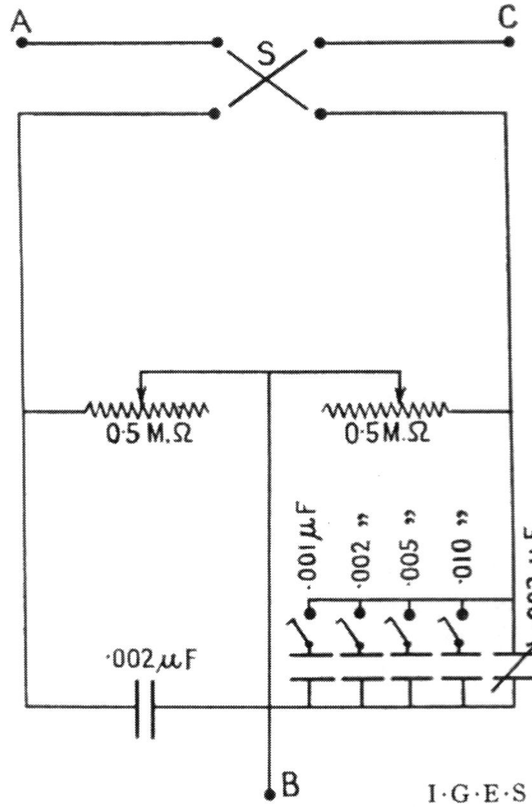

Fig. 205. Parallel ratio arm bridge

The terminals A, B, and C, are connected to the earthed electrodes in the manner shown in Fig. 201, from which it will be observed that resistances of unknown value are present in each arm, at the earth contacts A and C respectively. If very large, these resistances would seriously affect the accuracy of the observations, but although they may vary from a few hundred up to a thousand ohms in different types of ground, the errors arising from this source will be negligible if the impedances in the arms are kept sufficiently high. In practice a minimum impedance of about 50,000 ohms is maintained in each arm.

As an alternative to the series connected ratio arms an instrument was made, at the suggestion of the late Dr Bieler, in which the resistances and condensers in each arm were connected in parallel. The apparatus took the form shown in Fig. 205, and instead of being calibrated in ohms and capacity reactances, the corresponding reciprocal values (conductances and susceptances) were used. Dr Bieler's explanation of the instrument may be stated as follows:

"In the case of such an alternating current circuit with a resistance R and a capacity C connected in parallel, the total impedance Z in each arm will be given by

$$\frac{1}{Z} = \frac{1}{R} + \frac{1}{\frac{1}{j \times 2\pi fC}} = \frac{1}{R} + j.2\pi fC.$$

If $\frac{1}{Z} = y$, the admittance of the circuit,

$\frac{1}{R} = g$, the conductance of the circuit,

$2\pi fC = h$, the capacitative susceptance of the circuit,

we then have

$$y = g + jh.$$

"As already explained in the case of the series arm instrument, the same current i will flow through both arms when a balance is obtained. Therefore if V_1 is the potential drop between A and B we have

$$\frac{V_1}{i} = Z_1 = \frac{1}{y_1} = \frac{1}{g_1 + jh_1} \qquad \text{......(1)},$$

where Z_1 is the impedance, and y_1 the admittance of the arm AB.

"Similarly if V_2 is the potential drop between B and C,

$$\frac{V_2}{i} = \frac{1}{g_2 + jh_2} \qquad \text{......(2)}.$$

"Hence from (1) and (2) above, we get

$$\frac{V_2}{V_1} = \frac{g_1 + jh_1}{g_2 + jh_2}.$$

"In order to obtain the ratio of V_2 to V_1, $g_1 + jh_1$ and $g_2 + jh_2$ must be represented according to the convention established on p. 269. Then the ratio of the lengths of the two vectors representing the conductances is the ratio of the potential drops, and the angle between the vectors is the phase angle between the potential drops. In the case shown in Fig. 206

$$V_2 = \frac{OA}{OB} \cdot V_1$$

and V_2 leads V_1 by the angle θ.

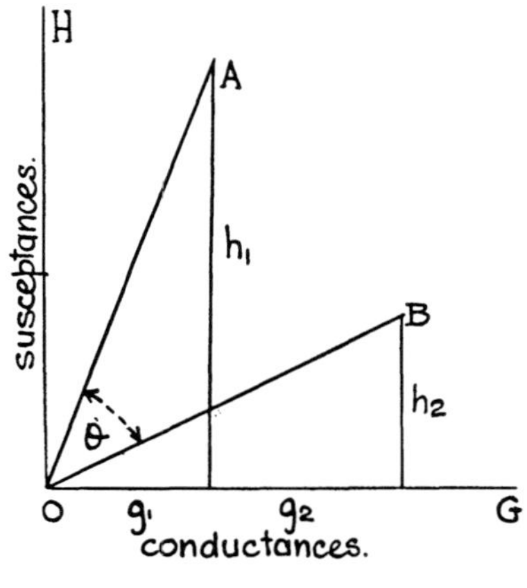

Fig. 206. Vector diagram, parallel ratio arm bridge

"In this apparatus the variable capacities and resistances are graduated so that the conductances g_1 and g_2 and the susceptances h_1 and h_2 are in the same units; the unit employed being the conductance of a 300,000 ohm resistance. When the frequency is

535 cycles per second this is equal to the susceptance of a condenser of a 0·001 microfarad capacity. With this unit the various portions of the variable capacity come out to be whole numbers, within an error of 1 per cent. at most."

To be able to correlate the resistances and capacities in this manner is a great advantage, and it was for this reason that the parallel connection of the condensers and resistances was suggested. Thus the condenser values in thousandths of a microfarad become individual units of the system and are additive, and the variable resistances are calibrated into the equivalent reciprocal ohms $\left(1 \text{ unit} = \frac{1}{300{,}000} \text{ reciprocal ohm}\right)$. This arrangement simplifies the calibration and reading of the instrument and facilitates the working out of the results[a]. In the case of the series connected ratio arms it is possible to graduate the instrument directly in resistances and reactances, but it is a rather difficult arrangement, since reactance varies inversely as the capacity.

Field Procedure. This has already been described on pp. 50–54, but those who may make use of the method will find the following suggestions of some practical value:

(i) The operator should stand at the central electrode *B* (Fig. 201), and to minimise errors due to any body capacity or other effects, both he and his instruments should be symmetrically disposed with reference to the three-electrode system.

(ii) The electrodes are conveniently made from short lengths of light drill-steel, pointed at one end, and fitted with a two-inch brass collar, soldered or brazed to the steel at a point about 6 in. below the striking end.

(iii) To the electrode ends of the three insulated wire leads large size battery clips are fastened, so that connection to the brass collars on the electrodes may be made or broken without difficulty.

(iv) It is convenient, and also good practice, to employ a fourth electrode which should always be put into position at the next station on the traverse, whilst the observations are being made at those immediately preceding it. In this way there is no need to move the two forward electrodes, after an observation has been made, since they are already in position to serve as the two rear electrodes for the next set up of the equipment. The operator and his assistants—one at each electrode—have merely to detach the battery clips and move forward, trailing the wire along the ground behind them.

(v) The adjustments should be so conducted that the total impedance in each arm, at the point of balance, is not too low; otherwise appreciable errors may arise owing to the electrode contact resistances, which are of unknown value. Even when dealing with large potential ratios it should be possible to maintain minimum impedances of about 50,000 ohms, which will be quite sufficient to guard against this danger.

[a] A further simplification is due to D. C. Gall who suggested that the potential variation curves need only be drawn for the in-phase components. In this case an instrument embodying the circuit shown in Fig. 205 may be made direct reading by substituting a fixed resistance in the one arm and removing the fixed condenser from the other arm of the bridge.

It is convenient to book the field observations in the manner indicated by the table shown below, which is intended for use with the series ratio arm instrument. In the left-hand columns R_1, R_2 and X_1 give the resistance and reactance values as read on the instrument in thousands of ohms; X_2, being the reactance of the fixed condenser, the value of which is entered at the head of the sheet. On the right-hand side of the table the columns *MR* and *PV* are the multiplied ratios and phase variation values respectively; the latter being obtained by addition of the phase angles. The potential ratios and differences in phase angle may then be determined graphically in the manner shown in Fig. 203; or use may be made of the mechanical device illustrated in

TABLE XIII. *Imperial Geophysical Experimental Survey. A.C. Potential Ratio Traverse*

$X_2 = 153$. *District*, Moonta, South Australia

Traverse, P. V. D. *Operator*, S. H. S. *Date*, 11th November, 1929					*Computer*, I. W. M. *Date*, 12th November, 1929							
Field observations					Office computations							
Stations	Switch D or R	R_1	X_1	R_2	V_1	θ_1	V_2	θ_2	V_2/V_1	*MR**	$\theta_2-\theta_1$	*PV*†
0	*D*	82	154	98	17·3	27° 30′	17·9	32° 0′	1·04		4° 30′	
1									*1·05*	1·00	+ *4° 15′*	0
2	*R*	94	157	78	18·1	30° 15′	17·0	26° 15′	1·06	1·05	4° 0′	+ 4° 15′
1	*R*	82	111	98	13·5	36° 0′	17·9	32° 0′	0·74		4° 0′	
2									*0·75*		+ *4° 8′*	
3	*D*	92	206	82	22·6	23° 0′	17·2	27° 15′	0·76	0·79	4° 15′	+ 8° 23′
2	*D*	96	360	96	37·9	13° 15′	17·9	31° 30′	0·47		18° 15′	
3									*0·47*		+ *18° 22′*	
4	*R*	62	60	84	8·2	47° 0′	17·3	28° 30′	0·47	0·37	18° 30′	+26° 45′
3	*R*	96	215	54	23·5	23° 15′	16·1	19° 0′	1·46		4° 15′	
4									*1·46*		+ *4° 30′*	
5	*D*	56	111	94	12·2	26° 0′	17·8	31° 0′	1·46	0·54	5° 0′	+31° 15′
4	*R*	96	255	64	27·4	19° 45′	16·9	22° 0′	1·62		2° 15′	
5									*1·63*		− *2° 8′*	
6	*D*	60	89	92	10·8	32° 15′	17·7	30° 15′	1·64	0·88	2° 0′	+29° 7′
5	*D*	74	126	96	14·6	30° 15′	17·9	31° 30′	1·23		1° 15′	
6									*1·24*		+ *1° 23′*	
7	*R*	100	185	76	21·0	27° 15′	16·9	25° 45′	1·24	1·09	1° 30′	+30° 30′
6	*R*	92	195	66	21·6	24° 30′	16·4	22° 30′	1·32		2° 0′	
7									*1·31*		+ *1° 30′*	
8	*D*	72	120	96	13·8	30° 30′	17·9	31° 30′	1·30	1·43	1° 0′	+32° 0′

* *MR* = multiplied ratio. † *PV* = phase variation.

Fig. 207. This consists of a thin wooden baseboard about 14×9 in., along the bottom of which a scale of capacity reactances is marked. The resistance values are engraved on a celluloid cursor *DE* which slides over the surface of the board at right angles to the scale *AB*. A second celluloid strip *AC*, graduated in any convenient scale, rotates about the point *A*, which occupies the zero division on the reactance scale, and also marks the centre point of a 90° protractor which is engraved on the baseboard. To operate the device, the cursor *DE* is adjusted to the reactance value X_1 on the scale *AB*, and the vector arm *AC* is moved beneath the cursor *DE* to a position *x*, such that its centre line passes through the resistance value R_1. The length *Ax* and the angle *xAB* are then read off and are entered in the columns V_1 and θ_1 of the table. As X_2 is fixed, the values of V_2 and θ_2 for any resistance R_2 can be read off directly from the table which is marked on the right-hand side of the baseboard. This device works very well, and can be operated by the computer as fast as the field values can be called out and the results entered in the proper columns.

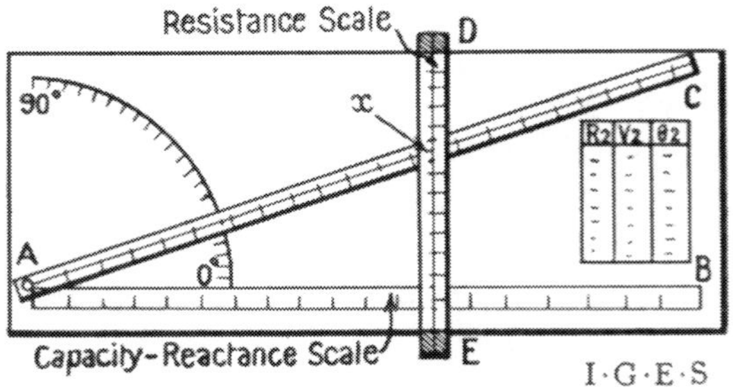

Fig. 207. Sliding scale device for obtaining vectors

The ratios of V_2 to V_1, and the successive multiplications of these ratios are carried out by slide rule and are entered in the columns marked V_2/V_1 and *MR* respectively. Finally the phase angles $\theta_2-\theta_1$ are entered up and are summed together in phase-variation column, *PV*. It will be noted in the table that there is close agreement between the values obtained for the direct and reversed positions of the switch and that the mean is used in each case. This double observation is a valuable check on the work as apart from errors of manipulation in the field and instrumental misbehaviour there is always a possibility of mistakes during the office computation.

The final values entered in the columns *MR* (multiplied ratios) and *PV* (phase variations) should now be plotted to a suitable scale at points midway between the pair of traverse stations to which they refer (see Fig. 36). The resulting curves will then show the potential and phase variations (per unit distance) along the line of traverse.

SECTION 5. ELECTROMAGNETIC METHODS

GENERAL PRINCIPLES

Introduction. In this section it is proposed to deal in greater detail with those electromagnetic methods which were used by the I.G.E.S. Descriptions are given of the instruments, of the field procedure, and of the methods of plotting results. Certain aspects of the underlying theory and its application in the specific interpretation of field results are also considered.

Induction Phenomena. The following system of symbols and units is used in this section unless otherwise stated:

Symbol	Unit	Quantity
$i_1 = I_1 \sin(\omega t)$	ampere	Primary current
$i_2 = I_2 \sin(\omega t - a)$	ampere	Secondary current
$h_1 = H_1 \sin(\omega t)$	gauss	Primary field
$h_2 = H_2 \sin(\omega t - a)$	gauss	Secondary field
$h_{2v} = H_{2v} \sin(\omega t - a)$	gauss	Vertical component of secondary field
$h_{2h} = H_{2h} \sin(\omega t - a)$	gauss	Horizontal component of secondary field
$f = \omega/2\pi$	cycle/sec.	Frequency
μ		Permeability
r_1, r_2, d, l, p, q, a	cm.	Distances
R	ohm	Resistances
ρ	ohm cm.	Resistivities
C	farad	Capacity
L, M	henry	Inductance

To illustrate the phenomena of electromagnetic induction fundamental to the methods under discussion the following special case may be examined.

Suppose two coaxial circles of wire are at a distance d apart (Fig. 208). The upper has a radius r_1, and represents the energizing loop carrying a current $i_1 = I_1 \sin \omega t$, where $\omega/2\pi = f$, the frequency in cycles per second.

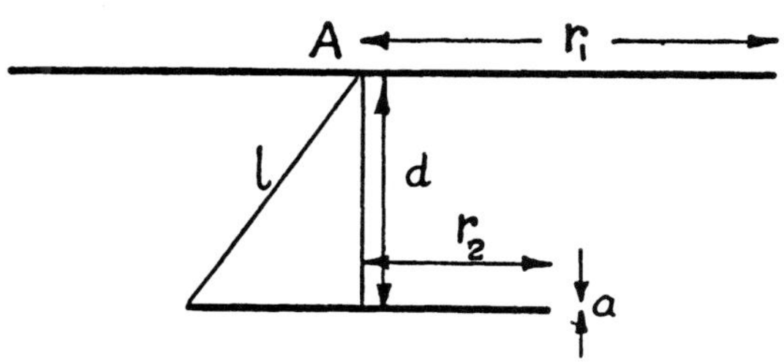

Fig. 208

If the mutual inductance between the circuits is M, then the E.M.F. induced into the lower circuit is

$$e_2 = \omega M I_1 . \sin\left(\omega t - \frac{\pi}{2}\right),$$

and if the impedance of the lower circuit is

$$Z = \sqrt{X^2 + R^2},$$

then the current that flows in it is

$$i_2 = \frac{\omega M I_1}{Z} \sin\left(\omega t - \frac{\pi}{2} - \phi\right),$$

where

$$\tan \phi = \frac{X}{R}.$$

At the point A, the field arising from the upper circuit or the primary field is given by

$$h_1 = \frac{\pi i_1}{5 r_1} = \frac{\pi I_1 \sin \omega t}{5 r_1},$$

and is vertical in direction. At the same point the secondary field is

$$h_2 = \frac{\pi i_2 d^2}{5\,(r_2^2 + d^2)^{\frac{3}{2}}} = \frac{\pi \omega M I_1 d^2}{5 l^3 \sqrt{X^2 + R^2}} \sin\left(\omega t - \frac{\pi}{2} - \phi\right),$$

where $l^2 = r_2^2+d^2$. Moreover it appears that the secondary field lags behind the primary by a phase angle $\left(\frac{\pi}{2}+\phi\right)$.

Thus at the point A, the ratio of the maximum values of the fields is given by

$$\frac{H_2}{H_1} = \frac{\omega M r_1 d^2}{l^3 \sqrt{X^2+R^2}}.$$

If the secondary circuit has self-inductance L and capacity C and the wire in it has a specific resistance ρ and a sectional radius a, then

$$H_2 = \frac{2\pi f M r_1 d^2 . H_1}{l^3 \sqrt{\left(2\pi f L - \frac{1}{2\pi f C}\right)^2 + \frac{4\rho^2 r_2^2}{a^4}}}.$$

This expression shows the various factors upon which the strength of the secondary field depends. It has been assumed that the medium between the two circuits is a perfect insulator but in practice there is generally a certain amount of absorption of the fields in the overburden. For substances whose permeability is not unity, the values of M and L will be somewhat changed.

Absorption of Electromagnetic Waves. The question of the extent to which electromagnetic waves penetrate the surface materials is of fundamental importance to the inductive methods. If the ground were a perfect insulator then there would be no absorption. Actually the ground conducts electricity to a considerable extent, owing to the presence of moisture and dissolved salts; and the magnetic field falls off according to an exponential law. An approximation to the actual state of affairs is afforded by the theoretical formula

$$H = H_0 \epsilon^{-2\pi d \sqrt{\frac{\mu f}{\rho}}},$$

where H is the intensity of the field at a depth d in a medium of specific resistance ρ E.M. unit and permeability μ_0. H_0 is the intensity at the surface and f is the frequency.

Comparatively high frequencies are often used in electrical prospecting with a view to inducing as large E.M.F.'s as possible in the buried conductors, but from the formula given above it is evident that the absorption also increases with the frequency. It becomes necessary then to determine at what frequency the greatest advantage can be obtained. There is not much published information upon this point, but from experiments carried out by Eve and Keys[a] in tunnels and caves, it would appear that even at the extremely high frequencies used by radio broadcasting stations signals can be received at depths of 300 ft. In the same group of experiments an energizing loop was laid on the ground above the caves, and of the frequencies from 20 to 110 kilocycles, the band ranging from 20 to 30 kilocycles gave the best results. It was also found that comparatively little absorption occurred when a frequency of 500 cycles was tried.

[a] *Applied Geophysics*, pp. 136, 137, University Press, Cambridge, 1929. *Nature*, vol. 124, p. 178, London, 1929. Technical Publication No. 316, A.I.M.E., New York, 1930.

Phase of the Secondary Field. In the simple case of the interaction between two circuits, considered on p. 275, it was shown that the secondary field lagged in phase behind the primary field by an angle $\left(\frac{\pi}{2}+\phi\right)$ where

$$\tan\phi = \frac{X}{R} = \frac{a^2\left(2\pi fL - \frac{1}{2\pi fC}\right)}{2\rho r_2}.$$

Thus in the first place a phase difference between the fields of $\frac{\pi}{2}$ is brought about by the fundamental law of induction. From the equation given above it is evident that the self-inductance, capacity, and resistivity, of the secondary circuit, which represents say an ore-body, also have a share in determining the out-of-phase conditions. The angle will be positive or negative according as the reactance of the body is predominantly inductive or capacitative. The frequency is also a factor and, moreover, the equation shows that by altering the frequency the phase may be altered. At high frequencies the inductance term predominates and ϕ will become $\pi/2$. In this case the phase difference between the primary and secondary fields is π and the phenomenon of elliptical polarization disappears.

Another set of factors is introduced by the fact that the velocity of the electromagnetic waves when travelling through the overburden is different from that when travelling through the air. Owing to this, the distances between the energizing circuit, the conducting body, and the search coil, all enter the problem.

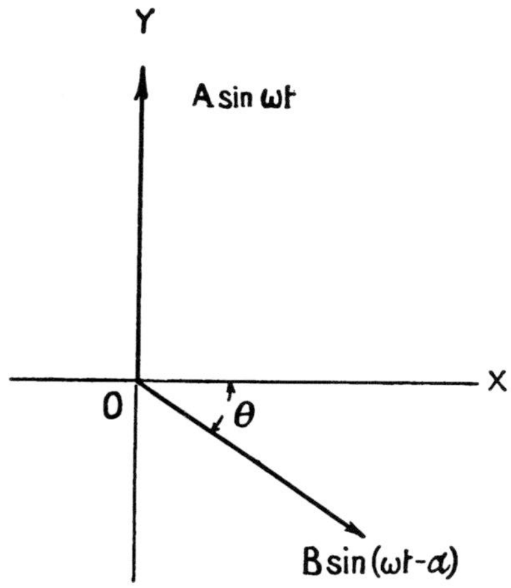

Fig. 209

Elliptical Polarization. Whenever the secondary field lags in phase with regard to the primary then the resultant field has an elliptically polarized structure[a]; for suppose that at a point O (Fig. 209) two fields are acting, differing in phase by α and making an angle of $(\pi/2+\theta)$ with each other, resolving gives

$$Y = A\sin\omega t - B\sin(\omega t-\alpha)\sin\theta,$$
$$X = B\sin(\omega t-\alpha)\cos\theta.$$

Eliminate ωt from these two equations and

$$X^2\left\{\frac{(A\cos\alpha - B\sin\theta)^2 + A^2\sin^2\alpha}{A^2B^2\cos^2\theta}\right\} - \frac{2(A\cos\alpha - B\sin\theta)}{A^2B\cos\theta}XY + \frac{Y^2}{A^2} = \sin^2\alpha.$$

This is of the form $MX^2+2HXY+NY^2 = 1$,

which is the general equation of an ellipse.

If the fields are at right angles to each other and the phase difference is $\frac{\pi}{2}$ then, putting $\theta = 0$ and $\alpha = \frac{\pi}{2}$,

[a] For a detailed discussion see Ambronn, *Gerlands Beitr. z. Geophysik*, Bd. XIX, Heft 1, 1928.

we obtain
$$\frac{X^2}{B^2}+\frac{Y^2}{A^2} = 1,$$
the simplest form for the equation of an ellipse.

Compounding of Elliptical Fields. In practice, at any point at which a measurement is being taken there are a great number of fields arising from current elements distributed throughout the area. These fields will differ in their intensities, directions, and phases, but it can easily be shown that the resultant field is still represented by a single vector whose extremity periodically traces out an ellipse.

At the point in question, any one vector can be replaced by components along three axes at right angles to one another:

$$X = A \sin(\omega t+\alpha),$$
$$Y = B \sin(\omega t+\alpha),$$
$$Z = C \sin(\omega t+\alpha).$$

If the same procedure is followed for all the other vector fields and the components along each axis are added, then

$$X = \Sigma\, A \sin(\omega t+\alpha),$$
$$Y = \Sigma\, B \sin(\omega t+\alpha),$$
$$Z = \Sigma\, C \sin(\omega t+\alpha),$$

and these may be re-written

$$X = A_1 \sin \omega t+A_2 \cos \omega t,$$
$$Y = B_1 \sin \omega t+B_2 \cos \omega t,$$
$$Z = C_1 \sin \omega t+C_2 \cos \omega t.$$

Compounding the three sine terms gives a single vector $H_1 = P \sin \omega t$ and the three cosine terms give another vector $H_2 = Q \cos \omega t$. Since these differ in phase by $\frac{\pi}{2}$ the resultant is a single rotating vector tracing out an ellipse.

This accounts for the well-known experimental fact that no matter how complicated the system of conductors may be, there is only one plane at any point in their vicinity in which a search coil can be placed so that it is not cut by any lines of force. This is the plane of the ellipse. In the determination of this plane two general cases may be considered. Let the major axis A of the ellipse lie in a vertical plane with the minor axis B perpendicular to that plane[a]. Then the horizontal component of the field will be an ellipse whose principal axes are B and $A \cos \theta$ where θ is the dip of the plane of the ellipse.

(1) $B > A \cos \theta$. A search coil lying in a vertical plane is rotated about its vertical axis until a minimum of intensity is obtained. The plane of the coil will then be perpendicular to the vector $A \cos \theta$ and will contain the vector B. If now the coil is rotated about its horizontal axis the major axis A will be brought into the plane of the coil, which will be indicated by silence in the telephones.

[a] For the sake of simplicity the B axis is assumed to be horizontal in these two cases.

(2) $B < A \cos\theta$. The preliminary search for a minimum in the horizontal field will place the plane of the coil so as to contain the vector $A \cos\theta$ and to be perpendicular to B. A further rotation through 90° about the vertical axis will now give the conditions obtained in case (1) and the coil may be placed in the plane of the ellipse by rotating about the horizontal axis as before. Case (2) is by far the more frequent in the field.

HORIZONTAL LOOP METHODS

The Field of the Loop. In order to measure in absolute units the secondary electromagnetic field in the neighbourhood of a conductor it is necessary, for the methods under discussion, to determine quantitatively the primary field due to the loop. This information is also required in deciding the most suitable size for the loop.

The field is vertical at all points in the plane of the loop. Actually the search coil is set up a few feet above the ground level, but for practical purposes the primary field may still be considered vertical, provided measurements are not taken within say 150 ft. of the cable.

The intensity of the field at any point due to the current flowing in the cable may be calculated from the formula $dH = \frac{ids \sin\theta}{r^2}$. At a point K, distant p and q from the sides of the loop as shown in Fig. 210 (a), the field is given by

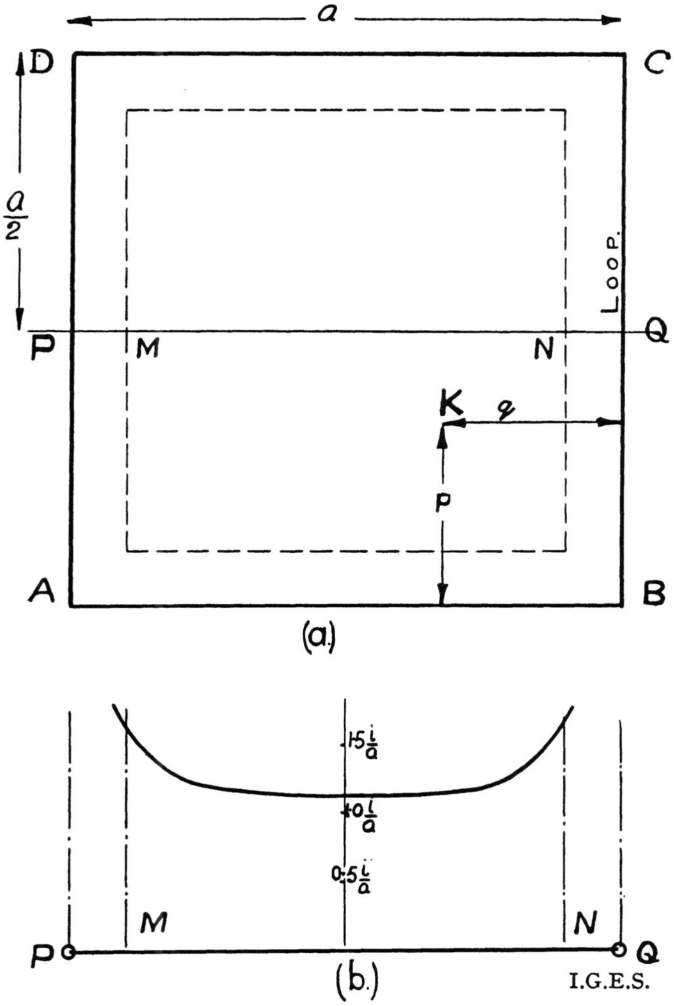

Fig. 210. The field of a horizontal loop

$$H = \frac{i}{10}\left\{\frac{\sqrt{p^2+q^2}}{pq} + \frac{\sqrt{p^2+(a-q)^2}}{p\,(a-q)} + \frac{\sqrt{q^2+(a-p)^2}}{q\,(a-p)} + \frac{\sqrt{(a-p)^2+(a-q)^2}}{(a-p)\,(a-q)}\right\} \text{ gauss.}$$

The way in which the intensity of the electromagnetic field varies from point to point along a line PQ (Fig. 210 (a)) is shown in Fig. 210 (b). As the cable is approached at each end, the intensity increases very rapidly giving an additional reason why readings are not taken close to the loop boundaries. Stations are only laid out on the portion of the traverse lying between M and N (Fig. 210 (b)) and since the curves for parallel traverses have a similar shape, the field has a fairly uniform intensity throughout the region investigated.

At the centre of the loop where $p = \frac{a}{2} = q$ the field is readily found to be

$$H_0 = \frac{8\sqrt{2}}{10} \cdot \frac{i}{a} = 1{\cdot}13\,\frac{i}{a} \text{ gauss.}$$

This is the minimum value of the field, but is close to the uniform value existing over the greater part of the region investigated. It is now possible to arrange the degree of amplification in the field set, and the current passed through the cable, to suit some selected size for the loop. If $i = 3$ amps and $a = 500$ m. then

$$H_0 = 6{\cdot}8 \times 10^{-5} \text{ gauss.}$$

An Elliptically Polarized Field. In this section the elliptically polarized field in the neighbourhood of a linear conductor will be examined[a]. This is a simple but fundamental case from which more complex ones may be built up, approximating to various types of problems met in the field. Further applications of the following theory will be found later in this chapter where the interpretation of field results is considered.

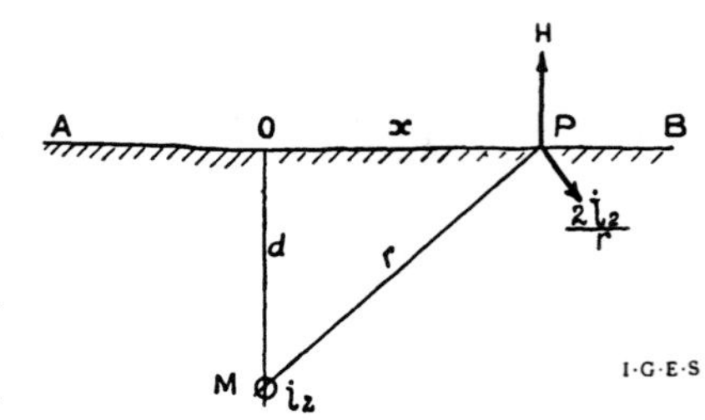

Fig. 211

Consider an infinitely long straight conductor M of small radius, shown in section in Fig. 211, at a depth d below the ground surface AB, and carrying an induced current i_2. The problem becomes two-dimensional and the origin of co-ordinates is taken at O, a point on the surface directly above the conductor. The nature of the electromagnetic field at any point P on the surface, distant x from O, will be investigated.

The primary field due to the current flowing in the loop is vertical, and may be denoted by

$$h_1 = H_1 \cos \omega t \qquad \text{......(1).}$$

An induced current i_2 flows along the conductor and lags in phase by an angle α behind the primary field. Then at P the secondary field has a value

$$h_2 = \frac{2I_2}{r} \cos(\omega t - \alpha)$$

$$= \frac{2I_2}{\sqrt{x^2+d^2}} \cos(\omega t - \alpha) \qquad \text{......(2),}$$

and has a direction making $\cos^{-1} \frac{d}{r}$ with the horizontal.

Resolving the field gives for the vertical component

$$Y = h_1 - h_2\, x/r$$

$$= H_1 \cos \omega t - \frac{2I_2 x}{x^2+d^2} \cos(\omega t - \alpha) \qquad \text{......(3),}$$

[a] The following calculations were made by the late Dr Bieler. They have been abstracted from his notes and included here as they are considered to be of practical value.

and for the horizontal component

$$X = h_2 . \frac{d}{r}$$

$$= \frac{2I_2 d}{x^2+d^2} \cos(\omega t - \alpha) \qquad \ldots\ldots(4).$$

Eliminating t from equations (3) and (4)

$$Y^2+2XY\left\{\frac{x}{d}-\frac{(x^2+d^2)}{2I_2 d}H_1\cos\alpha\right\}+X^2\left\{\frac{x^2}{d^2}-\frac{x(x^2+d^2)}{I_2 d^2}H_1\cos\alpha+\frac{(x^2+d^2)^2}{4I_2^2 d^2}H_1^2\right\}$$
$$= H_1^2\sin^2\alpha \qquad \ldots\ldots(5).$$

This is the equation of the magnetic field at the point P, which is elliptically polarized.

Before any information can be obtained from this equation it is necessary to obtain a relation between H_1 and I_2. The general relation will be rather cumbersome, involving the absorption of the primary field in the overburden and the electrical constants of the conductors. To illustrate the type of results to be expected, the case will be treated in which the amplitude of the primary field is equal to the secondary field at the point O, i.e.

$$H_1 = \frac{2I_2}{d},$$

and further that the two fields differ in phase by $\pi/2$. For this case the ratio of the principle axes of the ellipse will be determined and also the ratio of the out-of-phase horizontal component to the in-phase vertical component.

Equation 5 may now be written

$$Y^2+2XY\frac{x}{d}+X^2\left\{\frac{x^2}{d^2}+\left(\frac{x^2}{d^2}+1\right)^2\right\} = H_1^2,$$

or $$Y^2+2UXY+\{U^2+(U^2+1)^2\}X^2 = H_1^2,$$

where $$U = \frac{x}{d}.$$

The equation is of the type

$$AX^2+2HXY+BY^2 = 1,$$

which represents an ellipse whose axes are given by

$$\frac{1}{r^2} = \tfrac{1}{2}(A+B)\pm\tfrac{1}{2}\sqrt{(A-B)^2+4H^2},$$

and the angle θ between the major axis and the vertical is given by

$$\tan 2\theta = \frac{2H}{A-B}.$$

Hence for the case considered

$$\frac{1}{r^2} = \frac{1}{2H_1^2}\{(U^2+1)(U^2+2)\}\pm\sqrt{\{(U^2+1)^2+U^2-1\}^2+4U^2}$$

$$= \frac{1}{2H_1^2}(U^2+1)\{U^2+2\pm U\sqrt{U^2+4}\} \qquad \ldots\ldots(6),$$

and $$\tan 2\theta = \frac{2U}{(U^2+1)^2+U^2-1} = \frac{2}{U(U^2+3)} \quad \text{......(7).}$$

The computations made from formulae (6) and (7) are shown in Fig. 212. The shapes of the ellipses at various points are shown and also the ratio of the axes (curve II).

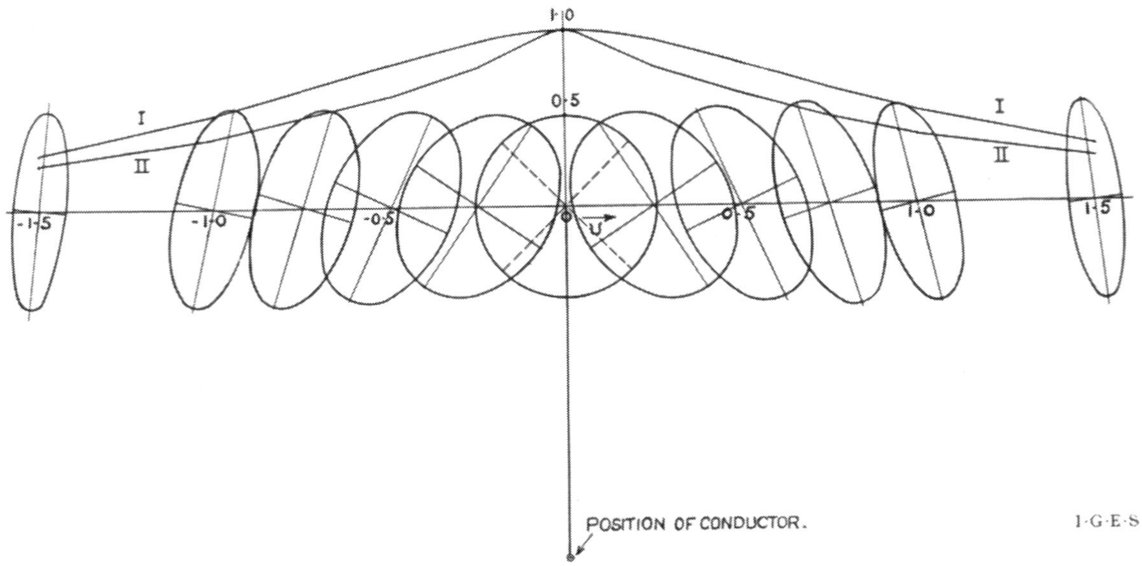

Fig. 212. Elliptically polarized field due to the combined effect of a primary uniform vertical field, and a secondary field due to a current in the conductor lagging in phase by $\frac{\pi}{2}$

Curve I. Ratio of out-of-phase horizontal component to the in-phase vertical field.
Curve II. Ratio of axes of representative ellipse.

Curve I represents the ratio of the horizontal out-of-phase component to the vertical in-phase component. This ratio is given by $1/(U^2+1)$, for the out-of-phase horizontal component has an amplitude

$$\frac{2I_2 d}{x^2+d^2},$$

and the in-phase vertical component is $\frac{2I_2}{d}$, and hence the ratio

$$= \frac{d^2}{x^2+d^2} = \frac{1}{1+U^2}.$$

It will be seen that the two curves bear a close resemblance to each other, a fact which is of great importance in the application of the horizontal loop method used by the I.G.E.S.

Description of Apparatus. In the field operations carried out by the I.G.E.S., the horizontal loop consisted of a single turn of three-strand cable connected by a non-inductive lead to a Bosch generator set up in some convenient position outside the loop. With an applied P.D. of 80 volts the current in the loop was about 3 amps. It is essential that the frequency of the generator be maintained constant at the selected value (500 cycles per second), and the current must be as free as possible from harmonics, which is attained by placing a wave filter between the generator and the loop.

The magnetic field set up is investigated by an apparatus built in two units[a]; a double coil for picking up the field, and a three-stage amplifier for making the small currents in the coils sufficiently intense to operate a pair of telephones.

A diagram of connections for the double coil is given in Fig. 213. The coils, which are permanently fixed at right angles to each other, consist of fine wire (0·018 in. diameter) wound in grooves and boxed over so that no wires are exposed. The vertical coil has 345 turns, and the horizontal coil 402. Further constructional details are given in the following list:

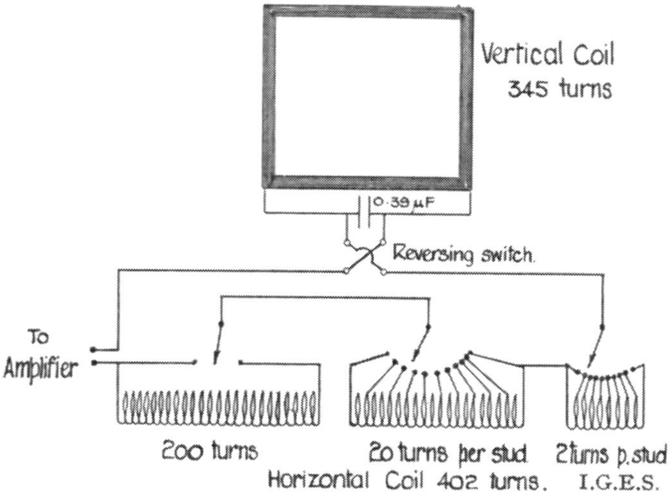

Fig. 213. Wiring diagram of the Bieler-Watson double coil[b]

	Vertical coil	Horizontal coil
Mean dimensions	23·38″ × 23·38″	11·38″ × 21·38″
Side section of winding ...	0·62″	0·62″
Total resistance	82Ω	67Ω
Capacity of condenser across coil	0·390 μF	—

A pointed staff fixed to the vertical coil takes the weight of the apparatus when in use, and a control box attached to the frame-work of the coils contains the condenser which tunes the vertical coil, a reversing switch, and three multipoint switches. From the diagram it can be seen that by the use of the reversing switch the current in the vertical coil can be made to aid or oppose that in the horizontal coil, and the multipoint switches allow any number of turns from 2 to 402 to be connected in the horizontal coil. A non-inductive lead makes connection with the amplifier.

The amplifier (Fig. 214) is of special construction, designed to filter out harmonics as far as possible. This is necessary as, otherwise, when the currents flowing in the coils are balanced for the 500-cycle frequency, they would not at the same time be balanced for frequencies which are multiples of that value. When the fundamental note is no longer present, the high harmonics would still be heard in the telephones making the determination of the balance a difficult matter. For this reason tuned transformer coupling is used in the amplifier, the tuning being adjusted to the 500-cycle frequency.

[a] Bieler, *Canad. Min. Bul.* No. 193, pp. 638–644, Canad. Inst. M.M., Montreal, 1928.
[b] Bieler. Australian Patent No. 15,536/28 with priority from 10th Sept. 1927.

Mullard valves (one PMIA and two PMILF) are used in each of the three stages, with filaments connected in series to a 6-volt battery; 60 volts being applied to the plate circuit. The filament current is automatically switched on when the telephones are plugged into the set. At 500 cycles per second the apparatus has a total amplification of about 4500.

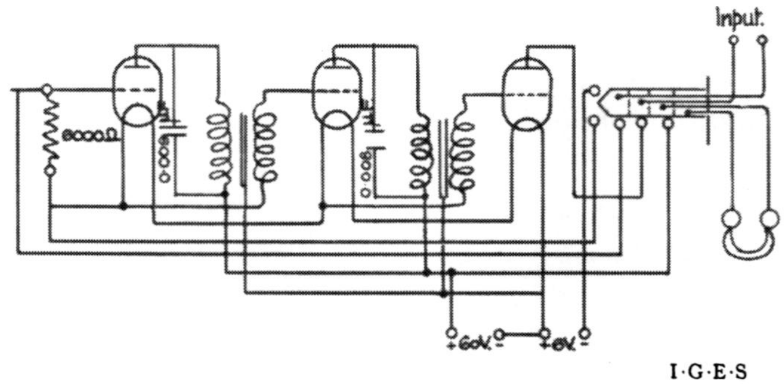

Fig. 214. Wiring diagram of tuned transformer coupled amplifier

The way in which a balance is obtained with the apparatus will be understood from the following considerations. Suppose that at any point of observation the apparatus is held with the vertical coil perpendicular to the minor axis of the representative ellipse; and the horizontal coil perpendicular to the major axis. Since the vertical coil is tuned to the frequency of the generator, the current flowing in it will be in phase with the E.M.F. induced by the magnetic field threading it, but the E.M.F. across the tuning condenser will lag in phase behind the current by 90°. The fields represented by the axes of the ellipse differ in phase by 90°, so that the E.M.F. across the condenser is either in phase or 180° out of phase with the E.M.F. across the horizontal coil. In the second case the reversing switch is thrown over and then the number of turns in the horizontal coil is varied until the E.M.F. due to it balances the E.M.F. across the condenser.

In practice it would take too long to search for the plane of the ellipse in order to take a measurement, and the usual procedure is to record at each station the number of turns required in the horizontal coil to obtain a balance; firstly with the plane of the vertical coil lying in a north-south, and secondly in an east-west direction. Compounding these two readings (Fig. 42) gives the number of turns of the horizontal coil which is proportional to the ratio of the fields represented by the axes of the ellipse, and this resultant is frequently plotted directly on the plan as being sufficient for purposes of interpretation. The ratio of the axes is approximately equal to the ratio of the out-of-phase horizontal field to the in-phase vertical field. Thus, knowing the primary field at any point, and the number of turns of the horizontal coil required for a balance, it is possible to calculate the horizontal out-of-phase component in absolute units at that point.

For if H_h is the amplitude of the horizontal field, $A_v N_v$ the area of the vertical coil multiplied by the number of its turns, and R and C are respectively the resistance of the vertical coil and the capacity across it; then, since the coil is tuned to resonance with the field, the current flowing in it is $\frac{\omega H_h A_v N_v}{R}$, and the E.M.F. across the condenser will be given by

$$\frac{H_h A_v N_v}{CR} \qquad \text{......(8).}$$

If H_v is the maximum value of the vertical field,
A_h is the area of the horizontal coil, and
N_h turns are inserted in it,

then the E.M.F. across the condenser due to the horizontal coil is

$$\omega H_v A_h N_h \qquad \text{......(9).}$$

If there is no sound in the telephones, then the E.M.F.'s due to the coils are balanced across the condenser, so that expressions (8) and (9) must be equal.

$$\therefore\ H_h = \omega CR.\frac{A_h}{N_v A_v}.N_h H_v = kN_h H_v,$$

where k is the instrument constant.

Substituting the values of the constants of the I.G.E.S. instruments, and taking $f = 540$ cycles per second,

$$H_h = 14 N_h H_v\ 10^{-5} \text{ gauss.}$$

Use of Double Coils when Secondary Fields are of Great Magnitude. When the secondary fields are exceptionally strong the maximum number of turns in the horizontal coil may be insufficient to obtain a balance. In this case a special procedure is adopted in order to find the ratio of the axes of the field ellipse. The coil system is mounted on a tripod and provision is made for reading its strike and dip. All the turns are cut out of the horizontal coil and the vertical coil is placed in the plane of silence. A fixed number of turns is then inserted into the horizontal coil and the whole is rotated about the vertical axis until silence is obtained.

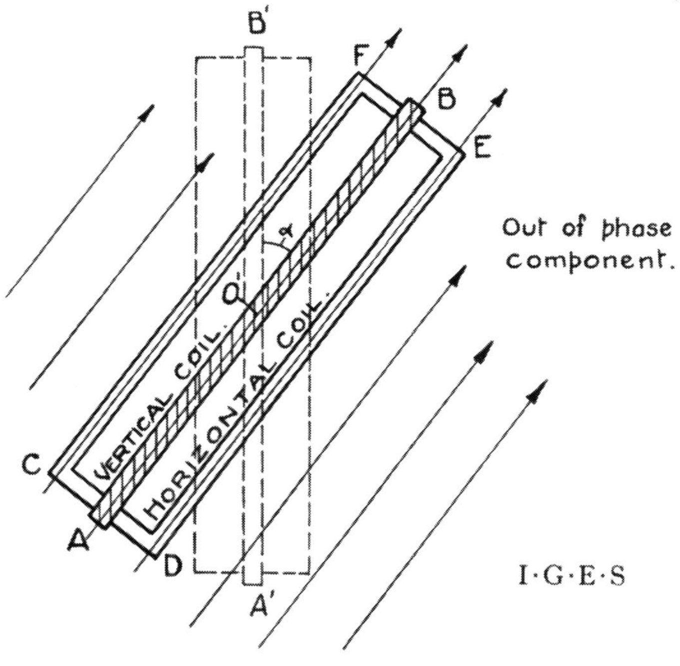

Fig. 215

In Fig. 215 it will be assumed that the plane of the ellipse is perpendicular to the plane of the paper and that the out-of-phase field, as shown by the arrows, lies parallel to the page. The coil system is placed so that the vertical coil AB lies in the plane of the ellipse and the horizontal coil $CDEF$ lies in the plane of the diagram. When the two coils are rotated about an axis passing through O and perpendicular to the plane of the paper the E.M.F. set up in the horizontal coil will remain constant and equal to

$$n\omega A_h H_v,$$

where n is the fixed number of turns in the horizontal coil. The E.M.F. set up across the vertical coil will change as the angle made with it by the out-of-phase field alters. If α is the angle (position $A'B'$ of coil) the E.M.F. across the condenser is

$$\frac{A_v N_v}{CR} H_h \sin \alpha .$$

If in this position there is no sound heard in the telephones the two expressions are equal and

$$H_h = \frac{n}{\sin \alpha} \cdot \frac{\omega A_h CR}{A_v N_v} . H_v = k\,(n/\sin \alpha)\, H_v .$$

If now the coils had been balanced by the normal procedure

$$H_h = k N_h H_v ,$$

whence

$$N_h = n \operatorname{cosec} \alpha .$$

It is obvious that two positions of silence will be found, and that they will be symmetrically disposed on each side of the plane of silence. The angle 2α between these two positions is determined experimentally and used in the above formula.

It would appear that the most accurate results would be obtained if n is kept as large as possible, but in practice it is found that the balances are difficult to determine when this is the case. An optimum number of turns is about 200, and this value has been used in the field work.

In practice the ellipse is not generally vertical as is supposed here, and errors will be introduced. For most purposes these are negligible, but if the dips of the planes of silence have been recorded in the field book, then a correction may be computed if thought necessary.

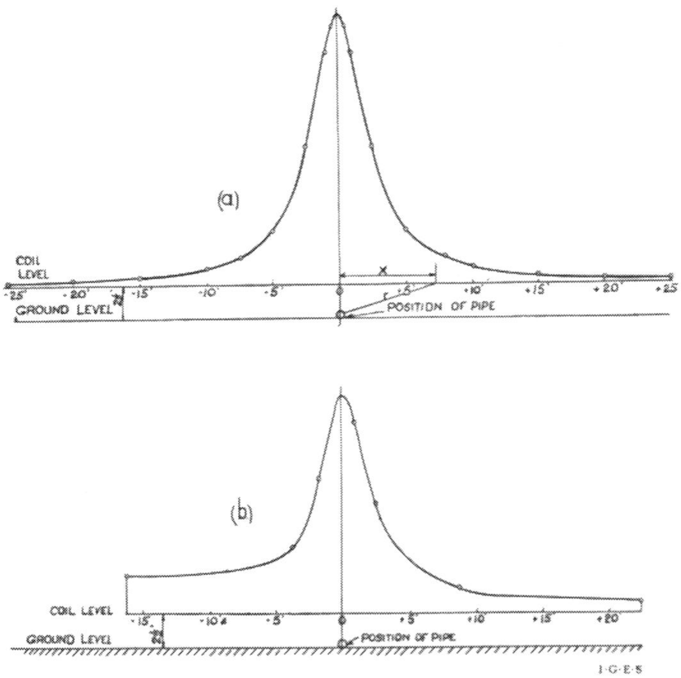

Fig. 216. Measured and calculated fields due to a pipe

Interpretation. It has been explained how the double coil is designed to compare the axes of the representative ellipse and that the readings taken in the field are proportional to this ratio. The readings are also approximately proportional to the horizontal out-of-phase component of the magnetic field. On the other hand it is possible to compute the ratio of the axes and the ratio of the horizontal out-of-phase to the vertical in-phase component at points on the surface, by making assumptions as to the shape of various combinations of buried conductors. By modifying these assumptions until the calculated results agree with the field results, much can be deduced concerning the position and dimensions of the conductive bodies. In many cases this work is not very laborious, as geological evidence will usually afford a fairly definite clue as to the assumptions which should be made. Although the axes of the ellipse can generally be computed, for given conditions, the horizontal out-of-phase component is obtained much more readily so that in most cases it is used for purposes of interpretation.

In Fig. 216 (*b*) is shown the section of a long pipe which is lying on the surface of the ground in the field of a large horizontal loop. The curve gives the results of actual measurements made with the double coil under these conditions. The ordinates are proportional to the horizontal out-of-phase field at points along a traverse above and at right angles to the pipe. For comparison the curve in Fig. 216 (*a*) has been calculated for the given conditions, and its ordinates are proportional to the horizontal field at any point P on the traverse, due to the induced current flowing in the pipe.

The maximum value of the horizontal out-of-phase field at the point P is given by

$$H_2 = \frac{2I_2}{r} \cdot \frac{d}{r} = \frac{2I_2 d}{x^2+d^2}.$$

In Fig. 217 a section is shown of a flat-lying conductor of which the length is large compared with either the width or depth. The current induced into it from a horizontal loop flows around its boundaries, so that the field set up is assumed to be that due to a horizontal circuit of infinite dimensions perpendicular to the plane of the page.

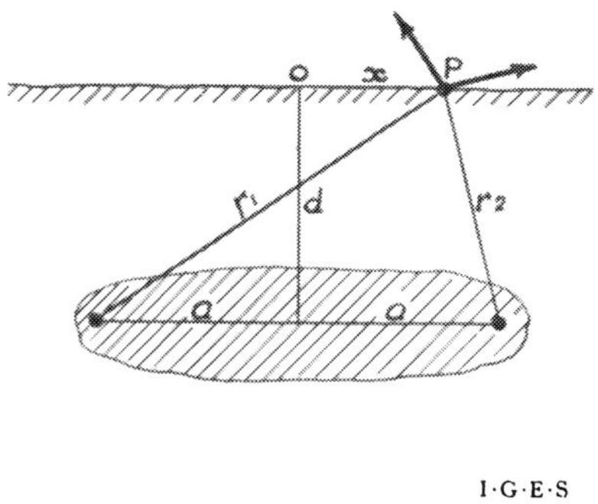

Fig. 217. Field of a circuit of infinite dimensions perpendicular to the plane of the paper

The width of the body (i.e. the distance between the current elements) is $2a$ and the horizontal plane in which the current is assumed to flow is at a depth d below the surface. O is a point on the surface equidistant from the current elements.

Then, at a point P, the amplitude of the horizontal component of the secondary field is given by

$$H_{2h} = 2I_2 \left(\frac{d}{r_2^2} - \frac{d}{r_1^2}\right)$$

$$= \frac{8I_2 xad}{(x^2+a^2+d^2)^2 - 4a^2 x^2},$$

and the vertical component of the secondary field is given by

$$H_{2v} = 2I_2 \left(\frac{a+x}{r_1^2} + \frac{a-x}{r_2^2}\right)$$

$$= \frac{4I_2 a (a^2+d^2-x^2)}{(x^2+a^2+d^2)^2 - 4a^2 x^2}.$$

Profiles computed from these expressions are usually sufficient to compare with the profiles of field results in an attempt to determine the depth and width of a body suspected to be lying flatly.

In such a case the ratio of the axes of the representative ellipse may also be readily obtained if required. Suppose that the primary field is given by

$$h_1 = H_1 \cos \omega t$$

and that the secondary current

$$i_2 = I_2 \cos (\omega t - \alpha).$$

For purposes of illustration it will be further assumed that $a = d$, and that the maximum value of the secondary field at O is the same as that of the primary field, whence it follows that

$$I_2 = \frac{H_1 d}{2}.$$

The horizontal component of the total field is

$$h_{2h} = \left(\frac{4d^3 x}{4d^4 + x^4}\right) H_1 \cos (\omega t - \alpha),$$

and the vertical component of the total field is

$$h_1 - h_{2v} = H_1 \cos \omega t - 2\left(\frac{2d^4 - d^2 x^2}{4d^4 + x^4}\right) H_1 \cos (\omega t - \alpha).$$

It is easily seen that by altering the time origin these expressions may be put in the form

$$x' = a \cos \omega t,$$

$$y' = b \cos \omega t + c \sin \omega t,$$

and eliminating ωt gives

$$\frac{1}{a^2 c^2}\{(b^2+c^2)\, x^2 - 2ab\, xy + a^2 y^2\} = 1,$$

which is the equation of the general ellipse

$$Ax^2 + 2Hxy + By^2 = 1,$$

where $$A = \frac{b^2+c^2}{a^2 c^2}; \qquad H = \frac{-ab}{a^2 c^2}; \qquad B = \frac{a^2}{a^2 c^2}.$$

The lengths of the semi-axes are given by

$$\frac{1}{r^2} = \frac{1}{2a^2 c^2}\{a^2+b^2+c^2 \pm \sqrt{(a^2+b^2-c^2)^2 + 4a^2 b^2}\}$$

and the inclination of the major axis to the vertical by θ where

$$\tan 2\theta = \frac{2ab}{a^2+b^2-c^2}.$$

Assuming values of d, x, and α, the values of the coefficients a, b, and c, are obtained and substituted in these expressions. The computed values of r and θ may then be compared with the values obtained in the field.

Similar methods of calculation are used for dipping bodies. In this case the angle of dip is another unknown quantity introduced into the calculations, but a clue to its value can often be obtained from known bodies in the district.

In Fig. 218 (*a*) the section of a lode[a] is shown as indicated by a drill hole, an adit, and an outcrop. A traverse was taken over this, and the profile of field results is given. For comparison the horizontal component of the field due to current induced in the body has been calculated and plotted in Fig. 218 (*b*). The position of the upper conductor, to which the greater part of the field is due, was fixed by consideration of the depth of oxidation usual in the area.

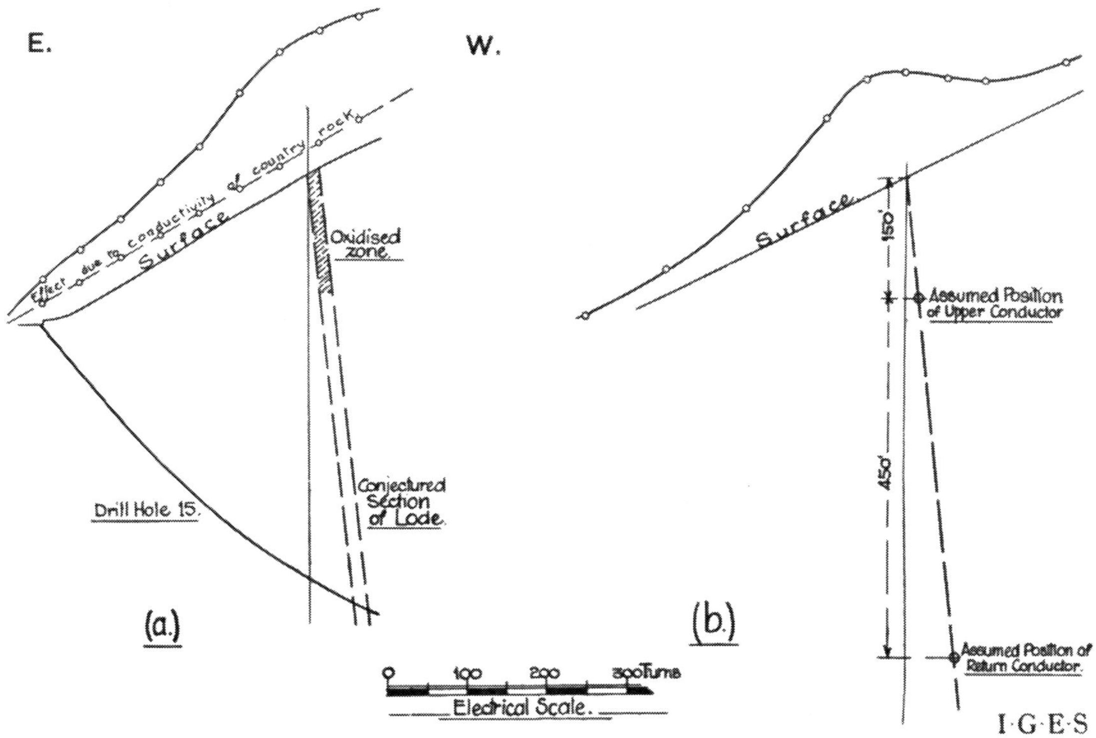

Fig. 218. Measured and calculated fields over a known lode

Finally the effects of general ground conductivity and of topography upon the interpretation of results will be briefly discussed.

Before prospecting an area the resistivities of samples of the surface materials as well as of representative mineral specimens should be determined. These measurements may be made in the laboratory, but a far better idea of the ground conductivity can be obtained from tests carried out in the field by one of the established methods[b]. If by these methods the area prospected is known to have a high general conductivity and large out-of-phase fields are measured, then it may become necessary to make an estimation of how much of the field is due to the general conductivity, and how much to possible bodies lying at depth. This may be roughly accomplished by laying out a loop in nearby ground having approximately the same conductivity, and known to be

[a] See p. 76. [b] See p. 24.

free from conductive bodies, and then making a reconnaissance survey. This will establish the order of magnitude of the vectors due only to the current flowing in the overburden. It may happen that the conductivity of the surface layers is so great that it is useless to attempt to apply the method. Under such conditions the fields due to a buried conductor will be extremely weak owing to the absorption of the energizing fields in the overburden, and they will also be obscured to a great extent by the intense fields set up by currents flowing in the surface layers.

When carrying out an electromagnetic survey in irregular country it is necessary to contour the area; say at intervals of 25 ft. The horizontal out-of-phase component can then be calculated at the ground surface so that topographical effects may be allowed for. It is a great advantage, however, to be able to take the measurements in the plane of the loop since it is then possible to see at a glance if the vector plan shows any anomalies. This condition is always striven for in practice by laying the loop on flat ground, or on hillsides sloping fairly uniformly[a].

VERTICAL LOOP METHODS

Work at Low Frequency (500 cycles per second). The loop originally used by the I.G.E.S. consisted of 10 turns of three-strand cable wound in the form of a triangle. The length of the base was 133 ft. and the height 45 ft., making a total area of 23,750 sq. ft. (2375 sq. ft. per turn).

The field due to this loop is horizontal, but its direction is not uniform as is the case with a large horizontal loop. Consequently its field was investigated mathematically and a table drawn up from which its direction at any point in the region could be read off.

This calculation may be made with sufficient accuracy in the following manner. Suppose the loop behaves as a magnetic doublet whose axis is perpendicular to the plane of the loop. Then at any point P (Fig. 219 a) the potential is given by

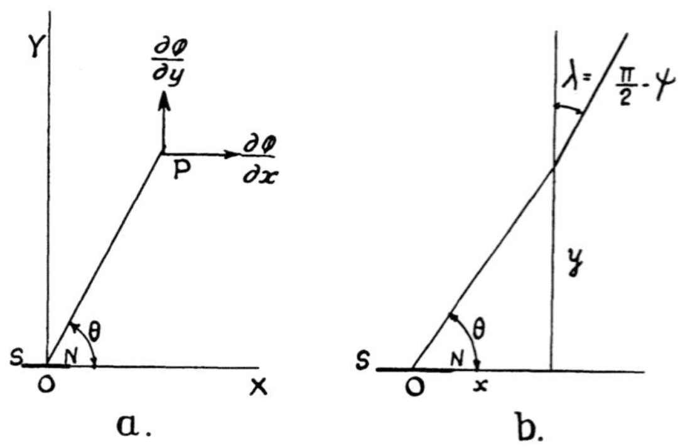

Fig. 219

$$\phi = \frac{-M\cos\theta}{r^2},$$

so that

$$\frac{\partial\phi}{\partial x} = \frac{M}{2r^3}(3\cos 2\theta + 1) \quad \text{and} \quad \frac{\partial\phi}{\partial y} = \frac{M}{2r^3}\,3\sin 2\theta.$$

The intensity of the resultant magnetic field is given by

$$\frac{M}{r^3}\sqrt{1+3\cos^2\theta},$$

[a] See p. 58.

and the angles ψ and λ which it makes with the x and y axes respectively are given by

$$\tan\psi = \left(\frac{\partial\phi}{\partial y}\right) \Big/ \left(\frac{\partial\phi}{\partial x}\right) = \frac{3\sin 2\theta}{3\cos 2\theta + 1},$$

and

$$\tan\lambda = \tan\left(\frac{\pi}{2} - \psi\right) = \cot 2\theta + \tfrac{1}{3}\operatorname{cosec} 2\theta.$$

The values of λ for a large number of values of θ corresponding to various values of x and y were computed and graphed.

Some idea of the structure of the magnetic field set up may be obtained from the section in Fig. 220. Directly above the conductor the primary and secondary fields are in the same horizontal plane and will combine to give a horizontal ellipse. On each side of this position the ellipses will be inclined to the horizontal.

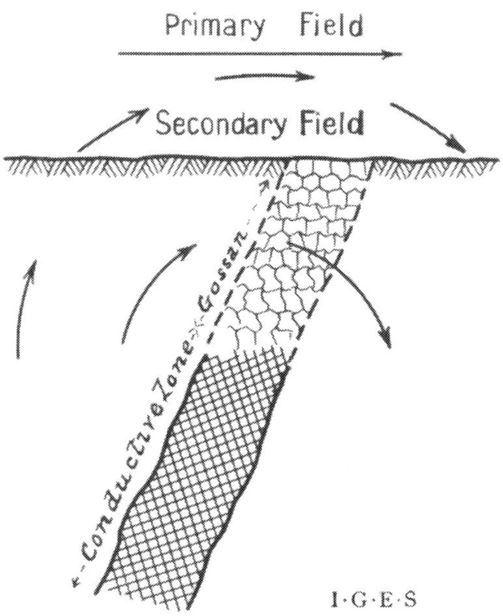

Fig. 220. Secondary field due to steeply dipping lode

To investigate the combined field, the double coil was set up on a tripod in such a way as to rotate about two axes, (1) a horizontal axis parallel to the common line of both coils, (2) a vertical axis in the axis of the tripod. Devices were fitted to permit the reading of the strike and dip of the coil.

To take a measurement at any station the plane at right angles to both coils is set in the direction of the primary field by referring to the graph that provides this information for any point within the area of the survey. The coils are rotated about the horizontal axis and the number of turns in the variable coil is altered until silence is obtained in the telephones. The angle made with the horizontal by the large coil and the number of turns in the small coil are then recorded.

This number of turns is a measure of the out-of-phase component at right angles to the primary field. Hence, directly above a conductor the position of the large coil for minimum sound in the telephones is horizontal, and the setting of the dial switch determining the number of turns in the small coil is zero for no sound. On each side of the conductor, when the coils are set for no sound, the large coil will not be horizontal and the setting of the dial switch will not be zero. It is important to note that the dial settings on the two sides of the body will be in opposite directions.

The results of the test are plotted by means of arrows whose lengths are proportional to the out-of-phase component at right angles to the primary field and whose directions are normal to that of the primary field. Thus, as a conducting body is crossed, the vectors are reversed in direction and vanish directly above it.

The vertical out-of-phase component may be computed approximately from the following considerations. Suppose that in Fig. 221 a long linear conductor is shown in section lying below a ground surface *AA*. If an induced current i_2 flows in the conductor, then at any point on the surface *P* the vertical out-of-phase component is

$$H_{2v} = \frac{2i_2}{r} \cdot \frac{x}{r} = \frac{2i_2 x}{x^2+d^2}.$$

The value of this expression for various values of *x* has been computed; and the manner in which the arrows change in direction and intensity as the conductor is crossed is shown in Fig. 221. The results are more conveniently shown by means of the curve *BB* of which the ordinate at any station is proportional to the length of the arrow there.

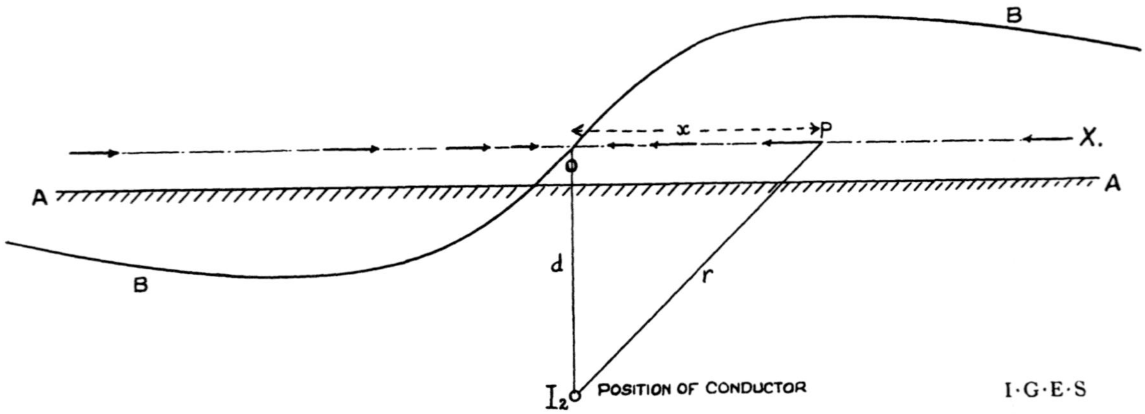

Fig. 221. Traverse across buried conductor showing change in vertical out-of-phase component

Unfortunately, owing to the weakness of the signals it was found impossible to take readings at distances greater than 500 ft. from the loop, and the vertical loop method described above was abandoned in favour of high frequency fields.

Principles of High Frequency Work. If a current of amplitude I_1 flows in an aerial of height *h* then, according to Dellinger[a], the strength of the magnetic component of the field at a distance *r* from the aerial is given by

$$H = \frac{2\pi h f I_1}{rc} \cos 2\pi f(t-r/c) + \frac{hI_1}{r^2} . \sin 2\pi f(t-r/c),$$

where *c* is the velocity of the waves and *f* is the frequency. The first is known as the induction term and the second as the radiation term. These are equal in value at points distant one wave length from the aerial but at shorter distances the induction term has the predominating effect. Since in the application of the method the maximum depths prospected are at most one-fiftieth of the wave length used the phenomena may be regarded as almost entirely inductive.

Introducing the absorption factor into the induction formula, it can be shown that

[a] *U.S.A. Bureau of Standards*, vol. 15 (1919–20).

the E.M.F. induced into a body at a depth *d* in ground of specific resistivity ρ e.m. unit and permeability μ is given by

$$E = 2\pi f M I_1 . \epsilon^{-2\pi\sqrt{\frac{f\mu}{\rho}}.d}.$$

If ρ and μ for the material of the surface formations in the region to be examined are measured, then the frequency at which induction in a body suspected to be lying at a depth *d* is a maximum may be computed from the formula. At a frequency of 60,000 cycles per second it was not expected that effective work could be done with bodies lying at depths greater than 150 ft. if the resistivity of the overburden were less than 10,000 ohm/cm. If, however, the resistivity were ten times this value then there would be possibilities of detecting conductors at depths of about 400 ft.

The high frequency method that has been described[a] consists essentially of finding the direction of an electromagnetic field. No attempt is made to determine the nature of the field ellipse, as has been done in the low frequency methods. If, however, the resultant field is elliptically polarized to such an extent that its directional characteristics can no longer be determined with sufficient accuracy, then the primary and secondary fields may be brought into phase, and this is most readily done by altering the frequency.

Several factors may introduce errors into the readings of dip and strike. Failure to set the plane of the loop towards the receiving coil or to direct the sighting axis of the coil towards the loop will obviously result in erroneous readings which must be specially guarded against in thickly wooded country. The distortion of the wave front, which both primary and secondary fields undergo, due chiefly to the fact that the electromagnetic waves travel with greater velocities in the air than in the ground, should also be kept in mind when interpreting field results. Thus in rough country, where there are hills and valleys between the energizing and receiving units, the distortion of the wave front of the primary field along the surface of the ground may give rise to considerable anomalies. These are minimized by setting up the transmitting aerial on as level a piece of ground as possible and working with the receiving apparatus at relatively short distances from it. When the secondary wave fronts emerge from the surface of the ground into the air, they may undergo considerable refraction which has the effect of making a conductive body appear nearer the surface than it actually is.

High Frequency Apparatus[b]. The loop consists of 10 turns of insulated wire spaced in the same plane at intervals of ½ in. and wound in the form of a square with 8 ft. sides. It is supported by a cross[c] of which the vertical member is 16 ft. high and shod at the base with a steel point, on which it can rotate as a pivot. At the top is a brass disc rotating on an iron shaft fixed to the wood to which four guys are attached. The cross arms are hinged to the lower section of the upright member so that they can be folded downwards for carrying. The loop is held in shape by means of cords attached near the ends of the arms in such a way as to form a square of nearly the same size as the loop.

[a] See p. 69.

[b] Designed by W. G. Baker, B.Sc., B.E. to I.G.E.S. requirements, and constructed at Sydney University, N.S.W. by kind permission of Professor Madsen.

[c] See photo on p. 71.

The constants of the transmitting loop circuit may be summarized as follows:

Consists of 10 turns of 0·063 in. diameter wire; 8 ft. square; pitch of turns 0·53 in. (0·75 in. on the diagonal).

Mean size of turn 91¼ in. square. Effective depth 4·77 in.

Inductance about $710 \mu H$ (Nagaoka formula).

Estimated resistance at 60,000 cycles per second 1·9 ohms (Butterworth formula).

Capacity made up of five condensers $0{\cdot}002 \mu F$ each.

Minimum frequency about 60,000 cycles per second.

Impedance of tuned circuit 37,700 ohms.

I.G.E.S.

Fig. 222. The transmitting set

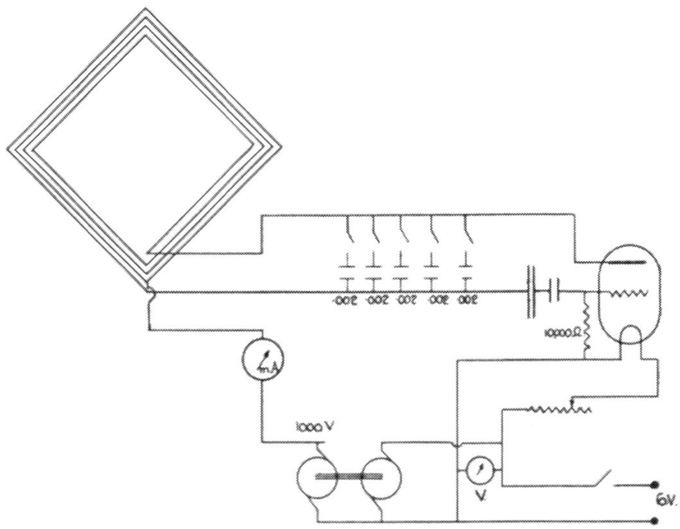

I·G·E·S

Fig. 223. Wiring diagram of oscillator

The oscillator is built into a frame of light brass angle bars which fits into a case as shown in Fig. 222. The circuit is of the Hartley type working directly on to the loop, and its connections are given in Fig. 223. The tuning condensers are each arranged with a single pole knife switch in order to allow the connection of different numbers into the circuit. A grid condenser is used (protected by a large high-voltage condenser in series) and a 10,000 ohm grid leak is connected directly from grid to filament thus supplying bias to the grid. A transmitting valve type A.T. 26[a] (anode dissipation 25 watts) is used, taking anode voltage up to 1000 and filament voltage not exceeding 5·5. The plate voltage is supplied by a small generator run by accumulators of large capacity, which also supply the filament current. Meters are provided for reading battery voltage and plate current.

[a] Army telegraph type supplied from ordnance stores.

The receiving coil was wound on a thin bakelite hoop, 2 ft. in diameter and about $2\frac{1}{2}$ in. in width. The whole was varnished and baked after winding to render waterproof and was strengthened by wooden cross-pieces. The coil was mounted on a theodolite head in place of the telescope so that its orientation could be found accurately. Provision was made for earthing the coil by connecting the frame of the theodolite to the metal points on the feet of the tripod legs.

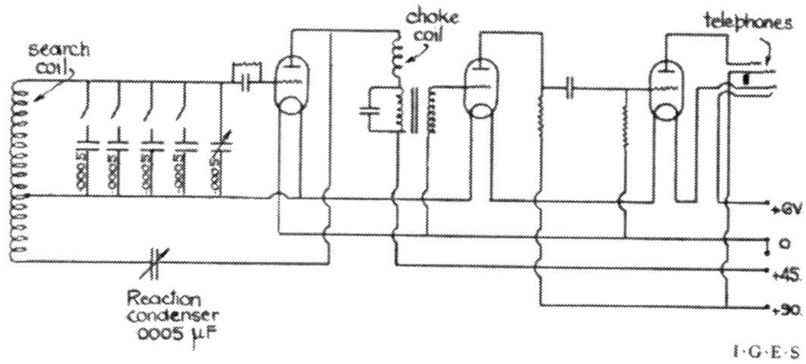

Fig. 224. Wiring diagram of receiving set

The constants of the receiving loop circuit are as follows:

Consists of 50 turns, 2 ft. in diameter, of 0·036 in. diameter enamelled wire.

Length of winding 2 in.

Inductance (by Nagaoka formula) $3{\cdot}225\mu H$.

Capacity made up of four fixed condensers each $0{\cdot}0005\mu F$ and one variable condenser of $0{\cdot}0005\mu F$.

Estimated resistance at 60,000 cycles per second 6 ohms.

Step up ratio of tuned circuit approximately 200.

In designing the apparatus it was first of all decided that it would be better to use unmodulated transmission and heterodyne reception than to modulate the transmitter, since in the former case the received signal would be proportional to the field strength, but in the latter case proportional to its square. Thus an autodyne detector is used followed by enough amplification at low frequency to make the set sufficiently sensitive.

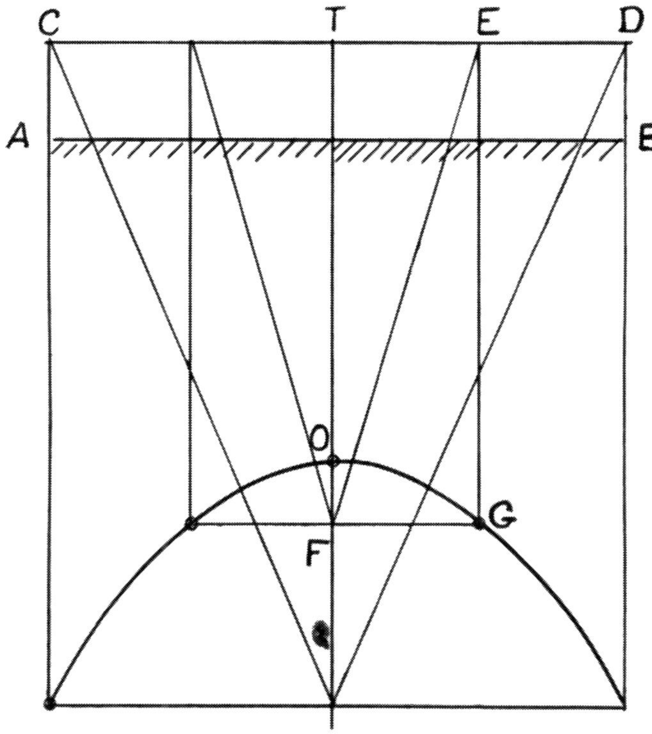

Fig. 225. Construction of index curve

The wiring diagram of the receiving circuit is given in Fig. 224. It is seen that oscillation is produced by taking a tap about one-third of the way along the coil and connecting this point to the filament. The ends of the coil are connected respectively to the grid and the reaction condenser. Switches of the filament type were used for putting the various fixed condensers in the circuit. The telephone jack is of the filament control type, lighting the filaments on being plugged in and extinguishing them on withdrawal. Mullard valves lit by small dry cells were used, the high tension being supplied by a 90-volt battery also fitted into the receiver case.

Interpretation. It has been described[a] how so-called dip and strike readings are taken in the field. For purposes of interpretation these may be plotted directly as curves along the traverse lines, the ordinate at any station being proportional to the reading of dip or strike taken there. These curves are of use in presenting a comprehensive view of the field results, and they immediately indicate points on the surface above conducting bodies.

More specific information about a conductor can be found by plotting an index curve as shown in Fig. 225. In the section, *AB* is the level of the ground and *CD* is the level at which the coil is set up. A cylindrical conductor *O* lies buried directly below the point *T*. If *E* is a station on the traverse then the normal to the plane of the coil is drawn through *E* and cuts the vertical through *T*, say at the point *F*. Through *F* a line is drawn perpendicular to *TF* and a point *G* is placed on it directly below *E*. The dips at other stations are similarly dealt with, and the curve *XOY* which is the locus of the point *G* is termed the index curve.

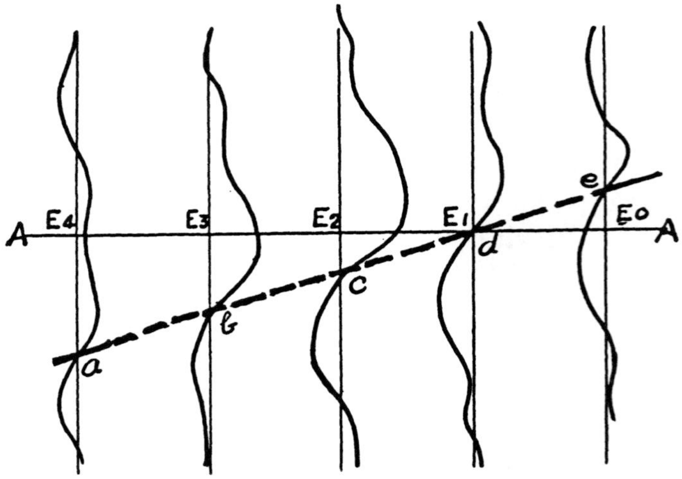

a Plan showing strike of Electrical Axis

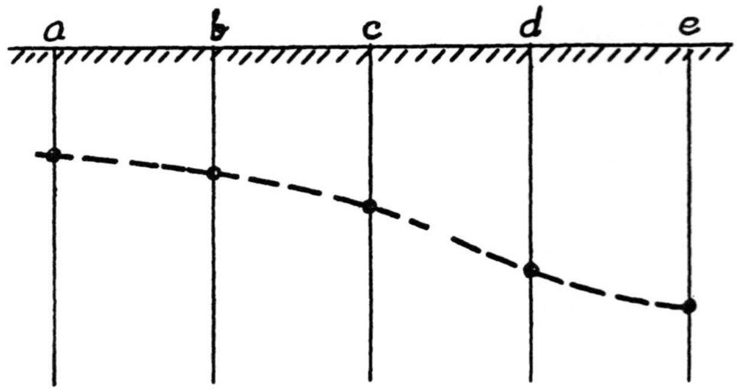

b Section showing pitch of Electrical Axis

I.G.E.S.

Fig. 226

It is seen from the figure that the apex of the curve coincides with the position of the conductor, and for the more complex cases encountered in practice an approximate idea of the depth of the body may often be obtained. The curve itself must of course be regarded as an arbitrary geometrical construction providing a concise representation of field results. Its value follows from the fact that for various typical cases it is possible to calculate the shape of the corresponding index curve so that the curve from the field results may sometimes be recognized as indicating a certain set of conditions. In a detailed interpretation, when the field curves

[a] See pp. 70 and 72.

are found to agree with the calculated curves, it may be possible to deduce the width, dip, and the depth of the body below the surface.

The preliminary results of a survey may be presented by means of plans similar to those shown in Fig. 226. In Fig. 226 *a* the traverse lines E_0–E_4 are set off perpendicular to the base line *AA* at intervals of 100 ft. The dip curves are plotted along the traverse lines, the ordinate at each station being proportional to the dip reading taken there. By this means the strike (*a*, *b*, *c*, *d*, *e*) and length of the conductive body are indicated. In Fig. 226 *b* a section is taken through *a*, *b*, *c*, *d*, *e* and on vertical lines dropped from these points the depths to the electrical axis of the conductor, as indicated by the index curves, are marked off. Joining these points gives the electrical axis of the body, and the way in which it pitches is clearly seen.

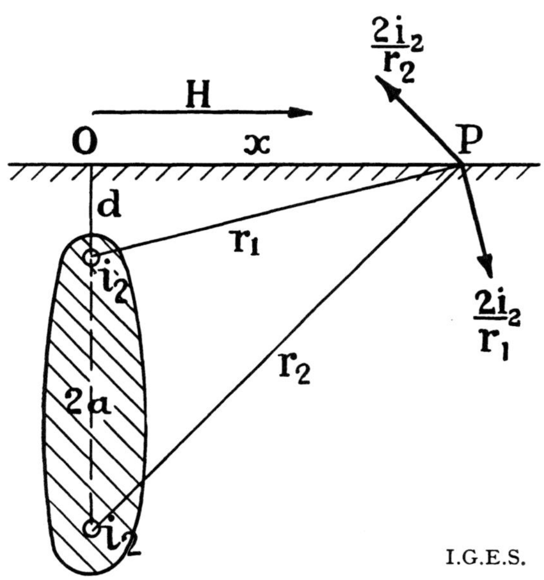

Fig. 227

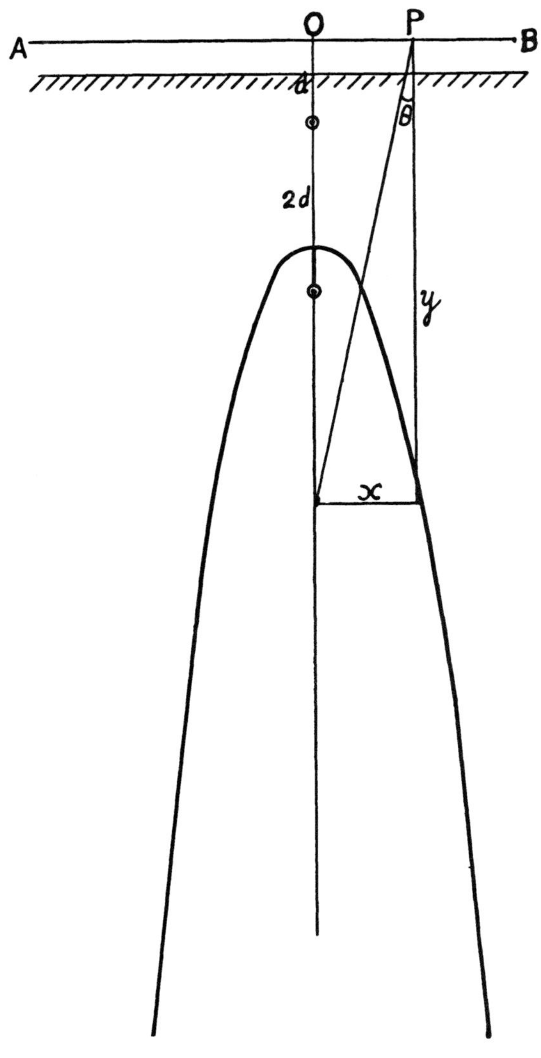

Fig. 228. Index curve calculated from Fig. 227

The shape taken by the index curve corresponding to a typical combination of conductors will be considered. Owing to the nature of some of the simplifying assumptions, however, the apex of the curve will not necessarily indicate the depth of the conductor. The overburden is supposed to be fairly shallow and to have a low conductivity.

In Fig. 227 an ore-body is shown in section with its upper surface lying at a uniform depth *d*. A horizontal field $h_1 = H_1 \cos \omega t$ induces in the body a current i_2 which flows along the top and bottom surfaces. Then at any point in the surface P the vertical field is given by

$$h_{2v} = 2I_2 x\left(\frac{1}{r_1^2}-\frac{1}{r_2^2}\right)\cos(\omega t-\alpha),$$

where α is the phase difference between primary and secondary fields and the horizontal field is

$$h_1+h_{2h} = H_1\cos\omega t+2I_2\left\{\frac{d}{r_1^2}-\frac{(2a+d)}{r_2^2}\right\}\cos(\omega t-\alpha).$$

If the phase difference is negligible and it is assumed that

$$a = d \quad \text{and} \quad I_2 = \frac{H_1 d}{3},$$

then the amplitudes of the components may be written

$$Y = \frac{16H_1 d^3 x}{3(x^2+d^2)(x^2+9d^2)},$$

$$X = \frac{H_1(3x^4+26d^2x^2+39d^4)}{3(x^2+d^2)(x^2+9d^2)}.$$

It follows that the angle of dip will be given by

$$\tan\theta = \frac{Y}{X} = \frac{16d^3 x}{3x^4+26d^2x^2+39d^4}.$$

The dips at different points along AB may be computed from this expression and the index curve obtained by graphical construction.

In simple cases such as the one under consideration it is possible to give the algebraic equation expressing the curve. If (x, y) is any point on the index curve then from Fig. 228 it is seen that the relation connecting these two quantities with the dip is

$$y = x\cot\theta.$$

Substituting the value for θ, obtained above, gives

$$y = \frac{3x^4+26d^2x^2+39d^4}{16d^3},$$

which is the equation of a parabola symmetrical about the y axis. This parabola has been plotted in Fig. 228.

CHAPTER VIII

GRAVIMETRIC METHODS

SECTION I. THEORY

Theory of the Gradiometer and Torsion Balance[a]. Take O, the origin of co-ordinates at the centre of the beam with OX, OY, OZ directed towards the north, the east, and vertically downwards respectively. The direction of the radial arm carrying the upper gradiometer mass, or the arm of the torsion balance carrying the lower mass, will be referred to as the directing arm of the beam, making an angle α with OX.

If U_0 is the value of the gravitational potential at the origin, the components of the field are given by

$$X_0 = \frac{\partial U_0}{\partial x} = 0; \quad Y_0 = \frac{\partial U_0}{\partial y} = 0; \quad Z_0 = \frac{\partial U_0}{\partial z} \quad \ldots(1).$$

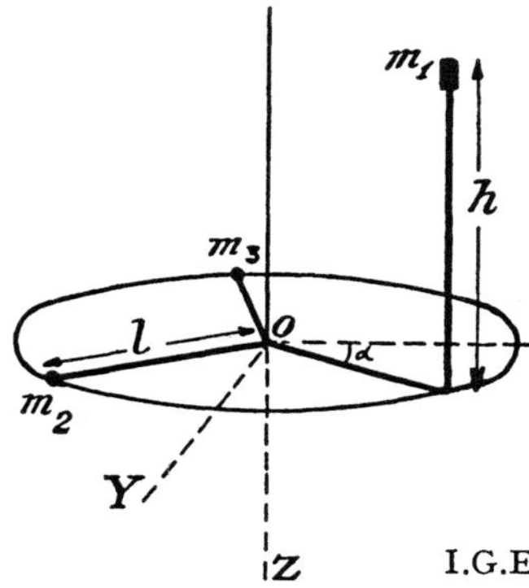

Fig. 229. Beam system of the gradiometer

For any neighbouring point (x, y, z) they are given by the expansion of a Taylor's series:

$$\left.\begin{aligned} X &= \frac{\partial X_0}{\partial x}.x+\frac{\partial X_0}{\partial y}.y+\frac{\partial X_0}{\partial z}.z+ \text{ terms of higher order} \\ Y &= \frac{\partial Y_0}{\partial x}.x+\frac{\partial Y_0}{\partial y}.y+\frac{\partial Y_0}{\partial z}.z+ \quad \text{"} \quad \text{"} \\ Z &= Z_0+\frac{\partial Z_0}{\partial x}.x+\frac{\partial Z_0}{\partial y}.y+\frac{\partial Z_0}{\partial z}.z+ \quad \text{"} \quad \text{"} \end{aligned}\right\} \ldots(2).$$

U_{xx}, U_{xy}, U_{xz}, etc. can be written for $\dfrac{\partial^2 U}{\partial x^2}, \dfrac{\partial^2 U}{\partial x.\partial y}, \dfrac{\partial^2 U}{\partial x.\partial z}$, etc. respectively and (2) may be represented as follows:

$$\left.\begin{aligned} X &= x.U_{xx}+y.U_{xy}+z.U_{xz} \\ Y &= x.U_{yx}+y.U_{yy}+z.U_{yz} \\ Z &= Z_0+x.U_{zx}+y.U_{zy}+z.U_{zz} \end{aligned}\right\} \quad \ldots\ldots(3),$$

where the higher terms in (2) are omitted, i.e. it is assumed that, within the small portion of the field occupied by the instrument, the second differentials of the potential function are constant.

The components X and Y produce on the beam a couple C given by

$$C = \int (xY-yX).dm \quad \ldots\ldots(4),$$

where (x, y, z) are the co-ordinates of an element of the beam of mass dm.

[a] Eötvös, *Ann. d. Physik*, vol. 59, pp. 354–400 (1896).

Substituting the values of X and Y given in (3),

$$C = \int \{(x^2-y^2)\, U_{xy}+xyU_{\Delta}+xzU_{yz}-yzU_{xz}\}\, dm \qquad \text{......(5)},$$

where U_{Δ} denotes $(U_{yy}-U_{xx})$.

In order to evaluate C when the directing arm is in any azimuth α the co-ordinates are transformed to another right-angled system, u, v, w, fixed with reference to the beam, with the u-axis horizontal and in the vertical plane through the upper mass m_1, and the w-axis vertically downwards.

The two systems of co-ordinates are connected by the equations

$$\left.\begin{aligned} x &= u \cos\alpha - v \sin\alpha \\ y &= u \sin\alpha + v \cos\alpha \\ z &= w \end{aligned}\right\} \qquad \text{......(6)}.$$

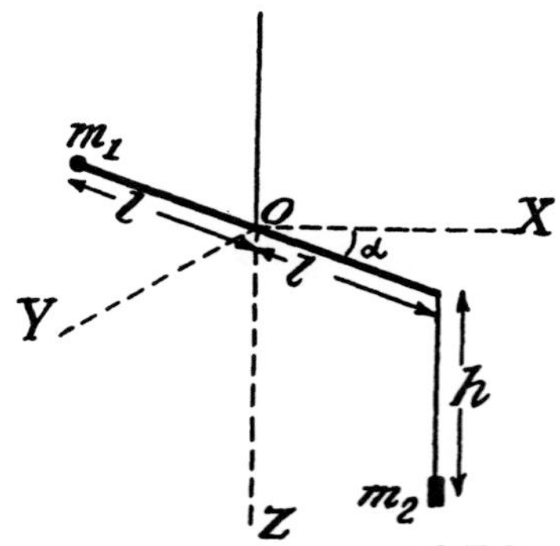

I.G.E.S.

Fig. 230. Beam system of the torsion balance

C is then given by

$$\begin{aligned} C = \int [U_{xy}\, &\{(u^2-v^2) \cos 2\alpha - 2uv \sin 2\alpha\} \\ &+\tfrac{1}{2}U_{\Delta}\, \{(u^2-v^2) \sin 2\alpha + uv \cos 2\alpha\} \\ &+U_{yz}\, (u \cos\alpha - v \sin\alpha) . w - U_{xz}\, (u \sin\alpha + v \cos\alpha) . w]\, dm \qquad \text{......(7)}. \end{aligned}$$

Case (i). *Gradiometer.*

For the upper mass m_1,

$$u = l, \quad v = 0, \quad w = -h;$$

and thus $\quad C/m = U_{xy}\, l^2 \cos 2\alpha + \tfrac{1}{2}U_{\Delta} l^2 \sin 2\alpha - hl\, (U_{yz} \cos\alpha - U_{xz} \sin\alpha) \qquad$ ......(8).

For the lower masses α is replaced by $(\alpha+120)$ and $(\alpha+240)$ respectively while $w = 0$.

For the circular beam and the spokes the value of the moment C is 0.

Adding, the total moment is given by

$$\begin{aligned} C/m = l^2 U_{xy}\, &\{\cos 2\alpha + \cos 2\, (\alpha+120) + \cos 2\, (\alpha+240)\} \\ &+\tfrac{1}{2}l^2\, U_{\Delta}\, \{\sin 2\alpha + \sin 2\, (\alpha+120) + \sin 2\, (\alpha+240)\} - hl\, (U_{yz} \cos\alpha - U_{xz} \sin\alpha) \text{......(9)}. \end{aligned}$$

The turning moment $C = \tau\theta$, where τ is the torsion constant of the wire and θ radians the deflection of the wire.

Thus $$\tau\theta = mhl\, (U_{yz} \cos\alpha - U_{xz} \sin\alpha) \qquad \text{......(10)}.$$

Case (ii). *Torsion Balance.*

For the upper mass m_1, $\quad u = -l, \quad v = 0, \quad w = 0.$

For the lower mass m_2, $\quad u = l, \quad v = 0, \quad w = h;$

and thus

$$C = \tau\theta = ml^2\, (2U_{xy} \cos 2\alpha + U_{\Delta} \sin 2\alpha) + mhl\, (U_{yz} \cos\alpha - U_{xz} \sin\alpha) \qquad \text{......(11)}.$$

There will also be a moment due to the beam itself equal to

$$(2U_{xy} \cos 2\alpha + U_{\Delta} \sin 2\alpha) \int r^2 dm,$$

where dm is an element of the mass of the beam at a distance r from its centre.

Now $\theta = (n_\alpha - n)/2D$,

where $n_\alpha =$ the scale reading for the azimuth α,
$n =$ the scale reading for the undeflected beam,
$D =$ the distance of the scale from the mirror expressed in scale divisions,
$K = 2ml^2 + \int r^2 dm$ is the moment of inertia of the beam about an axis through the torsion wire.

The equation for the torsion balance is thus

$$N_\alpha = (n_\alpha - n) = \frac{DK}{\tau}(U_\Delta \sin 2\alpha + 2U_{xy} \cos 2\alpha) + 2\frac{Dmhl}{\tau}(U_{yz} \cos \alpha - U_{xz} \sin \alpha) \qquad \ldots\ldots(12)$$

and the corresponding equation for the gradiometer is

$$N_\alpha = (n_\alpha - n) = 2\frac{Dmhl}{\tau}(U_{yz} \cos \alpha - U_{xz} \sin \alpha) \qquad \ldots\ldots(13).$$

To denote quantities associated with the second beam of the torsion balance, which is at 180° to the first one, dashes will be used. The solutions of the six equations of the above type obtained by substituting $\alpha = 0°, 120°, 240°$ are then as follows:

$$\left.\begin{aligned}
& N_0 + N_{120} + N_{240} = N_0' + N'_{120} + N'_{240} = 0 \\
\text{i.e.}\quad & n_0 + n_{120} + n_{240} - 3n = n_0' + n'_{120} + n'_{240} - 3n' = 0 \\
U_\Delta &= 2\{Q'(N_{240} - N_{120}) + Q(N'_{240} - N'_{120})\}/\sqrt{3}(PQ' + P'Q) \\
U_{xy} &= (N_0 Q' + N_0' Q)/(PQ' + P'Q) \\
U_{xz} &= \{P'(N_{240} - N_{120}) - P(N'_{240} - N'_{120})\}/\sqrt{3}(PQ' + P'Q) \\
U_{yz} &= (N_0 P' - N_0' P)/(PQ' + P'Q) \\
\text{where}\quad P &= 2KD/\tau, \quad Q = 2Dmhl/\tau \\
P' &= 2K'D'/\tau', \quad Q = 2D'm'h'l'/\tau'
\end{aligned}\right\} \qquad \ldots\ldots(14).$$

The curvature values U_Δ and U_{xy} depend upon the departure of the equipotential surface from the spherical shape. If the equipotential surface through any point A were spherical, then the sections made by all planes through the normal at A to this surface would be circles with the same centre and radius as those of the sphere. In the general case, where the surface through A is not spherical, the sections are other curves. Of these planes through the normal at A there are two principal sections, one having the greatest and the other the least radius of curvature. The radii of these principal sections are known as the principal radii of curvature.

If r_1 and r_2 are the least and greatest radii respectively, and λ the angle that the principal section with the greatest radius of curvature makes with the meridian, then

$$\tan 2\lambda = -2U_{xy}/U_\Delta, \text{ and } g\left(\frac{1}{r_1} - \frac{1}{r_2}\right) = -U_\Delta \sec 2\lambda.$$

The quantity $\left(\frac{1}{r_1} - \frac{1}{r_2}\right)$ is a measure of the departure of the equipotential surface from the spherical shape; for a sphere $r_1 = r_2$ and this quantity $= 0$. Also $g\left(\frac{1}{r_1}-\frac{1}{r_2}\right)$ is sometimes termed the horizontal directing tendency, or H.D.T.

Values of Gradients due to Assumed Bodies. These have been published in various places in the literature; some of the more useful are given here. They are for two-dimensional bodies which extend with a uniform vertical cross-section to infinity on both sides of the reference plane ZOX, parallel to the axis OY which is perpendicular to the paper.

(i) *Elementary cylinder* (Fig. 231) of cross-section da and the radius vector r to this element makes an angle ϕ with OX.

The gradient at O is given by

$$U_{yz} = 0,$$

$$U_{xz} = \frac{2Gs}{r^2}\sin 2\phi da \qquad \text{......(15)},$$

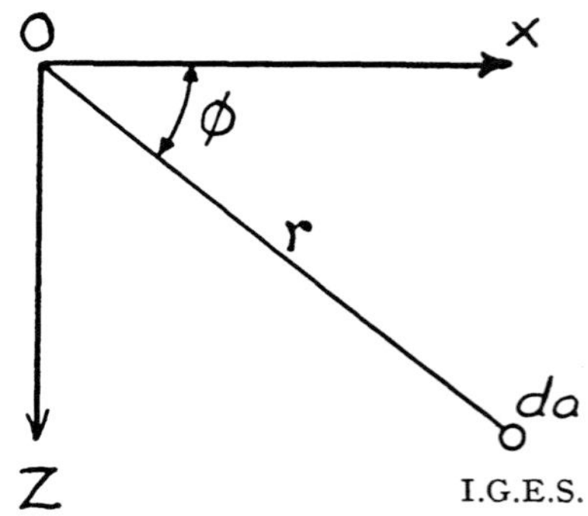

Fig. 231. Gradient due to elementary cylinder

where G is the gravitational constant, and s the density difference between the material of which the body is composed and the surrounding material.

(ii) *Infinite block with sloping edge* (Fig. 232).

(iii) *Infinite sloping dyke with horizontal upper surface* (Fig. 233).

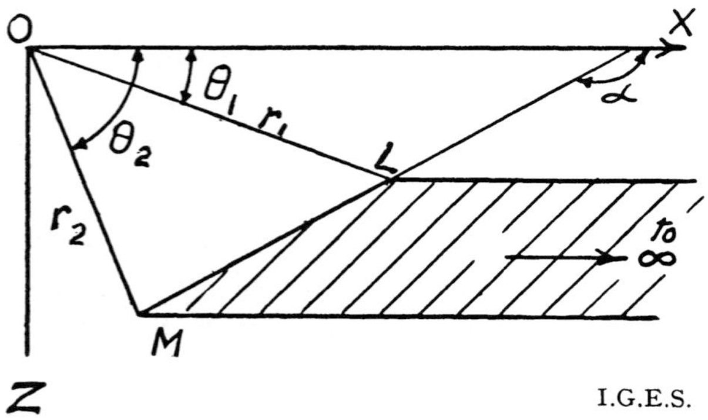

Fig. 232. Gradient due to infinite block with sloping edge

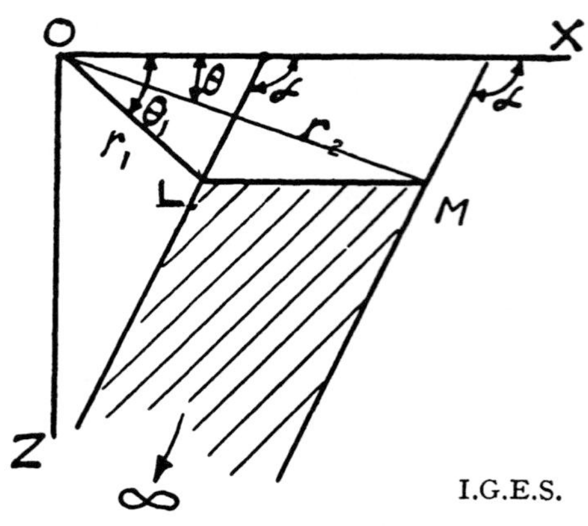

Fig. 233. Gradient due to infinite sloping dyke with horizontal upper surface

In cases (ii) and (iii) the gradient at O is given by

$$U_{yz} = 0,$$

$$U_{xz} = Gs\{2\sin^2\alpha . \log . r_2/r_1 - (\theta_2-\theta_1)\sin 2\alpha\} \qquad \text{......(16)},$$

where the angles are all measured in a clockwise direction from that of OX, and r_1, r_2, at angles θ_1 and θ_2 to OX, are the distances of L and M respectively in each case.

If the edge of the block or the dyke is vertical, $\alpha = \pi/2$ so that

$$U_{xz} = 2Gs . \log . r_2/r_1.$$

Graphical method of obtaining the Values of the Gradients due to any Two-dimensional Body. The method used is that suggested by Jung[a].

Equation (15) may be written

$$U_{xz} = \frac{2Gs}{r^2} \sin 2\phi . r . d\phi dr \qquad \text{......(17)},$$

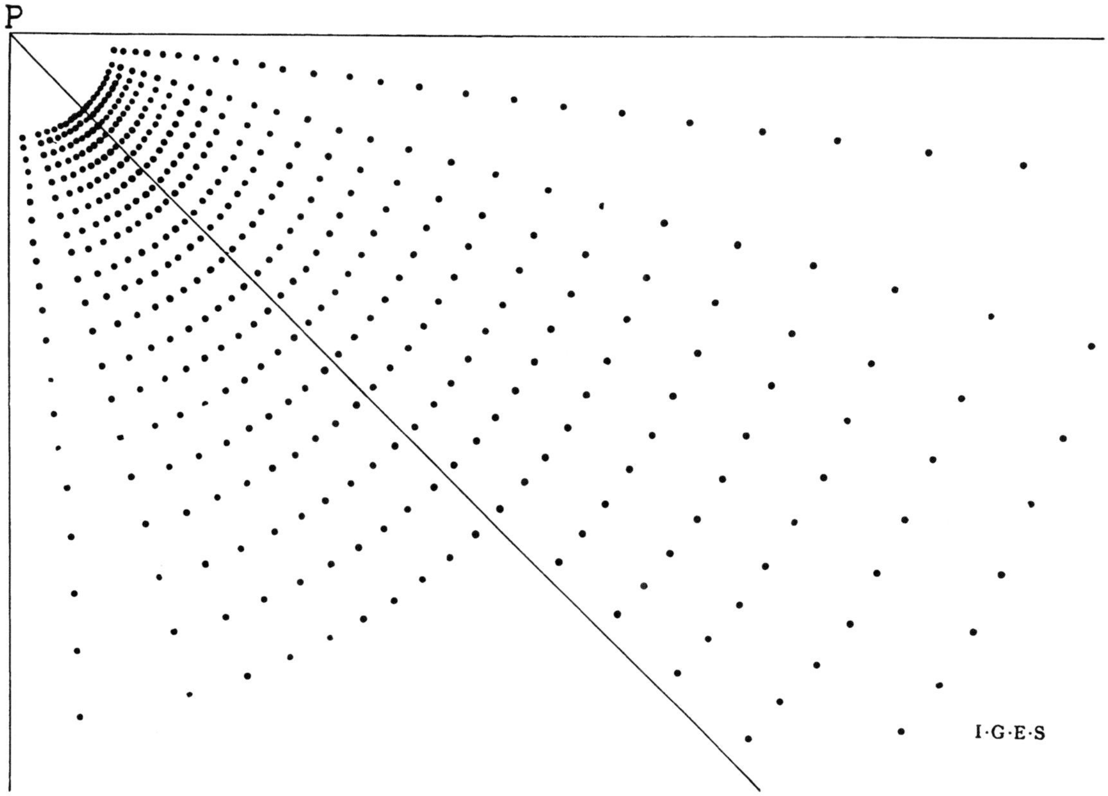

Fig. 234. Graticule for determining gradients due to two-dimensional bodies

and the gradient due to the two-dimensional body is the integral of this over its total cross-section, i.e.

$$U_{xz} = 2Gs \iint \frac{\sin 2\phi}{r} d\phi\, dr \qquad \text{......(18)},$$

and the gradient due to the part of the body with cross-section included between r_1 and r_2 and ϕ_1 and ϕ_2 is given by

$$U_{xz} = 2Gs \int_{r_1}^{r_2} \int_{\phi_1}^{\phi_2} \frac{\sin 2\phi}{r} dr\, d\phi$$

$$= -Gs\,(\cos 2\phi_2 - \cos 2\phi_1) . \log . r_2/r_1 \qquad \text{......(19)}.$$

The principle of the method is to divide the cross-section of the body into a series of zones such that the quantity $(\cos 2\phi_{n+1} - \cos 2\phi_n) . \log . r_{m+1}/r_m$ is constant.

[a] *Zeits. f. Geophysik*, vol. 3, pp. 210–212 (1927).

Fig. 234 shows a diagram which can be used for the purpose, where the zones are bounded by a series of circles and of radial lines whose radii and directions respectively are given by

$$\left.\begin{aligned} r_m &= r_1\, e^{m/10} \\ 2\phi_n &= \cos^{-1} n/10 \end{aligned}\right\} \qquad \ldots\ldots(20),$$

so that

$$\left.\begin{aligned} \log . r_{m+1}/r_m &= 0{\cdot}1 \\ (\cos 2\phi_{n+1} - \cos 2\phi_n) &= 0{\cdot}1 \end{aligned}\right\} \qquad \ldots\ldots(21).$$

and

These circles and lines are not shown, but the effective centre of each zone is marked by a dot.

This diagram is drawn on tracing cloth, and, when in use is placed with its origin P over the point O, at which it is required to calculate the gradient due to the two-dimensional body of cross-section F, and either of the bounding lines of the diagram is placed along OX. The number of dots n falling in the area F is counted, and G, the gradient at O, is given by

$$\text{Gradient} = Gsn/100 \qquad \ldots\ldots(22),$$

or in Eötvös units the gradient is approximately $2ns/3$.

In using the diagram, the scale on which the cross-section is drawn is only subject to the limitation that no part of the cross-section falls in the blank area between P and the innermost ring of points. A convenient diagram for use has $r_1 = 1$ in.; thus the scale must be such that the distance between O and the nearest part of F is not less than 1 in. The diagram could be extended radially in either direction by calculating additional radii less than r_1 and greater than the outermost radius by the use of the formulae in (20). It is not of very much use for certain types of two-dimensional body, particularly that shown in Fig. 115, where the angle α is nearly 180°. Other graphical methods for determining two-dimensional effects have been described by Haalck[a] and Barton[b].

Derivation of the Isogams. The value of $\int \frac{dg}{ds}\, ds$ (where $\frac{dg}{ds}$ is the gradient of gravity along the direction of s) taken round any closed path should $= 0$. If the closed path is in the form of a rectangle, with sides in the north and south direction and at right angles to it, then $\frac{dg}{ds}$ is the value of U_{xz} and U_{yz} respectively at points along those sides. In making the calculations for the preparation of the isogam plan at Gelliondale, Fig. 112, two such rectangles were chosen, one in the north-east, and one in the south-east of the field. The stations in both areas are sufficiently close together to justify the assumption that the components of the gradients change linearly between any two adjacent stations, i.e. the value of the integral $\int \frac{dg}{ds}\, ds$ between two adjacent stations A, B is equal to the distance AB multiplied by half the sum of the components at the stations of the gradient in the direction AB.

[a] *Zeits. f. Geophysik*, vol. 4, pp. 209–219 (1928).
[b] *Geophysical Prospecting 1929*, pp. 488–491, A.I.M.E., New York, 1929.

The residual values of the integrals for the two rectangles were $7{\cdot}8 \times 10^{-4}$ and $1{\cdot}5 \times 10^{-4}$ cm. sec.$^{-2}$ respectively; these are quite small compared with the sums of the magnitudes of the integrals along each side, viz. $123{\cdot}4 \times 10^{-4}$ and 119×10^{-4} gm. cm.$^{-2}$ respectively. This justifies the assumption as to linear change of the gradients and also is a check on the accuracy with which these gradients were measured.

The procedure of tracing out the isogams was as follows:

The residual value of the integral was divided evenly between all the stations on the rectangle, and the value of the gravity anomaly, taken as zero at some arbitrary point, was plotted for each of the sides of the rectangles. From these curves the positions along each of the lines where the anomaly was 0, $\pm 0{\cdot}5$, $\pm 1{\cdot}0$, etc. $\times 10^{-}$ cm. sec.$^{-2}$ were obtained and marked on the gradient plan; in each case with a short line at right angles to the gradient direction, which was usually obtained by interpolation from the directions of the adjacent gradients. These short lines form portions of the isogams for a short distance in the neighbourhood of their positions. Other suitable lines were then taken parallel to the sides of these rectangles, and, if necessary, the values of the gravity anomaly adjusted so that the value of the integral around the enclosed rectangle was zero, and the crossing points of the isogams were then marked on these lines.

When a suitable number of such crossing points had been obtained, those corresponding to the same value of the gravity anomaly were joined together to form isogams, the intermediate portions being drawn as nearly as possible at right angles to the directions of the gradient arrows lying in their path. This method of adjustment is only approximate, but a complete and accurate method would be very laborious.

SECTION 2. FIELD PROCEDURE

Reduction of Field Observations. Theoretically, readings of the torsion balance in three azimuths, e.g. 0°, 120°, and 240°, give sufficient equations for a unique solution for the six unknown quantities, n and n' the scale readings corresponding to the undeflected positions of the beams, two gradient components, and two curvature terms. (In the case of the gradiometer, readings in three azimuths give n and the two components of the gradient.) In general, however, the value of n is not constant, and the change in its value must be followed by taking additional readings.

The gradients are then usually calculated according to the following tabular system, which applies to each of the beams of the torsion balance and to the gradiometer[a]. In the three lower columns the blank spaces are filled by the value of N derived from the other two values in the same row by using the equation

$$N_0 + N_{120} + N_{240} = 0.$$

As soon as the successive rows are in good agreement (i.e. differing by 0·0 or 0·1 scale divisions for the gradiometer), the instrument may be taken as ready for removal to the next site. With the gradiometer, in the majority of cases, the I.G.E.S. party found

[a] For a numerical example see the specimen record sheet, Fig. 239, in which, for convenience, N_0, N_{120}, and N_{240} have been replaced by N_1, N_2, and N_3.

Azimuth a	Reading (n_a)	Value of n	Value of N, i.e. of $(n_a - n)$
0°	n_0		
120°	n_{120}	$\frac{1}{3}(n_0+n_{120}+n_{240})$	$n_{120}-\frac{1}{3}(n_0+n_{120}+n_{240})$
240°	n_{240}	$\frac{1}{3}(n_{120}+n_{240}+{}'n_0)$	$n_{240}-\frac{1}{3}(n_{120}+n_{240}+{}'n_0)$
0°	$'n_0$	etc.	etc.
120°	$'n_{120}$		
240°	$'n_{240}$		

1	2	3
	N_{120}	N_{240}
N_0		N_{240}
N_0	$'N_{120}$	
	$'N_{120}$	$'N_{240}$
etc.	etc.	etc.

that this occurred after five or six settings in azimuth. In many cases further readings were taken which merely served to confirm the values already obtained. If the fourth setting (e.g. $'n_0$) differs from the first by 0·0 or 0·1 scale division, it is also reasonable to assume the instrument ready for removal. With the torsion balance, if photographic registration is used, the seven readings recorded on the plates are utilized. This gives four rows of N's, so that if one of the readings n is lost, which is usually due to sticking of the beam, there may still be sufficient readings to obtain satisfactory values of the gradient. If more than six settings have to be taken, which may occur if the values of n change irregularly, it is usually preferable to take for N the mean of the values in each column.

The gradients, expressed in Eötvös units, are then given by the formulae on p. 141. If one of the systems of the torsion balance is out of action, the gradients and curvature terms could be obtained from readings on one beam in five azimuths; with the Oertling torsion balance this is not feasible, as the observation windows in the metal cover are at angles of 120°. Photographic recording can be used, by taking two plates, one for the three usual azimuths, and one for the azimuths 180°, 300°, and 60°, by setting up with the compass needle reversed. Then, for either beam,

$$U_{xz} = \frac{1}{2\sqrt{3}Q}(N_{240}+N_{300}-N_{60}-N_{120}),$$

$$U_{yz} = \frac{1}{2Q}(N_0-N_{180}),$$

where Q has the value $\frac{2Dmhl}{\tau}$.

Laboratory Testing and Field Calibration. A useful test for the torsion balance, both in the laboratory and in the field, is to measure a gradient both with normal and with reversed azimuths (see above). The value of the gradient is then derived independently from each system, and naturally these two determinations should agree. If they do not, a lead sphere should be used in the manner described by Eötvös[a] in order to check the values of the constants of the torsion wires. The centre of the lead sphere is placed level with the centre of gravity of the lower mass, so that the line joining these points is perpendicular to the vertical plane through the beam. The value of the torsion constant τ is then given by

$$\frac{\tau\,(n'-n)}{2D} = 2\,\frac{Mm}{r^2}.Gl.\frac{1}{\sqrt{1+\lambda^2/4r^2}},$$

where $n'-n$ = the difference of the readings when the sphere is moved from one side of the weight to the other (a sufficient time for the beam to come to its equilibrium position must be allowed),

M gm. = the mass of the lead sphere,

λ cm. = the length of the lower weight,

r cm. = the mean distance[b] between the centre of the sphere and the centre of gravity of the lower mass,

and τ, m, D, l have the meanings previously assigned to them.

If, using the values of τ thus obtained, and the values of D, m, h, and l given by the makers, the gradients and curvatures obtained with the two beams do not agree, a complete calibration is required, for which laboratory accommodation is necessary. For m a simple measurement of mass is required, and for h and l of length, while D is obtained by measuring the angles through which the beam turns for measured changes in the scale-reading of the instrument. These angles are measured by observing the beam mirror with a telescope and scale; the procedure is described in standard text-books of practical physics. For the gradiometer, D can be determined in a similar manner, but a small auxiliary mirror must be attached to the beam stem with soft wax, in a vertical plane at right angles to that of the beam mirror.

On account of the construction of the gradiometer the calculation of the value of τ from the deflection due to a sphere is not simple, but it can be done to a sufficient degree of accuracy. In the I.G.E.S. determination the sphere was supported with its centre at the same height as the lower weights and on a radius through the torsion wire at right angles to that carrying the upper rod, so that the surface of the sphere was just clear of the outer casing of the instrument[c]. The attractions between the

[a] Eötvös, *loc. cit.* p. 345 (1908).

[b] The value of r is usually taken as half the distance between the two positions of the centre of the sphere, or as the distance between the centre of the sphere and the axis of the tube in which the lower mass hangs. This, however, disregards the fact that the beam turns through an angle of $(n'-n)/2D$ radians when the deflecting sphere is changed from one to the other of its positions. The value of r as obtained above is thus too large by $l\,(n'-n)/4D$ cm. If this effect is neglected, the value of τ as obtained above may be too low by a few per cent.

[c] *Referee's comment.* It would have been preferable to place the sphere with its centre at the same height as the upper mass, thus reducing the effect due to any lack of symmetry in other portions of the suspended system, and simplifying the calculation.

sphere and the following parts of the beam of the gradiometer were calculated separately:

(*a*) The metal plug at the lower end of the rod carrying the upper mass, treated as a point mass.

(*b*) The rod carrying the upper mass, treated as a uniform cylinder[a].

(*c*) The small metal sliding piece on this rod.

(*d*) Each of the three masses, treated as point masses.

(The error involved in treating these as point masses instead of cylinders is less than 1 part per 1000.)

This calculation neglects the effects of the radial arms of the beam, but, owing to the symmetrical disposition of the latter, these are small compared to the term *d*. If the calibration is always carried out with the same sphere in the same position relative to the beam, the deflection (obtained from a series of successive readings with and without the sphere) is inversely proportional to the value of the torsion constant of the gradiometer wire; the constant of proportionality is determined from the above calculation. For example, with one wire in the instrument, the deflection due to a lead sphere of mass 14,711 gm. and of mean radius 13·64 cm. was $7{\cdot}75 \pm 0{\cdot}05$ scale divisions (obtained as a mean of 13 determinations). The value of τ calculated from this was 0·0874, which may be compared with the figure given by the makers, 0·0864, and the value, obtained by oscillation experiments by the I.G.E.S. party, of 0·0879.

Corrections. 1. *Normal effect.* Owing to the shape and rotation of the earth the value of the acceleration g, due to gravity, is greatest at the poles and least at the equator. In the southern hemisphere there is thus a gradient directed to the south, and in the northern hemisphere to the north; its magnitude can be calculated from Helmert's formula[b] for the variation of g with latitude

$$\text{Gradient} = 8{\cdot}15 \sin 2\phi.$$

Since the gradient components are referred to the magnetic meridian through the point concerned, they are given in Eötvös units by

$$U_{xz} = +8{\cdot}15 \cos\lambda \sin 2\phi,$$
$$U_{yz} = -8{\cdot}15 \sin\lambda \sin 2\phi,$$

where ϕ is the latitude and λ the magnetic declination which is positive when the magnetic north is east of geographical north. These equations hold for both hemispheres if the latitude is taken positive in the northern and negative in the southern hemisphere. Similar normal effects occur also with the curvature terms.

[a] The gravitational attraction on a thin uniform cylinder of mass M gm., of height h cm., with the centre of its base at a distance r from the centre of the attracting sphere of mass m, and with its axis perpendicular to the line joining the centres, is equal to

$$\frac{GMm}{r^2\sqrt{1+h^2/r^2}}.$$

[b] $g = 980{\cdot}616 - 2{\cdot}5928 \cos 2\phi + 0{\cdot}0069 \cos^2 2\phi - 0{\cdot}0003086\, H$ gm. cm.$^{-2}$, where H metres is the height above sea level.

2. *Terrain and topographical effects*. The original method of Eötvös[a] is only applicable to a locality such as the Hungarian plain, where the surface is flat and regular.

Various other methods have been given by Nikiforov[b], Schweydar[c], Numerov[d], Jung[e], Haalck[f], Ansel[g], Heiland[h], and Lancaster Jones[i]. Of these, Nikiforov, Schweydar and Ansel give analytical, Jung and Haalck graphical, methods for the determination of terrain and topographical effects; Numerov gives a graphical method for topographical and an analytical method for terrain, and Heiland an analytical method for topographical effects. The disadvantage of analytical methods is that, in general, one must carry out the levelling according to a fixed system (for which the constants used in the calculation have been determined), and cannot adapt the procedure to the type of surface met with in the different parts of the field. This is not a serious objection when dealing with terrain effects, for which it can be obviated to a certain extent (see below), but for topographical effects a graphical method involving the preparation of a contoured plan of the locality is preferable to the carrying out of the routine levelling required for their analytical determination (except perhaps in certain types of country such as a uniformly sloping plain).

The I.G.E.S. party used Schweydar's method to correct for terrain, and Numerov's for topographical, effects.

In Schweydar's first paper he expresses the heights on the various circles with the station as centre in terms of Fourier's Series, the constants of which are calculated from the heights; the general expression for the difference of height z_n between the station and points on any circle is

$$z_n = a_n + b_n \sin \alpha + c_n \cos \alpha + d_n \sin 2\alpha + e_n \cos 2\alpha + \dots \text{ etc.} \qquad \dots\dots(1).$$

If r, α = the polar co-ordinates of an element of the terrain material referred to the station point and the magnetic north,

s gm. cm.$^{-3}$ = the density of the material,

h cm. = the height of the centre of gravity of the instrument above the station point,

z cm. = the height of the element above the station point,

G = the gravitational constant,

then the gradient, U_{xz}, due to this element is given by

$$U_{xz} = 3Gs \frac{r^2 \cos \alpha \, (h-z) \, d\alpha \, dr dz}{\{r^2 + (h-z)^2\}^{\frac{5}{2}}} \qquad \dots\dots(2),$$

[a] *Loc. cit.* pp. 357–362 (1908).

[b] "L'anomalie de la gravité dans la région de Kursk," *Bull. de l'Inst. Phys.-Math. de l'Ac. de Sc. de Russie*, vol. 1, pp. 35–45 (1922).

[c] *Zeits. f. Geophysik*, vol. 1, pp. 81–89 (1924); vol. 3, pp. 17–23 (1927).

[d] *Ibid.* vol. 1, pp. 367–371 (1925); vol. 4, pp. 117–134 (1928).

[e] *Ibid.* vol. 3, pp. 201–212 (1927). [f] *Ibid.* vol. 4, pp. 161–178 (1928).

[g] *Lehrbuch der Geophysik*, pp. 533–544, Borntraeger, Berlin, 1926.

[h] *Geophysical Prospecting 1929*, pp. 544–560, A.I.M.E., New York, 1929.

[i] *Monthly Notices of Roy. Astron. Soc.* pp. 121–140 (1929).

and the total gradient due to the terrain is given by

$$U_{xz} = 3Gs \int_0^{2\pi}\int_0^r\int_0^z \frac{r^2 \cos \alpha\,(h-z)\,d\alpha\,dr\,dz}{\{r^2+(h-z)^2\}^{\frac{5}{2}}} \qquad \ldots\ldots(3).$$

Now $\dfrac{z^2-2hz}{r^2+h^2}$ is small, so that we can put approximately

$$U_{xz} = 3Gs \int_0^{2\pi}\int_0^r \frac{r^2 \cos \alpha\,d\alpha\,dr}{(r^2+h^2)^{\frac{5}{2}}}\,(hz_r-\tfrac{1}{2}z_r{}^2) \qquad \ldots\ldots(4),$$

where z_r refers to the value of z for the circle of radius r. A similar expression is obtained for U_{yz} by replacing $\cos\alpha$ by $\sin\alpha$.

Now, if the differences of height are small, we can neglect the term $\frac{1}{2}z_r{}^2$ in the bracket in comparison with hz_r. Then, since we have

$$\left.\begin{aligned} \int_0^{2\pi} z_n \cos\alpha\,d\alpha &= \pi c_n \\ \int_0^{2\pi} z_n \sin\alpha\,d\alpha &= \pi b_n \end{aligned}\right\} \qquad \ldots\ldots(5),$$

the only constants of the Fourier's Series required for the determination of the gradients are b_n and c_n; similarly for the curvatures only d_n and e_n are required. If the levelling is done in eight azimuths (0°, 45°, 90°, etc.) and the differences of height on any circle are $z_1, z_2, \ldots z_8$ (see Fig. 235), we obtain, by substituting in equation (1), the following values:

$$\left.\begin{aligned} 4b_n &= \frac{1}{\sqrt{2}}(z_2-z_6+z_4-z_8)+z_3-z_7 \\ 4c_n &= \frac{1}{\sqrt{2}}(z_2-z_6-z_4+z_8)+z_1-z_5 \\ 4d_n &= z_2-z_4+z_6-z_8 \\ 4e_n &= z_1-z_3+z_5-z_7 \end{aligned}\right\} \qquad \ldots\ldots(6).$$

In irregular country the levelling is done in 16 azimuths for such of the circles as have the greatest irregularity, and the values are given by

$$\left.\begin{aligned} 8b_n &= z_5-z_{13}+0{\cdot}383\,(z_2-z_{10}+z_8-z_{16})+0{\cdot}707\,(z_3-z_{11}+z_7-z_{15})+0{\cdot}924\,(z_4-z_{12}+z_6-z_{14}) \\ 8c_n &= z_1-z_9+0{\cdot}383\,(z_4-z_{12}-z_6+z_{14})+0{\cdot}707\,(z_3-z_{11}-z_7+z_{15})+0{\cdot}924\,(z_2-z_{10}-z_8+z_{16}) \\ 8d_n &= z_3-z_7+z_{11}-z_{15}+0{\cdot}707\,(z_2+z_{10}+z_4+z_{12}-z_6-z_{14}-z_8-z_{16}) \\ 8e_n &= z_1-z_5+z_9-z_{13}+0{\cdot}707\,(z_2+z_{10}-z_4-z_{12}-z_6-z_{14}+z_8+z_{16}) \end{aligned}\right\} \qquad \ldots\ldots(7).$$

The assumption is then made that the variation of height from one circle to the next in any given azimuth is proportional to the radial distance from the station. The value of the integral in equation (4) obtained on this assumption is of the form

$$\left.\begin{aligned} U_{xz} &= (K_1c_1+K_2c_2+\ldots) \\ U_{yz} &= (K_1b_1+K_2b_2+\ldots) \\ -U_\Delta &= (L_1e_1+L_2e_2+\ldots) \\ 2U_{xy} &= (L_1d_1+L_2d_2+\ldots) \end{aligned}\right\} \qquad \ldots\ldots(8),$$

where b_1, c_1, d_1, e_1 refer to the first, b_2, etc. to the second circle, and so on, and the K's and L's are constants which depend only on the values of h and of the successive radii.

Schweydar computed these values for a value of $h = 90$ cm., and for radii of 1·5, 3, 5, 10, 20, 30, 40, 50, 70, 100, etc. metres.

For levellings in feet at distances of 5, 10, 15, 20, 30, 50, and 100 ft. and $h = 3$ ft., the values of the above constants have been recalculated by the I.G.E.S. and agree with those given by McLintock and Phemister[a].

$$\left.\begin{aligned} U_{xz} &= (35{\cdot}6c_1+9{\cdot}14c_2+2{\cdot}79c_3+1{\cdot}48c_4+0{\cdot}916c_5+0{\cdot}435c_6+0{\cdot}142c_7)\\ 2U_{xy} &= (49{\cdot}8d_1+27{\cdot}1d_2+13{\cdot}2d_3+10{\cdot}1d_4+9{\cdot}2d_5+7{\cdot}3d_6+4{\cdot}35d_7)\end{aligned}\right\} \quad \ldots(9).$$

U_{yz} is obtained by replacing the c's by b's, and $-U_\Delta$ by replacing the d's by e's.

Unless the differences of height are small we cannot neglect the term $\frac{1}{2}z_r^2$ in comparison with hz_r in equation (4), and the results calculated for some of the circles may be subject to a very large error and may even have the wrong sign. By the above method of calculation the terrain effect would increase continuously with increase of z. Actually this is not the case, because the gradient due to terrain material for which z is greater than h is in the opposite direction to that in which z is less than h.

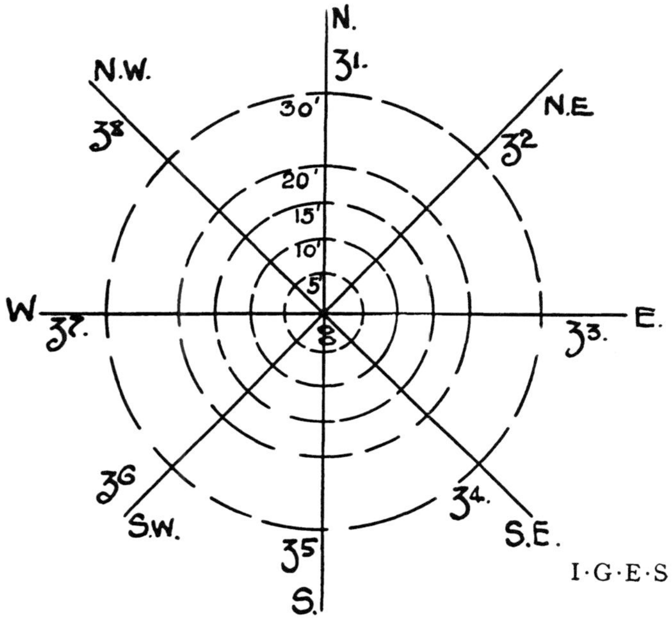

Fig. 235. Layout for terrain levelling

In a later paper Schweydar shows how the effect due to this z^2 term can be calculated; this effect is an additional one to be subtracted from those given in equations (8) and (9). The squares of the height differences on any circle are expressed in terms of Fourier's Series; the constants, B_n and C_n, of these series are calculated in exactly the same manner as are b_n and c_n in equations (6) and (7).

Thus $$4B_n = z_3^2 - z_7^2 + \frac{1}{\sqrt{2}}(z_2^2+z_4^2-z_6^2-z_8^2) \text{ etc.} \quad \ldots\ldots(10),$$

and similarly for levellings in 16 azimuths.

The integration for the terrain is carried out as before, and the effect of the z^2 terms is given by an expression of the form

$$\left.\begin{aligned} M_1C_1+M_2C_2+\ldots\\ M_1B_1+M_2B_2+\ldots\end{aligned}\right\} \quad \ldots\ldots(11),$$

to be subtracted from the expression given in (8) for U_{xz} and U_{yz} respectively.

[a] *A Gravitational Survey over the Swynnerton Dyke, Yarnfield, Staffordshire.* Mem. Geological Survey, Summary of Progress for 1927, Part II, p. 5 (1928).

Schweydar calculated the values of the constants M for a metric system of radii; the values used by the I.G.E.S. have been calculated by the writer and are in agreement with those given by McLintock and Phemister[a], the z^2 term being given by

$$U_{xz}=(7{\cdot}805\,C_1+0{\cdot}500\,C_2+0{\cdot}634\,C_3+0{\cdot}130\,C_4+0{\cdot}205\,C_5+0{\cdot}0506\,C_6+0{\cdot}031\,C_7)\quad\ldots(12),$$

and a similar term for U_{yz} in which B's replace the C's. No z^2 term is required for the curvature values, as it does not occur in the integral for these values corresponding to that for the gradients given in (4).

The calculation of terrain effects may be very much expedited by the use of suitable forms. A specimen form used by the I.G.E.S. is shown in Fig. 239; the products of the integers and the constants in (9), i.e. the factors F, being obtained from tables when working out the results. The forms may be used when it is necessary to calculate the z^2 term effects; a convenient method is to write down in red the squares of the height differences (in feet) below the corresponding figure for the height difference and then to carry through the calculation in exactly the same way, except that in the column for the factor F the appropriate value given in (12) is inserted. In most cases the bulk of the terrain effects occur in the circles of radii 5, 10, 15 and 20 ft., and for these the z^2 term is nearly always negligible. For the 100 ft. circle it is very often greater than the simple term. It was pointed out that the value of the constants b_n, c_n depend on the height h of the centre of gravity of the instrument. If h is increased, these constants for the nearer circles decrease; and this fact has been used in other parts of the world where the thickness of the tropical vegetation makes the necessary levelling extremely laborious. The instrument is set up on a platform 10 or more feet from the ground, so that the constants for the nearer circles are so small that the terrain effects can be neglected. In practice, however, the time thus saved in terrain levelling would usually be less than the time required for the preparation of a sufficiently stable support for the instrument.

Topographical Effects. The method of Numerov[b] has been described by Shaw and Lancaster Jones[c], who explain the use of the diagrams for these effects.

The effect due to the prism of mean height z, the cross-section of which is included between the radii r_n and r_{n+1}, which are in azimuths θ_{m+1} and θ_m, is obtained by integrating the expression on the right-hand side of (3) between the limits θ_{m+1}, θ_m, and r_{n+1}, r_n, where the assumption is made, as the value of r is large (over 100 ft.), that $(h-z)^2$ can be neglected in comparison with r^2.

The gradient terms are given by

$$\left.\begin{aligned} U_{xz} &= 3Gsz\left(h-\frac{z}{2}\right)\left(\frac{1}{2r^2_n}-\frac{1}{2r^2_{n+1}}\right)(\sin\theta_{m+1}-\sin\theta_m)\\ U_{yz} &= -3Gsz\left(h-\frac{z}{2}\right)\left(\frac{1}{2r^2_n}-\frac{1}{2r^2_{n+1}}\right)(\cos\theta_{m+1}-\cos\theta_m)\end{aligned}\right\}\quad\ldots\ldots(13).$$

The principle of the method is to divide the area round the station into zones such that the quantity $\left(\frac{1}{2r^2_n}-\frac{1}{2r^2_{n+1}}\right)(\sin\theta_{m+1}-\sin\theta_m)$ is constant. Shaw and

[a] *Trans. Roy. Soc. Edinburgh*, vol. 56, Part 1, p. 146 (1929).
[b] *Loc. cit.*
[c] *Mining Magazine*, vol. 36, pp. 339–340 (1927).

Lancaster Jones prepared a diagram for this purpose on a scale of 1 in. to 100 ft., such that the radii and their directions are given by

$$\left.\begin{aligned} \left|\sin\theta_{m+1}-\sin\theta_m\right| &= 0{\cdot}1 \\ \frac{1}{r^2_n}-\frac{1}{r^2_{n+1}} &= 1\times10^{-5} \end{aligned}\right\} \qquad \text{......(14)},$$

where r is measured in feet.

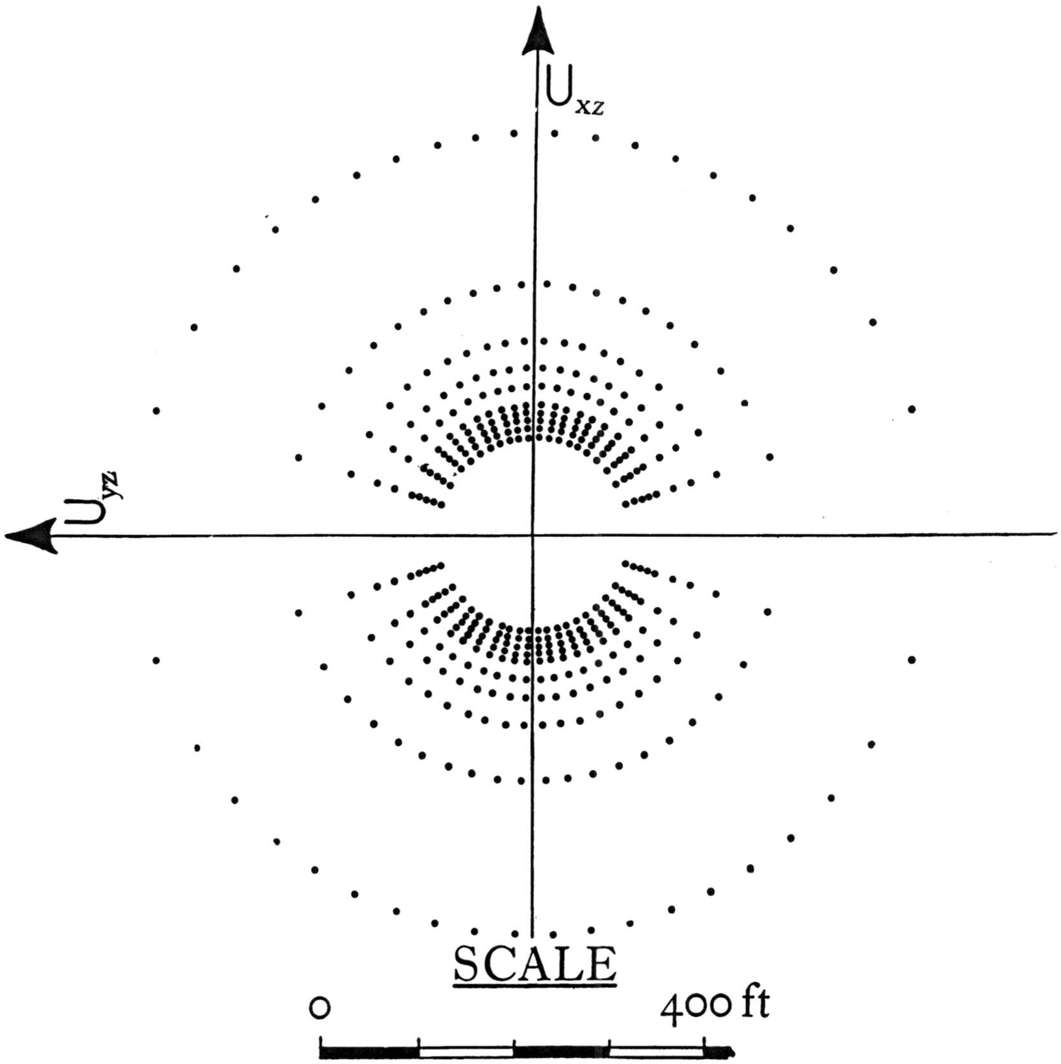

Fig. 236. Graticule for determination of gradients due to topographical features (after Shaw and Lancaster Jones)

The diagram does not show the actual radii and circles, but the effective centre of each area is marked by a dot; the radii and angles which intersect in these dots are given by

$$\left.\begin{aligned} \pm\theta &= 2{\cdot}9^\circ, 8{\cdot}6^\circ, 14{\cdot}5^\circ, 20{\cdot}5^\circ, 26{\cdot}7^\circ, 33{\cdot}4^\circ, 40{\cdot}5^\circ, 48{\cdot}6^\circ, 58{\cdot}2^\circ, \text{ and } 71{\cdot}8^\circ \\ r &= 103, 108, 115, 124, 135, 149, 169, 200, 258, \text{ and } 447 \text{ ft.} \end{aligned}\right\} \ldots(15).$$

The value of U_{xz} expressed in E corresponding to each point is given by

$$U_{xz} = sz\left(h - \frac{z}{2}\right) \times 10^{-4} \qquad \text{......(16).}$$

The I.G.E.S. party found that such a diagram (Fig. 236) was unsatisfactory in very rough country, particularly when any very marked features occurred between 500 ft. and 1000 ft. from the station site[a]. For these cases a second diagram (Fig. 237)

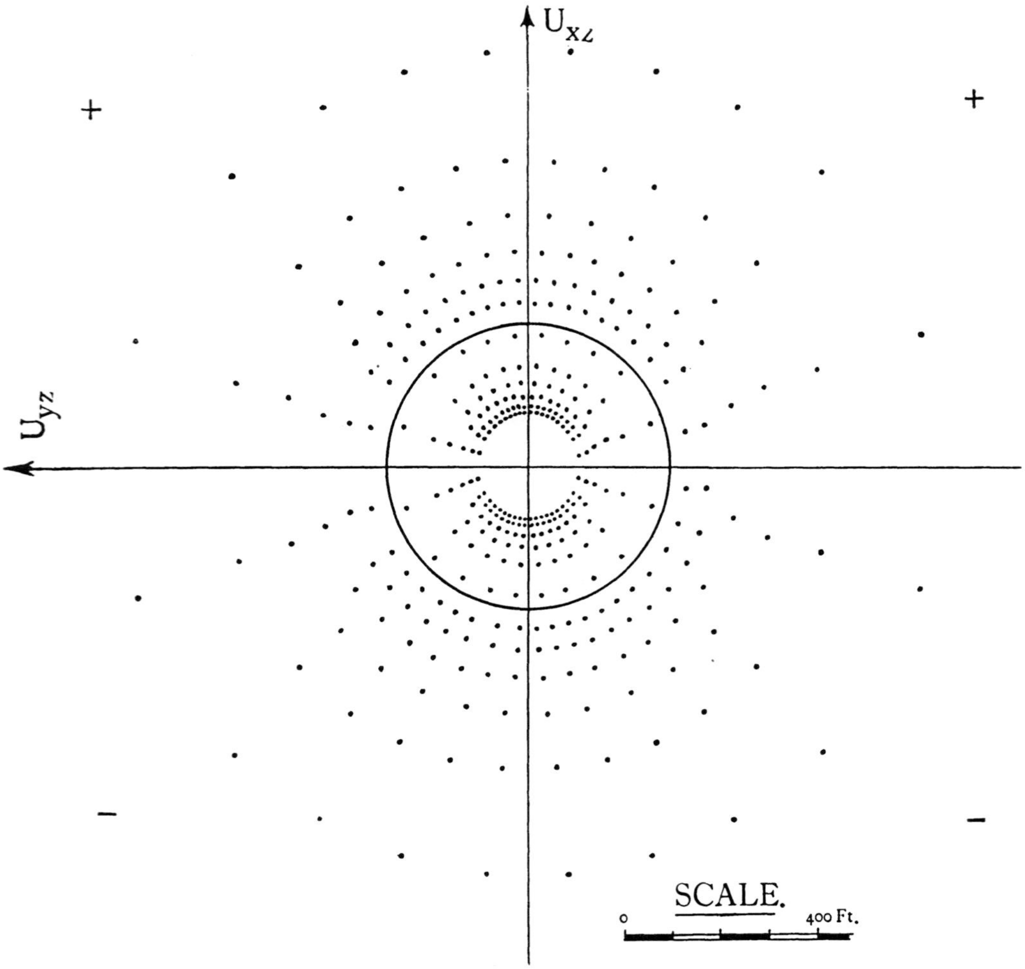

Fig. 237. Graticule for determination of gradients due to topographical features (after Numerov)

was constructed similar to the one described by Numerov in his second paper[b]. In this diagram the value of

$$\left(\frac{1}{r^2_n} - \frac{1}{r^2_{n+1}}\right)(\sin \theta_{m+1} - \sin \theta_m)$$

[a] For instance, at one station at Lakes Entrance the topographical effects as determined by the graticule shown above were $-5\ E$ and $-1\ E$ for U_{xz} and U_{yz} respectively; the values, viz. $-7\ E$ and $-5\ E$ respectively, found by using the Numerov graticule are probably much more accurate. The difference is largely due to the effect of a deep gully at a distance of about 500 ft. from the station.

[b] *Loc. cit.* pp. 129–133.

is taken as 2×10^{-6} for each area inside, and 2×10^{-7} for each area outside the full circle. It is preferable to choose the values of the separate terms in the above product so that each area may approximate as nearly as possible to a square. In this diagram the following series of radii and angles were used, and points were placed in the middle of the areas formed by them.

r	*m*
100 ft.	0°, 5·7°, 11·5°, 17·5°, 23·6°, 30·0°, 36·9°, 44·4°, 53·1°, 64·2°, 90·0°
112	0°, 5·7°, 11·5°, 17·5°, 23·6°, 30·0°, 36·9°, 44·4°, 53·1°, 64·2°, 90·0°
129	0°, 7·2°, 14·5°, 22·0°, 30·0°, 38·7°, 48·6°, 61·0°, 90·0°
151	0°, 9·6°, 19·5°, 30·0°, 41·8°, 56·4°, 90·0°
177	0°, 9·6°, 19·5°, 30·0°, 41·8°, 56·4°, 90·0°
224	0°, 11·5°, 23·6°, 36·9°, 53·1°, 90·0°
316	0°, 5·7°, etc.
354	0°, 5·7°, etc.
408	0°, 7·2°, etc.
476	0°, 9·6°, etc.
559	0°, 9·6°, etc.
707	0°, 11·5°, etc.
1000	

A plan of the district, contoured at 5 ft. intervals, within a radius of about 1000 ft. of each station is required. As the topographical effects commence at 100 ft. from the station this plan must be on a fairly large scale, say 1 in. to 200 ft., or to 100 ft. The graticule used must be on the same scale. In many cases, for the gradients, the effects beyond 500 ft. may be neglected, and thus the surveying is considerably reduced. Further, beyond 200 ft. or 300 ft. from the station, contours every 10 ft. are sufficient.

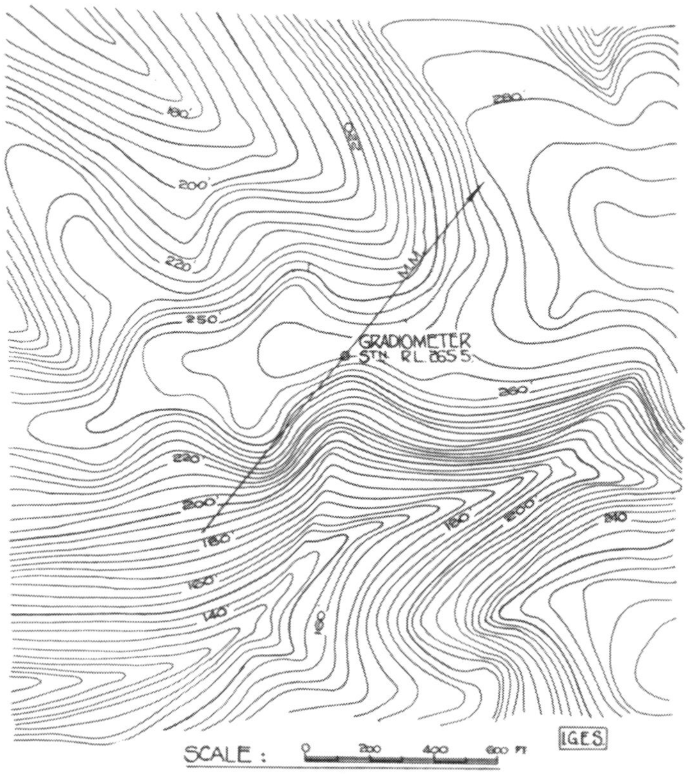

Fig. 238. Contoured plan of district surrounding a gradiometer station, Lakes Entrance, Victoria

Fig. 238 is reduced from the contoured plan prepared for the work at Lakes Entrance on the scale of 1 in. to 200 ft. As the country here was rather rugged the Numerov graticule was used. In using either graticule, its centre is placed on the station point, and the arrow marked U_{xz} or U_{yz} is placed along the direction of magnetic north. The points that fall between the successive contour bands are then counted, those falling in the parts of the plan marked + and − being taken as positive and negative respectively. The figures for the station shown on Fig. 238 are given in Table XIV. In this table n_x and n_y are the number of points counted for U_{xz} and U_{yz} respectively. z is the value obtained by subtracting the height of the station from the mean height of the contour band considered. Points falling within the circle count 10, those outside it as 1. As a check on the accuracy of the counting, the sum of the + and − points are taken; it should be 990 for each column. The difference in the values of the sums shown in Table XIV shows that the point counting has not been done with absolute accuracy; the discrepancies, however, are so small that they may be neglected[a].

For topographical effects in the case of the curvature terms a new graticule is required. It is used in the same way as is the one for the gradients[b].

TABLE XIV. *Example of the determination of topographical corrections for the gradients U_{xz} and U_{yz}*

Height of Station, 265·5 ft. Value of s = 1·8 gm. cm.$^{-3}$.

Contour band	n_x		n_y		n_x	n_y	z	$h-\frac{z}{2}$	$n_xz\left(h-\frac{z}{2}\right)$		$n_yz\left(h-\frac{z}{2}\right)$	
	+	−	+	−					+	−	+	−
ft.							ft.	ft.	$10^3\times$	$10^3\times$	$10^3\times$	$10^3\times$
130–140	—	−1	1	—	−1	1	−130	68	8·7	—	—	8·7
140–150	—	−1	—	—	−1	—	−120	63	7·6	—	—	—
150–160	—	−3	2	—	−3	2	−110	58	19·1	—	—	12·8
160–170	—	−2	3	—	−2	3	−100	53	10·6	—	—	15·9
170–180	—	−8	8	—	−8	8	−90	48	34·5	—	—	34·5
180–190	1	−11	11	−1	−10	10	−80	43	34·4	—	—	34·4
190–200	—	−17	7	−4	−17	3	−70	38	45·1	—	—	8·0
200–210	4	−16	14	−5	−12	9	−60	33	23·8	—	—	17·8
210–220	5	−35	37	−9	−30	28	−50	28	42·0	—	—	39·2
220–230	12	−43	54	−14	−31	40	−40	23	28·5	—	—	36·8
230–235	8	−74	55	−10	−66	45	−33	20	43·6	—	—	29·7
235–240	7	−103	107	−9	−94	98	−28	17	44·5	—	—	46·5
240–245	40	−129	100	−28	−89	72	−23	14	28·5	—	—	23·1
245–250	50	−64	95	−25	−14	70	−18	12	3·0	—	—	15·1
250–255	97	−76	55	−63	21	−8	−13	9½	—	26·0	1·0	—
255–260	207	−66	96	−154	141	−58	−8	7	—	7·9	3·2	—
260–265	400	−143	122	−311	257	−189	−3	4½	—	3·5	2·5	—
265–270	130	−200	200	−350	−70	−150	2	2	—	0·3	—	0·6
270–280	13	—	8	−1	13	7	10	−2	—	0·3	—	0·1
280–290	3	—	1	—	3	1	20	−7	—	0·4	—	0·1
290–300	2	—	1	—	2	1	30	−12	—	0·7	—	0·4
Sums	979	−992	977	−984	—	—	—	—	373·9 39·1 334·8	39·1	6·7	323·7 6·7 317·0

$U_{xz} = 3{\cdot}6\times 3{\cdot}34 = 12{\cdot}0\,E.$ $U_{yz} = 3{\cdot}6\times -3{\cdot}17 = -11{\cdot}4\,E.$

Effect of uncertainty in Density of Terrain. Theoretically, by levelling in a sufficient number of azimuths for terrain, and making a sufficiently detailed contoured plan for topographical and cartographical effects, they could be accurately determined

[a] A similar check can be made when using the other graticule (Fig. 236). Here the sums should each equal 200.

[b] Shaw and Lancaster Jones, *loc. cit.* Numerov, *loc. cit.*

IMPERIAL GEOPHYSICAL EXPERIMENTAL SURVEY

TORSION BALANCE GRAVITY GRADIENTS

Station No 59 Date 27-28/8/1928 Instrument O.T.B Observer N.B.L. photog. Surveyor A.G.B.

TERRAIN CORRECTION.

Level Differences in hundredths of a foot — Computation for Gradients — S = Soil Specific Gravity = 1.0

Dist. \ Azimuth	0 / 180	45 / 225	90 / 270	135 / 315	a 0 − 180	b 45 − 225	c 90 − 270	d 135−315	e ·707(b−d)	f ·707(b+d)	a + e	c + f	Factor F	U_{xz} = F×(a+e)	U_{yz} = F×(c+f)
5	12 8	−2 0	8 −7	5 −6	4	−2	15	11	−9	6	−5	21	35·6	−178	749
10	25 11	10 2	−1 32	8 −9	14	8	−33	17	−6	18	8	−15	9·14	73	−137
15	−8 6	1 5	6 8	18 −6	−14	−4	−2	24	−20	14	−34	12	2·79	−95	33
20	−3 6	4 −1	9 4	17 −8	−9	5	5	25	−14	21	−23	26	1·48	−34	−38
30	8 28	5 −17	4 16	16 9	−20	22	−12	7	11	20	−9	8	0·916	−8	7
50	22 −5	14 −36	2 24	33 6	27	50	−22	27	16	54	43	32	0·435	19	14
100													0·141		
200													0·0315		

Time	Azimuth	A BEAM (Left) N	N_0	N_1	N_2	N_3	B BEAM (Right) N′	N'_0	N'_1	N'_2	N'_3	Temp. Inside Out
21·30	N	102·7					491·2					
22·30	120	97·6	108·2		−10·6		509·2	485·3		23·9		
23·30	240	124·2	107·4			16·8	455·6	484·5			−28·9	
0·30	N	100·5	107·4	−6·9			488·8	483·5	5·3			
1·30	120	97·4	107·8		−10·4		506·2	483·4		22·8		
2·30	240	125·6	108·8			16·8	455·2	483·2			−28·0	
3·30	N	103·5					488·1					
			MEANS	−6·6	−10·3	16·9		MEANS	5·2	22·8	−28·0	

	U_{xz}	U_{yz}	$2U_{xy}$	U_Δ
TOTALS	−223	628	−621	−853
S/400	0·0025		0·0025	
Terrain Correction	−0·6	1·6	−1·6	−2·1
Topographical Correction	~	~	~	~
Special Correction	~	~	~	~
Normal Correction	−7·9	1·1	1·7	6·0
Total Correction	−8·5	2·7	0·1	3·9
Observed Value	120	−34	−36·5	−42
Corrected Value	129	−37	−37	−46

LOCATION OF STATION

Co-ordinates

−1195′, 815′

Plate No. C5

OBSERVED GRADIENT VALUES.

$U_{xz} = 1{\cdot}99\,(N_3 - N_2) - 1{\cdot}31\,(N'_3 - N'_2)$

$= 1{\cdot}99\,(27{\cdot}2) - 1{\cdot}31\,(-50{\cdot}8)$

$= 54{\cdot}1 + 65{\cdot}7 = \underline{120}$ E

$U_{yz} = 3{\cdot}45 \times N_1 - 2{\cdot}27 \times N'_1$

$= -22{\cdot}9 - 11{\cdot}8 = \underline{-34}$ E

REMARKS

Beam Released at 20·30

Calm clear night.

Rather warm.

Imperial Geophysical Experimental Survey

CURVATURE VALUES

Station No 59

TERRAIN CORRECTION.

Computation for Curvatures

$2U_{xy}$ = F′×(b′−d′)	U_Δ = F′×(c′−a′)	a′ 0+180	b′ 45+225	c′ 90+270	d′ 135+315	c′−a′	b′−d′	Factor F′	Distance
−50	−948	20	−2	1	−1	−19	−1	49·8	5
353	−136	36	12	31	−1	−5	13	27·11	10
−79	212	−2	6	14	12	16	−6	13·24	15
−61	101	3	3	13	9	10	−6	10·11	20
−340	−147	36	−12	20	25	−16	−37	9·19	30
−444	65	17	−22	26	39	9	−61	7·27	50
								4·34	100
								1·806	200

OBSERVED CURVATURE VALUES.

$U_\Delta = 6{\cdot}7\,(N_3 - N_2) + 4{\cdot}4\,(N'_3 - N'_2)$

$= 6{\cdot}7\,(27{\cdot}2) + 4{\cdot}4\,(-50{\cdot}8)$

$= 182 - 224 = \underline{-42}$ E

$2U_{xy} = 11{\cdot}6 \times N_1 + 7{\cdot}7 \times N'_1$

$= -76{\cdot}5 + 40 = \underline{-36}$ E

$\tan 2\lambda = -\frac{33}{43} = -0{\cdot}767$

$\therefore\ 2\lambda = 142{\cdot}5^\circ$ or $322{\cdot}5^\circ$

$\lambda = 161^\circ$

H.D.T. = 43 sec 2λ = 54

Fig. 239. Specimen sheet for recording torsion balance or gradiometer results

even for the most irregular types of country. Apart from the very extensive surveying required, a limit is set on the accuracy of the above determinations by uncertainty as to what value should be taken for the density, *s*. This is usually determined by digging a small hole of measured dimensions and calculable volume and weighing the material taken out of it. The I.G.E.S. party at Gulgong used a block of the terrain material resulting from a buried explosion made by the seismic operations. This was sufficiently coherent to be shaped roughly into some definite geometrical form and weighed directly. The method of packing terrain material into a rectangular brass box of measured dimensions, and weighing it, is not nearly so accurate, in general giving too low a value for *s*. In most cases the terrain is not homogeneous; for example, there may be a covering of light soil on top of a denser layer of sand. In most cases the value of *s* is only known to within 0·1 or 0·2; thus for a terrain correction of 60 *E*, which was occasionally obtained at Lakes Entrance, there may be an uncertainty of more than 6 *E* in the correction. In these cases the accuracy could probably be increased by raising the instrument farther above the gound[a].

Special Effects. In general the site of the station can be so chosen that these effects are negligible. If not they must be calculated, and in most cases this can be done by determining the effect for the ideal geometrical body to which the feature concerned most closely approximates, using for the purpose one of the set of formulae compiled by Lancaster Jones[b]. Alternatively, it is usually a simple matter to calculate how far the station must be distant from the feature concerned in order that the gradient caused by it is negligible. For example, at Lakes Entrance the gradients due to trees were calculated on the assumption that they could be treated as uniform cylinders extending from the ground (i.e., 3 ft. below the centre of gravity of the instrument) to a height of 103 ft., and on this assumption it was found that the gradient due to a tree of radius *r* ft. of specific gravity 0·9, at a distance *R* ft. from the station, was approximately given by

$$\text{Gradient} = \frac{180r^2}{R^2} \text{ if } R \text{ was more than 10 ft.}$$

Thus a tree, 2 ft. in diameter, at a distance 10 ft. from the station gives a gradient of about 2 *E*; if it were 20 ft. from the station the gradient would be less than half a unit. It can thus be seen that, except in very heavily wooded country, the stations can be so chosen that the gradients due to trees are negligible.

SECTION 3. BEHAVIOUR OF INSTRUMENTS

Discrepancy between Torsion Balance and Gradiometer Results. During the progress of the work at Gelliondale, certain marked discrepancies were observed in the gradient measured at the same station with the torsion balance and the gradiometer. It was found that the values had an approximately constant ratio, so that it was not thought advisable at the time to interrupt the field work, but to make a careful calibra-

[a] See p. 312.

[b] *Geophysical Prospecting 1929*, pp. 505–529, A.I.M.E., New York, 1929.

tion and comparison of the two instruments in Melbourne at the close of field operations. It was further decided not to replace the gradiometer torsion wire in the field, as the values of repeated measurements with the gradiometer agreed, and as only one spare wire was at the time available. The results of the calibration and comparison are summarized as follows:

(i) The two instruments were compared in a basement before proceeding to the field in June 1928, and no disagreement was detected.

(ii) From the results obtained at Gelliondale, at the 23 stations where both instruments were used, it was concluded that unit gradient on the torsion balance = 1·36 on the gradiometer. The relation between the readings is shown in Fig. 240.

(iii) The values of D, m, h, and l were re-determined, and found to agree with the maker's figures.

(iv) The values of τ for the two wires in the torsion balance were determined by the deflection experiment as described on p. 307, giving 0·651±0·002 dyne cm./radian (mean of 8 observations) and 0·468±0·004 (mean of 17 values) compared with the maker's values of 0·651 and 0·410 respectively. The second value is 14 per cent. higher than the maker's figure, and the results for this wire were always less consistent than those for the first one. The field results for the torsion balance were re-calculated, using the value of 0·468. The gradients obtained at the station where the two beams were compared as described on p. 307 were then as follows:

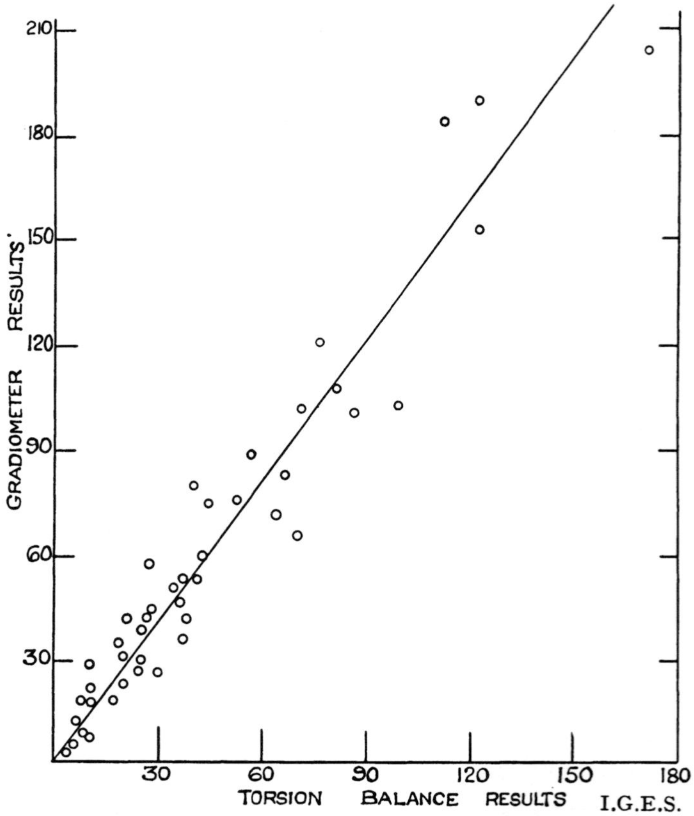

Fig. 240. Graph connecting torsion balance and gradiometer results at Gelliondale. The slope of the straight line is the mean (1·36) of the ratio (gradiometer reading) : (torsion balance reading)

	First Beam	Second Beam	Ratio
U_{xz}	−78·6	−76·9	0·98
U_{yz}	35·9	37·0	1·03

(v) The values of τ for the gradiometer wires were calculated from the time of oscillation of an inertia bar when suspended by the wire, the weight of the bar being approximately that of the gradiometer beam. In order to minimize the effect of convection currents the oscillating system was placed inside a vessel consisting of two concentric cylinders of sheet metal separated by granulated cork, the oscillations of the bar being observed through a narrow mica-covered slot cut out of each cylinder. The moment of inertia of the carefully turned brass cylinder about its axis of suspension was calculated from its dimensions, and corrections were applied for the moment of inertia of the wire cradle carrying the bar, and of the mirror attached to it.

The results for one wire were as follows:

Wire no. B. 4005

Date	Range of amplitude of oscillation	Mean time of oscillation	Range of times
		(sec.)	(sec.)
9. ii. 29	500°–150°	157·10	156·6–157·6
11. ii. 29	60°–10°	157·06	156·4–157·6
11. ii. 29	10°–1⅓°	157·11	156·6–157·6

Mean time = 157·1 sec. Value of τ = 0·0851. Maker's value = 0·0862.

For the wire that had been in use in the gradiometer, B. 4018, widely varying times were obtained[a] and no definite value of τ could be deduced. It is evident that this wire, when loaded with the gradiometer beam, did not behave as an elastic body. Nevertheless the results obtained with it in the field were fairly consistent, and did not show any alteration with time, as will be seen from Table XV, where the results at the repeated stations are given. When a considerably lighter inertia bar was used, its behaviour became normal and very consistent times were obtained. These results are summarized below:

Wire no. B. 4018

Weight of inertia bar	Value of τ
(gm. wt.)	
9·1	0·0586
24·2	0·0645
44·2	0·0765
84·0	—
86·0 (approx. weight of beam)	0·0881 (maker's figure)

The discrepancies between the two instruments were then ascribed to the faulty gradiometer wire, and another wire (B. 4020) was put in the instrument. It was found to be satisfactory; the value of τ previously found by oscillation experiments (0·0879) agreed with the maker's value (0·0864).

(vi) A comparison of the results of the two instruments in a deep basement gave little information about their relative behaviour on account of the marked variations in gradient from point to point in the basement.

[a] For instance, with the inertia bar used the range of the observed times of oscillation was from 524–570 sec.

TABLE XV. *Comparison between values of gradients obtained with the gradiometer at Gelliondale at repeated stations*

Station no.	Observed values		Date
	U_{xz}	U_{yz}	
3	47	−61	23. vii. 28
	42	−59	24. vii. 28
5	35	−4	19. vii. 28
	34	2	24. vii. 28
89	4	−45	29. viii. 28
	−2	−34	29. viii. 28
	8	−46	30. viii. 28
94	−55	−16	31. viii. 28
	−55	−19	31. viii. 28
111	99	15	16. vii. 28
	103	11	17. vii. 28
	101	18	7. ix. 28
	104	9	7. ix. 28
127	9	−26	17. ix. 28
	9	−30	18. ix. 28
112	−55	−9	11. ix. 28
	−51	−13	12. ix. 28
114	−73	0	12. ix. 28
	−71	12	13. ix. 28
133	−55	−14	19. ix. 28
	−57	−3	20. ix. 28
151	−119	7	28. ix. 28
	−125	19	5. x. 28
195	156	−178	27. x. 28
	155	−166	29. x. 28
243	61	−92	19. xi. 28
	57	−91	20. xi. 28
250	60	−61	21. xi. 28
	64	−51	22. xi. 28
290	68	−7	8. xii. 28
	39	3	10. xii. 28
59	190	−50	16. viii. 28
	193	−42	3. i. 29 day
	180	−47	3–4. i. 29 night
	183	−42	4. i. 29 day

(vii) The gradiometer, with the new wire in it, was taken back to Gelliondale and the gradients at certain stations re-determined. The results, which are given in Table XVI, were consistently lower than those previously obtained using the other wire, but were very much closer to them than to the torsion balance figures[a].

Behaviour of Gradiometer Wires. (i) B. 4020. This wire behaved very well at first, but towards the end of the work at Lakes Entrance the fluctuations in zero reading

[a] It would appear from the preliminary tests in the basement at Melbourne that the gradiometer wire was in a satisfactory condition when received, and its subsequent erratic behaviour was probably due to a strain which it received subsequent to the preliminary tests.

The torsion coefficient of a wire varies considerably with the load when the latter is large, so that it is necessary to determine the torsion constant of the wire under the same load as the wire carries in the balance.

had become very much more marked. Its torsion constant was periodically checked, with the following results:

Wire no. B. 4020

Method	Place	Date	Value of τ
Makers	London	—	0·0881
Oscillation	Melbourne	Jan. 1929	0·0879
Deflection	Melbourne	May 1929	0·0874
Deflection in field	Gelliondale	May 1929	0·091
Deflection in field	Lakes Entrance	June 1929	0·088
Deflection in field	Lakes Entrance	Aug. 1929	0·087

(ii) B/40/105/1. This wire was used at Gulgong and behaved satisfactorily throughout the work. The value of its torsion constant, 0·108, determined there by the deflection experiment was rather lower than the maker's figure of 0·116.

The value of the gradients at repeated stations, using this wire, are given in Table XVII.

(iii) B/40/105/2. This was the first wire tried with the new gradiometer and could not be used at all. The torsion head could not be adjusted so that the readings of the instrument came on the visible part of the scale, as the zero readings were much too variable.

TABLE XVI. *Comparison between gradiometer and torsion balance at Gelliondale*

Station no.	Value of U_{xz}			Value of U_{yz}'		
	Gradiometer		Torsion balance	Gradiometer		Torsion balance
	Old wire	New wire		Old wire	New wire	
3	47, 42	36	27·5	−61, −59	−56	−42
55	204	189	171	−72	−75	−64
75	153	145	122	−76	−73	−52
59	190, 193, 180, 183	175	112, 120	−50, −42, −47, −42	−48	−27, −34
83	26	27	30	−42	−47	−38

(iv) P/04/01/7. This was used to replace the above wire in the new gradiometer. For the first week the zero was very variable. After that the instrument could be used to measure a gradient, but the readings were very unreliable. When it was first set up on the new station the readings apparently were quite consistent, but gave completely erroneous values for the gradient. Usually when it had been on the station for a day or more it gave concordant results which agreed fairly well with those of the other instrument. The explanation of this behaviour seemed to be that the zero was very much displaced when the beam was unclamped at the new station, and that it needed rather a long time to return to its normal value. The value of the torsion constant, obtained by the deflection experiment, was 0·154; the maker's figure was 0·159.

TABLE XVII. *Values of gradients at repeated stations at Gulgong*

Station no.	Date	U_{xz}	U_{yz}
6	21. x. 29	−19	8
	1. xi. 29	−24	8
25	5. xi. 29	−30	− 5
	12. xi. 29	−27	− 6
24	6. xi. 29	−24	− 7
	8. xi. 29	−22	− 5
30	16. xi. 29	−32	−15
	18. xi. 29	−31	−14

The values of the gradients, obtained when the results of the two instruments were compared, are given in Table XVIII.

Conclusions. It will thus be seen that the results from the two gradiometers agree moderately well with each other and that consistent values can be obtained with the gradiometer either by day or by night, and even if the instrument is used in a climate subject to large diurnal variations of temperature. The results from the torsion balance are consistently lower than those from the gradiometer by an amount far exceeding the probable error of either instrument and, despite a thorough calibration of both instruments, this difference remains unexplained.

TABLE XVIII. *Comparison of gradiometers at Gulgong*

Station no.	Value of U_{xz}		Value of U_{yz}	
	No. 1	No. 2	No. 1	No. 2
1	−26	−18, −20	−15	−23, −25
3	−14	−13	0	3
6	−19, −24	−23	8, 8	3
22	−19	−13	− 8	− 7
25	−30, −27	−26	− 5, − 6	0
27	−30	−27	−41	−35
32	−55	−61	−32	−33
44	−59	−51	− 7	− 7

The I.G.E.S. has not come across any cases in the literature in which the results obtained with different types of instrument are compared and concludes that while reliable results are readily obtained for relative values of gradients, the determination of the absolute value is much more difficult.

CHAPTER IX

MAGNETIC METHODS

SECTION I. THEORY OF INSTRUMENTS

Theory of the Magnetic Balance for Measuring Vertical Intensity[a]. In Fig. 241, let SN represent the magnetic system of moment M e.m. unit and mass m gm. inclined at an angle θ radian to the horizontal in a plane at right angles to the magnetic meridian in the *southern* hemisphere; and let G represent its centre of gravity at a distance e cm. from the knife edge O and inclined at an angle ω radian to ON.

If the vertical intensity of the earth's magnetic field is Z, the couple, C, acting on the system in an anticlockwise sense is

$$C = MZ \cos \theta.$$

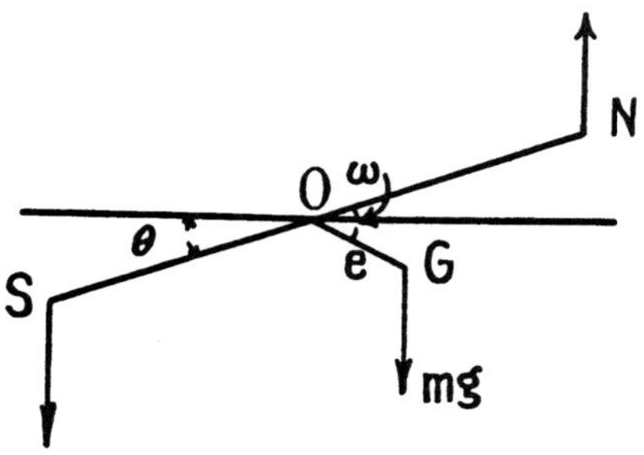

Fig. 241. Vertical magnetic balance

If the system is in equilibrium, this couple is balanced by the couple due to the weight of the system acting at G, so that

$$MZ \cos \theta = mg.e \cos (\omega - \theta)$$
$$= mg (a \cos \theta + d \sin \theta),$$

where $a = e \cos \omega$, and $d = e \sin \omega$,

$$\therefore \quad \tan \theta = \frac{MZ - mg.a}{mg.d}.$$

For the Schmidt balance let S_0 = the scale reading in the horizontal position;
S = the scale reading in the deflected position;
and f scale divisions = focal length of the object lens in the optical system.

Then $$\frac{S - S_0}{f} = \tan 2\theta = 2 \tan \theta \text{ nearly, if } \theta \text{ is small.}$$

Therefore $$\frac{S - S_0}{2f} = \frac{MZ - mga}{mgd} \quad \text{......(1).}$$

If the system is adjusted to balance horizontally in the normal field Z_0, then

$$\tan \theta = 0;$$

and $$MZ_0 = mga;$$

so that $$\frac{S - S_0}{2f} = \frac{M(Z - Z_0)}{mgd} = \frac{M \Delta Z}{mgd},$$

where ΔZ denotes the increment of intensity causing the deflection.

[a] For a general theory, see Heiland, *Geophysical Prospecting 1929*, p. 268, A.I.M.E., New York, 1929.

Therefore $$S - S_0 = \frac{2fM}{mgd}\,\Delta Z = \frac{\Delta Z}{\epsilon} \qquad \ldots\ldots(2),$$

where $\epsilon = \frac{mgd}{2fM}$ is the scale value, and is a constant for any given adjustment.

For the Thalen-Tiberg magnetometer θ is observed directly. When adjusted to balance horizontally in a normal vertical field Z_0

$$\tan\theta = 0,$$

$$\therefore \quad MZ_0 = mga,$$

so that $$\tan\theta = \frac{MZ - MZ_0}{mgd} = \frac{M\Delta Z}{mgd},$$

or $$\Delta Z = \frac{mgd}{M}\tan\theta = K\tan\theta \qquad \ldots\ldots(3),$$

where K is the constant of the instrument and can be varied by altering the value of d.

Theory of Deflection Magnetometer. Two methods are available for the measurement of horizontal intensity in terms of the moment of a standard magnet: (*a*) by the tangent magnetometer, Fig. 242, and (*b*) by the sine magnetometer, Fig. 243.

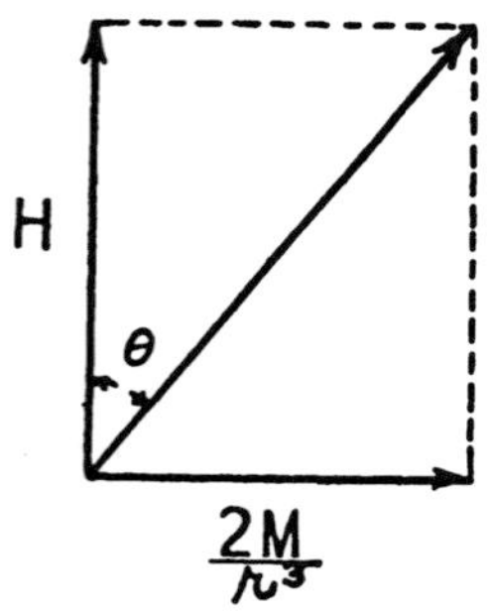

Fig. 242. Deflection magnetometer (tangent method)

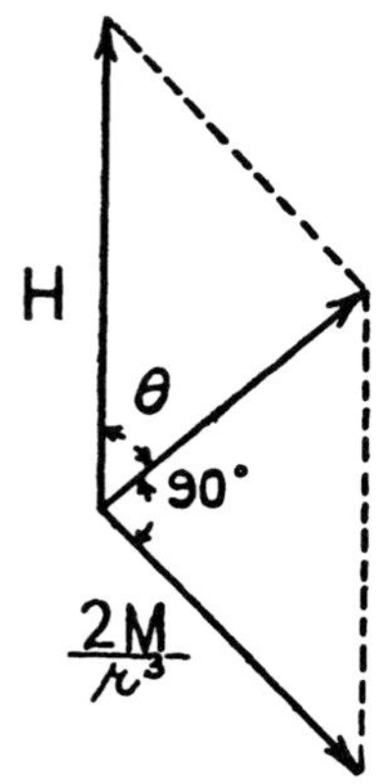

Fig. 243. Deflection magnetometer (sine method)

In (*a*) the standard magnet is placed to the magnetic E or W of the magnetometer, whose magnet is acted upon by two fields at right angles, namely H gauss, the horizontal component of the earth's field, and $\frac{2M}{r^3}$ gauss, the field of the deflecting magnet of moment M at a distance r cm. from the magnetometer magnet. The magnetometer places itself in the direction of the resultant of these two fields and is deflected from the NS direction through an angle θ radian given by

$$\tan\theta = \frac{2M}{r^3}\cdot\frac{1}{H} \text{ approx.} \qquad \ldots\ldots(4).$$

In (*b*) the deflecting (standard) magnet is moved round the magnetometer magnet until the axes of these two are at right angles. In this case

$$\sin\theta = \frac{2M}{r^3}\cdot\frac{1}{H}\text{ approx.} \qquad \text{......(5)}.$$

Both these relations are approximations, inasmuch as the field of the deflecting magnet at a distance r along its axis is only approximately $\frac{2M}{r^3}$ gauss. For a discussion of this, see any standard treatise on magnetism.

SECTION 2. ADJUSTMENT OF SCHMIDT VERTICAL BALANCE

Adjustment of Magnetic System. When purchasing a magnetometer, the manufacturer should be informed of the vertical intensity of the earth's field for which the magnetic system needs to be adjusted. On taking the instrument into the field, however, further adjustment is usually needed to compensate any small difference in vertical intensity from that for which the balance was previously adjusted. Some part of the area which, from geological considerations and tests at a few points, is deemed to be free from large magnetic anomalies, is selected for making this adjustment, which can be done by altering the position of the centre of gravity of the magnetic system, by means of the screw provided for the purpose, until the scale reads a suitable value (10 to 30).

Determination of Scale Value. The scale value may be defined as the change in magnetic field, measured in γ[a], which produces a deflection of one scale division. It is determined by measuring the deflection due to the reversal of a known magnetic field, which can be produced conveniently by means of the auxiliary magnets[b] and the special attachment which fits into the tripod head. Scale readings, S_1 and S_2, are taken first with the *N*-pole and then with the *S*-pole upwards, while a note is made of the temperature and the distance r of the auxiliary magnet from the magnetic system of the instrument.

The scale value is then given by[c]

$$\epsilon = \frac{4M\left(1+\frac{K}{r^2}\right)}{r^3\,(S_1-S_2)},$$

where $K = \frac{1}{2}L^2 - \frac{3}{4}l^2$; the makers give $\frac{3}{4}l^2 = 51{\cdot}7$.

L cm. = the distance between the poles of the deflecting magnet which is about 5/6ths of its overall length.

l cm. = the distance between the poles of the deflected magnet of the instrument.

M e.m. unit = the magnetic moment of the deflecting magnet.

[a] $1\gamma = 10^{-5}$ gauss.

[b] This field can be produced more conveniently by a coil system through which a known current is passed.

[c] Heiland and Duckert, *Zeits. f. angew. Geophysik*, vol. 1, p. 300 (1924).

The value of the term $1+\frac{K}{r^2}$ is very close to unity, and varied only from 0·97 to 0·9998 in the case of the Schmidt balances used by the I.G.E.S.

It is advisable to redetermine the scale value of the instrument periodically, as it is inversely proportional to the magnetic moment of the beam system[a], which usually decreases in the course of time. The following table shows the alteration in scale value of the instruments used by the I.G.E.S.

Scale Values in γ per Division			
Measured at	Date	Instrument	
		No. 88,057	No. 88,059
Potsdam	March, 1928	32·8	33·7
Gelliondale	November, 1928	35·8	28·3
Gulgong	November, 1929	54·9*	31·9

* This high value is partly due to the fact that the observer altered the position of the knife edge with respect to the beam.

Maximum Range. The scale value of the instrument has been shown on p. 324 to be equal to $\frac{mgd}{2fM}$. The only factor in this expression which can be varied conveniently is d, the vertical distance of the centre of gravity of the beam system below the knife edge. This dimension can be altered simply by moving the vertical brass screw up or down, thus varying the scale between the approximate limits of 15 and 100γ per scale division[b]. As the scale has a range of 80 divisions, the maximum range of intensity is therefore about 8000γ, which can be further increased by about 10,000γ by using the auxiliary magnets.

Magnetic Moment of Auxiliary Magnet. The magnetic moments of all the auxiliary magnets are given by the makers, but as they vary with temperature and time it is desirable to check these values at suitable intervals; they can be determined by standard methods in a laboratory[c]. This was done at Melbourne and the values for five of the magnets agreed to within 1 per cent. and the sixth to 4 per cent. of the makers' figures, determined ten months earlier at Potsdam.

Field Manipulation. A few modifications to the prescribed[d] field procedure which save time are given below:

(i) *Orientation.* Instead of orientating the tripod head to the magnetic north by means of the compass and subsequently turning it through 90° to bring it to the east, the tripod head may be adjusted in one operation by setting the compass needle to 270°.

[a] See formula (2), p. 324.

[b] By using a gold screw a value of 300γ per division is obtainable; Heiland, Henderson and Malkovsky, *U.S. Bureau of Mines Tech. Paper* 439, p. 19 (1929).

[c] See standard textbooks of practical physics; e.g. Watson or Kohlrausch.

[d] Directions for Use; loc. cit.

(ii) *Repeated readings.* A further saving of time can sometimes be effected by taking only one reading in each azimuth instead of three, as the error of reading the instrument is often negligible compared with the magnitude of the anomaly. It is only necessary to repeat the readings when the instrument has an inferior knife edge or when the anomalies are small.

(iii) *Temperature variations.* The effect of temperature on the balance readings depends upon the temperature gradients inside its outer case, as well as on the actual temperature[a]. It is important, therefore, to allow the instrument to acquire the temperature of the surrounding air before any readings whatever are taken. On one traverse made at Gelliondale, when this precaution was not taken, the discrepancy on repeating at the base station was as high as 48γ after making corrections for temperature and diurnal variation.

(iv) *Magnetic material.* The observer must be absolutely free from magnetic material. Even a wrist-watch causes an appreciable deflection. The obvious way to decide whether any doubtful article can be carried by the observer is simply to move it about near the instrument and observe the deflection, which should be less than 0·1 div. for accurate work. Boots, belts, trouser-buttons, etc., should be tested in this way. Iron wire fences are common in Australia but their effect was found to be quite negligible at distances greater than 50 ft.

SECTION 3. BEHAVIOUR OF INSTRUMENTS

The behaviour of the Schmidt balances during the 18 months they were used in Australia was quite satisfactory. Possibly the levels of the instrument could be improved, as it was found that they got out of adjustment very easily. Of the two instruments belonging to the I.G.E.S. one had a much better knife edge than the other, the probable error in the reading of the better instrument being about 1/10 scale division, while for the other it was about twice as large.

The Thalen-Tiberg magnetometer was not so satisfactory as the workmanship of the vital parts was poor. No provision has been made for locking the sliding brass weight in position on the magnetic needle after making an adjustment, so that it is liable to slip unless handled with the greatest care. Considerable difficulty was often experienced from this cause in trying to obtain satisfactory repeats. There also seems to be too much friction at the pivot, and this may be partly due to the severe grinding to which it is subjected by the frequent clamping and releasing. The levelling head is well designed, but an adjustable bubble instead of a fixed one would be an improvement. Another satisfactory feature of the instrument is the tripod which, although light and portable, is sufficiently rigid.

[a] *Geophysical Prospecting 1929*, p. 285, A.I.M.E., New York, 1929.

CHAPTER X

SEISMIC METHODS

SECTION I. PROPAGATION OF SEISMIC DISTURBANCES

Paths of Elastic Waves in Seismic Prospecting. In prospecting over an interface between a surface formation and a lower formation of higher velocity[a], the conventional "refraction" method assumes the existence of seismic waves, or pulses, following the two different paths shown in Fig. 244:

a. a pulse which travels along the surface with the velocity of the longitudinal wave in the upper material;

b. a pulse which (i) first travels to the interface at the critical angle, with the velocity of the longitudinal wave in the upper material; (ii) then travels along the interface[b] with the velocity of the longitudinal wave in the lower material; and (iii) finally travels from the interface at the critical angle, back to the surface, with the velocity of the longitudinal wave in the upper material.

The real existence of these pulses has not been completely explained theoretically, and is not consistent in all respects with the experience of seismologists.

The ordinary theory of *reflection* and *refraction* does not allow either of these paths. Thus, firstly, with respect to the surface pulse *a*, it may be shown[c] that, when a longitudinal wave is incident on a free surface at grazing incidence, no reflected longitudinal wave appears, but the energy goes entirely into a reflected transverse wave; so that a longitudinal wave (and similarly, a transverse wave) cannot be propagated along a free surface. Secondly, with respect to the refracted wave *b*; when incidence is at the critical angle as for *b* i, a negligible part of the energy is refracted; while from the wave *b* ii, travelling in the interface, a negligible part of the energy is refracted to generate a wave such as *b* iii. Tentative mathematical investigations, however, have shown that the transmission of these pulses, *a* and *b*, may be explicable in terms of *diffraction*, but a full mathematical treatment has not yet been presented.

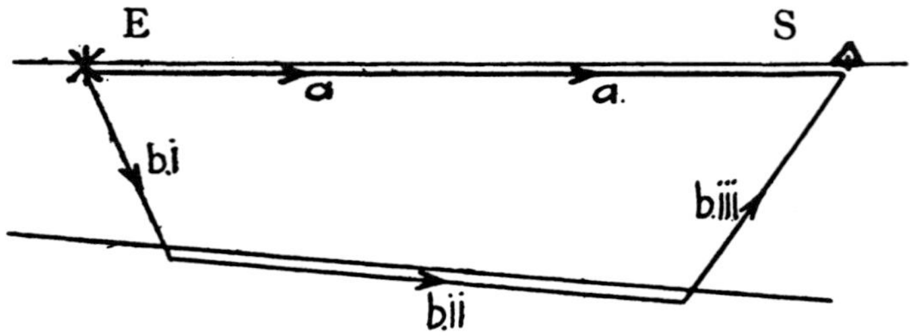

Fig. 244. Paths of waves in seismic prospecting
E, Explosion. *S*, Seismometer. *a*, Surface wave.
b i, *b* ii, *b* iii, Refracted wave

The existence of a longitudinal wave *a* travelling along the surface, Fig. 244, receives no support from tectonic seismology. The longitudinal and transverse waves

[a] See p. 196, note b.
[b] Or, where the interface is convex upwards, by the shortest path through the lower material.
[c] Walker, *Modern Seismology*, pp. 43–45, Longmans Green and Co., London, 1913.

recorded in natural earthquakes both travel along the chord of the geoid from the earthquake focus[a] to the seismometer. More complex waves, known as Rayleigh waves and Love waves, follow the arc of the earth's surface; but neither longitudinal nor transverse waves do so.

The anomalous phenomena pointed out by A. Mohorovičić[b] in 1909, for "near" earthquakes, i.e. at distances up to about 500 miles from the focus, have, however, been explained by the assumption of paths, both for longitudinal and transverse waves, similar to that assumed for the "refracted wave" *b*, Fig. 244, in seismic prospecting. Fig. 245 shows the paths of the six different waves recorded in near earthquakes, involving refraction, at the critical angle, at three main boundaries known to exist in the structure of the earth's crust[c]. The real existence of these wave-paths has been established by calculation from earthquake records.

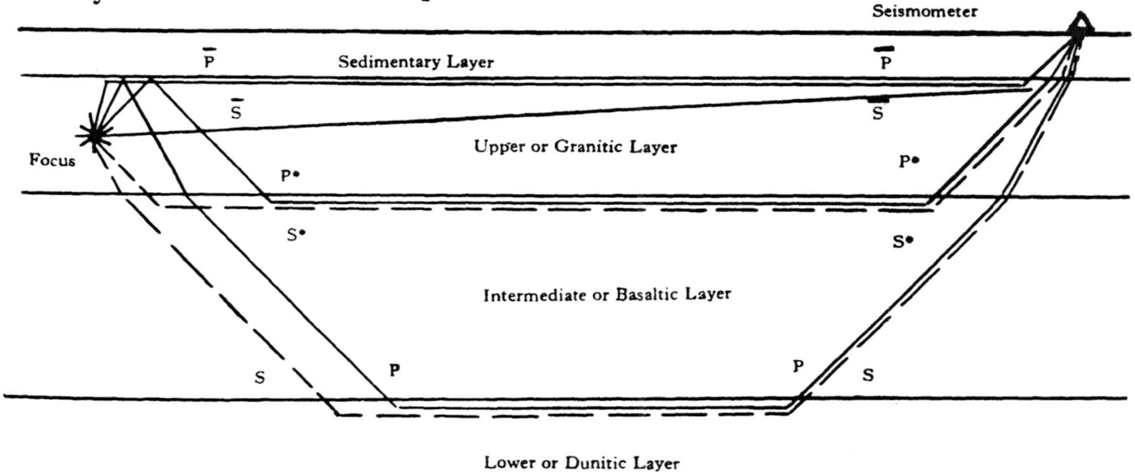

Fig. 245. Paths of waves in near earthquakes (after Jeffreys[d]). Approximate thickness of upper layer, 10 km.; of intermediate layer, 20 km. (*Note*. The transitions may be gradual and not sudden. The *P* waves may be generated at the upper surface, and not at the base of the sedimentary layer)

The position with regard to the conventional assumptions of paths for longitudinal waves in seismic prospecting is therefore as follows:

Path	Velocity	Theory of simple reflection and refraction	Seismology
a Along surface	Upper material	Surface longitudinal wave impossible	Not observed
b (i) Incident upon boundary at critical angle	Upper material	Negligible part of energy refracted	Observed in near earthquakes
b (ii) Along boundary	Lower material		
b (iii) Leaving boundary at critical angle	Upper material		

[a] The point of origin of an earthquake is known as the *focus*; the surface point directly above the focus is known as the *epicentre*.

[b] *Jahr. d. Meteor. Obs. Agram.* (*Zagreb*), 1909.

[c] Jeffreys, *The Earth*, 2nd ed., pp. 95–116, University Press, Cambridge, 1929.

[d] *Ibid.* p. 100.

The calculations in seismic prospecting have for the most part been based upon the simple assumption of the two pulses following paths *a* and *b* (Fig. 244), and have been confirmed by boring and other tests. The conventional theory thus gives results which are at least approximately[a] true. Without further theoretical and precise practical investigation, however, it cannot be admitted as a full and accurate account of the waves and wave-paths used in seismic prospecting. A fuller discussion will be given in the following paragraphs.

Chief Elastic Waves in Homogeneous Isotropic Solids. It is necessary to give a more general account of the propagation of seismic disturbances than has been done in pp. 195–199. The properties of the more important waves are set out in the following table:

Name	Seismological designation	Boundary conditions	Velocity	Nature of displacements	Direction of displacements	Type of strain	Wave front*
Longi-tudinal	Primary *P*	Remote from boundaries	$v_L = \sqrt{\frac{k+\frac{4}{3}n}{\rho}}$	Linear	Parallel to direction of propagation	Dilatational; irrotational	Sphere
Trans-verse	Secondary *S*	Remote from boundaries	$v_T = \sqrt{\frac{n}{\rho}}$	Linear	Perpendicular to direction of propagation	Distortional; equivoluminal	Sphere
Rayleigh	*L*	Near to a surface	$v_R = c\sqrt{\frac{n}{\rho}}$ $[c = f(k_1 n) \simeq 0{\cdot}9]$	Elliptical (approximately)	In vertical plane through direction of propagation	Complex	Cylinder†
Love	*M*	In surface layer of lower velocity, for transverse waves, over medium of higher velocity	Velocity intermediate between transverse wave velocities in overburden and bed-rock	Linear	Parallel to surface and perpendicular to direction of propagation	Distortional	Cylinder‡

* Wave front, for point source of disturbance.
† Cylinder with vertical axis, increasing in depth with distance from source; amplitude decreasing exponentially with depth.
‡ Cylinder with axis perpendicular to slab.

Elastic Constants. For a homogeneous isotropic solid two moduli of elasticity are required, e.g. incompressibility k, and rigidity n; or Lamé's constants λ, μ; or Young's modulus E, and Poisson's ratio σ. The interrelations are:

$$k = \lambda + \tfrac{2}{3}\mu = \frac{E}{3(1-2\sigma)}, \qquad \lambda = k - \tfrac{2}{3}n = \frac{\sigma E}{(1+\sigma)(1-2\sigma)}, \qquad E = \frac{\mu(3\lambda+2\mu)}{\lambda+\mu} = \frac{9kn}{3k+n},$$

$$n = \mu = \frac{E}{2(1+\sigma)}, \qquad \mu = n = \frac{E}{2(1+\sigma)}, \qquad \sigma = \frac{\lambda}{2(\lambda+\mu)} = \frac{3k-2n}{6k+2n}.$$

For an incompressible solid, $\sigma = \frac{1}{2}$, $k = \infty = \lambda$, $n = \mu = \frac{1}{3}E$. It is found for the

[a] Schweydar and Reich, *Gerlands Beitr. z. Geophysik*, vol. 17, p. 121 (1927), followed by Ambronn, *Elements of Geophysics*, p. 251, McGraw Hill, New York, 1928, state that calculations are more accurate if it is assumed that incidence is normal to the boundary and not at the critical angle. See also p. 200.

materials of the earth's crust that the value of $\sigma = 0{\cdot}29$[a]. This approximates to "Poisson's condition" (derived from assumptions as to intermolecular forces), that

$$\sigma = \tfrac{1}{4}, \quad \text{or} \quad \lambda = \mu = n = \tfrac{3}{5}k = \tfrac{2}{5}E.$$

The Longitudinal and Transverse Waves. Poisson, in 1828, showed that two types of wave could be propagated in an isotropic solid, one a longitudinal disturbance, and one a transverse disturbance; and Stokes, in 1849, showed that the former wave was one of irrotational dilatation, the latter of equivoluminal distortion. Their velocities are respectively

$$v_L = \sqrt{\frac{k+\frac{4}{3}n}{\rho}} = \sqrt{\frac{\lambda+2\mu}{\rho}} = \sqrt{\frac{E}{\rho}\left(1+\frac{2\sigma^2}{1-\sigma-2\sigma^2}\right)};$$

and

$$v_T = \sqrt{\frac{n}{\rho}} = \sqrt{\frac{\mu}{\rho}} = \sqrt{\frac{E}{\rho}\left(\frac{1}{2(1+\sigma)}\right)}.$$

The boundary of an elastic disturbance, in an isotropic solid, will advance with one of these two velocities, v_L or v_T; if the strain is pure compression, then the velocity is necessarily v_L; if the strain is pure shear, then the velocity is necessarily v_T. In an aelotropic solid, at any point there are three velocities in a given direction; there are thus three wave surfaces arising from a point disturbance. In an isotropic solid, these wave surfaces are all spheres, two of them coinciding, so that there are in effect two spherical wave surfaces, advancing with velocities v_L and v_T. If the initial disturbance is a compression, then only the former sphere will exist; if a shear, then only the latter; in general, however, both wave surfaces will be developed. These will be inner and outer boundaries of a shell within which the disturbance is confined[b].

For $\sigma = \frac{1}{4}$, the velocities become

$$v_L = \sqrt{\frac{3n}{\rho}}; \qquad v_T = \sqrt{\frac{n}{\rho}}; \qquad \text{so that} \qquad v_L/v_T = \sqrt{3}.$$

From an earthquake the longitudinal waves are normally the first to arrive, and so they are known to seismologists as *Primary*, or *P* waves; and the transverse, as *Secondary*, or *S* waves. At the surface, the latter may be denoted by *SH* or *SV*, according as the displacement is horizontal or vertical. Although they travel primarily along the chord of the geoid, arrivals corresponding to one or more reflections at the surface are also recorded.

Reflection. At a plane boundary an incident wave produces, in general, four new waves. An incident longitudinal wave gives not only a reflected and a refracted wave of the original longitudinal type, but also a reflected and a refracted wave of the transverse, *SV*, type. Similarly, an incident transverse, *SV*, wave gives a reflected and a refracted wave of a longitudinal type, as well as of the original transverse, *SV*, type. An incident transverse wave of the *SH* type, however, produces no waves of transformed type,

[a] Adams and Coker, *An Investigation into the Elastic Constants of Rocks* (Publ. Carnegie Inst., Washington, 1906), state that they found the value of σ to range from 0·198 to 0·29.

[b] Love, *The Mathematical Theory of Elasticity*, 4th ed., ch. XIII, University Press, Cambridge, 1927.

but a reflected and a refracted wave of the original type only. The distribution of energy depends in each case upon the angle of incidence; and the effect analogous to total internal reflection may prohibit the existence of one or both of the refracted waves. As indicated above, the amplitude of a wave—if diffraction effects are neglected—is zero when refracted into a plane interface; and neither longitudinal nor transverse waves can travel in a free surface.

The Rayleigh Wave. In 1885 Rayleigh[a] found, however, that a certain wave, combining compression and distortion, could travel along a free surface, and was the only type of *harmonic* wave possible at such a surface. This is known as the *Rayleigh*, or *L* wave. The phase is independent of the depth, but the amplitude decreases exponentially, and rapidly, with depth, becoming negligible at a depth equal to the wave length. The velocity is

$$v_R = c\sqrt{\frac{n}{\rho}},$$

where c is a function of the elastic moduli; being 0·9954 when $\sigma = \frac{1}{2}$ (that is, $\lambda = \infty = k$), and 0·9194 for the usual case, $\sigma = \frac{1}{4}$ (that is, $\lambda = \mu$). The ratio of maximum vertical and horizontal displacements is 1·9 in the former case, 1·5 in the latter. Actually the path of the displaced particle is a pure ellipse only for a plane wave front, being slightly modified for a diverging wave.

Rayleigh waves are found in earthquake records; in fact, since the energy in the Rayleigh wave follows the arc of the earth's surface and attenuates in two dimensions only, it constitutes the bulk of the shock received at a distant point. The axes of the displacement ellipse are found to have a ratio smaller than the theoretical ratio of 1·5; usually from 0·9 to 1·0[b].

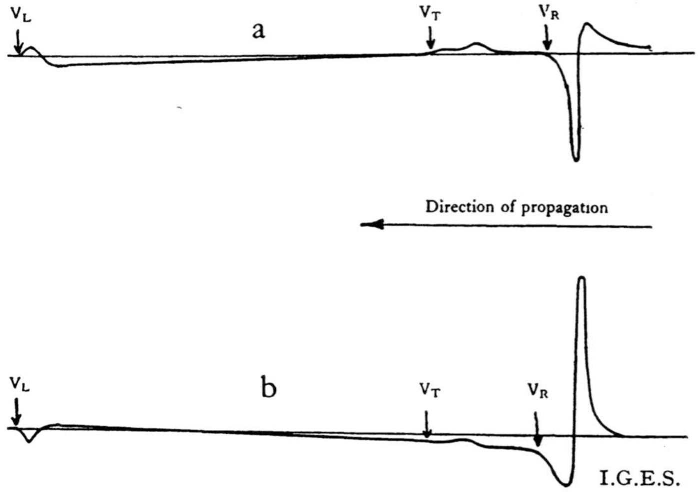

Fig. 246. Instantaneous values of the displacement of particles in a given region of the surface, propagated from a distant point by application of a vertical impulse to the surface

a, Displacements in line of propagation. *b*, Displacements in vertical plane.

v_L, v_T, v_R, Phases travelling with the velocities of longitudinal transverse, and Rayleigh waves respectively.

Diffraction. The curvature of the wave front is often sufficiently large, relative to the wave length, to introduce diffraction effects at a boundary. In this case a longitudinal or transverse, *SV*, wave can no longer be reflected at a

[a] *Lond. Math. Soc. Proc.* vol. 17, p. 4 (1885–6). Rayleigh, *Scientific Papers*, vol. 2, p. 441, University Press, Cambridge, 1900.

[b] Mainka, *Phys. Zeits.* vol. 16, p. 117 (1915).

free surface as a combination of longitudinal and transverse waves alone; but the reflection must involve the production of a Rayleigh wave.

Analysis of the diffraction of elastic waves was initiated by Lamb[a] in 1904, who investigated the surface disturbance resulting from an impulse applied at a surface point. At a distant point on the surface, he finds that the disturbance commences after a time corresponding to propagation with the velocity of the longitudinal wave; that a second phase of the disturbance commences after a time corresponding to propagation with the velocity of the transverse wave; and that a third phase, of much greater amplitude, commences after a time corresponding to propagation with the velocity of the Rayleigh wave.

Fig. 246 shows the horizontal and vertical components of the complete disturbance, the former being in the direction of propagation; there is no horizontal component at right angles to the direction of propagation.

Nakano[b], 1925, following Lamb's work, showed the following effects at a surface point, distant x from the epicentre of an impulsive disturbance at a focus of depth h, comparable with the wave length (Fig. 247):

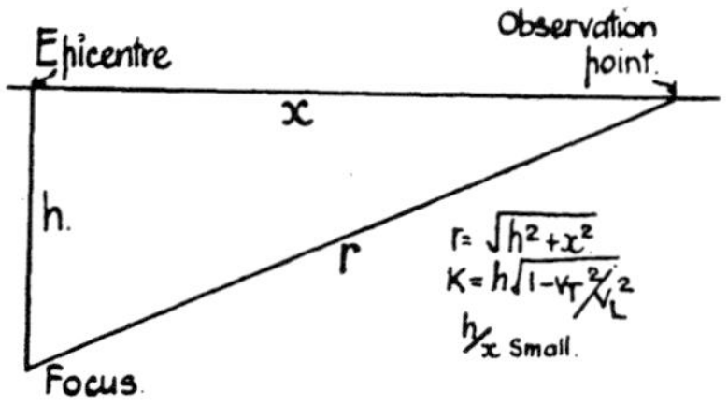

Fig. 247. Seismic disturbances from shallow focus

Initial disturbance: Pure compression

Longitudinal, P, wave	Arrives at time r/v_L
Transverse, SV, wave	Builds up to a maximum at time x/v_T
Rayleigh, L, wave	Builds up to a maximum at time x/v_R

Initial disturbance: Pure shear

Horizontal, P, wave	Begins at time $k/v_T + x/v_L$
Transverse, S, wave	Begins at time x/v_T
Rayleigh, L, wave	Builds up to a maximum at time x/v_R

The portion of the incident energy transformed into the Rayleigh wave is, for $\sigma = \frac{1}{4}$ (that is, $\lambda = \mu$), 41·7 per cent. in the former case, 88·5 per cent. in the latter case.

Effect of Heterogeneous Stratification. Love[c], in 1908, investigated the case of a horizontal stratum overlying an extended medium of higher velocity. He showed that certain transverse waves, namely, the *SH*, of velocity intermediate between those proper to transverse waves in the two media, could be transmitted in the stratum. These are known as *Love*, or *M*, waves. Difficulty has been experienced in distinguishing Love waves in earthquake records, but there is no doubt of their real existence[d].

[a] *Roy. Soc. Phil. Trans.* A, vol. 203, p. 1 (1904).
[b] *Jap. J. of Astron. and Geophysics*, vol. 2 (1925).
[c] Love, *Some Problems in Geodynamics*, ch. XI, University Press, Cambridge, 1911.
[d] See for example the table of Love wave velocities collected by Gutenberg, *Phys. Zeits.* vol. 25, pp. 377–381 (1924), and records of trans-pacific Love waves published by Matuzawa, *Bull. Earthquake Research Inst. Tokyo*, vol. 6, p. 213 (1929).

Sezawa, 1927–28, has shown that for such a structure the Rayleigh wave will have a velocity intermediate between those of the Rayleigh waves in the two media, and that there is a similar effect upon the velocity of Rayleigh waves in a thin stratum of one medium, with a faster medium overlying and underlying it.

Dispersion. The velocities of longitudinal, transverse, and Rayleigh waves, in a homogeneous solid, are independent of wave length, but Love waves and Rayleigh waves, in stratified structures, have velocities which are a function of the ratio of wave length to depth of stratum. It has also been shown[a] that the velocities of Rayleigh waves on curved surfaces are a function of the ratio of wave length and radius of curvature. Accordingly, trains of waves of definite frequencies are always found in earthquake records.

As a consequence of dispersion effects, the velocities observed are in reality group velocities rather than true wave velocities, and caution must be exercised in this respect in the interpretation of seismic records in terms of wave propagation.

Angle of Emergence. Considerable importance is attached by some writers to a determination of the angle of emergence. The ratio, given by simultaneous records of x and z components, will be the tangent of the angle of emergence. This will, however, be only the apparent angle, θ', including the effect of the reflected waves of both types. The true angle, θ, is given by

$$\cos\theta = v_L/v_T \sqrt{\tfrac{1}{2}(1-\sin\theta')}.$$

The I.G.E.S. party, using the Schweydar seismometer, found that, owing to lack of sufficient damping, the records of vertical and horizontal components were not sufficiently comparable to enable the angle of emergence to be readily and accurately estimated. It was possible, however, to decide whether the initial movement was primarily one of vertical or of horizontal displacement. Thus the refracted and the surface waves were both found to be longitudinal displacements; but the former, affecting only the vertical-component pendulum, reached the surface approximately vertically; the latter, affecting the horizontal-component pendulum, at grazing incidence.

Surface-formation Disturbances found by the I.G.E.S. To study disturbances propagated in a surface formation a set of explosions was carried out at Gulgong in an alluvial overburden between 150 and 250 ft. deep, Fig. 170. The records were made with the Schweydar seismometer, the explosive being placed on the surface[b] at a succession of points spaced at intervals of 50 ft. up to a distance of 1200 ft. Records obtained at distances of 800 and 1200 ft. have been shown in Figs. 167, 168, and discussed briefly on pp. 221–222. At these distances the refracted disturbance was the first-arrival, emerging approximately vertically, as stated in the last paragraph. Fig. 248 shows the records

[a] *Bull. Earthquake Research Inst. Tokyo*, vol. 3, p. 1 (1927); vol. 5, p. 85 (1929).
[b] For convenience and rapidity of working at these small distances.

obtained at 400 ft., where the surface disturbance was the first-arrival[a]. In this case the horizontal components appeared before the vertical, and the corresponding point could also be found in the records for greater distances, at which it was not that of the first-arrival. After this point of the record, vertical displacements gradually appeared; but the next pronounced feature was the sudden commencement of motion with large horizontal component, without appreciable increase in vertical displacements. At a later stage of the record, a phase of large vertical displacements suddenly appeared; an increase in horizontal displacements was discernible at the same point, but the vertical were thereafter the greater.

As appears in Fig. 167 the amplitude of the transverse, *SH*, component never became very large.

Three characteristic features were thus selected throughout the set of records: (*a*) the first-arrival of the surface wave; (*b*) the commencement of large horizontal displacements; (*c*) the commencement of large vertical displacements. The corresponding points have been marked *A*, *B*, *C*, on the records reproduced in Figs. 167, 168, 248. A time-distance graph from the complete set of records is given in Fig. 249. The points corresponding to the arrival of these three phases fall on three straight lines: *OA*, velocity = 3600 ft./sec.; *OB*, velocity = 2500 ft./sec.; *OC*, velocity = 1750 ft./sec.; indicating that surface waves are concerned.

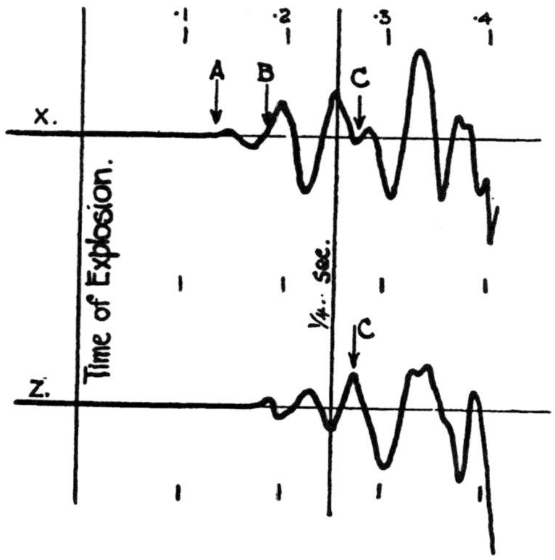

Fig. 248. Records of surface disturbances obtained with Schweydar seismometer. *X*, Horizontal component. *Z*, vertical component. Charge 0·1 lb. blasting gelatine on sandy clay

Conclusions as to Seismic Prospecting. Some geophysicists[b] have found that, at a sufficient distance from the explosion point the first-arrival disturbance travelling through the surface formation is of longitudinal type, and arrives vertically from below. This may be accounted for by a progressive increase of velocity with depth, as explained on p. 198; arising, for example, from an increase with depth of the degree of consolidation of sedimentary formations. Under such conditions the theoretical objections to the propagation of a longitudinal disturbance along the surface do not arise.

In the Gulgong experiments of the I.G.E.S., however, the first-arrival disturbance was of longitudinal type, but at grazing incidence, even at a distance of 1200 ft. from the explosion. The only theoretical support for this seems to be the work of Lamb and Nakano referred to above, from which it appears that longitudinal and transverse

[a] *Referee's comment.* Although this figure certainly seems to indicate the absence of the onset *A* in the vertical record, the evidence for the complete absence on all occasions is not convincing (see p. 221). It is not in agreement with the experience of other investigators.

[b] See p. 198, note b, and *Technical Publication*, No. 267, A.I.M.E., New York, 1930.

waves travelling along the surface might be maintained by continual diffraction, provided that the radius of curvature of the wave front were of the order of the wavelength[a]. The I.G.E.S. records are naturally more complex than the form of the disturbance deduced by Lamb for the simple conditions which he assumed. They do not show the arrival of any marked phase with the velocity of the transverse wave; but this phase is not very conspicuous in Lamb's curves. The "*C*" phase of the I.G.E.S. records, with large vertical components, is, of course, that which travels with the Rayleigh wave velocity. It is evident that further investigation, both theoretical and experimental, will be required before there is an adequate knowledge of the "surface" disturbance of seismic prospecting.

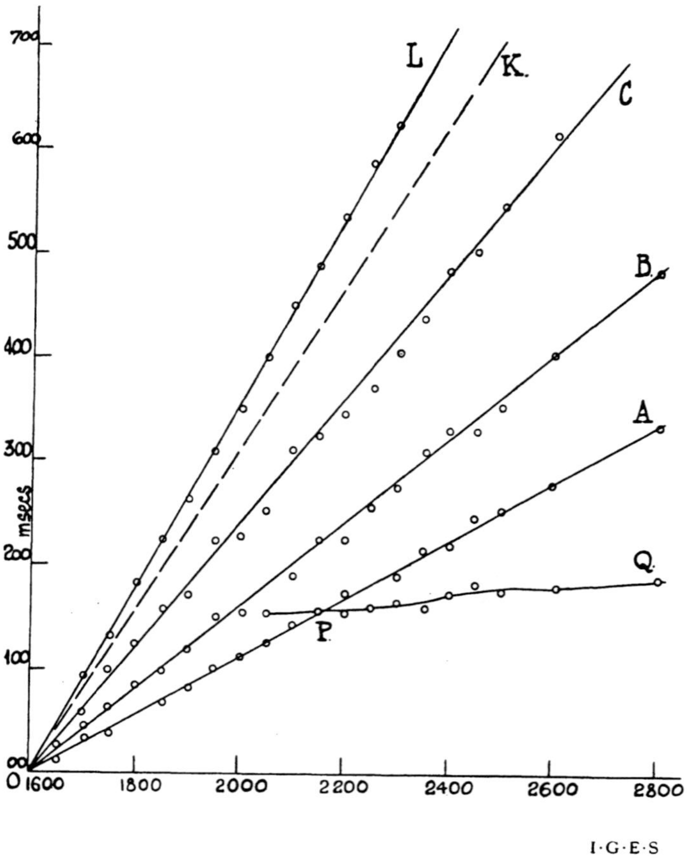

Fig. 249. Time-distance graph of surface disturbances *OA*, *OB*, *OC*, Surface waves. *OK*, *OL*, Sound waves. *PQ*, Refracted wave.

Similarly, the "refracted" disturbance of seismic prospecting and the phenomena of near earthquakes both require further investigation. To throw light upon the anomalous distribution of energy in the latter case, Jeffreys[b] has taken the special case of a boundary between non-rigid media. Arising from a compressional disturbance, he finds, at a point on the surface, a pulse arriving at a time which would correspond with a path and velocities such as are assumed in explaining the times of arrival of waves in a near earthquake. He also finds that the pulse possesses a proportion of energy of the order found for the wave-paths in near earthquakes. It is to be presumed that these phenomena of seismic prospecting and near earthquakes may be explained by the operation of diffraction, as in the work of Lamb and Nakano on propagation at a free surface. A fuller investigation of the nature of the diffraction and an extension to media possessing rigidity, are not yet available.

[a] With a frequency of 30 per second and a velocity of 3000 ft./sec. (e.g. in alluvium), the wave length would be 100 ft.; with the same frequency, and a velocity of 21,000 ft./sec. (e.g. in a granite), the wave length would be 700 ft.

[b] *Cambridge Phil. Soc. Proc.* vol. 23, p. 472 (1925).

SECTION 2. INTERPRETATION OF FIELD OBSERVATIONS

Relation between Slope of Boundary and Slope of Graph. A fundamental formula, for the refracted wave from a single boundary, is that connecting the slope of the graph with the slope of a boundary between homogeneous formations. Suppose that, in the vertical plane through the direction of shooting[a], the boundary between the surface and lower formations is inclined at a constant angle α to the ground surface, which is horizontal.

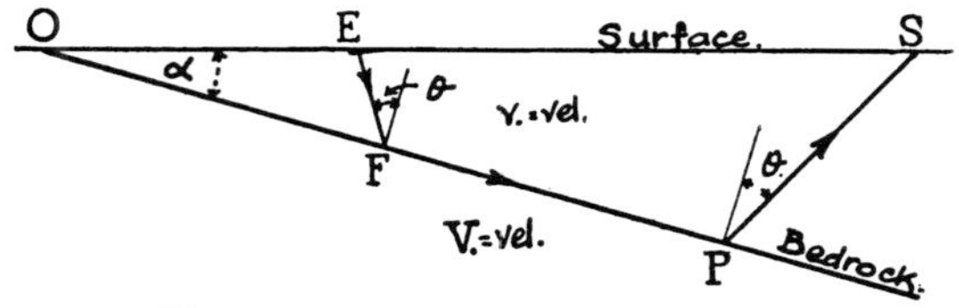

Fig. 250. Disturbance refracted at sloping boundary

Let the velocities in the upper and lower formations be v and V respectively.

The critical angle is given by

$$\sin\theta = v/V \qquad \text{......(1).}$$

With the notation of Fig. 250 and with the explosion point E, and seismometer point S

$$OE = x_0, \qquad OS = x,$$
$$OF = x_0.\cos(\theta-\alpha)/\cos\theta, \qquad OP = x.\cos(\theta+\alpha)/\cos\theta,$$
$$EF = x_0.\sin\alpha/\cos\theta, \qquad SP = x.\sin\alpha/\cos\theta.$$

The time of travel from E to S by the path $EFPS$ is

$$t = x_0.\sin\alpha/v.\cos\theta+[x.\cos(\theta+\alpha)-x_0.\cos(\theta-\alpha)]/V.\cos\theta+x.\sin\alpha/v.\cos\theta.$$

Then
$$v.dt/dx = v.\cos(\theta+\alpha)/V.\cos\theta+\sin\alpha/\cos\theta$$
$$= [\sin\theta.\cos(\theta+\alpha)+\sin\alpha]/\cos\theta$$
$$= \sin(\theta+\alpha).$$

Thus, if $U\left(=\dfrac{dx}{dt}\right)$ is the "apparent velocity" in the lower formation for shooting in this direction as derived from the time-distance curve,

$$\sin(\theta+\alpha) = v/U \qquad \text{......(2).}$$

For shooting in the reverse direction, but with the same slope as positive, the graph will give a different apparent velocity in the lower formation, U', from which

$$\sin(\theta-\alpha) = v/U' \qquad \text{......(3).}$$

From (2) and (3) may be found the values of θ and α; while the true bed-rock velocity, V, follows from (1) $\quad V = v/\sin\theta.$

Depth of Boundary. This may be determined by using the distance from shot-point to the critical point of the graph; that is, the point at which the graphs of arrivals of surface and refracted disturbances intersect.

[a] It must be remembered that in practice the problem is always three-dimensional, and the method is one of first-arrival. The shortest time path is not necessarily in the vertical plane through the surface line: for example, in the case considered, if the boundary between the formations is a plane, its line of greatest slope must lie beneath the line of shooting if the theory given is to hold.

The time of travel of the surface wave from E to S is

$$T = (x-x_0)/v.$$

The point at which the graphs of arrivals of surface and refracted disturbances intersect will be found by putting $t = T$, or

$$(x-x_0)/v = (x+x_0).\sin\alpha/v.\cos\theta+[x.\cos(\theta+\alpha)-x_0.\cos(\theta-\alpha)]/V.\cos\theta;$$

whence, using (1)

$$x-x_0 = x.\sin(\theta+\alpha)-x_0.\sin(\theta-\alpha),$$

which leads to

$$x-x_0 = x_0.(2.\cos\theta.\sin\alpha)/[1-\sin(\theta+\alpha)].$$

If the depth to the boundary be H_0 at E, then

$$\begin{aligned} H_0 &= x_0.\tan\alpha \\ &= (x-x_0).[1-\sin(\theta+\alpha)]/(2.\cos\theta.\cos\alpha). \end{aligned}$$

Thus, with a slight change of notation, if X be the distance from the explosion point to the critical point of the graph, we have as formulae for the depth at shot-point

$$H = X\frac{1-\sin(\theta+\alpha)}{\cos(\theta+\alpha)+\cos(\theta-\alpha)} \qquad \text{......(4)},$$

when shooting in the direction of positive slope; and

$$H' = X'\frac{1-\sin(\theta-\alpha)}{\cos(\theta+\alpha)+\cos(\theta-\alpha)},$$

when shooting in the direction opposite to that of positive slope.

Calculation of Boundary Profile. For uniform slope of boundary surface and for homogeneous media two shots, giving values for v, U, U', X, X', will be sufficient for the calculation of the values of θ, α, V, H, H'.

When the slope is not uniform the same fundamental equation, $\sin(\theta+\alpha) = v/U$, will hold. The values of $\sin(\theta+\alpha)$ and $\sin(\theta-\alpha)$, given by values of v/U and v/U' at a surface point, are, however, not for the same point of the boundary, since the path of the disturbance in the upper formation will have opposite inclinations to the vertical, for the two cases. The horizontal coordinate of the corresponding points in the boundary will not, in fact, be known until the depth is known. If the slope is not very irregular it will be sufficiently accurate to combine such pairs of values, and so find the variation of α (and of V, if the velocity in the lower formation varies also) from point to point of the boundary.

The form of the bed-rock profile may be traced out by such a method. If the bed-rock velocity be known, then shooting in one direction will be sufficient.

Calculation of Depth. When the conditions are not uniform estimates may be made of the depths at two points, from which shots have been carried out in opposite directions over the intervening line, by drawing the ideal graph of two straight lines (as for uniform conditions) which would correspond most closely to the points actually plotted on the graph. This will give average values for U and U', and corresponding values of X and X', which may be used for calculation. From the depths thus calculated at shot-points, a straight line may be drawn which will represent the average boundary profile between the two points. It must be remembered that the true profile will intersect this line at intermediate points, and not, in general, beneath the shot-points. Moreover, care must be exercised that the values taken for U and U' are average values for the whole of the interval between the shot-points. For example, the graphs shown in Fig. 251, over an anticline, would give no indication of the anticline, but the effect of the structure would merely be to give a value for the velocity in the lower formation, which would be greater than the true value. With similar graphs available, the effect of a syncline would be to indicate too low a value for the velocity in the lower formation. In such a case it would be necessary that the remainder of the interval between the points should be covered by graphs for shooting, in each direction, from more distant shot-points.

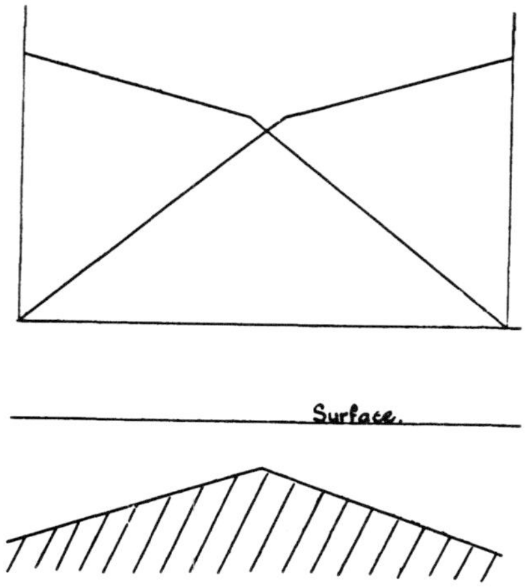

Fig. 251. Shooting over an anticline. Two shots such as these would not indicate the structure

Method of Differences. The following method has at times been used by the I.G.E.S. party for the determination of depths. Shots are fired successively at two out of three points in a line, and in each case the time of arrival of the refracted wave is found at the remaining two points. The spacing is so chosen that the refracted wave will arrive before the surface wave. Thus, taking the notation of Fig. 252, times of travel may be found for the paths *ALMB*, *BNPC*, and *ALPC*. Subtracting the third from the sum of the other two would give a remainder which would be approximately twice the time of travel over *BM* or *BN*. Knowing velocity in the upper formation, the depth at *B* follows at once. Continuing with a fourth point, *D*, the depth at *C* could be found, and so on for further points.

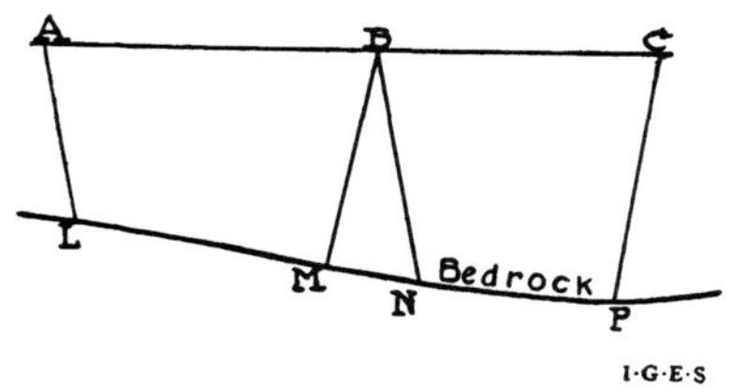

Fig. 252. Depth by method of differences. The depth at the central point is found by shooting along the three paths shown

With the times for the distance *BN* and *PC* known, the time occupied for the distance *NP* follows by subtraction, whence an approximate value for velocity in the lower formation may be found. Once this is known, the correction for the distance *MN*, neglected in the subtractions, may be estimated. The second approximations, given by the application of this correction, will in general be sufficiently accurate.

Examples from Gulgong Experiments. The following examples of calculations are taken from the experiments at the line of bores at Gulgong (see p. 225). In the following table the figures represent the time in m.sec. (milliseconds) for the first-arrival at the point indicated at the beginning of the row[a]. The points are denoted by their distance in feet from one end of the traverse line, as in Fig. 171.

Seismometer point \ Explosion point	00	400	800	1200	1600	2000	2400	2800	3200
00	—	140	178	210	237	267	288	308	331
400	140	—	136	183	210	240	260	279	302
800	178	136	—	132	183	212	232	251	275
1200	210	183	132	—	122	181	201	220	244
1600	237	210	183	122	—	113	171	190	215
2000	267	240	212	181	113	—	111	160	185
2400	288	260	232	201	171	111	—	98	146
2800	308	279	251	220	190	160	98	—	102
3200	331	302	275	244	215	185	146	102	—

Application of Method of Differences. Time from 00 to 800 = 178; from 800 to 1600 = 183; from 00 to 1600 = 237.

Time between surface and interface, beneath 800,

$$= \tfrac{1}{2}.[(178+183)-237] = 62.$$

The times for other points are found similarly, giving the following table:

Point	800	1200	1600	2000	2400
Time in upper formation ...	62 m.sec.	62	61	60	51
Velocity in upper formation*	3·0 ft./m.sec.	3·2	3·4	3·6	3·8
Depth to interface	186 ft.	198	207	216	194

* A fairly regular variation in velocity was found along this line.

The time in the interface between 800 and 1600 will then be 183−(62+61) = 60, which gives a velocity of $\frac{800}{60}$ = 13·3 ft./m.sec. Similarly the velocity in the interface between 1600 and 2400 is calculated as 13·6 ft./m.sec.

To a sufficient degree of approximation the correction for the distance in the interface between the feet of the inclined paths in the upper formation, at 800, is then

$$= 2\times 186\times \frac{3\cdot 0}{13\cdot 3} = 84 \text{ ft.}$$

The time correction will be $= \frac{1}{2}\times \frac{84}{13\cdot 3} = 3$ m.sec. This must be added to the time of 62 m.sec. found above, giving a corrected value of 65 m.sec., for the time in the upper formation beneath 800. Similar corrections may be applied for the other points.

The corrected depths will be, to the nearest 5 ft.

Point	800	1200	1600	2000	2400
Depth... ...	200 ft.	210	220	230	200

[a] Since the time of travel between two points is the same for shooting in either direction, the table is symmetrical about a diagonal; the rows might equally well correspond to explosion points, and the columns to seismometer points.

Calculation of Boundary Profile. The calculation of the profile of the boundary is embodied in the following table:

Distance ft.	Δt m.sec.	$\Delta t'$ m.sec.	ΔT m.sec.	$\sin(\theta+a)$ $\frac{v}{V}=\frac{\Delta t}{\Delta T}$	$\sin(\theta-a)$ $\frac{v}{V'}=\frac{\Delta t'}{\Delta T}$	θ	a	V ft./m.sec. $v/\sin\theta$	v ft./m.sec. $\frac{400}{\Delta T}$	ΔH ft. 400 tan a
00										
	31	28	140	0·221	0·200	12° 7′	0° 35′	13·8	2·9	4
400										
	31	28	136	0·228	0·206	12° 30′	0° 41′	13·4	2·9	4
800										
	32	31	132	0·242	0·235	13° 48′	0° 12′	12·5	3·0	1
1200										
	27	30	122	0·221	0·246	13° 30′	−0° 44′	14·1	3·3	− 5
1600										
	30	30	113	0·265	0·265	15° 22′	0° 00′	13·2	3·5	0
2000										
	20	39	111	0·180	0·351	15° 28′	−5° 05′	13·5	3·6	−36
2400										
	19	40	98	0·194	0·408	17° 38′	−6° 27′	13·5	4·1	−45
2800										
	25	34	102	0·245	0·333	16° 50′	−2° 39′	13·4	3·9	−18
3200										

Δt, $\Delta t'$ m.sec. = Difference between time of arrival of refracted disturbance at successive points, for forward and backward shots, respectively.
ΔT m.sec. = Difference between time of arrival of surface waves at successive points.
ΔH ft. = Increase in depth of boundary from one point to the next.

Adopting a value of 220 ft. for the depth at the point 1600 (as given by the method of differences) the following are the depths to the nearest 10 ft. given by the values found for "ΔH."

Point ...	00	400	800	1200	1600	2000	2400	2800	3200
Depth ...	220 ft.	220	220	220	220	220	180	140	120

These agree, within 10 per cent., with the depths found by the method of differences.

SECTION 3. DESCRIPTION OF EQUIPMENT

The Einthoven Galvanometer System. The layout of the Einthoven galvanometer and camera is shown in Fig. 156. To reduce the width of the film necessary to accommodate the images of the six strings, the images of the latter are turned by individual reflection at six silvered prisms. The six patches of light reflected from the prisms, each with a central image of a string, form a continuous line of light on the film in the camera, behind a cylindrical lens. This arrangement enables images of the six strings, at intervals of 1 mm., magnified 30-fold, to be accommodated on a film whose width is 35 mm. The characteristics of each string are: diameter 0·02 mm., effective length 150 mm., resistance 70 ohms. They are mounted in a "harp," shown in Fig. 253. The deflection for a current of 0·3 m.amp. is 1 mm. after magnification, but is varied by altering the string tension. The sensitivity was checked at regular intervals and the strings were periodically cleaned and annealed. The timing system was also checked.

The field coils, lamp and timing system are all designed for a 12-volt supply, which gives a current of 2 amps. in the field coils, and an approximately saturated magnetic field. The whole system could be run from a single 12-volt battery; but it was found preferable to have a second, for a separate supply to the timing system, in order to avoid reaction on the tuning-fork circuit from the galvanometer circuit; and a third, to double the voltage on the field coils and lamp at the time of the explosion.

An automatic developing camera was used, shown in Fig. 254. This was also run from one of the 12-volt batteries, and included a magnetic clutch for feeding the photographic paper and a magnetic control of developer flow. The camera, supplied by the Cambridge Scientific Instrument Company with the galvanometer, had been designed for military "sound-ranging" work. It had a film speed from 1 to 2 in./sec., which was found insufficient for the purpose to which it was now adapted. A hand turned camera, without automatic development, was therefore substituted for it. This camera had a film speed of about 8 in./sec.

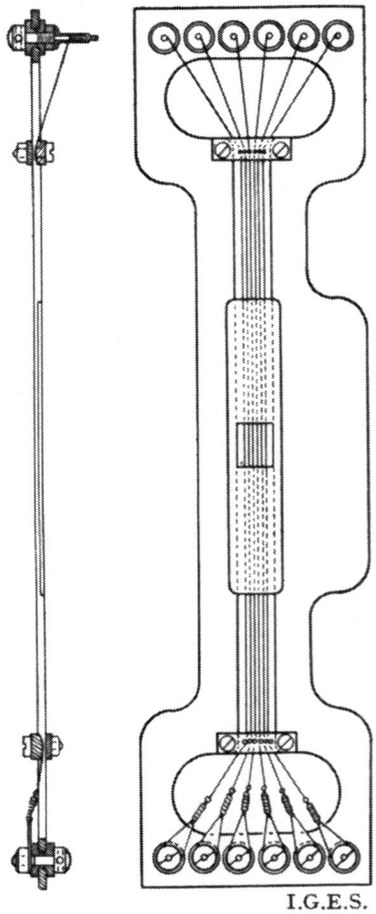

Fig. 253. Harp of six strings for Einthoven galvanometer. The harp fits tightly between the pole pieces of the electromagnet. The strings are joined to the external circuits by sliding contacts

The Schweydar Seismometer. The optical system of the Schweydar seismometer is shown in plan and elevation in Fig. 255. The lamp whose extinction recorded the

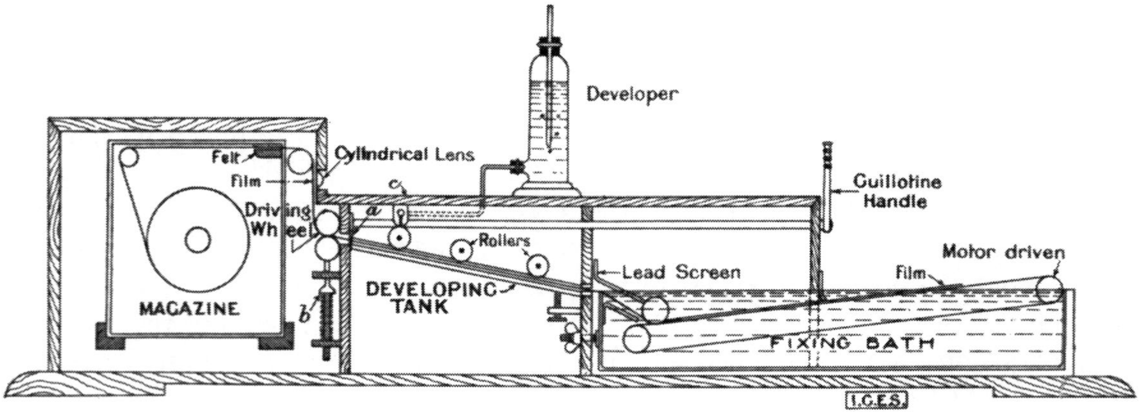

Fig. 254. Automatic developing camera. *a*, guillotine; *b*, spring holding clutch; *c*, prism bench mounted here. The magnetic clutch and developer release are not shown

moment of explosion was placed behind the mirrors, and sent a beam between them. Other workers have used a galvanometer mirror for this purpose, but the system described was found to be more convenient and was probably more accurate. The

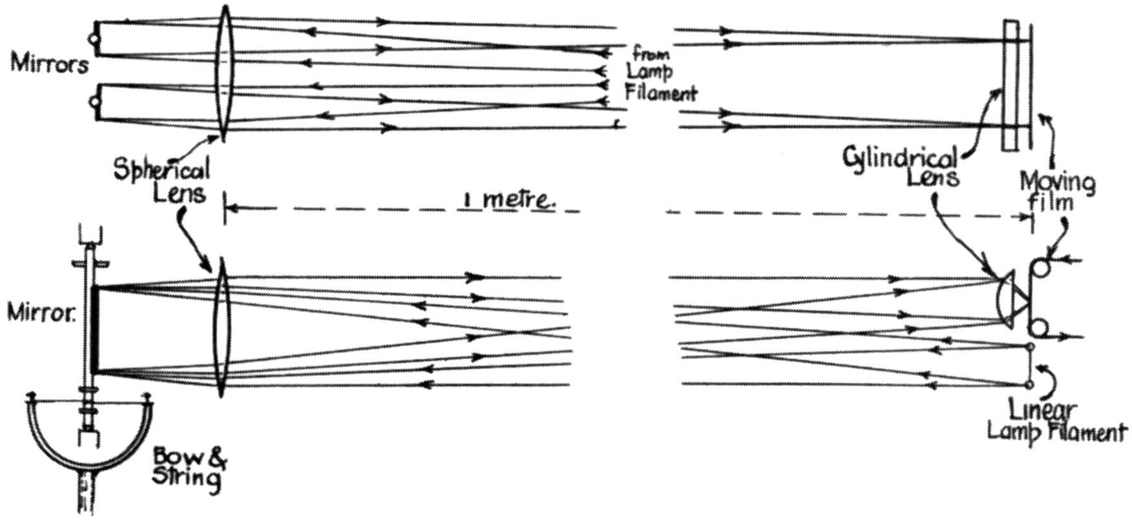

Fig. 255. The optical system of the Schweydar seismometer

camera contains a guillotine and special magazine which enable removal of exposed film in full daylight. Apart from preliminary adjustments of the position of mirror images the seismometer may be used in ordinary daylight, but for this adjustment it is a great advantage to have a dark tent.

Fig. 256. The head of the Schweydar seismometer

Adjustments. The chief difficulty experienced with the Schweydar seismometer was in the adjustment of the mirror images to fall on the cylindrical lens. The mechanism with which this is done appears in the photographs of the seismometer head, Figs. 160, 256. Both vertical and horizontal adjustments of each image can be made by turning milled knobs at the back of the head. The vertical adjustment is by a pair of springs and pistons, and presents no difficulty—except that, in the instrument investigated, it was found to affect the horizontal position of the image. For the horizontal adjustment the mirror-spindle has to slip in its encircling bow-string. A detailed diagram is given in Fig. 257. A pin is run through the spindle transversely, lying within the walls of a slot in a cover which surrounds the upper bearing. The milled knob for this adjustment is connected through two universal joints with a worm and pinion mechanism which turns the cover until one wall of the slot engages the pin so that the spindle is turned until the bow-string slips. The cover is then rotated slightly in the reverse direction, since in the final adjustment it is of course essential that the pin be not touching the walls of the slot, but free to vibrate.

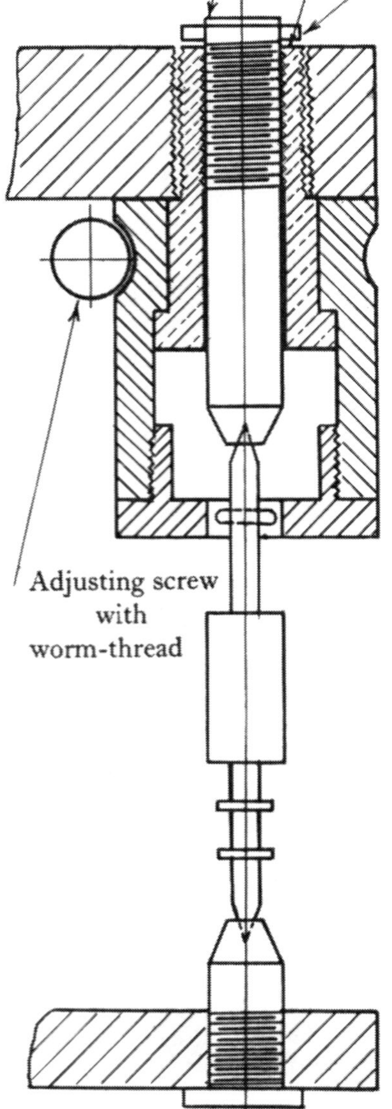

Fig. 257. Mechanism for adjustment of mirror orientation, Schweydar seismometer

Hair Tension. The bow-string is a human hair; and the possibility of adjustment depends upon the hair tension being sufficiently small to allow slipping, rather than a stretching of the hair and a straining of the thin cone on which the bow is mounted. On the other hand, the tension must be sufficient to overcome the stiffness of the hair, so that it grips the spindle sufficiently to transmit small vibrations without slip. To clamp the hair with this optimum tension needs some skill; the tension advised by the makers is 2 gm. wt. Unless the correct tension is established, however, difficulty and loss of time are experienced at every adjustment of the mirror images. The latter operation is necessary after each occasion on which the inertia masses are clamped; or after a shock sufficiently severe to knock the pin of the mirror-spindle against the walls of the slot; while even minor vibrations, unavoidable when the masses are left on suspension, tend to allow the spindle to creep.

I.G.E.S.

Fig. 258. Interior of hut

G. Einthoven galvanometer. *C*. Hand camera.
A. Automatic camera. *B*. Battery balance board.
T. Transformers (replacing battery balance board)

Time Markings. The rate of travel of the record is about 10 cm. per sec., but was found to vary slightly from day to day, with extremes of 9·6 and 10·4 cm. per sec.; so that it had to be measured afresh from the time markings for each record. These time intervals are marked, not at the line of focus of the cylindrical lens, but at a distance of 2·06 cm. from it, corresponding to a time (measured on the film) of 0·21 sec. after the recording of the earth vibration. Thus it is essential that the paper be running steadily, otherwise difficulty is experienced in reading times with accuracy.

Free Period. There is some doubt as to the natural frequency of the pendulum systems. The makers state that the period is 0·282 sec., whereas the typical record provided by them for this seismometer shows the pair of periods, 0·031 and 0·063. The I.G.E.S. party found that a large number of different methods of excitation developed only a pair of periods, 0·04 and 0·08 sec., and a high-frequency vibration of period 0·008 sec.

Damping. The introduction of the smaller frequency in the pair of periods given above is an effect of the type of damping system employed. The record of an impulsive shock commences with the full period and, after a greater or smaller time, depending upon the viscosity of the damping liquid used, this becomes transformed into the half-period. With heavy damping, e.g. with castor oil, the transformation is complete after a few vibrations; with the damping chambers empty, the half-period does not develop. Thus a consequence of the method of damping is to complicate the record, making it much more difficult to analyse than if the natural vibration were purely sinusoidal.

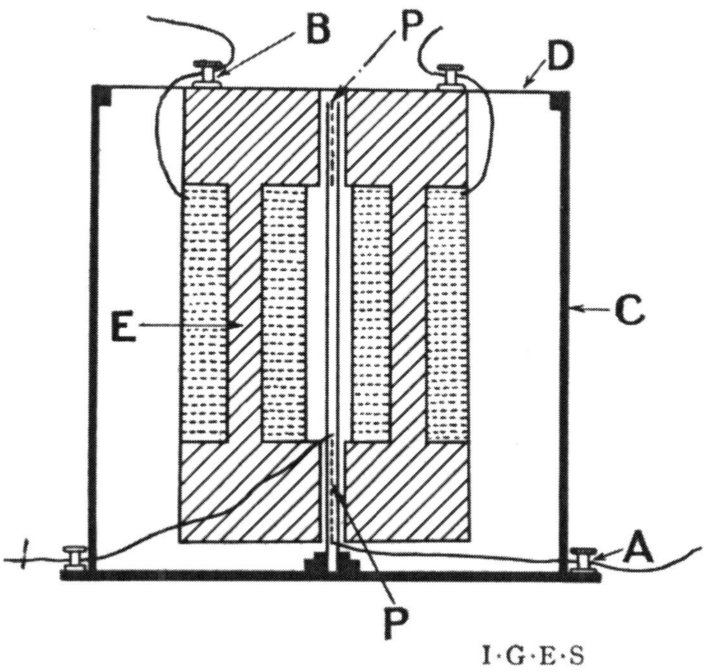

Fig. 259. The electromagnetic seismometer

The Electromagnetic Seismometer. The electromagnetic seismometer was made at the University of Sydney, and is illustrated in Fig. 259. The principles upon which it operates have been explained on p. 210. It consists of a cylindrical brass box *C*, with a thin diaphragm of steel *D* as lid, from which an electromagnet *E* is suspended; the latter acts as the inertia mass. Attached to the base, and passing between the poles of the suspended magnet, is a thin, flat coil *P* of insulated copper wire, held between two thin plates of mica. There are thus two circuits; *B* to excite the electromagnet, and *A* from the coil either direct to the Einthoven string, or to the primary of one of the transformers. A simple calculation shows that this seismometer is unsuitable, without amplification, for low-frequency disturbances.

The Hot-Wire Seismometer[a]. The construction of this seismometer is seen from Fig. 260, and the principles upon which it operates have been explained on pp. 210, 211.

In practice, the instrument finally adopted, of which a set of five was used by the I.G.E.S. field party, had an iron casting as base plate and rim, of 9 in. internal diameter and of $\frac{3}{4}$ in. internal depth. The inertia mass was a lead cylinder of 3 lb. wt., pressed and riveted to a no. 36 gauge brass sheet, which was then pressed and retained in position as a diaphragm by a circular iron ring, locking it with the iron base plate rim. The whole seismometer weighed 20 lb.; it had a natural frequency of vibration of about 30 per sec.

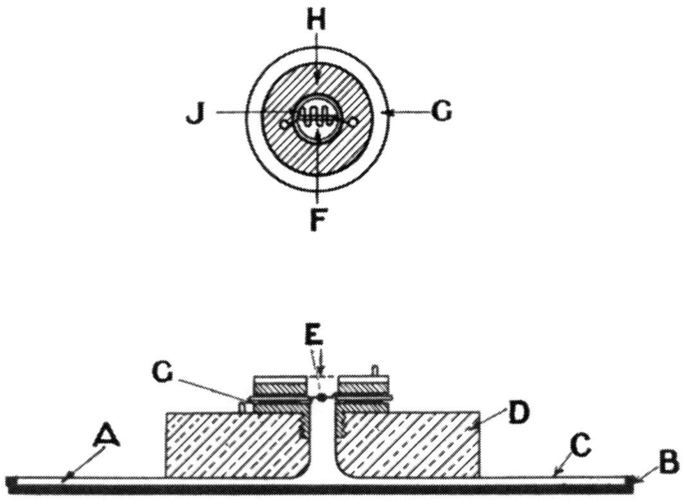

Fig. 260. The hot-wire seismometer

A, Air space. *B*, Iron plate. *C*, Brass diaphragm. *D*, Inertia mass. *E*, Gauze cover. *F*, Hot wire microphone. *G*, Mica disc. *H*, Silver discs on each side of mica. *J*, Fine platinum wire on porcelain rod.

The seismometer as illustrated detects vertical movements only: a horizontal motion may rock the inertia mass, but does not materially alter the volume of air enclosed. It may, however, be mounted with the base in a vertical plane, so as to detect a horizontal component; or in an inclined plane, so as to be sensitive to both horizontal and vertical components, like the Western Electric microphone.

With a wire of diameter of 5×10^{-3} mm., the hot-wire seismometer approached the sensitivity of the Western Electric microphone; it was steady unless externally disturbed, and was well damped. Placed with the base horizontal, it recorded the same time of arrival for refracted disturbances as did the carbon-granule seismometer; but for disturbances propagated through the surface formation, in which the first-arrival had horizontal component of displacement only, it was not excited until the vertical component appeared. Thus a ratio of 1·52 was found between velocities of such first-arrival surface disturbances when recorded by the carbon-granule and hot-wire seismometers, respectively.

The hot-wire seismometer in its present form was not available until the close of field work.

Typical Records with the Hot-Wire Seismometer. Two typical records with the hot-wire seismometer are shown in Fig. 261 (i) and (ii); the former being from 0·1 lb. of blasting gelatine at 100 ft., and consequently showing surface velocities only; the latter being from 30 lb. at 1000 ft., and showing the refracted wave arriving first. The over-

[a] Australian Commonwealth Provisional Patent, No. 23,284, by Major Edgar Booth.

burden, of argillaceous clay, was known to be about 150 ft. in depth. With the hot-wire seismometers at 100 ft. (strings 1 to 5) was also a carbon-granule seismometer (string 6). In each case, string 1 was the moment-of-explosion string; the break is denoted by *a*.

The carbon-granule seismometer, sensitive to both horizontal and vertical displacements, records a surface wave *b*, Fig. 261 (i), of velocity 3460 ft./sec. The hot-wire seismometers, sensitive only to vertical displacements, record first the arrival of a wave at *c*, of velocity 2270 ft./sec.; this is also detected by the carbon-granule seismometer; for which, in fact, the string movement is lost, and the string thereafter remains in continual agitation. The next major arrival, *d*, is of velocity 1390 ft./sec. These velocities, 3500 ft./sec., 2300 ft./sec., and 1400 ft./sec. are typical. The ratio $\frac{3500}{2300} = 1{\cdot}52$ is apparently the ratio of the longitudinal to the transverse wave velocity, under the given conditions.

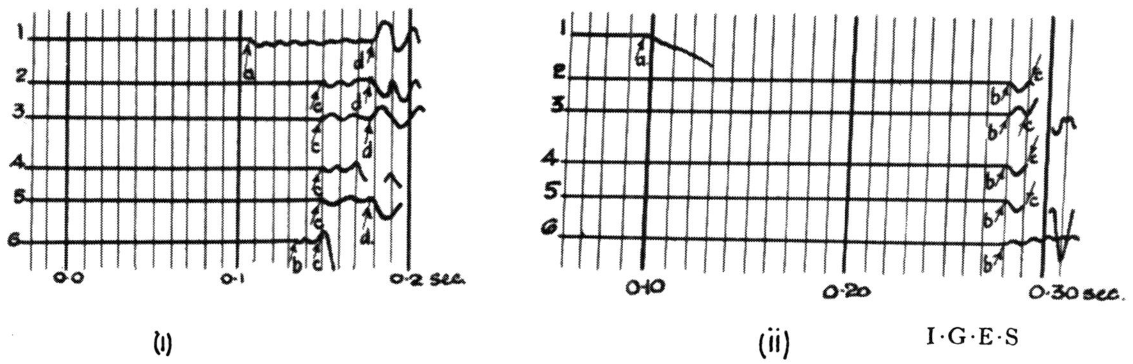

Fig. 261. Typical records with the hot-wire seismometer
(i) Charge 0·1 lb.; distance 100 ft. (ii) Charge 30 lb.; distance 1000 ft.

In Fig. 261 (ii) the refracted wave is picked up at *b* by all the instruments, for it has a predominant vertical component and therefore affects the hot-wire microphones. This time of first-arrival agrees with that determined by other types of seismometer for the same explosion and seismometer points. It is followed by a deflection at *c* so great that the string movements thereafter were not readable, and surface wave arrivals could not be picked out.

SECTION 4. CONCLUSIONS

Accuracy of Timing. The time recorded for first-arrival disturbances varied with the sensitivity of the seismometer. This was traced to two causes. In the succession of disturbances of different type which arrive at the seismometer the energies were found to be successively greater. An insensitive seismometer would therefore miss the first disturbance, and would record a later disturbance[a] as an apparent first-arrival. More-

[a] For example, the arrival *B* instead of the arrival *A*, p. 335.

over, the energy travelling in a given type of disturbance was not confined to the first onset, but it was found that, in general, there was a building up on the record to successively greater amplitudes. Thus, even though a seismometer might record the first-arrival disturbance, the record of the arrival might show a gradual "emersion" from the initial undisturbed record, so that it would be difficult to pick out the moment of first-arrival, and the tendency would be to decide upon a time which was later than the true time. With a sufficiently sensitive seismometer, however, there was a sharp "break" on the record at the moment of first-arrival; and no increase in sensitivity disclosed an earlier "break."

A more violent explosion will produce a larger zone of pulverization, in which the shock travels with abnormally high velocity. It is well known that disturbances of extreme amplitude, in the immediate neighbourhood of an explosion point, appear to be propagated with abnormal velocities. It is therefore to be expected that the velocities recorded in seismic prospecting would vary somewhat with the violence of the explosion. Under the conditions of the I.G.E.S. experiments, however, it was found that the decrease in recorded velocity with decrease in size of charge was usually due, to an overwhelming degree, to the effect of seismometer sensitivity, discussed in the last paragraph: for a weak explosion the seismometer will not show the true first-arrival. It is necessary, therefore, to see that the charge of explosive is sufficient to suit the sensitivity of the seismometer, the distance of detection, and the nature of the intervening media. Using carbon-granule seismometers, whose sensitivity is liable to vary over a wide range, no precautionary measures available were sufficient to ensure full sensitivity at the moment of explosion, and a time which was abnormal with respect to the time recorded by other seismometers at neighbouring points was looked upon with suspicion, and the observation was repeated if sufficiently important.

Similarly, any discrepancy in times recorded by different seismometers was not a true one, but was due (apart from difference of sensitivity) to the fact that some seismometers (for example, the form of hot-wire seismometer available to the I.G.E.S. field party) responded to one component of vibration only. With the Schweydar seismometer—which recorded horizontal and vertical components separately—the first-arrival was shown, under some conditions by the horizontal component, and under others by the vertical component. The type of carbon-granule seismometer used was responsive to both components, without separation; it recorded the same time of first-arrival as did the Schweydar seismometer (no matter for which component this appeared), provided that it was sufficiently sensitive, when the disturbance reached it.

Development of Apparatus—Component to be Detected. It seems desirable that detectors should be capable of recording horizontal and vertical movements either conjointly, or independently as in the case of the Schweydar seismometer. Any alteration to, or redesigning of, apparatus should therefore embody this principle. Admittedly, this is not the general practice. It was the experience of the I.G.E.S. that a vertical component seismometer recorded the first-arrival of a disturbance in the surface formation *after* the horizontal-component Schweydar seismometer, or after a carbon-granule seismometer susceptible to both components of displacement (see Fig. 261). Certainly the first-arrival of the refracted disturbance had a vertical component of displacement only, and all instruments recorded it at the same time. For small scale operations, however, to rely upon the vertical-component record of the seismometer alone would have given wrong results. The justification for the employment by geophysicists of vertical component seismometers only is the practical experience that the first-arrival disturbance through the surface formation has a predominantly vertical component; but, as shown on pp. 335–336, this was not the experience of the I.G.E.S.

Amplification. The addition of valve amplifiers complicates the apparatus, thereby increasing the risk of failure, and adds to field difficulties. In view of statements that an amplification of "several million"[a] has been employed with the electromagnetic detector, a low frequency amplifier of 10^3 voltage-amplification was built, but it was not found practicable to shield the instrument and leads from stray variable fields even with screened leads, and screened amplifier box. From this experience it would seem most undesirable to depend upon high amplification; rather, the seismometers should be designed to be sufficiently sensitive to give direct magnification of the degree desired.

The degree of magnification is governed by the microseism level; with little interference from microseisms it is possible to use smaller shocks and to record them with greater magnification. It should be mentioned here, as the point has been brought out in support of accelerometers[b], that the microseisms which were disturbing to the work of the I.G.E.S. were not of low frequency, but of frequency of the order 100 per second, due to wind in the trees and undergrowth. Such interference would not be overcome by the use of an accelerometer.

Extension of Procedure to Larger Scale Work. If, as suggested on p. 232, radio-electrical control is used, thus eliminating all connecting wires, then independent stations can be employed; each station being self-contained on a motor car, under the charge of a trained mechanic. Alternatively, the equipment could be delivered at the required point and set up in a tent, if rate of working were not the primary consideration. In the latter case, one skilled man could set up each station and leave it ready to record, employing an unskilled man to start the light and film-feed on receipt of a cautionary wireless signal. Such a procedure has been adopted with mechanical seismometers; even dispensing with the unskilled attendant by starting the camera by radio-electrical signal, and stopping it automatically after the passage of a set time.

[a] Barton, *Geophysical Prospecting 1929*, p. 601, A.I.M.E., New York, 1929.

[b] Ambronn, *Elements of Geophysics*, pp. 241–242, McGraw Hill, New York, 1928; Ambronn, *Engineering and Mining Journ.* vol. 128, p. 93 (1929).

CHAPTER XI

GENERAL CONCLUSIONS

The experience of the Survey has shown that the outstanding applications for geophysical work in Australia are:

Electrical surveys for base metal ores (principally sulphides) in Tasmania and in other parts of the Commonwealth where the salinity of the surface and underground waters does not preclude the use of electrical methods.

Electrical surveys in saline water bearing districts with a view to determining the distribution and character of the underground water supplies.

Gravimetric surveys in the brown-coal fields of Victoria and South Australia.

Magnetic surveys over the sub-basaltic tin and gold deep-leads in New South Wales, Victoria, and Tasmania.

Seismic and gravimetric surveys in sub-alluvial gold and tin deep-lead areas.

ELECTRICAL METHODS

The practical applications of the electrical methods are largely concerned with the search for metalliferous deposits of substantial dimensions and of relatively high electrical conductivity. Under favourable circumstances they will prove of great value in this connection but, as a rule, they are ineffective for the direct location of disseminated masses, such as auriferous quartz reefs. In the cases of metalliferous lodes or lenticular bodies having pronounced angles of dip there can be very few mineral fields in which the expense of an electrical survey would be justified, unless substantial ore-bodies might be expected to occur within 300 ft. of the surface. The detectability of such deposits diminishes with increasing depth at a most disconcerting rate and, in general, one cannot expect to locate a lode unless its strike extent is at least double its depth beneath the surface. In the cases of flat-lying deposits of great thickness and lateral extent it may be possible to carry out effective electrical surveys at greater depths but, unfortunately, it is rarely the case that deposits of this kind occur under sufficiently favourable conditions.

Several of the electrical methods have practical applications other than metalliferous prospecting. They may be used with effect in a variety of geological investigations; for example, under certain circumstances it is possible to make satisfactory determinations of the depth to the water table, or to bed-rock in country covered by recent and unconsolidated formations. The value of such measurements in some engineering problems will be apparent. It is also believed that this branch of applied geophysics has an important future in Australia in connection with water supply investigations.

The suitability of an area for electrical prospecting operations depends on many factors but largely upon the electrical conductivity of the country rocks in which the mineral deposits are expected to occur. Preliminary tests of ground conductivity, and of its variations with increasing depth from the surface, must be made on the ground *in situ*; special attention being paid to the conductivity of the overburden, surface soils, and surface and underground waters. Tests carried out in the laboratory upon samples sent from the district in question are quite inadequate, although they may yield information of value.

The electrical methods are numerous, and success in a new district largely depends on the selection of those which are most suited to the local conditions. A wide field experience under a variety of geological conditions is perhaps of greater importance in the electrical than in any other of the geophysical methods of prospecting. Whilst gravimetric and seismic surveys are usually concerned with large scale geological structures, the reverse is the general case when prospecting for metalliferous deposits by electrical methods. Whereas the results obtained by the former methods can be submitted to a full mathematical treatment this is rarely possible in an electrical survey, the interpretation of which is largely empirical.

The Spontaneous Polarization Method is unique amongst electrical methods in that it is concerned entirely with naturally occurring phenomena. Although its practical applications are limited it will occasionally provide a very simple and expeditious means of locating sulphide ores.

The Resistivity Methods are of great importance in connection with structural and water investigations of the kind already indicated. Although they have comparatively few direct applications to metalliferous prospecting they are used as a preliminary to such operations in order to determine the local ground conductivity conditions. Resistivity apparatus is an essential part of the equipment of every electrical prospecting party, irrespective of the nature of the problems to be investigated.

The Equipotential Line Method can be recommended for general reconnaissance purposes when prospecting for metalliferous deposits of a normal lode or lenticular type. Excepting the cases of shallow deposits, the indications are ill-defined and as a rule they have to be subjected to more detailed examinations by electrical methods of greater precision. The advantages of the method are the simplicity of the apparatus and the speed of operation.

The A.C. Potential Ratio Method. This was developed by the I.G.E.S. for making detailed investigations of the surface potentials and the phase displacements that arise in the neighbourhood of conductive bodies. Although the field investigations carried out in Australia during 1928–29 were quite sufficient to establish the value of the method the authorized period of the Survey came to a close before certain of its special applications could be fully explored.

Apart from its use in carrying out independent surface potential surveys the ratiometer is of particular value for making detailed examinations of indications obtained by other methods; for example, by the standard equipotential line method which, although of admitted value for reconnaissance work, is too crude for many purposes.

Amongst the special uses of the A.C. potential ratio (ratiometer) method[a] are:

1. The sub-division of the base line for an equipotential line survey, so that the lines may be commenced at equal potential intervals or, the calibration of a survey already completed, so that values may be assigned to the lines and potential profiles constructed from them.

2. Determinations of the depths to underground discontinuities by the electrode probe method (p. 29).

3. The investigation of bore-holes and the correlation of bore-hole records. This application is a modification of the electrode probe method in which a potential or resistivity profile is constructed along the length of the bore-hole; the necessary measurements being made by lowering the three electrode contacts down the bore-hole to a series of pre-arranged positions.

4. The complete determination of surface potentials, i.e. of the surface ellipses.

Since the close of the I.G.E.S. an entirely new method of conducting A.C. surface potential surveys has been developed. It is based upon an appreciation of the fact that, just as the primary field (approximately in phase with the applied current) is represented by equipotential lines (E.P. lines) in the well known manner, so can the out-of-phase field be depicted by an analogous series of *equiquadrature*, or E.Q. lines. The latter can actually be traced out over the ground surface and the ratiometer is the most convenient instrument for the purpose. Since the values of the quadrature components are the same at all points on an E.Q. line it follows that when all three electrodes, connected to the ratiometer, are situated on the same E.Q. line it is possible to balance the in-phase components by a simple adjustment of resistance only or capacity only. In practice advantage is taken of this fact and with three electrodes and a simplified ratiometer equiquadrature lines may be followed much in the same way that equipotential lines are usually traced out.

The ratiometer is also used for preparing the equiquadrature survey in such a manner that successive lines in each series (E.P. and E.Q.) are spaced apart at equal potential intervals, and so that the potential interval of the one series is known in terms of that of the other. In this way, taking the time relations of the rotating vectors into account, the surface potential conditions for the entire area may be fully determined[b] and shown upon a plan in a convenient diagrammatic form.

The first details of the A.C. potential ratio method and of the ratiometer were published early in 1931[c], just over a year after the system was last used in Australia. In two subsequent publications[d] the outlines of a very similar method and apparatus

[a] Patent applications covering the A.C. potential ratio method, and the special use of the ratiometer for locating equiquadrature lines, are controlled by the Empire Marketing Board, 2 Queen Anne's Gate Buildings, London, S.W. 1. Requests for particulars should be addressed to the Secretary.

[b] The surface ellipse at any point is determined by the directions of the E.P. and E.Q. lines at that point and by the spacing of the lines of each series in its immediate neighbourhood.

[c] Broughton Edge, *Nature*, January 3, 1931. Macmillan & Co., London.

[d] Lundberg and Zuschlag, *Technical Publication*, No. 415, A.I.M.E., New York Meeting, February 1931 and *Bul. Canad. Inst. M.M.* March 1931.

are given and it is satisfactory to observe that the authors hold a high opinion of the importance of this new method of electrical prospecting.

The Electromagnetic Methods used by the Survey employed both horizontal and vertical energizing loops. It was found that the former are more effective in prospecting for flat lying deposits and the latter for steeply dipping lodes. A notable advantage of electromagnetic fields is that they will readily penetrate non-conductive surface layers. This becomes a matter of importance in arid or frozen country where difficulty may be experienced with the surface potential methods, which require good electrical contacts with the ground. Other features in favour of inductive methods are that they take greater advantage of the large conductivity ratios that are sometimes met with in practice, and that the magnetic fields produced are more amenable to complete and accurate measurements. An adverse feature is the high degree of absorption of electromagnetic fields by conductive overburden. Although the surface potential methods also suffer from an analogous screening effect, in most cases they do so to a less serious degree.

The relative merits of the surface potential and the electromagnetic methods, from the point of view of the depths to which they are effective in locating conductive lode deposits, depend on several factors the individual effects of which cannot be fully determined. The experience of the Survey has been that the electromagnetic methods employed have not always confirmed distinctive surface potential indications obtained over known deposits, but contrary cases might well have been recorded had the electromagnetic methods been used more extensively. Nevertheless, since it is the integral magnetic field that is measured in the latter methods, it is to be expected that the effects of a deep-seated body will be masked rather readily by those due to high surface conductivity, or to other factors which may have a predominating influence on the resultant field above the ground surface.

Electrical Surveys in Australia. In many of the arid and semi-arid parts of Australia the electrical methods of prospecting for minerals are at a serious disadvantage on account of the high saline content of the underground waters. Salt water is a good conductor of electricity and when the ground is permeated by it the conductivity ratio between ore and country rock, upon which the successful operation of the electrical methods depends, is usually reduced below the effective limits. In order to determine the extent to which the various electrical methods are affected by this adverse factor the Survey carried out investigations in two of the affected districts: at Moonta in South Australia, where the underground waters have a mean resistivity of 45 ohm cm. and in the Northampton district of Western Australia, where the values range between 135 and 350 ohm cm. These resistivity values are representative of considerable portions of Australia, including several of the well-known mining fields (e.g. Wiluna, Kalgoorlie, Broken Hill). The general conclusion reached by the Survey was that there is little prospect of success by any of the known electrical methods in areas where the ground is permeated by water having a resistivity of less than 200 ohm cm. In cases where the value lies between 200 and 500 ohm cm., it may be possible to obtain satisfactory results by employing special methods (p. 114). Other factors have to be considered, however, in determining the possibilities of success in any particular area.

In Tasmania, and in the high rainfall areas of the Eastern States where the surface and underground waters are fresh, the conditions for electrical prospecting are satisfactory, although even in such districts it was found that a relatively high ground conductivity is the general rule. In Australia, as in other parts of the world, the principal applications are in connection with base metal sulphide ores, such as those which are so widely distributed along the west coast of Tasmania, and in certain parts of Queensland and New South Wales.

As already stated one of the most promising fields for the employment of electrical methods is in connection with water supply, which in Australia is a matter of equal or even greater importance than the search for base metals. With a view to investigating the possibility of using the resistivity method for this purpose, a short series of tests was carried out in the Mallee district of Victoria. Although the instruments available were not well suited to the low ground resistivities met with in this district (100–1000 ohm cm.), the results obtained were sufficient to indicate that in saline areas the method has important possibilities as a means of determining the character and depths of underground water levels. It is strongly recommended that the investigations commenced by the I.G.E.S. should be continued in other parts of Australia where the underground water conditions afford suitable opportunities.

GRAVIMETRIC METHODS

Gravimetric surveying, by comparison with other geophysical methods, is a relatively slow and expensive procedure and only occasionally do opportunities arise for employing it on metalliferous mining fields. Unquestionably it is more suited to investigations of geological structure, such as may be required in coal and oil fields, than to the direct location of ore-bodies. To a great extent it has now been superseded by the seismic method but some of the work carried out in Australia shows that there are certain types of problem which can be dealt with more effectively by the torsion balance or gradiometer than by the seismic or any other geophysical method. In particular this is the case where a detailed investigation of small scale structures is required.

The gravimetric survey carried out in Victoria over a portion of the Gelliondale brown-coal field illustrates very well the class of work which is possible over this type of deposit. The general character and modes of occurrence of the brown coal in the States of Victoria and South Australia are favourable to the gravimetric method and it is to be expected that, from time to time during the development and exploitation of these immense deposits, important problems will arise which will be capable of a satisfactory solution by this means. The Gelliondale survey affords a good example of a case in which the gravimetric method was definitely superior to the seismic, since it was able to delimit the actual coal boundary; had the seismic method been employed, it would not have been possible to distinguish the coal from the surrounding formations (p. 173).

Two distinct types of instrument were used by the I.G.E.S. in conducting gravity surveys: the large torsion balance constructed by Oertling, and the gradiometer, which is a more portable instrument made by the same maker. The former did not prove to be satisfactory for use under Australian climatic conditions and on this account its use was discontinued. The gradiometer, on the other hand, was reliable, and on account of its greater portability was at a definite advantage in a country in which transport is one of the major difficulties.

MAGNETIC METHODS

The magnetic method can be applied qualitatively in a great variety of geological problems and in many cases it will provide some information of value towards their solution. The observations can be carried out rapidly and cheaply and it is surprising that their importance as an auxiliary to geological survey work has received so little attention until within the last few years.

The obvious applications of the method are connected with the discovery of highly magnetic ores, such as magnetite and pyrrhotite, and an excellent example of work of this kind is provided by the magnetic survey carried out at Renison Bell in Tasmania, where striking results were obtained over bodies of tin-bearing pyrrhotite. There are many indirect applications, however, and a magnetic survey will frequently yield information of substantial value in cases where the presence of distinctive magnetic anomalies might not at first be suspected. Remarkable results can sometimes be obtained when investigating the extent and distribution of igneous rocks, and even in the case of sedimentary formations, which have low magnetic susceptibilities, it is occasionally found that recognizable magnetic anomalies are associated with the major structural features. An example of this is provided by the magnetic survey at Gellion-dale, Victoria, where an interesting relationship was observed between the gravimetric and the magnetic anomalies, which correspond respectively to the greater density and greater magnetic susceptibility of the jurassic rocks of that area (p. 159).

The trial magnetic survey carried out over the sub-basaltic gold deep-leads at Gulgong, N.S.W. has shown that the method has important applications in areas in which basalt flows are concealed by alluvium. The boundaries of the flows may be determined, and in some cases the course of the sub-basaltic lead itself may be inferred by studying the magnetic anomaly due to the basalt that fills it. In view of the fact that important deposits of gold and tin have been mined in the past beneath the basalt sheets of New South Wales, Victoria, and Tasmania, and that in some instances work has only been stopped in consequence of the loss of the lead, it is believed that magnetic surveying has an important future in these parts of the Commonwealth. The magnetic instruments used by the I.G.E.S. were two Schmidt vertical balances and one Thalen-Tiberg magnetometer. The former were quite satisfactory during the eighteen months they were used in Australia. The Thalen-Tiberg instrument, although of good design, gave considerable trouble owing to the workmanship of the vital parts being poor. The application of the magnetic methods of prospecting is comparatively simple and it was found that the magnetometers could be used in the field by anyone accustomed to handling instruments.

SEISMIC METHODS

The seismic method is suited to the investigation of large scale geological structures of a simple character; in particular to conditions of extended horizontal stratification, for which the gravity variometer method fails. Under favourable circumstances, depths from the surface, slopes (dips), and certain other features may be determined for concealed structures, such as buried topographies, anticlines, and synclines. Although electrical methods are now being applied to such investigations to an increasing extent, it has not yet been shown that they are effective where the depths exceed a thousand feet from the surface. In the case of the seismic methods there is well authenticated evidence that, in some parts of the world, convincing results have been obtained at depths of the order of several thousand feet.

The experience of the I.G.E.S., although limited to a relatively short series of investigations, is sufficient to permit the statement that the seismic method has been found to be applicable to simple structures in which one reasonably homogeneous formation overlies another; the velocity of the elastic waves in the latter being the greater. An example of such conditions is afforded at Gulgong, N.S.W. (p. 225), where a high-velocity bed-rock is overlain by several hundred feet of unconsolidated alluvium. Under these favourable conditions the form and depth of the lower formation can be found, and if the latter changes in character over a large area this fact will also be disclosed, provided the velocities in the adjacent materials differ sufficiently (e.g. extensive granite masses within slate country). With some prior knowledge of the local rock types and their respective velocities it is thus possible to establish the nature as well as the depth and form of a hidden formation. In some cases an investigation may be made of three superimposed formations (e.g. alluvium, slate, granite) and assuming that the major structural features are simple they could be identified and their depths given.

Repeated observations made by the I.G.E.S. show that the surface velocity will sometimes vary greatly in different parts of what may be presumed on geological evidence to be the same material, and it follows that there may be appreciable inaccuracies in determining the minor features of an underlying formation (e.g. a buried topography). For this reason alone it is considered that a claim for an order of accuracy greater than 10 per cent. would have to be justified by special conditions—i.e. simplicity of structure, and marked homogeneity of the formations concerned.

The experience of the I.G.E.S. Seismic Section was largely gained with electrically recording instruments, although one mechanical seismometer was used for purposes of comparison. The Survey's opinions regarding the advantages of electrical over mechanical recording are given on pp. 232 and 349.

GEOPHYSICAL METHODS USED IN COMBINATION

One of the duties of the Survey was to report on the relative merits of the several geophysical methods and to determine the practical advantages to be gained by using them in combination. To draw close comparisons between the methods is hardly possible, however, in view of their specialized nature and since most of their practical applications are so different in character. For example, during the course of the work in Australia, no case was found in which the gravimetric and the electrical methods might usefully be employed together or as a check upon each other. There were occasions when the deductions from a gravimetric survey could be supported by magnetic observations made over the same ground (e.g. Gelliondale Survey), and it was also found that the electrical and the magnetic methods were compatible in a number of cases (e.g. Renison Bell and Gulgong Surveys). In general it is recommended that every gravimetric and electrical prospecting party should have access to magnetic instruments, and in view of the speed and small expense with which magnetic observations can be made, there should be no difficulty in complying with this requirement. Although there are many instances in which seismic and gravimetric surveys would afford valuable checks on each other, it is not often the case that circumstances would permit the provision of instruments and personnel for both methods. On the other hand the supplementary evidence to be obtained by the second method may be of such value (p. 145) as to justify the additional expense.

WORKING COSTS

Although the field work carried out by the I.G.E.S. was largely of an experimental character, it is possible to give the following account of operating costs for the several methods employed. They are approximately correct for the conditions under which the I.G.E.S. parties were working in Australia, but it should be understood that they do not include administrative and other overhead charges, or such items as travelling expenses. Although each party was an independent unit, working under the charge of an experienced officer, the general direction and ultimate responsibility for the work was in the hands of a senior member of the Survey who, as a rule, would spend from three to five days with each party during every period of six weeks. The cost of this general supervision is also omitted from these statements.

Although the initial expenditure on apparatus and field equipment is given for each of the geophysical methods these items are not allowed for in the operating costs. Owing to the rapid developments in the design of geophysical instruments it will in general be wise to allow for their depreciation at the rate of 40 per cent. per annum. Even in the cases of the gravimetric and some of the electrical methods which have become more standardized it is advisable that these charges should be fully written off within a period of three years.

Electrical Methods. As a rule the electrical methods are concerned with investigations of a more detailed character and on a much smaller scale than is the case with the gravimetric or seismic methods. Observations have to be made at a great number of stations which in many cases will only be spaced 15 or 20 yards apart. The costs and speed of work vary greatly according to the nature of the country; especially with respect to the topography and the thickness of the vegetation.

The following estimates of working costs are based on Australian conditions and for a minimum personnel consistent with efficient working; namely the electrical operator in charge and a junior surveyor, whose combined salaries would total about £80 per month, and three labourers who would cost £56 per month. The inclusion in the party of a second operator involves the employment of another labourer, but although the rate of work is substantially increased it is not actually doubled. For large scale work where the time factor is important and where overhead expenses are heavy the employment of the second operator is desirable. In all cases, salaries and wages are the heaviest items, expenditure on stores and maintenance of equipment being of minor importance.

Reliable figures are given below for the equipotential line and the Bieler-Watson methods, both of which were extensively used in Australia. In the cases of the A.C. potential ratio, spontaneous polarization, and the high frequency methods, the figures available are less exact, since these methods were employed principally for detailed or confirmatory work, rather than for systematic surveys of large areas.

Equipotential Line Method. Taking an area of 320 acres (half a square mile) as the standard, the following estimates are arrived at for a reconnaissance survey:

Local conditions	Speed of survey (acres per day)*	Cost per acre (£)	Cost per 320 acres† (£)
Open or lightly timbered country—damp ground	28	0·46	147
,, ,, ,, —dry ground	16	0·62	200
Dense bush—damp ground	14	0·70	225
,, ,, —dry ground	12	0·78	250

* Excluding time spent on layout and assuming reasonable weather conditions.
† These costs would be relatively higher for a smaller area and *vice versa*.

Thick vegetation and rugged country are the principal obstacles in equipotential work, but it is also sometimes hindered in regions of high ground conductivity where phase displacements are large and the equipotential points difficult to locate. As a rule, however, this only occurs in areas of limited extent, and the speed of the survey as a whole is not materially affected.

Bieler-Watson Method. The speed at which readings can be taken depends to a very important extent upon the skill of the operator, but it is reasonable to assume a rate of 10 acres per day in normal country, inclusive of time spent in preliminary and final operations. There are fewer figures upon which an estimate of the cost of this method may be based than in the case of the equipotential line method, but under normal conditions the cost per acre was approximately £0·5.

Other Methods[a]. From the data available the cost per acre for a detailed survey may

[a] No figures can be given for resistivity investigations as the I.G.E.S. had no opportunities for conducting systematic surveys by this method.

be estimated at £0·7 for the high-frequency method, £0·8 for the A.C. potential ratio method, and £0·4 for the spontaneous polarization method, inclusive of all surveying costs but not allowing for special electrode layouts for the potential ratio method.

Since most of the electrical apparatus used by the I.G.E.S. was made privately in University Workshops and Government Research Stations it is difficult to say what its actual cost would be if obtained through ordinary channels. There is little doubt, however, that the whole of the equipment required for the five methods discussed above, as well as for resistivity surveys, could be obtained for a sum not exceeding £600.

Gravimetric Method[a]. From the experience gained by the I.G.E.S. Gravimetric Section the following summary of operating costs per station can be given for the two extreme types of country represented by Gelliondale and Lakes Entrance in Victoria.

(*a*) *Favourable country*. It was found that in flat open country, with stations not more than 400 ft. apart, a party consisting of an experienced operator, two surveyor-assistants, and three labourers could determine the gradients at 6 stations per day when working with two gradiometers and three portable huts (i.e. 120 stations per month). The operating costs per month may be stated as follows:

Salary and wages	£ per month	
O.C. Section, salary and allowance	55	
Surveyor-assistants, 2 at £35 per month ...	70	
Labourers, 1 at £5 per week	23	
„ 2 at £4 per week	37	
	—	185
Operating expenses		
General expenses	10	
Local transport, etc.	5	
	—	15
Total operating costs per month		£200

Cost per station £1·66.

(*b*) *Unfavourable country*. At Lakes Entrance in Victoria gradients were measured at 63 stations in a period of eight weeks, using one gradiometer. The surveyor spent an extra three months in the field and a further month was required for office work and computations. In the case of this survey the costs were as follows:

	£
Salaries	330
Wages	150
Transport and stores	50
	£530

Cost per station £8. 10*s*. 0*d*.

[a] Barton states that the total cost, including depreciation, is $1000–1200 per month for a single torsion balance doing 50 stations, i.e. $20–24 per station. *Geophysical Prospecting 1929*, p. 476, A.I.M.E., New York, 1929.

The initial expenditure on instruments and field equipment used on the surveys discussed above was as follows:

	£
Gradiometers, 2 at £375	750
Spare wires for gradiometer, 8 at £5	40
Huts for gradiometer, 3 at £100	300
Surveying instruments, etc.	150
	£1240

Magnetic Method. To give reliable figures for the cost of magnetic surveying is difficult since there were no occasions in Australia on which the attention of a field party was entirely confined to this method. The work was always carried out in conjunction with surveys by other methods; in many cases along traverse lines that had previously been laid out for other purposes. It should be explained, however, that this subordination of the magnetic work is no reflection on the merits of the method; it was due to the fact that the procedure is already so well established and so quickly applied that a special personnel was not considered to be necessary. The general experience obtained was as follows:

The speed at which the field measurements can be made depends upon the weather, the type of country, the accuracy required, and the distance between stations. Two types of magnetometer were used by the I.G.E.S. for determining vertical intensity, the Schmidt vertical field balance for accurate observations and the Thalen-Tiberg magnetometer where large anomalies existed. For each of these instruments ten stations per hour is the rate obtained by averaging the speeds of different observers in various types of country in which the station intervals did not exceed 200 ft. The rate of work with the Thalen-Tiberg magnetometer would have been greater but for certain defects in this instrument.

In the case of a magnetic survey such as that conducted by the I.G.E.S. at Gulgong, N.S.W. one observer and one surveyor-assistant can work comfortably at the rate of 250 stations per week. With traverses spaced 400 ft. apart and observation stations along them at 200 ft. intervals, as at Gulgong, there would be less than 250 stations per $\frac{1}{2}$ square mile; even making ample allowance for closer spacing of the stations at critical parts of the survey. Apart from the salaries of the two men the operating costs are negligible. £25 per week (i.e. per $\frac{1}{2}$ square mile) would cover both salaries and all incidental expenses excepting depreciation of the magnetometer, and of the simple surveying appliances that are required. The purchase prices of the Schmidt and Thalen-Tiberg magnetometers were £128 and £16 respectively.

Seismic Method. In the case of the Seismic method it is possible to give the following summary of working costs based upon the experience gained in New South Wales. These investigations extended over a period of several months and were concerned with a special type of structural problem.

Salary and wages	£ per month	
O.C. Section, salary and allowances	55	
Surveyor-assistant, salary and allowances	35	
Shot-firers, 2 at £20 per month	40	
Labourers, 2 at £4. 2s. 6d. per week	36	
	——	166
Expendable stores, etc.		
Explosives, 20 shots per week at £2. 10s. 0d. ...	200	
Films, chemicals, etc.	10	
Transport running expenses	10	
Incidental	14	
	——	234
Total operating costs per month		£400

With a field party constituted as above and engaged on a detailed survey, four shots per day can be fired for five days in the week, representing a backwards and forwards shooting of two sections of traverse per day. With each section 2600 ft. in length this rate of progress gives a *run* of about four miles per week.

On a close survey of an area for deep-leads, or other buried topographical features where careful small scale work is necessary, with detector points 400 ft. apart along traverses at 800 ft. intervals (a very close reticulation), this represents only $2\frac{1}{2}$ square miles per month. Precise small scale work of this character would only be conducted where positive results of considerable importance were being obtained. In the case of a survey by the fan method (p. 219), a deep-lead can sometimes be followed quite readily once it has been picked up. In such cases the survey is more of a linear character than one of area, and under favourable conditions the lead might be followed at the rate of over a mile per week.

From the statement given above it will be observed that the expenditure on explosives represents half of the total operating costs. The outlay on the apparatus, field equipment, and transport (motor truck) used by the I.G.E.S. amounted to £600, but for a completely mobile unit, with the recording station mounted on a second truck, this figure would have approached £1000. In general it would be a good policy to write off the whole of this amount in two years since the transport is subjected to very rough conditions and the equipment would largely be out-of-date at the end of this period.

An estimate of the costs for large scale work, or for long periods, would be without the backing of practical experience, but even in such cases it is likely that explosives would still account for half the necessary expenditure.

INDEX

Personal names, and page numbers in Part II, are shown in italics. References to footnotes are indicated by "n" after the page number

46-2

Lightning Source UK Ltd.
Milton Keynes UK
UKOW011108140812

197503UK00001B/2/P

9 781107 612020